Artificial Intelligence in Agriculture

Artificial Intelligence in Agriculture

Guest Editor

Arslan Munir

MDPI

Basel • Beijing • Wuhan • Barcelona • Belgrade • Novi Sad • Cluj • Manchester

Guest Editor
Arslan Munir
Department of Electrical
Engineering and Computer
Science
Florida Atlantic University
Boca Raton, FL
USA

Editorial Office
MDPI AG
Grosspeteranlage 5
4052 Basel, Switzerland

This is a reprint of the Special Issue, published open access by the journal *AI* (ISSN 2673-2688), freely accessible at: https://www.mdpi.com/journal/ai/special_issues/AI_Agriculture.

For citation purposes, cite each article independently as indicated on the article page online and as indicated below:

Lastname, A.A.; Lastname, B.B. Article Title. *Journal Name* **Year**, *Volume Number*, Page Range.

ISBN 978-3-7258-7100-1 (Hbk)
ISBN 978-3-7258-7101-8 (PDF)
https://doi.org/10.3390/books978-3-7258-7101-8

Cover image courtesy of Arslan Munir

Contents

About the Editor

Arslan Munir

Arslan Munir is currently a Professor in the Department of Electrical Engineering and Computer Science at Florida Atlantic University. He was a postdoctoral research associate in the Electrical and Computer Engineering (ECE) department at Rice University, Houston, Texas, USA, from May 2012 to June 2014. He received his M.A.Sc. in ECE from the University of British Columbia (UBC), Vancouver, Canada, in 2007 and his Ph.D. in ECE from the University of Florida (UF), Gainesville, Florida, USA, in 2012. From 2007 to 2008, he worked as a software development engineer at Mentor Graphics Corporation (now Siemens) in the Embedded Systems Division. Munir's current research interests include embedded and cyber–physical systems, secure and trustworthy systems, parallel computing, quantum computing, artificial intelligence, and computer vision. Munir has received many academic awards, including the doctoral fellowship from the Natural Sciences and Engineering Research Council (NSERC) of Canada. He has earned gold medals for best performance in electrical engineering, and gold medals and academic roll of honor for securing rank one in pre-engineering provincial examinations (out of approximately 300,000 candidates). His research has been sponsored by the U.S. Department of Energy (DOE), U.S. Department of Defense (DOD), Air Force Office of Scientific Research (AFOSR), Air Force Research Laboratory (AFRL), National Aeronautics and Space Administration (NASA), U.S. Department of Agriculture (USDA)/National Institute of Food and Agriculture (NIFA), National Science Foundation (NSF), and Semiconductor Research Corporation (SRC). He has been awarded Faculty Fellowships at AFRL, Information Directorate (RI) in 2019, 2020, 2021, and 2023. His name is also included in the list of the World's Top 2% Scientists compiled by experts at Stanford University.

AI MDPI

Editorial

Artificial Intelligence and Machine Learning for Smart and Sustainable Agriculture

Arslan Munir

Department of Electrical Engineering and Computer Science, Florida Atlantic University, Boca Raton, FL 33431, USA; arslanm@fau.edu

1. Introduction

Agriculture is entering a profound period of transformation, driven by the accelerating integration of artificial intelligence (AI), machine learning, computer vision, autonomous sensing, and data-driven decision support [1]. These technological advances are reshaping the ways in which crops are monitored, soils are assessed, environmental conditions are forecasted, disease symptoms are detected, yields are predicted, and even animal welfare is evaluated [2]. What once required extensive manual labor, specialized expertise, and repeated field visits is increasingly supported or in some cases replaced by intelligent systems capable of delivering real-time insights and precision-guided interventions.

The contributions assembled in this edited volume capture the current momentum and future promise of AI for agriculture. Collectively, they illustrate how modern learning paradigms are being brought into the demanding context of agricultural environments, where variability in lighting, occlusion, weather, biological diversity, and resource limitations challenge the adaptability and robustness of conventional algorithms. The works included here demonstrate not only scientific innovation but also a strong emphasis on real-world applicability, highlighting robotic systems, mobile applications, edge devices, spectral-analysis pipelines, and federated learning networks [3] engineered for seamless deployment in agricultural settings.

2. Organization

To provide a coherent perspective on this broad and rapidly evolving landscape, the chapters in this edited volume are organized into six thematic clusters, each representing a major axis of contemporary research. These clusters progress naturally from field robotics and perceptual intelligence to soil and environmental modeling, plant stress and disease analytics, yield prediction and spatial modeling, smart hydroponics and edge-enabled computing ecosystems, and finally to livestock welfare assessment through advanced bioacoustic analysis. Figure 1 provides an overview of this organization, illustrating the structure of the volume across its six thematic clusters. Collectively, these contributions offer a comprehensive view of how artificial intelligence is reshaping agriculture from soil to canopy and from crops to livestock.

Cluster 1—Robotic Perception, Object Detection, and Scene Understanding in Agricultural Environments

Field robotics and automated sensing systems must contend with visual environments that are significantly more complex than those represented in standard computer vision benchmarks. Agricultural settings are marked by heterogeneous background textures, low-contrast crop canopies, foliage-induced occlusion, and highly dynamic illumination, all of which complicate robust perception. The works in this cluster advance the state-of-the-art

 https://doi.org/10.3390/10.3390/ai7010012

by developing perceptual methods capable of meeting these challenges and maintaining reliable performance under real-world field conditions.

Figure 1. Thematic map of clusters in AI and ML for smart and sustainable agriculture.

- **Spatial AI for Robotic Wheat Detection and Navigation** (Gunturu et al.) presents a stereo-vision robotic platform designed for autonomous navigation in dense wheat fields. By fusing depth perception with efficient object detection architectures, the system performs real-time wheat plants detection and collision avoidance, illustrating how spatial AI can support field scouting and future robotic field operations.
- **Model Size and Image Augmentation for Object Detection in Complex Background Scenes** (Sangha and Darr) studies the influence of different photometric image augmentation methods on model performance for standard one-stage and two-stage detectors. The work conducts a rigorous analysis of how detector size and photometric augmentation influence performance in the visually demanding context of agriculture. The work provides valuable guidance for designing models intended for deployment in the field rather than the laboratory.
- **Detection and Counting of Olive Tree Fruits in Olive Groves** (Osco-Mamani et al.) compares several deep learning models for olive tree fruits detection using imagery collected directly from olive orchards. Results indicate that YOLOv8m achieves highly accurate detection and strong correlation with manual counts, demonstrating that deep detection models are now sufficiently mature for practical yield-monitoring applications.

Cluster 2—Soil, Water, and Environmental Intelligence for Sustainable Crop Production

Sustainable crop production depends on a precise understanding of soil condition, water availability, and the complex interactions between plant physiology and the surrounding environment. As climate variability intensifies drought risk and amplifies nutrient–water coupling effects, growers and researchers increasingly rely on data-driven methods to monitor resource dynamics and anticipate stress before yield losses occur. The chapters in this cluster advance this goal by integrating proximal and remote sensing with meteo-

https://doi.org/10.3390/10.3390/ai7010012

rological observations and predictive machine learning. Together, they demonstrate how spectral analytics and forecast-informed models can support scalable estimation of soil attributes, improve soil moisture prediction for irrigation planning, and strengthen crop stress characterization under interacting environmental and nutrient constraints.

- **Dimensionality Reduction Models for Soil Attribute Prediction** (Wei et al.) demonstrates that dimensionality reduction statistical models such as principal components regression (PCR) and least absolute shrinkage and selection operator (LASSO) can derive meaningful predictions directly from raw spectral data without complex preprocessing. This study provides a promising direction for scalable, low-cost soil sensing in diverse agricultural contexts.
- **Machine Learning for Topsoil Moisture Forecasting** (Martínez et al.) demonstrates highly accurate soil moisture predictions by integrating soil probe measurements with meteorological and forecast data. Such models are essential for proactive irrigation management, especially in drought-prone regions where efficient water resource management is crucial for ensuring sustainable crop production.
- **Comparison of Nitrogen and Water Dual-Stress Effects Through Hyperspectral Imaging** (Maki et al.) examines how nitrogen availability and water stress jointly affect maize physiology. By leveraging hyperspectral imaging, the authors demonstrate that dual-stress modeling improves the robustness of relative water content prediction, while maintaining reliable nitrogen content estimation, which is an essential advancement for climate-adaptive crop breeding and precision management.

Cluster 3—Vision-Based Crop Stress, Disease Detection, and Agricultural Product Health Assessment

Vision-based analytics play a central role in enabling timely, scalable, and non-invasive assessment of crop condition across the agricultural lifecycle. From early detection of nutrient stress and pest infestation in the field to post-harvest evaluation of produce quality, computer vision and deep learning provide powerful tools for translating visual cues into actionable agronomic insights. The contributions in this cluster illustrate how modern image-based learning frameworks, ranging from convolutional neural networks and vision transformers to ensemble and neural architecture search-driven models, can reliably characterize plant health, diagnose disease, and assess product quality under both pre-harvest and post-harvest conditions. Together, these works demonstrate the maturation of vision-driven agricultural intelligence, emphasizing robustness, generalization, and deployability across diverse crops, environments, and stages of production.

- **Deep Learning for Nitrogen Stress Detection in Maize Using RGB Imaging** (Ghazal et al.) leverages different deep learning models to predict nitrogen deficiency in maize crops using a field-acquired RGB imaging dataset. The authors benchmark several convolutional neural networks (CNNs) alongside a custom-built Vision Transformer. Among the tested models, EfficientNetB0 achieves the best accuracy. This work demonstrates how cost-effective imaging solutions can support nutrient-stress monitoring at scale.
- **Enhancing Pest Detection on Tomato Plants Using Ensemble Techniques** (Giakoumoglou et al.) shows that ensemble techniques, such as Non-Maximum Suppression (NMS), Soft Non-Maximum Suppression (Soft NMS), Non-Maximum Weighted (NMW), and Weighted Boxes Fusion (WBF) significantly improve the detection of pest Tuta absoluta. This work contributes towards the development of more effective pest management strategies in tomato production.
- **Neural Architecture Search for Soybean Disease Recognition** (Omole et al.) proposes AgriNAS, a framework that employs neural architecture search to automatically

discover optimal architectures for soybean disease classification. By incorporating adaptive convolution and spatial–temporal augmentation, the approach outperforms conventional CNNs, demonstrating the promise of automated architecture design for pest and disease detection in different agricultural settings.

- **Detection of Leaf Diseases in Banana Crops Using Deep Learning Techniques** (Jiménez et al.) explores deep learning techniques for the detection of leaf diseases, such as Black Sigatoka and Cordana, in banana crops. The authors compare and evaluate the performance of various deep learning models (e.g., ResNet50, Efficient NetB0, VGG19) for disease detection in banana crops. Further, this work describes the implementation of deep learning models in a mobile application, which is an important step toward practical, in-field disease diagnostics for growers.
- **Machine Learning Model for Assessing Fruit Quality** (Apostolopoulos et al.) extends vision-based agricultural health assessment beyond the field and into the post-harvest stage by introducing a generalizable machine learning framework for fruit quality evaluation. Using deep visual representations learned through vision transformers, the authors demonstrate that a single, unified model can accurately distinguish between high-quality and defective produce across a wide range of fruit types, without reliance on fruit-specific training or handcrafted features. Extensive experiments across sixteen fruit categories show that the proposed approach achieves accuracy comparable to, and in many cases exceeding, that of dedicated per-fruit models. By framing fruit quality inspection as a transferable visual recognition problem, this work advances scalable and deployable solutions for automated grading, sorting, and quality control in agricultural supply chains, complementing earlier in-field vision-based health monitoring approaches within this cluster.

Cluster 4—Yield Prediction, Synthetic Data Generation, and Spatial Analytics

Accurate yield prediction is fundamental to agricultural planning, market forecasting, and risk management, yet it is often hindered by sparse, imbalanced, or spatially heterogeneous data. These limitations are especially pronounced in field-scale and smallholder contexts, where comprehensive historical records are rarely available. The contributions in this cluster address these challenges by introducing data-centric innovations, including synthetic data generation and spatially aware learning frameworks. Together, they demonstrate how generative modeling and neighborhood-informed analytics can improve predictive robustness and accuracy, even under constrained data conditions.

- **Data Synthesis for Alfalfa Biomass Yield Estimation** (Vance et al.) explores the use of generative models to synthesize datasets to accurately predict the Alfalfa crop biomass. By using Conditional Tabular Generative Adversarial Network (CTGAN) and Tabular Variational Autoencoder (TVAE) to augment limited yield datasets, the authors significantly improve prediction accuracy and introduce a practical tool Predict Your CropS (PYCS pronounced like Pisces) accessible to end users. Their work illustrates how generative models can compensate for the scarcity of labeled agricultural data.
- **Spatial-Lagged Machine Learning for Corn Yield Prediction** (Noa-Yarasca et al.) proposed an approach to crop yield prediction by incorporating spatially lagged spectral data (SLSD) from neighboring pixels through the spatial-lagged machine learning (SLML) model. The results indicate that incorporation of SLSD through SLML outperforms traditional vegetation-index-based methods in predicting crop yield. This research highlights the significance of spatial context and neighborhood information in improving corn yield prediction and emphasizes the need for optimizing spatial parameters, feature selection, and neighborhood size to enhance model accuracy for crop yield prediction.

https://doi.org/10.3390/10.3390/ai7010012

Cluster 5—Controlled-Environment Agriculture, IoT-Enabled Monitoring, and Edge/Federated Intelligence

Controlled and semi-controlled agricultural environments, such as hydroponic farms, grain storage facilities, and distributed field-deployed sensing networks, offer powerful opportunities for automation, continuous monitoring, and closed-loop decision support. Yet these settings also impose practical constraints: systems must operate reliably under variable illumination and environmental conditions, run on resource-limited devices, and scale across distributed deployments without excessive energy or connectivity demands. The contributions in this cluster address these challenges by coupling AI with Internet of Things (IoT) and edge computing, and by introducing federated learning strategies that reduce energy consumption and communication overhead while preserving performance. Together, these works illustrate how intelligent monitoring and optimization pipelines can be translated into deployable systems for controlled-environment agriculture, post-harvest protection, and large-area sensing networks.

- **Parameters Optimization for Smart Hydroponic Systems** (Shareef et al.) conducts a systematic literature review to determine the optimal growth parameters for leafy green vegetables, herbs, and spices cultivated in hydroponic systems. The work analyzed numerous research papers to identify the optimal ranges for pH, electrical conductivity (EC), temperature (nutrient solution and ambient), aeration/dissolved oxygen (DO), growing media/substrate, lighting/artificial illumination, relative humidity, and CO_2 dosing, as well as plant-specific parameters. The work further highlights the increasing role of AI and automation in controlled-environment agriculture.
- **Machine Learning for Insect Monitoring in Grain Facilities** (Mendoza et al.) addresses a critical but often underrepresented part of the agricultural pipeline, that is, post-harvest protection and stored-product security, by developing an affordable, camera-based insect monitoring system for grain and food facilities. The proposed setup couples a manual-focus camera with an NVIDIA Jetson Nano to run a trained deep learning detector on a live video stream, enabling real-time identification of key stored-product pests. The system is evaluated under multiple illumination conditions (white LED, yellow LED, and no dedicated lighting), demonstrating both the feasibility of low-cost automation and the practical sensitivity of machine-vision monitoring to lighting variability. The study further details an end-to-end deployment pipeline, from data collection and annotation to model training (SSD-MobileNet), model conversion (ONNX), and accelerated inference (TensorRT), highlighting the engineering considerations required for operational edge deployment. By translating deep learning-based detection into a deployable facility-scale monitoring workflow, this work contributes an important step toward scalable, continuous pest surveillance that can support integrated pest management decisions and reduce post-harvest losses.
- **FedBirdAg: A Low-Energy Federated Learning Platform for Bird Detection** (Benhoussa et al.) proposes an energy-efficient federated learning framework for bird detection in crop fields using a compact smart camera network. The results reveal that the proposed federated learning-based approach results in energy savings of eight times as compared to a centrally trained model. Their approach demonstrates how decentralized intelligence can support wildlife monitoring and crop protection at scale.

Cluster 6—Livestock Bioacoustics and Welfare Assessment Through AI

As agricultural intelligence expands beyond crop-focused applications, AI is increasingly being applied to the monitoring and management of livestock systems. Among emerging modalities, bioacoustic sensing offers a non-invasive and continuous means of assessing animal condition and behavior. The contribution in this cluster explores the use

https://doi.org/10.3390/10.3390/ai7010012

of transformer-based acoustic modeling to extract meaningful representations from animal vocalizations, highlighting a novel pathway for advancing livestock welfare assessment within precision livestock farming.

- **Adapting a Large-Scale Transformer Model to Decode Chicken Vocalizations** (Neethirajan) explores the feasibility of repurposing a large-scale Transformer-based model, OpenAI's Whisper, originally developed for human speech recognition, to the decoding of chicken vocalizations. The work records chicken vocal data under diverse experimental conditions, including healthy versus unhealthy birds, pre-stress versus post-stress scenarios, and quiet versus noisy environments. Processing these chicken vocal data recordings through Whisper produces text-like outputs which are further processed through natural language processing (NLP) sentiment analysis tools to obtain "negative" and "positive" scores with close correspondence to documented changes in vocal intensity associated with stress events and altered physiological states. The work reveals that the repurposed transformer-based architectures can learn meaningful representations of non-human acoustic signals. This work opens the door to non-invasive, real-time monitoring of poultry welfare.

3. Conclusions

The research presented in this volume highlights both the scientific sophistication and practical relevance of AI innovations across the agricultural domain. Several themes resonate throughout the contributions: the push toward field-ready robustness; the emphasis on model efficiency and energy-aware computation; the growing use of multi-modal sensing; and the extension of AI from crop-centered applications to soil, environment, and livestock systems.

Together, these works illustrate a field that is advancing rapidly toward intelligent, resilient, and sustainable agricultural ecosystems. As Guest Editor, I extend my deepest appreciation to the authors, reviewers, and editorial team whose efforts made this collection possible. It is my hope that this volume serves not only as a snapshot of current progress but also as a catalyst for future innovation at the nexus of agriculture and artificial intelligence.

Conflicts of Interest: The author declares no conflicts of interest.

List of Contributions:

1. Gunturu, S.; Munir, A.; Ullah, H.; Welch, S.; Flippo, D. A Spatial AI-Based Agricultural Robotic Platform for Wheat Detection and Collision Avoidance. *AI* **2022**, *3*, 719–738. https://doi.org/10.3390/ai3030042.
2. Sangha, H.S.; Darr, M.J. Influence of Model Size and Image Augmentations on Object Detection in Low-Contrast Complex Background Scenes. *AI* **2025**, *6*, 52. https://doi.org/10.3390/ai6030052.
3. Osco-Mamani, E.; Santana-Carbajal, O.; Chaparro-Cruz, I.; Ochoa-Donoso, D.; Alcazar-Alay, S. The Detection and Counting of Olive Tree Fruits Using Deep Learning Models in Tacna, Perú. *AI* **2025**, *6*, 25. https://doi.org/10.3390/ai6020025.
4. Wei, M.C.F.; Canal Filho, R.; Tavares, T.R.; Molin, J.P.; Vieira, A.M.C. Dimensionality Reduction Statistical Models for Soil Attribute Prediction Based on Raw Spectral Data. *AI* **2022**, *3*, 809–819. https://doi.org/10.3390/ai3040049.
5. Zambudio Martínez, M.; Silveira, L.H.M.d.; Marin-Perez, R.; Gomez, A.F.S. Development and Comparison of Artificial Neural Networks and Gradient Boosting Regressors for Predicting Topsoil Moisture Using Forecast Data. *AI* **2025**, *6*, 41. https://doi.org/10.3390/ai6020041.
6. Maki, H.; Lynch, V.; Ma, D.; Tuinstra, M.R.; Yamasaki, M.; Jin, J. Comparison of Various Nitrogen and Water Dual Stress Effects for Predicting Relative Water Content and Nitrogen Content in Maize Plants through Hyperspectral Imaging. *AI* **2023**, *4*, 692–705. https://doi.org/10.3390/ai4030036.

https://doi.org/10.3390/10.3390/ai7010012

7. Ghazal, S.; Kommineni, N.; Munir, A. Comparative Analysis of Machine Learning Techniques Using RGB Imaging for Nitrogen Stress Detection in Maize. *AI* **2024**, *5*, 1286–1300. https://doi.org/10.3390/ai5030062.
8. Giakoumoglou, N.; Pechlivani, E.-M.; Frangakis, N.; Tzovaras, D. Enhancing Tuta absoluta Detection on Tomato Plants: Ensemble Techniques and Deep Learning. *AI* **2023**, *4*, 996–1009. https://doi.org/10.3390/ai4040050.
9. Omole, O.J.; Rosa, R.L.; Saadi, M.; Rodriguez, D.Z. AgriNAS: Neural Architecture Search with Adaptive Convolution and Spatial–Time Augmentation Method for Soybean Diseases. *AI* **2024**, *5*, 2945–2966. https://doi.org/10.3390/ai5040142.
10. Jiménez, N.; Orellana, S.; Mazon-Olivo, B.; Rivas-Asanza, W.; Ramírez-Morales, I. Detection of Leaf Diseases in Banana Crops Using Deep Learning Techniques. *AI* **2025**, *6*, 61. https://doi.org/10.3390/ai6030061.
11. Apostolopoulos, I.D.; Tzani, M.; Aznaouridis, S.I. A General Machine Learning Model for Assessing Fruit Quality Using Deep Image Features. *AI* **2023**, *4*, 812–830. https://doi.org/10.3390/ai4040041.
12. Vance, J.; Rasheed, K.; Missaoui, A.; Maier, F.W. Data Synthesis for Alfalfa Biomass Yield Estimation. *AI* **2023**, *4*, 1–15. https://doi.org/10.3390/ai4010001.
13. Noa-Yarasca, E.; Osorio Leyton, J.M.; Hajda, C.B.; Adhikari, K.; Smith, D.R. Leveraging Spectral Neighborhood Information for Corn Yield Prediction with Spatial-Lagged Machine Learning Modeling: Can Neighborhood Information Outperform Vegetation Indices? *AI* **2025**, *6*, 58. https://doi.org/10.3390/ai6030058.
14. Shareef, U.; Rehman, A.U.; Ahmad, R. A Systematic Literature Review on Parameters Optimization for Smart Hydroponic Systems. *AI* **2024**, *5*, 1517–1533. https://doi.org/10.3390/ai5030073.
15. Mendoza, Q.A.; Pordesimo, L.; Neilsen, M.; Armstrong, P.; Campbell, J.; Mendoza, P.T. Application of Machine Learning for Insect Monitoring in Grain Facilities. *AI* **2023**, *4*, 348–360. https://doi.org/10.3390/ai4010017.
16. Benhoussa, S.; De Sousa, G.; Chanet, J.-P. FedBirdAg: A Low-Energy Federated Learning Platform for Bird Detection with Wireless Smart Cameras in Agriculture 4.0. *AI* **2025**, *6*, 63. https://doi.org/10.3390/ai6040063.
17. Neethirajan, S. Adapting a Large-Scale Transformer Model to Decode Chicken Vocalizations: A Non-Invasive AI Approach to Poultry Welfare. *AI* **2025**, *6*, 65. https://doi.org/10.3390/ai6040065.

References

1. Ghazal, S.; Munir, A.; Qureshi, W.S. Computer vision in smart agriculture and precision farming: Techniques and applications. *Artif. Intell. Agric.* **2024**, *13*, 64–83. [CrossRef]
2. Muqaddas, S.; Qureshi, W.S.; Jabbar, H.; Munir, A.; Haider, A. A Comprehensive Deep Learning Approach for Harvest Ready Sugarcane Pixel Classification in Punjab, Pakistan Using Sentinel-2 Multispectral Imagery. *Remote Sens. Appl. Soc. Environ.* **2024**, *35*, 101225. [CrossRef]
3. McMahan, B.; Moore, E.; Ramage, D.; Hampson, S.; Arcas, B.A. Communication-Efficient Learning of Deep Networks from Decentralized Data. In Proceedings of the 20th International Conference on Artificial Intelligence and Statistics (AISTATS) 2017, Fort Lauderdale, FL, USA, 20–22 April 2017; Proceedings of Machine Learning Research (PMLR): Cambridge, MA, USA, 2017; pp. 1273–1282.

https://doi.org/10.3390/10.3390/ai7010012

AI

MDPI

Article

A Spatial AI-Based Agricultural Robotic Platform for Wheat Detection and Collision Avoidance

Sujith Gunturu [1], Arslan Munir [1,*], Hayat Ullah [1], Stephen Welch [2] and Daniel Flippo [3]

[1] Department of Computer Science, Kansas State University, Manhattan, KS 66506, USA
[2] Department of Agronomy, Kansas State University, Manhattan, KS 66506, USA
[3] Department of Biological and Agricultural Engineering, Kansas State University, Manhattan, KS 66506, USA
* Correspondence: amunir@ksu.edu

Abstract: To obtain more consistent measurements through the course of a wheat growing season, we conceived and designed an autonomous robotic platform that performs collision avoidance while navigating in crop rows using spatial artificial intelligence (AI). The main constraint the agronomists have is to not run over the wheat while driving. Accordingly, we have trained a spatial deep learning model that helps navigate the robot autonomously in the field while avoiding collisions with the wheat. To train this model, we used publicly available databases of prelabeled images of wheat, along with the images of wheat that we have collected in the field. We used the MobileNet single shot detector (SSD) as our deep learning model to detect wheat in the field. To increase the frame rate for real-time robot response to field environments, we trained MobileNet SSD on the wheat images and used a new stereo camera, the Luxonis Depth AI Camera. Together, the newly trained model and camera could achieve a frame rate of 18–23 frames per second (fps)—fast enough for the robot to process its surroundings once every 2–3 inches of driving. Once we knew the robot accurately detects its surroundings, we addressed the autonomous navigation of the robot. The new stereo camera allows the robot to determine its distance from the trained objects. In this work, we also developed a navigation and collision avoidance algorithm that utilizes this distance information to help the robot see its surroundings and maneuver in the field, thereby precisely avoiding collisions with the wheat crop. Extensive experiments were conducted to evaluate the performance of our proposed method. We also compared the quantitative results obtained by our proposed MobileNet SSD model with those of other state-of-the-art object detection models, such as the YOLO V5 and Faster region-based convolutional neural network (R-CNN) models. The detailed comparative analysis reveals the effectiveness of our method in terms of both model precision and inference speed.

Keywords: spatial AI; deep learning; depth sensing; robotic platform; collision avoidance; agriculture; embedded computing

1. Introduction and Motivation

Wheat (*Triticum*) is one of the most important staple foods in the temperate world, of which the United States produces 8% of the world's total [1–3]. Thus, there is a great need to conduct research on its growth and development in field plot studies conducted as breeding program wheat performance trials. This helps wheat breeders predict plant traits (phenotypes), such as yield, based on their genetic constitutions (genotypes). One of the most important aspects of wheat research is to understand the relationship between wheat growth and the soil properties (for instance, soil moisture) of the fields where it is grown. One measuring device is the Geophex Ltd. Gem-2 electromagnetic induction soil electrical conductivity sensor. It weighs over 15 lbs. and is typically carried manually. To best support digital agricultural research, soil conductivity would be measured several times per week. However, a breeding trial that is 50 m × 75 m can take four hours, and should be done at specific times of day.

To facilitate breeding trials, we used a robot to carry the sensor through a field manually guided via a remote control. The robot saves investigators' time and energy while trying to acquire the soil's properties. While remote control provides an easy solution, it is imperative not to inadvertently harm the wheat in any of the three hundred (300) or more plots, which are separated by very constricted aisles (9 to 12 inches). As the field size increases, human control from greater distances becomes hard and the possibility that the robot will run over the wheat increases. One way to avoid this is to make the robots intelligent enough to distinguish the wheat from the aisles, turn its wheels to stay aligned with the aisles (autonomous navigation), and to halt, if, for any reason, that becomes impossible. This requires not only the detection of wheat in the field of view but also its distance from the robot. If the robot recognizes wheat far enough ahead, there is time to align its wheels and avoid collision. However, if the wheat is closer than some permitted threshold distance (for instance, if a surface irregularity unexpectedly deflects the wheels toward the wheat a few feet or inches away at the edge of the aisle), there will not be time to do anything but stop before running over the wheat. This entails the use of spatial artificial intelligence (AI).

Spatial AI is an ability of an AI system to reason based on not just what it is looking at but also how far away things are located. Spatial AI applies AI to not only identify the object but also provide information on where the object, in this case wheat, is in 3D space. When the robot makes the informed decision to halt, the operator and/or the robot itself can turn its wheels, while remaining in place, and then, when they are pointed in the correct direction, restart the acquisition of soil properties.

Our main contributions in this article are as follows:

- Design of an agricultural robotic platform for autonomous navigation in crops while avoiding collisions. The platform uses spatial AI and deep learning models for collision avoidance and crop (wheat) detection.
- Training of different state-of-the-art deep learning models, such as MobileNet single shot detector (SSD), YOLO, and Faster region-based convolutional neural network (R-CNN) with ResNet-50 feature pyramid network (FPN) backbone, for object (wheat) detection.
- Performance comparison of the state-of-the-art deep learning models for wheat detection on different computing platforms.
- Evaluation of the trained deep learning models for wheat detection through various metrics, such as accuracy, precision, and recall.

The remainder of this article is organized as follows. Section 2 summarizes the previous works related to object detection and deep learning, mainly focusing on CNNs. Section 3 discusses the proposed framework and its technical components. Following this, Section 4 discusses training evaluation and detailed experimental results. Finally, Section 5 concludes the article with its limitations and future directions.

2. Related Work

Deep learning has played an important role in various fields, such as biology, medicine, agriculture, and agronomy. This section discusses previous works in the literature related to computer vision and the use of computer vision in agriculture.

Redmon et al. [4] presented YOLO, a new approach in object detection. YOLO treats object detection as a regression problem, which is the main difference between YOLO and prior object detection models. YOLO is very fast, processing images at a higher rate than other object detection approaches, and gives the best performance in real-time object detection [5]. As illustrated in Redmon et al. [4], YOLO reasons globally about the image when making predictions and learns generalizable representations of objects. Results in [4] reveal that YOLO produces only half the number of background errors as compared to Fast R-CNN. These reasons led us to use YOLO as our deep learning model for wheat detection.

Inspired by the work of Ren et al. [6], we also implemented Faster R-CNN with an FPN backbone for real-time wheat detection in the field. Faster R-CNN has achieved state-of-the-art object detection accuracy on PASCAL VOC 2007, PASCAL VOC 2012 [7,8], and MS COCO [9] datasets with only 300 proposals per image. These exceptional results led us to use Faster R-CNN with a ResNet-50-FPN backbone model in our study. Faster R-CNN and RPN have been used by several entries in many competitions [10]. It should also be observed that R-CNN with a ResNet-50-FPN backbone not only provides a cost-efficient solution for practical usage, but also helps improve the accuracy of object detection. Faster R-CNN is composed of two modules. The first module is a deep fully convolutional network and proposes regions, and the second module is the fast R-CNN detector that uses the proposed regions [11]. In this article, the accuracy of YOLO is compared against that of Faster R-CNN with ResNet-50-FPN [12]. This comparison helped us decide which of the two models, that is, Faster R-CNN with ResNet-50-FPN or YOLO, is more suitable for our final implementation in the robotic platform. Results (Section 4.8) reveal that we were able to attain almost equal accuracy using both the models; however, YOLO is faster than Faster R-CNN. A more detailed discussion on the accuracies of the two models is presented in Section 4.10.

Liu et al. [13] have proposed a new approach named single shot detector (SSD), which discretizes the output space of bounding boxes into a set of default boxes with different aspect ratios and scales per feature map location. They observed that the SSD model is relatively simple to train. The SSD model eliminates proposal generation and the subsequent pixel or feature resampling stage and encapsulates all computations in a single network. The work done by He et al. [14] is most relevant to our work. They presented a residual learning framework to ease the training of networks that are substantially deeper than those used in other state-of-the-art models. The extremely deep representations also result in good generalization performance on other recognition tasks, which led them to win 1st place in ImageNet [15] detection, ImageNet localization, and COCO detection competitions. This state-of-the-art performance inspired us to adopt a ResNet backbone in our Faster R-CNN models for wheat detection.

Mosley et al. [16] performed experiments on detecting and counting sorghum head through parameter tuned SSD from the images obtained from unmanned aerial vehicles (UAVs). Their approach involves parameter tuned anchor boxes, which achieves an out-of-sample mean average precision of 0.95. However, their proposed system is unable to work in a real-time environment due to its high computational complexity. Ghosal et al. [17] explored machine learning-based approaches such as deep convolutional neural networks for efficient object detection. Their proposed methodology involves a weakly supervised deep learning framework inspired by active learning for sorghum head detection. They utilized an object detection framework called RetinaNet with ResNet-50. The backbone network adopted in this study was the feature pyramid network (FPN). The FPN was built on top of ResNet-50, and it has shown good results. However, due to complex architecture of ResNet-50, their method cannot be adopted for real-time object detection applications. Velumani et al. [18] have investigated the effect of ground sampling distance (GSD) on detection stage using a Faster R-CNN object detection algorithm for maize plants. Results have shown promising plant detection and counting with a root mean square error of 0.08. However, their method uses a computationally complex ResNet-50 feature extraction backbone in the Retina-Net architecture, which makes their method unsuitable for real-time applications.

Gonzalo-Martín et al. [19] implemented test time augmentation (TTA) to overcome the challenges of differences in shape and color of the sorghum head in UAV imagery. Results indicate that the detection achieved by TTA outperforms detection based on individually transformed testing sets. Xue et al. [20] proposed a velocity control strategy for autonomous agricultural vehicles based on the moment state, hazard severity, and distance between the object and vehicle. Results indicate that their collision avoidance strategy predicts collisions in real-time with an average detection time of 0.2 s. Shutske et al. [21] have developed a microwave sensor and control system to mitigate the probability of a fast-moving vehicle colliding with a slow-moving vehicle from the rear. The sensor unit which is outfitted on the back of a vehicle senses the distance and velocity of a vehicle moving closer from the rear.

Prior works have not leveraged spatial AI and deep learning models for wheat detection and collision avoidance in real-time. Furthermore, previous works did not compare the performances of the state-of-the-art deep learning models for wheat detection on different edge computing platforms. This work filled the void in previous works by devising a real-time wheat detection and collision avoidance system and evaluating the performance of proposed system on a variety of edge computing platforms.

3. Proposed Spatial AI System for Collision Avoidance in Wheat Fields

In this section, we present a detailed discussion of our proposed spatial AI-driven collision avoidance system and its technical components. The workflow of our proposed system was divided into three distinct phases, as shown in Figure 1.

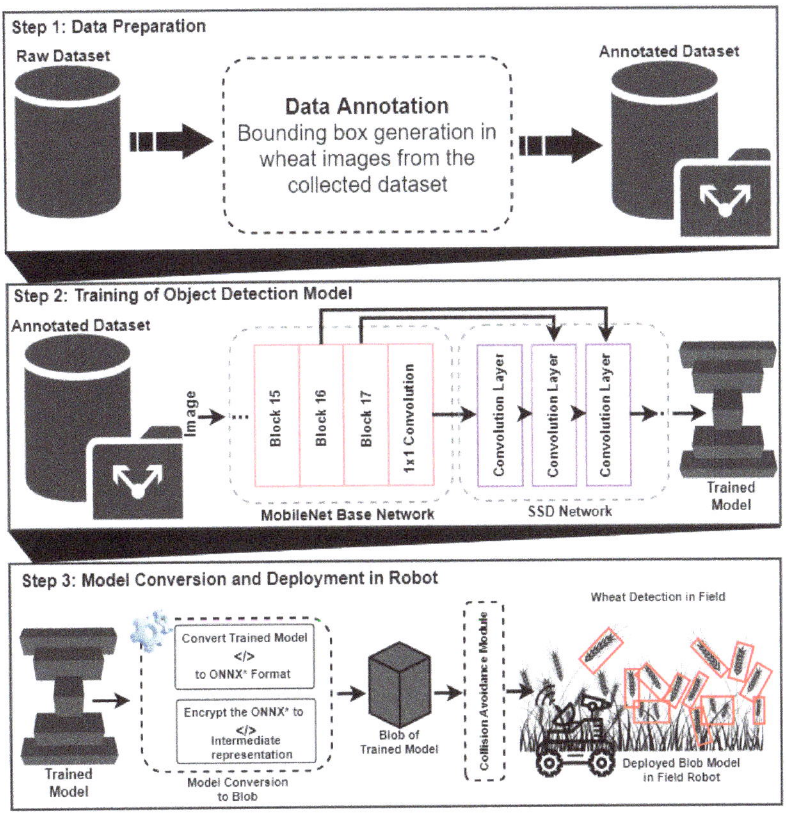

Figure 1. The visual overview of our proposed spatial AI system for collision avoidance in wheat field.

This first phase was the collection of image data for training from different sources, including farm sites and online repositories, such as Google Images and Kaggle. The collected image data were then manually annotated for training by generating bounding boxes on the wheat images. The second phase was the training phase where a computationally efficient yet robust object detector, such as MobileNet SSD, was trained on our prepared annotated dataset. Lastly, the third phase converted the trained wheat detection model to a blob format and then deployed it on the robot for practical use in a real-time application in wheat fields.

3.1. Data Preparation

In any computer vision problem, the performance of a deep learning model mainly depends on the quality of the dataset, which is, in turn, nothing but the quantity and quality of the images. For training our deep learning model, our dataset comprised images from multiple sources, including farm sites where the robots are used. The data were collected from the field at Ashland bottoms (KSU Agronomy Research Farm) spread over a latitude and a longitude from (39.128, −96.6157) to (39.1284, −96.6164) in the city of Manhattan, Kansas, United States. Most of the images in our collect training dataset were obtained on the KSU Agronomy Farm with a digital camera, as shown in Figure 2. The images were taken considering the view of the robot; that is, the training images are like the images that are to be encountered and predicted by the robot.

Figure 2. Research site (KSU Agronomy farm) in Manhattan, Kansas.

One of the important requirements of our deep learning model is that the image recognition must be done in all stages of wheat crop; that is, the wheat color can be brown or green depending on the season or time of operation of the robotic platform. Considering such variety in training data, we collected both brown and green wheat image data from the wheat fields. For ease of understanding, sample images of both brown and green wheat are depicted in Figure 3a,b, respectively. The image shown in Figure 3a was taken on 4 July 2021, and represents the stage of wheat growth called the reproductive stage. This is the stage where the wheat heads have fully emerged from the stem. Pollination is very quick and takes only 3 to 5 days to complete. Thus, the images were taken in quick succession, forming a dataset comprising 1200 images of pollinated wheat with emerged heads. The image shown in Figure 3b was taken on 22 June 2021. This stage of wheat growth is called the maturity stage, which is also known as hard dough. In this phase, the plant turns a straw color and the kernel becomes very hard. The focus of collected training images is to get view of robot and make sure that wheat is clearly visible and can be annotated for training. In addition, we also collected images from online public repositories, such as Google Images and Kaggle. From Google Images, we downloaded non-copyrighted images, including both brown and green wheat images. From the Kaggle repository, we acquired more than 3000 images, and this amount was later increased by performing data augmentation before training the model. After collecting the dataset, we manually annotated the collected images by labeling them with bounding boxes (rectangles) having four coordinates that specify the locations of wheat in the given image. A bounding box specifies the region of interest by (x, y) coordinates of the upper-left corner and the (x, y) coordinates of the lower-right corner, which together define a diagonal representation of a rectangle. The sample annotation of bounding boxes over a wheat image is depicted in Figure 4.

(a) (b)

Figure 3. The visual overview of wheat at different states. (**a**) A sample image from the KSU Agronomy Farm when wheat is in reproductive state. (**b**) A sample image from the KSU Agronomy Farm when wheat is in mature state.

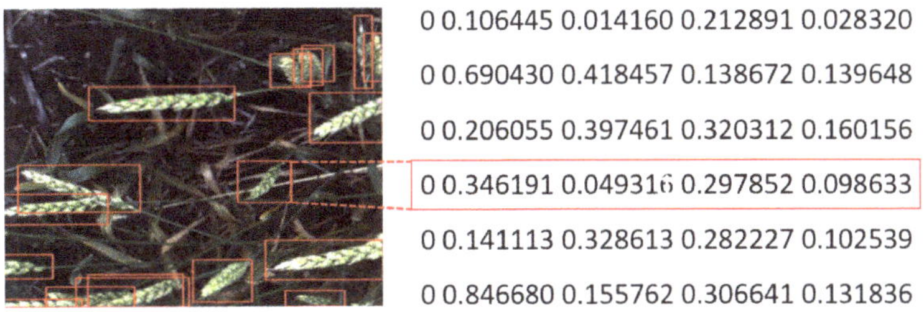

Figure 4. The visual depiction of sample annotated image from training data.

3.2. MobileNet SSD Architecture for Wheat Detection

In this research study, we explored different state-of-the-art object detection models, including YOLO v5, Faster R-CNN, and MobileNet SSD, and compared their performances for wheat detection. Based on the obtained performance, we determined that the MobileNet SSD is better than other models in terms of objection detection accuracy, model complexity, and time complexity. Therefore, we chose MobileNet SSD as the object detection architecture in our proposed framework (Figure 1) for wheat detection in the field.

A detailed discussion on the proposed MobileNet SSD architecture is provided in the subsequent subsection.

3.2.1. Architectural Details of MobileNet SSD

This section provides the technical details of the MobileNet SSD architecture employed in our proposed spatial AI system for wheat detection and collision avoidance. The MobileNet SSD architecture is an encapsulation of two modules that include a backbone feature extraction module and an SSD module (containing extra object-specific feature extraction layers), as depicted in Figure 5. The MobileNet V1 backbone feature extractor uses depthwise separable convolutions instead of standard convolutions, where each depthwise separable convolution layer consists of depthwise and pointwise convolutions, which greatly reduces the overall complexity of model. The standard MobileNet V1 architecture starts with a standard convolution layer, followed by 13 depth convolution layers. After the depthwise separable convolution layers, the obtained features maps are converted to a fully connected layer and pooled by a maxpooling layer, and finally the softmax layer generates the probabilities for a predefined number of classes based on the extracted features. Since here our objective is to use only the learned representation from the MobileNet V1 architecture, we froze the last three layers (fully connected layer, maxpooling layer, and softmax layer) and used the output from the last depthwise convolutional layer, and fed that to the SSD detector module as an input.

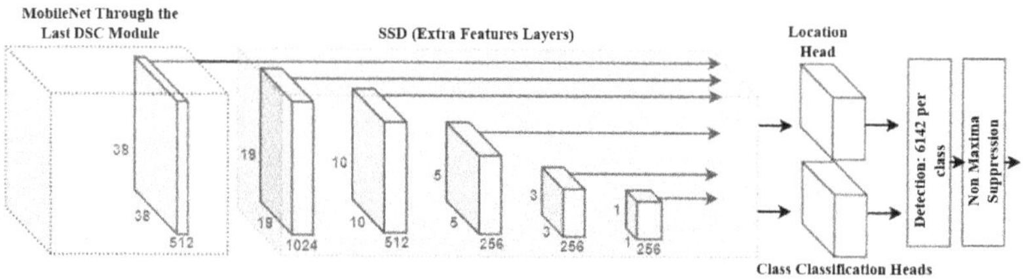

Figure 5. The visual overview of the MobileNet SSD architecture employed in our proposed framework.

The SSD detector module first applies a set of boxes to each cell of the feature maps learned from the MobileNet V1 architecture. Next, it predicts a score for each candidate class in the corresponding feature map cell. Consequently, for each map, SSD generates $(C_{candidate} + 4)/kwh$ results, where $C_{candidate}$ denotes number of classes, k represents the number of default bounding boxes, and w and h represent the width and height of feature map, respectively. The SSD architecture uses several feature maps having different resolutions to take advantage of both low-level and high-level features. Based on the utilized feature maps in the SSD architecture, the scale of the default box S_k can be mathematically expressed as follows:

$$S_k = S_{min} \frac{S_{max} - S_{min}}{m - 1} \times (k - 1) \quad (1)$$

where S_{min} and S_{max} are the scales of the lowest and highest feature map, respectively, and m represents the number of features maps. The five common aspect ratios of bounding boxes can be expressed as $A_R \in \{1, 2, 3, 0.5, 0.33\}$, where the width and height of the box can be computed using $S_k\sqrt{A_R}$. Similarly, the center of a bounding box can be estimated as $\left(\frac{i+0.5}{|f_k|}, \frac{j+0.5}{|f_k|}\right)$, where $|f_k|$ is the size of feature maps and $i, j \in [0, |f_k|]$. Further, SSD uses a Jaccard index metric to compute the matches, where the Jaccard value ≥ 0.5 between the ground-truth and the predicted bounding boxes is considered to be a match box. Mathematically, the Jaccard index metric can be expressed as follows:

$$Jaccard = \frac{A_o}{A_u} \quad (2)$$

where A_o denotes the area of overlap (i.e., the common area between the ground-truth and the predicted bounding box) and A_u denotes the area of union (i.e., combined area of the ground-truth and the predicted bounding box), respectively.

The SSD architecture uses the joint loss function which is the summation of localization loss L_{loc} and classification loss L_{cls}. The localiztaion loss is smooth $L1$ loss ($L1_{smooth}$), which can be mathematically expressed as follows:

$$L_{loc}(x,l,g) = \sum_{i \in Pos}^{N} \sum_{m \in \{cx,cy,w,h\}} x_{ij}^{k} L1_{smooth}(l_i^m - \hat{g}_j^m),$$

where (3)

$$\hat{g}_j^{cx} = \frac{g_j^{cx} - d_i^{cx}}{d_i^w}, \quad \hat{g}_j^{cy} = \frac{g_j^{cy} - d_i^{cy}}{d_i^h}, \quad \hat{g}_j^w = log\left(\frac{g_j^w}{d_i^w}\right), \quad \hat{g}_j^h = log\left(\frac{g_j^h}{d_i^h}\right).$$

Here, N is the number of correct matches, and the terms d, g, and l represent the default bounding box, ground-truth bounding box, and predicted bounding box, respectively. The symbols cx and cy represent the x and y coordinates, respectively, of the center location of the default bounding box, and w and h denote the width and height of the default bounding box, respectively.

The second type of loss that SSD uses is classification loss L_{cls}, which is the loss function related to the prediction of the object type by the predicted bounding box. The classification loss of SSD can be mathematically expressed as follows:

$$L_{cls}(x,c) = -\sum_{i \in Pos}^{N} x_{ij}^{p} log(\hat{c}_i^p) - \sum_{i \in Neg} log(\hat{c}_i^0) \quad (4)$$

where the term $\hat{c}_i^p = \frac{exp(c_i^p)}{\sum_p exp(c_i^p)}$ represents the model's predicted confidence score for class p. The term $\hat{c}_{i0}$ represents the confidence value for the negative match of the bounding box. Similarly, Pos and Neg represent the positive and negative matches of bounding boxes, respectively. The term x_{ij}^p is an indicator variable which verifies the match of the ith default bounding box and the jth ground-truth bounding box for class p. Together with both loss functions, the joint loss function $Loss_{total}$ can be calculated as follows:

$$Loss_{total} = \frac{1}{N}(L_{cls}(x,c) + \alpha L_{loc}(x,l,g)) \quad (5)$$

where N denotes the total number of correct matches and α represents the weight factor for the localization loss.

3.2.2. Motivation of Using MobileNet SSD for Wheat Detection

The selection of a suitable model for the problem under consideration is very challenging, particularly when the computing resources onboard the robot are limited. This section briefly discusses the reasons why we have chosen a MobileNet SSD architecture over YOLO V5 and Faster R-CNN architectures for our spatial AI-based framework. Several factors, such as model complexity, model accuracy, and model inference time, need to be considered while dealing with resource-constrained computing platforms and their usage for practical industrial applications, such as agriculture. To choose a suitable optimal object detection model, we have conducted extensive experiments and evaluated MobileNet SSD, YOLO V5, and Faster R-CNN models based on the aforementioned criteria. Based on the detailed model evaluation experiments (Section 4), we chose MobileNet SSD for our proposed framework, as MobileNet SSD model balances tradeoffs between model accuracy, model complexity, and inference time. Furthermore, MobileNet SSD is feasible for computing platforms with limited memory and resources, such as LattePanda, Raspberry Pi, and Intel Neural Compute Stick.

3.3. Model Conversion and Deployment on Field Robot

In this section, we discuss the conversion of the trained model to a blob format and its deployment on a robot for real-time wheat detection in the field. The proposed model conversion module is three-fold, which converts network binaries at three levels and obtains the final blob representation of the trained model, as shown in Figure 6. The model conversion module starts with the input trained model and converts it into an open neural network exchange (ONNX) format. The model conversion module encode the model's metadata in protocol buffer format (having extension .proto), which provides only the data type information (e.g., float32) of the trained model layers and learned knowledge in a JSON file. The next level of conversion is an intermediate representation, which is an ad hoc extraction of the ONNX information designed to facilitate its conversion to the final blob format. To enable the use of custom-trained models by DepthAI [22], converting them into a blob file format optimizes the best inference on Myriad X processor. The first three steps in model conversion pipeline are coded using open source modules publicly available in Python; the last step (also programmed in Python) is done via an online API call to software provided by Luxonis [23].

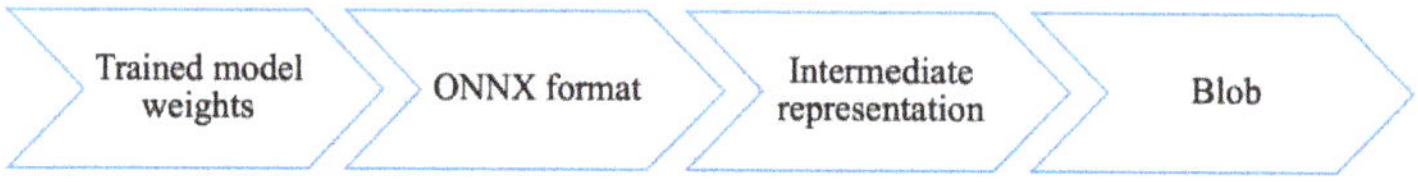

Figure 6. The visual overview of workflow for model conversion to blob format.

3.4. Workflow for Robot Operation and Collision Avoidance

In this section, we discuss the decision-making and collision avoidance workflow of a robot for wheat detection in field. The robot used in this research was equipped with a stereo camera [24] and the trained model. For better understanding, the operations of our proposed robotic system are put into two categories, namely, *wheat detection* and *collision avoidance*. The first category of robot operation is wheat detection, where the mounted stereo camera, along with the trained model, helps the robot to not only detect the wheat in the field but also supervises it to drive on the right path in the field to avoid collisions. The second category of robot operation is to avoid collisions with wheat via depth sensing and communication between the robot and the embedded computing device using the PySerial API. Both collision avoidance through depth sensing and communicating via PySerial API are discussed in detail the following subsections.

3.4.1. Collision Avoidance via Depth Sensing

The above section described how the object detection is achieved. However, the final decision on when to stop the robot is solely based on the distance between wheat and the robot. Thus, the depth information is crucial. To acquire depth sensing, left and right stereo cameras outfitted on the OpenCV AI kit [24] were utilized, and the object detection was accomplished by the red, green, blue (RGB) camera outfitted at the center of the OpenCV AI kit. An OpenCV AI kit was put on the robot. The OpenCV AI camera neither uses weights of the model nor the model directly to perform object detection; rather, it uses a blob which was obtained from the OpenVINO model. The blob conversion is accomplished using open source software provided by Luxonis [23], as discussed in Section 3.3. The obtained blob is then used in detecting wheat, and left and right stereo cameras of the OpenCV AI kit are used in estimating the distance of wheat from the camera (and thus the robot).

3.4.2. Communicating with the Robot Using PySerial

To stop the robot when it detects wheat at less than the threshold distance, a connection needs to be established between the robot and the embedded controller board (LattePanda [25] in the case of our implementation). LattePanda has a Universal Serial Bus (USB) port, which is connected to the Arduino controller of the robot. When the deep learning model detects wheat, it sends a signal to the USB port, via which the signal is sent to the robot's speed controller, which adjusts the robot's speed. A flowchart of this process is shown in Figure 7, and the related code snipped for robot control is shown in Figure 8. This code snippet is embedded into depth-sensing code for communicating with the robot. In the code, "conf" represents the confidence of object detection, whose threshold can be adjusted by the designer depending on the performance of the model.

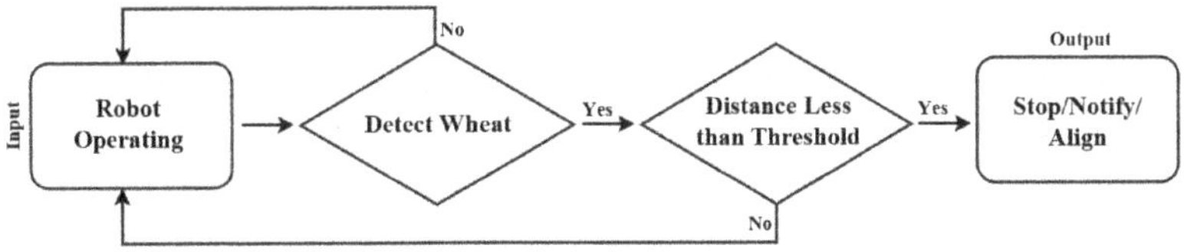

Figure 7. Workflow for robot's operation and collision avoidance.

```
import os, serial, time, sys

def write_read(x):
    arduino = serial.Serial(port = 'COM6', baudrate = 9600, timeout = 5)
    arduino = write(bytes(x, 'utf-8'))
    time.sleep(.05)
    data = arduino.readline()
    arduino.close()
    return data

wheatheads_count = 0

if conf.item() > 0.90 and distance < THRESHOLD_DISTANCE:
    wheatheads_count += 1
    if wheatheads_count > 1:
        num = 'H'
        value  = write_read(num)
        wheatheads_count = 0
        option = input("enter L to start AND Q to Quit")
        if option == 'Q':
            num = 'L'
            value = write_read(num)
            sys.exit()
        value = write_read(option)
```

Figure 8. Code snippet to perform collision avoidance.

4. Training Evaluation and Experimental Results

This section provides a detailed discussion on the training phase of this research study. First, we discuss data augmentation as a preprocessing step. Next, we present the embedded computing platforms with which we performed feasibility analysis to determine their suitability for deployment on our robotic system. We then briefly discuss the evaluation metrics for performance evaluation and comparison, followed by training details of MobileNet SSD and other comparative object detection models (YOLO V5 and Faster R-CNN).

4.1. Data Augmentation

Image augmentation is the process of taking the images that are already present in the training dataset and manipulating them to create various altered versions of the same image. Augmentation provides more images to train and gives a different viewpoint to the classifier. Different viewpoints can represent changes in the saturation, color, crop, and horizontal and vertical flips. To create augmented data, we used a PyTorch orientation module named Albumentations. Albumentations is a module where all the necessary augmentation steps for augmentations are provided in predefined functions. These functions are applied to the image dataset to create manipulated images, which are then used for training the deep learning model. A few examples of augmentations applied are:

SmallestMaxSize: Rescales an image so that the minimum size is equal to the given maximum size, while keeping the aspect ratio of the initial image.

ShiftScaleRotate: Randomly assigns transforms, such as translate, scale, and rotate, to the input images.

RandomCrop: Crops a random part of the image.

RGBshift: Randomly shifts values for each channel of the input RGB image.

RandomBrightnessContrast: Randomly changes brightness and contrast of the input image.

The transformations of training image after data augmentations are visually depicted in Figure 9.

Figure 9. Sample images from the augmented training dataset.

4.2. Embedded Computing Platforms

We benchmarked the deep learning models on various embedded computing platforms that are suitable for integration with the robotic platform for wheat detection. The embedded computing platforms that we leveraged in our experimentation and benchmarking included LattePanda, Intel Neural Compute Stick (NCS), and OpenCV AI kit. Here, we summarize the main characteristics of these embedded computing platforms.

4.2.1. LattePanda

LattePanda Alpha 864s [25] is a high-performance, palm-sized embedded board that runs Windows 10 and has low power consumption. It is being widely utilized in edge computing, vending, advertising machines, and industrial automation. This palm-sized machine was outfitted onto our autonomous robot and was used for running the deep learning model(s), communicating with the operator, and communicating with the robot's integral components. Key features of LattePanda Alpha 864s are an Intel Core M3-8100Y dual-core processor operating at 1.1–3.4 GHz, an Intel Ultra High Definition (UHD) Graphics 615, 8 GB Memory, and an integrated Arduino ATMEL 32U4 co-processor.

4.2.2. Intel Neural Compute Stick

Intel NCS [26] is an accelerator for deep learning to enable the deep learning model to recognize objects at a high rate of frames per second. It can support heterogeneous execution across computer vision accelerators implemented on a central processing unit (CPU), graphics processing unit (GPU), vision processing unit (VPU), and field-programmable gate array (FPGA). It supports the Intel OpenVINO toolkit, and also supports various operating systems, including Windows, Mac, and Ubuntu. It integrates an Intel Movidius Myriad X VPU. It supports various machine learning frameworks, including TensorFlow, Caffe, Apache MXNet, Open Neural Network Exchange (ONNX), PyTorch, and PaddlePaddle via an ONNX conversion.

4.2.3. OpenCV AI Kit

OpenCV AI Kit with Depth (OAK-D) [24] is a spatial AI platform that can simultaneously run advanced neural networks while providing depth from two 1 megapixel (MP) global shutter-synchronized stereo cameras, one on the left and one on the right, and color information from a single 4K 12 MP RGB camera in the center. OAK-D is essentially a smart camera with neural inference and depth processing capability. OAK-D can be used with any host operating system that OpenVINO supports. It supports 4K/30 fps H.265, JPEG, H.264, and H.265/HEVC encodings [27].

4.3. Evaluation Metrics

In the following, we discuss the evaluation metrics we have utilized for evaluating our deep learning models.

4.3.1. Precision

Precision P measures the accuracy of a model's prediction; that is, precision quantifies the percentage of correct predictions [28].

$$P = \frac{TP}{TP + FP} \tag{6}$$

where TP denotes *true positives*, that is, predicted as positive correctly; and FP denotes *false positives*, that is, predicted as positive, incorrectly.

4.3.2. Recall

Recall R measures how well a model finds all the positives [28]:

$$R = \frac{TP}{TP + FN} \tag{7}$$

where TP signifies *true positives*, that is, predicted as positive correctly, and FN represents *false negatives*, that is, predicted as positive, incorrectly.

4.3.3. Intersection over Union (IoU)

An object detection system's predictions are characterized by a bounding box and a class label [28]. For each bounding box, the measure of correctness is defined by an IoU metric, which measures the overlap between the predicted bounding box and the ground truth bounding box; that is,

$$IoU = \frac{A_o}{A_u} \tag{8}$$

where A_o and A_u are area of overlap and the area of union, respectively. For object detection tasks, precision P and recall R are calculated using the IoU for a given IoU threshold. For example, if IoU prediction is greater than the IoU threshold, then we classify that prediction as true positive (TF). If the IoU value for a prediction is less than 0.5, say 0.4, we classify that prediction as a false positive (FP). This implies that for a given prediction, we may get different binary TP, FP, and false negative (FN) values, and thus different P and R values, by changing the IoU threshold.

4.3.4. Mean Average Precision (mAP)

The average precision (AP) for a given class is obtained by calculating the area under the precision–recall (PR) curve for the object detections. The mAP is the average of AP over all classes and/or overall IoU thresholds for a set of detections. Often, interpolated AP, in particular, 11-point interpolated AP, is used for calculating mAP.

4.4. Training YOLO

The final and the most crucial part of creating a real-time deep learning model is to train the model on training images. YOLO performs supervised training for object detection. We performed the training of our YOLO model in Google Collaboratory using Colab Pro subscription. Google Collaboratory provides Nvidia Tesla K80 GPUs that have a dual-GPU design with 4992 CUDA cores (2496 CUDA cores per GPU). Nvidia Tesla K80 has a 24 GB of GDDR5 memory and has a PCI Express (PCIe) interface. For every cycle of data collected from the KSU Agronomy Farm, YOLO was trained on approximately 2000+ images for 7.679 h of GPU time. The training was performed on a pretrained model with available weights of a large YOLO model. As the new images were added to the dataset, training was performed on the new images to update the model weights. The training, validation, and testing splits for different data sources are given in Table 1. Training was done by setting image size to 1024 × 1024, batch size to 16, and the number of training epochs to 100.

Table 1. Training, validation, and testing splits (%) for different data sources.

Image Source	Training	Validation	Testing
Kaggle	80	10	10
Google Images	80	10	10
KSU Agronomy Farm	70	10	20

4.5. Training Faster R-CNN with ResNet-50-FPN

We have used stochastic gradient descent (SGD) as an optimizer in training this model. The learning rate (LR) scheduler used in training this model was step LR. The loss function used in the Faster R-CNN was binary cross-entropy in the first state of the region proposal network (RPN), and the classification loss used was normal cross-entropy [6]. Training was performed in Google Collaboratory. The initial training of Faster R-CNN with ResNet-50-FPN took 8 h of GPU time for 2000+ images. Object loss comparison between YOLO and Faster R-CNN with ResNet-50-FPN while training is shown in Figure 10. To insure the timely convergence and avoid over-fitting, we used an "early stop" strategy that allows one to stop the training if the model converges well-enough and there is no room for further improvement in training and validation losses.

4.6. Training MobileNet SSD

The anticipation and classification of bounding box positions in SSD architecture was done in a single pass by a single convolutional network in the SSD architecture. The network consists of a base architecture followed by several convolution layers. Originally, MobileNet SSD model was trained on a benhmark object detection dataset called "COCO dataset". In this research, we used the pre-trained MobileNet SSD model and retained it on our wheat images dataset using a transfer learning approach. We utilized Tensorflow object detection API and Model Zoo resources for training MobileNet SSD and other two object detection models, including YOLO V5 and Faster R-CNN. The MobileNet SSD, due to its single shot approach to recognizing several objects in the image, is faster as compared to two-shot RPN based approaches, such as R-CNN. TensorFlow Object Detection API is a framework designed to solve object detection problems. Model Zoo consists of pretrained computer vision models on the COCO dataset and the KITTI dataset.

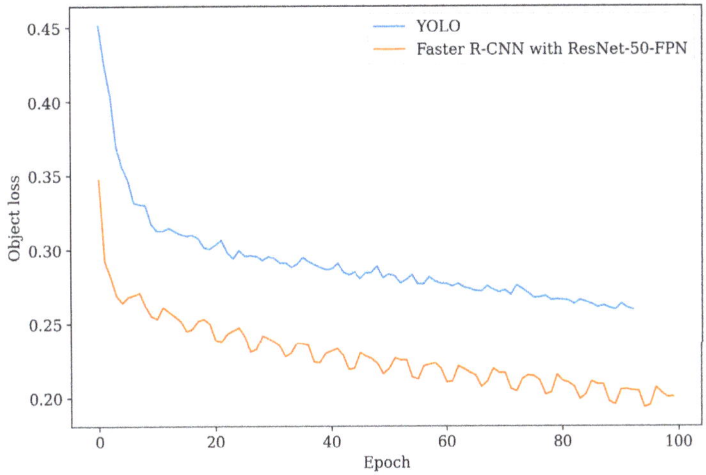

Figure 10. Object loss for YOLO and Faster R-CNN with ResNet-50-FPN.

4.7. Testing YOLO

To test the performance of YOLO, the model was run on test images from the dataset. The performance of the model P_m is defined by the following equation.

$$P_m = \frac{N_D}{N_T} \tag{9}$$

where N_D denotes the total number of wheat heads detected and N_T denotes the total number of wheat heads present in the image. Our YOLO model achieved a performance of 0.93 on the test dataset. The weights of this model were saved for future training and inference in the field.

4.8. Testing Faster R-CNN with ResNet-50-FPN

Our Faster R-CNN with ResNet-50-FPN model achieved a performance of 0.90 (using Equation (9)) in identifying the wheat heads. Results indicate that the performance of YOLO is 3.33% higher as compared to the Faster R-CNN with ResNet-50-FPN. Testing was done on LattePanda [29], which is the actual embedded computing platform used for real-time detection in the field.

4.9. Testing MobileNet SSD

We have evaluated the accuracy of our MobileNet SSD model using our images collected from the Agronomy farm (Figure 2). Results indicate that MobileNet SSD attained accuracy close to those of Faster R-CNN with ResNet50 FPN and YOLO. Results show that MobileNet SSD caused a slight decrease in detection accuracy: 5% and 3% as compared to YOLO and Faster R-CNN, respectively. Since the major concern in this study was to attain high speed inference on embedded platforms, we deployed MobileNet SSD on the OpenCV AI kit. Results reveal that the model achieved 12–13 fps on 600 × 600 images and 18–23 fps on 300 × 300 image.

4.10. Comparing YOLO, Faster-R-CNN, and MobileNet SSD

After training both YOLO and Faster R-CNN with ResNet-50-FPN on the training dataset, results revealed that the models attained almost identical performance results in terms of wheat head detection (Equation (9)). Results further indicate that the speed of YOLO is three times faster than that of the Faster R-CNN with ResNet-50-FPN. Figure 11 shows time taken in seconds by the three deep learning models to perform inferences for various numbers of images.

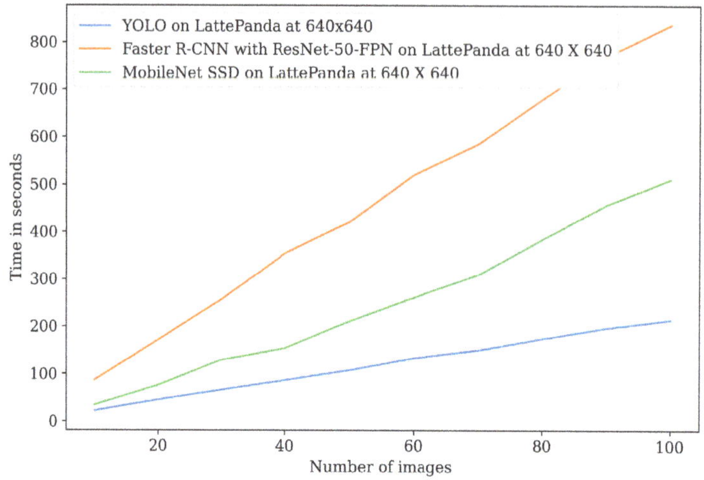

Figure 11. Time taken to perform wheat detection.

4.11. Benchmarking Deep Learning Models on Various Embedded Computing Platforms

Figure 12 and Table 2 depict the time taken in seconds to perform inferences for various number of images on different embedded computing platforms. The deep learning models that we benchmarked included YOLO, Faster R-CNN with ResNet-50-FPN, and MobileNet SSD. The embedded computing platforms on which these models were run included Intel stick, OpenCV AI kit, and LattePanda. Results indicate that MobileNet SSD on the OpenCV AI kit at an image size of 300×300 achieved the lowest inference time, whereas Faster R-CNN with ResNet-50-FPN at an image size of 640×640 resulted in the highest inference time among the compared models.

Table 2. Time taken to perform wheat detection on various platforms.

Images	YOLO on LattePanda at 640 × 640	Faster R-CNN on LattePanda at 640 × 640	MobileNet SSD on LattePanda 640 × 640	YOLO on Intel Stick 640 × 640	Faster R-CNN on Intel Stick 640 × 640	YOLO on OpenCV AI Kit at 640 × 640	MobileNet SSD on OpenCV AI Kit at 640 × 640	MobileNet SSD on OpeCV AI Kit at 300 × 300
10	22.13	86.33	34.33	15.49	60.43	2.4	0.93	0.52
20	45.19	171.74	75.85	31.63	128.80	5.64	1.88	0.52
30	65.68	256.17	128.74	45.32	181.88	8.21	2.77	1.55
40	86.31	353.87	153.94	63.71	254.79	9.69	3.76	2.06
50	107.91	420.85	211.5	84.24	303.01	12.40	4.38	2.56
60	133.38	520.19	261.43	93.36	374.53	13.47	5.85	3.31
70	150.23	585.90	309.73	102.70	433.56	15.02	6.48	3.64
80	174.49	680.53	384.76	132.62	496.79	17.44	7.65	4.14
90	197.60	770.64	456.66	146.22	562.57	17.96	8.35	4.63
100	214.36	837.08	511.13	160.97	627.81	23.84	9.37	5.15

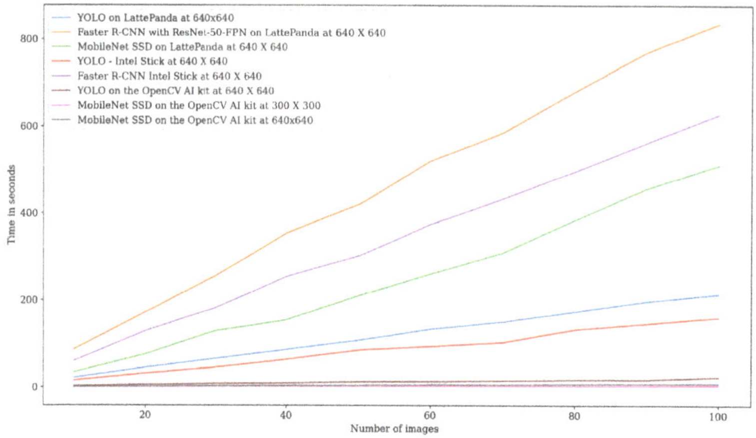

Figure 12. Time taken to perform wheat detection on various platforms.

4.12. In-Field Real-Time Object Detection and Depth Sensing

The performances of our object detection and depth-sensing models were tested in the KSU Agronomy Farm (Figure 2) where the robot was deployed. As indicated above, MobileNet SSD was used for real-time object detection in the field. The results of object detection are shown in Figure 13, and the results of depth sensing are shown in Figure 14. The effectiveness of the wheat recognition can be increased by varying the threshold/confidence level of the model.

Figure 13. Real-time detection of wheat heads in the field.

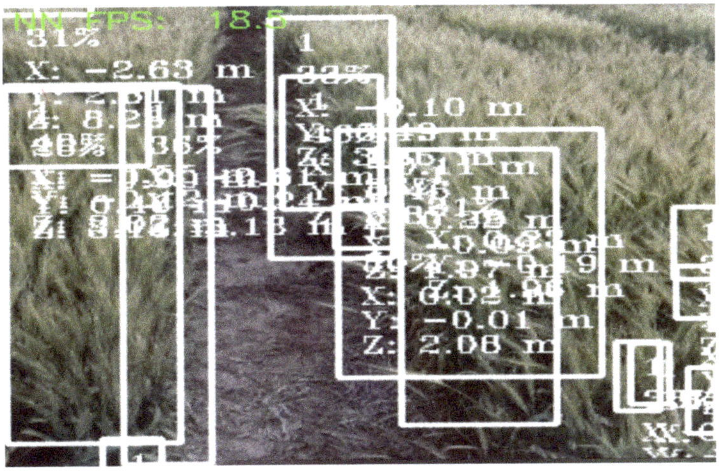

Figure 14. Real-time depth sensing of wheat in the field.

5. Conclusions

In this study, we designed an autonomous robotic platform that performs collision avoidance while navigating in crop rows using spatial AI. We explored and compared various deep learning models to determine the models that can provide high accuracy and inference speed on relatively low-cost embedded devices, such as LattePanda, Intel Neural Compute Stick, and OpenCV AI Kit, which are suitable for integrating on the robotic platform. We trained a MobileNet SSD architecture and other comparative object detection models, YOLO V5 and Faster R-CNN, on our prepared wheat images dataset. The experimental results revealed that the MobileNet SSD model attained the best detection performance on LattePanda for wheat detection with the runner-up inference speed (YOLO V5 was fastest), thereby dominating Faster R-CNN model in terms of both model detection accuracy and inference speed. Thus, MobileNet SSD achieves a better trade-off between model accuracy and time-complexity. Furthermore, results indicate that MobileNet SSD outperforms YOLO V5 on OpenCV AI Kit in terms of both model accuracy and inference speed. These results indicate that the MobileNet SSD model is a suitable candidate for real-time applications on resource-constrained computing platforms by providing stable accuracy and fast inference speeds. After these experimental evaluations, the trained MobileNet SSD model was combined with the stereo depth sensing on OpenCV kit mounted on our robotic platform to detect the distances of the objects (i.e., wheat) from the camera to avoid collision with wheat.

The current work had limitations, such as (i) not making the robot fully autonomous, and (ii) not identifying alleys present in the field. These limitations can be eliminated by using segmentation techniques to detect alleys in the field and making the robot fully autonomous using path planning algorithms, such as the dynamic window collision approach. Path planning algorithms can be combined with deep learning models to achieve fully autonomous driving of the robot. In the future, we plan to include a segmentation-based approach for path identification, which will make our robot more intelligent while moving in the field for wheat detection.

Author Contributions: Conceptualization, A.M., S.W., and D.F.; methodology, S.G., A.M., H.U., S.W., and D.F.; software, S.G.; validation, S.G.; formal analysis, S.G. and A.M.; investigation, A.M. and S.W.; resources, A.M., S.W., and D.F.; data curation, S.G.; writing—original draft preparation, S.G. and A.M.; writing—review and editing, A.M., H.U., and S.W.; visualization, S.G. and A.M.; supervision, A.M.; project administration, A.M., S.W., and D.F.; funding acquisition, S.W. and D.F. All authors have read and agreed to the published version of the manuscript.

Funding: This research was funded in part by the NSF/EPSCoR grant #1826820 to Kansas State University. Any opinions, findings, and conclusions or recommendations expressed in this material are those of the author(s) and do not necessarily reflect the views of the NSF.

Institutional Review Board Statement: Not applicable.

Informed Consent Statement: Not applicable.

Data Availability Statement: The data presented in this study are available on request from the corresponding author on reasonable request. The data are not publicly available due to proprietary reasons.

Acknowledgments: This research was supported in part by NSF/EPSCoR grant #1826820 to Kansas State University (KSU). Any opinions, findings, and conclusions or recommendations expressed in this material are those of the author(s) and do not necessarily reflect the views of the NSF. The authors would also like to acknowledge Calvin Dahms (cmdahms@ksu.edu) from KSU for his support for the robotic platform and help with taking the images from the robotic platform used in this work.

Conflicts of Interest: The authors declare no conflict of interest.

References

1. Asseng, S.; Ewert, F.; Martre, P.; Rötter, R.P.; Lobell, D.B.; Cammarano, D.; Kimball, B.A.; Ottman, M.J.; Wall, G.; White, J.W.; et al. Rising temperatures reduce global wheat production. *Nat. Clim. Chang.* **2015**, *5*, 143–147. [CrossRef]
2. Tack, J.; Barkley, A.; Nalley, L.L. Effect of warming temperatures on US wheat yields. *Proc. Natl. Acad. Sci. USA* **2015**, *112*, 6931–6936. [CrossRef] [PubMed]
3. Ihsan, M.Z.; El-Nakhlawy, F.S.; Ismail, S.M.; Fahad, S.; Daur, I. Wheat phenological development and growth studies as affected by drought and late season high temperature stress under arid environment. *Front. Plant Sci.* **2016**, *7*, 795. [CrossRef] [PubMed]
4. Redmon, J.; Divvala, S.; Girshick, R.; Farhadi, A. You only look once: Unified, real-time object detection. In Proceedings of the IEEE Conference on Computer Vision and Pattern Recognition, Las Vegas, NV, USA, 27–30 June 2016; pp. 779–788.
5. Srivastava, S.; Divekar, A.V.; Anilkumar, C.; Naik, I.; Kulkarni, V.; Pattabiraman, V. Comparative analysis of deep learning image detection algorithms. *J. Big Data* **2021**, *8*, 1–27. [CrossRef]
6. Ren, S.; He, K.; Girshick, R.; Sun, J. Faster R-CNN: Towards real-time object detection with region proposal networks. *Adv. Neural Inf. Process. Syst.* **2015**, *28*, 91–99. [CrossRef] [PubMed]
7. Everingham, M.; Van Gool, L.; Williams, C.K.; Winn, J.; Zisserman, A. The pascal visual object classes (voc) challenge. *Int. J. Comput. Vis.* **2010**, *88*, 303–338. [CrossRef]
8. Everingham, M.; Eslami, S.A.; Van Gool, L.; Williams, C.K.; Winn, J.; Zisserman, A. The pascal visual object classes challenge: A retrospective. *Int. J. Comput. Vis.* **2015**, *111*, 98–136. [CrossRef]
9. Lin, T.Y.; Maire, M.; Belongie, S.; Hays, J.; Perona, P.; Ramanan, D.; Dollár, P.; Zitnick, C.L. Microsoft coco: Common objects in context. In *European Conference on Computer Vision*; Springer: Cham, Switerland, 2014; pp. 740–755.
10. ImageNet Large Scale Visual Recognition Challenge 2015 (ILSVRC2015). Available online: https://www.image-net.org/challenges/LSVRC/ (accessed on 16 August 2022).
11. Girshick, R. Fast r-cnn. In Proceedings of the IEEE International Conference on Computer Vision, Santiago, Chile, 7–13 December 2015; pp. 1440–1448.
12. Tan, M.; Le, Q. Efficientnet: Rethinking model scaling for convolutional neural networks. In Proceedings of the 36th International Conference on Machine Learning, Long Beach, CA, USA, 9–15 June 2019; ML Research Press: Maastricht, The Netherlands, 2019; pp. 6105–6114.
13. Liu, W.; Anguelov, D.; Erhan, D.; Szegedy, C.; Reed, S.; Fu, C.Y.; Berg, A.C. SSD: Single shot multibox detector. In Proceedings of the European Conference on Computer Vision, Munich, Germany, 8–14 September 2016; Springer: Cham, Switerland, 2016; pp. 21–37.
14. He, K.; Zhang, X.; Ren, S.; Sun, J. Deep residual learning for image recognition. In Proceedings of the IEEE Conference on Computer Vision and Pattern Recognition, Las Vegas, NV, USA, 27–30 June 2016; pp. 770–778.
15. Krizhevsky, A.; Sutskever, I.; Hinton, G.E. Imagenet classification with deep convolutional neural networks. *Adv. Neural Inf. Process. Syst.* **2012**, *25*, 1097–1105. [CrossRef]
16. Mosley, L.; Pham, H.; Bansal, Y.; Hare, E. Image-based sorghum head counting when you only look once. *arXiv* **2020**, arXiv:2009.11929.
17. Ghosal, S.; Zheng, B.; Chapman, S.C.; Potgieter, A.B.; Jordan, D.R.; Wang, X.; Singh, A.K.; Singh, A.; Hirafuji, M.; Ninomiya, S.; et al. A weakly supervised deep learning framework for sorghum head detection and counting. *Plant Phenomics* **2019**, *2019*, 1525874. [CrossRef] [PubMed]
18. Velumani, K.; Lopez-Lozano, R.; Madec, S.; Guo, W.; Gillet, J.; Comar, A.; Baret, F. Estimates of maize plant density from UAV RGB images using Faster-RCNN detection model: Impact of the spatial resolution. *arXiv* **2021**, arXiv:2105.11857.
19. Gonzalo-Martín, C.; García-Pedrero, A.; Lillo-Saavedra, M. Improving deep learning sorghum head detection through test time augmentation. *Comput. Electron. Agric.* **2021**, *186*, 106179. [CrossRef]

20. Xue, J.; Xia, C.; Zou, J. A velocity control strategy for collision avoidance of autonomous agricultural vehicles. *Auton. Robot.* **2020**, *44*, 1047–1063. [CrossRef]
21. Shutske, J.M.; Gilbert, W.; Morgan, S.; Chaplin, J. Collision avoidance sensing for slow moving agricultural vehicles. *Pap.-Am. Soc. Agric. Eng.* **1997**, 3. Available online: https://www.researchgate.net/publication/317729198_Collision_avoidance_sensing_for_slow_moving_agricultural_vehicles (accessed on 16 August 2022).
22. Luxonis. DepthAI's Documentation. 2022. Available online: https://docs.luxonis.com/en/latest/ (accessed on 4 August 2022).
23. Luxonis. Luxonis-Simplifying Spatial AI. 2021. Available online: https://www.luxonis.com/ (accessed on 4 August 2022).
24. OpenCV. OpenCV AI Kit: OAK-D. 2021. Available online: https://store.opencv.ai/products/oak-d (accessed on 4 August 2022).
25. LattePanda. LattePanda Alpha 864s. 2021. Available online: https://www.lattepanda.com/products/lattepanda-alpha-864s.html (accessed on 4 August 2022).
26. Intel. Intel Neural Compute Stick. 2021. Available online: https://www.intel.com/content/www/us/en/developer/tools/neural-compute-stick/overview.html (accessed on 4 August 2022).
27. Naushad, R. Introduction to OpenCV AI Kits (OAK-1 and OAK-D). 2021. Available online: https://medium.com/swlh/introduction-to-opencv-ai-kits-oak-1-and-oak-d-6cdf8624517 (accessed on 4 August 2022).
28. Yohanandan, S. mAP (mean Average Precision) Might Confuse You! 2020. Available online: https://towardsdatascience.com/map-mean-average-precision-might-confuse-you-5956f1bfa9e2 (accessed on 7 December 2021).
29. LattePanda Alpha 864s (Win10 Pro activated)—Tiny Ultimate Windows/Linux Device. Available online: https://www.dfrobot.com/product-1729.html (accessed on 16 August 2022).

AI

MDPI

Article

Influence of Model Size and Image Augmentations on Object Detection in Low-Contrast Complex Background Scenes

Harman Singh Sangha * and Matthew J. Darr

Department of Agricultural and Biosystems Engineering, Iowa State University, Ames, IA 50011, USA; darr@iastate.edu

* Correspondence: hssangha@iastate.edu; Tel.: +1-785-370-9472

Abstract: Background: Bigger and more complex models are often developed for challenging object detection tasks, and image augmentations are used to train a robust deep learning model for small image datasets. Previous studies have suggested that smaller models provide better performance compared to bigger models for agricultural applications, and not all image augmentation methods contribute equally to model performance. An important part of these studies was also to define the scene of the image. Methods: A standard definition was developed to describe scenes in real-world agricultural datasets by reviewing various image-based machine-learning applications in the agriculture literature. This study primarily evaluates the effects of model size in both one-stage and two-stage detectors on model performance for low-contrast complex background applications. It further explores the influence of different photo-metric image augmentation methods on model performance for standard one-stage and two-stage detectors. Results: For one-stage detectors, a smaller model performed better than a bigger model. Whereas in the case of two-stage detectors, model performance increased with model size. In image augmentations, some methods considerably improved model performance and some either provided no improvement or reduced the model performance in both one-stage and two-stage detectors compared to the baseline.

Keywords: deep learning; computer vision; agriculture; image augmentations

1. Introduction

The introduction of electronic systems and computers in agriculture has helped improve crop production and the overall agricultural experience for farmers [1,2]. An increase in crop yield has been observed partly due to the employment of smart farm input management techniques, which are heavily reliant on modern computational methods [3,4]. These applications can vary from high throughput phenotyping of crops, remote sensing, etc., to precise machinery controls and field application [5–7]. To advance the technology further, various studies are being conducted to explore the use of machine learning and artificial intelligence in agricultural applications [8]. Artificial intelligence and some machine learning applications take advantage of deep learning methods due to their ability to learn important features in complex datasets. Furthermore, with advancements in imaging and manufacturing technologies, in recent years, camera sensors have become inexpensive and lightweight. This has enabled camera sensors on platforms, such as unoccupied aerial vehicles (UAV), field robots, etc., to collect agricultural image data. These images are used in various applications including, but not limited to, plant phenotyping, disease monitoring, drought monitoring, pest control, in-row navigation, weed management, precision planting, fruit picking, crop harvesting, and crop handling [9–17].

Using machine learning with images further enables researchers to optimize agricultural systems and increase productivity. A computer can learn important features in an image and present them in the form of feature identification, feature detection, feature characteristics, and feature segmentation. Among these applications, deep learning-based object detection methods have gained considerable traction in the last decade and have been tested in many agricultural applications [9,13,15,16,18–28].

It is often observed in typical object detection deep neural networks that the bigger and more complex the model, the better its ability to learn difficult features in the images. In one of the studies in the literature, the authors used a YOLO model to detect lemon fruits in orchards [27]. The image scene was complex in nature with fruits having similar color space as the background. The authors stated that for a complex background environment, a smaller or less complex model had a better performance compared to a more complex or bigger model. Furthermore, agricultural datasets are often limited in the number of images or training samples present for deep learning model training. Moreover, it is always advised to use image augmentation methods to improve the model's ability to generalize to unknown features that might be present but are not captured by the current image dataset when the size of the data is small. While working towards developing a new version for YOLO, the authors also found during testing that not all image augmentations were contributing towards improving the model performance [29].

In the literature, it was observed that many researchers have attempted to define real-world agricultural dataset scenes. Terms like contrast and complex background were repeated multiple times to explain the aesthetic quality of the images in the dataset. However, each definition was different and addressed different aspects of the research the study was focused on. Hence, one of the goals of this research study is to develop an all-encompassing definition of real-world agricultural dataset scenes that can be used to describe the nature of images in real-world agricultural deep learning applications. A universal definition will assist in performing more focused studies and developing easily generalized deep learning models for agricultural applications specific to low-contrast complex backgrounds. The second objective of this study is to evaluate the effects of model size in both one-stage and two-stage detectors on model performance for low-contrast complex background applications. And thirdly, an objective is to gauge the influence of different photo-metric image augmentation methods on model performance for standard one-stage and two-stage detectors.

2. Defining Characteristics of Agricultural Image Datasets

To achieve the research goal, the literature was investigated using the keywords "agriculture", "background", and "deep learning". The Web Of Science Core Collection journal database was searched using the mentioned keywords on 2 November 2022, and was limited to only research articles. The database search yielded 358 research articles mentioning "agriculture", "background", and "deep learning". To further filter the pool of papers, only open-access articles were considered, resulting in 210 articles in total.

The selected articles were further screened based on the following inclusion criteria:

1. Articles published during and after the year 2018.
2. Articles referring to the subject area of "Plant Science", "Agricultural Engineering", "Agronomy", "Computer Science", "Electrical Engineering", "Artificial Intelligence", or "Remote Sensing".
3. Articles published in the English language.

This reduced the number of articles to 126. A careful reading of the abstracts and articles allowed us to eliminate those that did not describe the environment in the scene for images, used lab-generated imagery, or were unavailable articles within the selected database. Only 63 articles remained after sorting. A detailed reading of these articles allowed us to identify common words and phrases used to describe the scene in agricultural image datasets. Based on the findings, a global definition was presented. Refs. [30–32] influenced the method used to search and develop a definition for this topic. A year-wise list of the selected articles is provided in Table 1.

Table 1. Year-wise distribution of reviewed articles.

2022	2021	2020	2019	2018
[33]	[34]	[26]	[35]	[36]
[37]	[38]	[39]	[40]	[41]
[42]	[43]	[44]	[45]	-
[46]	[47]	[48]	-	-
[49]	[50]	[51]	-	-
[52]	[53]	[54]	-	-
[55]	[56]	[57]	-	-
[58]	[59]	[60]	-	-
[61]	[62]	[63]	-	-
[64]	[65]	[66]	-	-
[67]	[68]	-	-	-
[69]	[70]	-	-	-
[71]	[72]	-	-	-
[73]	[74]	-	-	-
[75]	[76]	-	-	-
[77]	[78]	-	-	-
[79]	[80]	-	-	-
[81]	[82]	-	-	-
[83]	[84]	-	-	-
[85]	[86]	-	-	-
[87]	-	-	-	-
[88]	-	-	-	-
[89]	-	-	-	-
[90]	-	-	-	-
[91]	-	-	-	-
[92]	-	-	-	-
[93]	-	-	-	-

2.1. Image Scene

The features of a scene in an image are described using two-level descriptors. The first level includes features such as color and texture. The second level contains information regarding the objects present in the scene, such as a car and house, and the interaction of these objects. For example, the scene in Figure 1 can be described as "A red car in front of a grey house on a cloudy day". Here, the low-level descriptors are red, grey, and cloudy, conveying information about the colors and textures present in the scene. The high-level feature descriptors are car, house, and in front of, characterizing the objects in the scene and how they interact.

Figure 1. An image scene containing a red car in front of a grey house on a cloudy day. The description of the scene in the image uses both low-level descriptors (red, grey, and cloudy) and high-level descriptors (car, house, and in front of) (Google Image Search: 'car in front of a house').

Similarly, various low-level and high-level descriptors explain the scene in the agricultural datasets. The descriptions in the literature commonly used shadow, sunlight, type of crop, background, color, occlusions, and other biological material terms to create a textual reference for the different types of scenes present in the images for a particular agricultural dataset. There are cases where the images are staged to create a real-world reference in the lab to decrease the data collection time (Figure 2) [94,95]. Here, the real-world descriptors are used to mimic the hue, saturation, lighting, etc., in the lab. The environment present in real-world image data is very complex. Various small and large features compose a scene. Creating an accurate description is very important. The following section discusses the various attempts at describing the scene found in agricultural image data and an effort to find common descriptors to establish a global definition for real-world agricultural datasets.

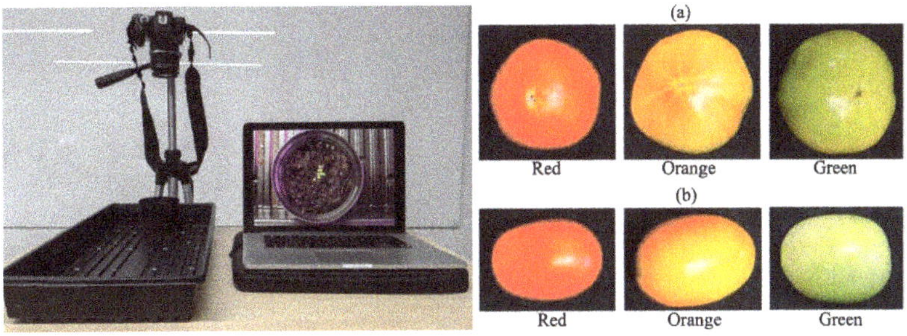

Figure 2. Examples for staged images used for machine learning applications [94,95]. Maturity levels in (**a**) aroma tomato and (**b**) pear tomato

2.2. Scene Descriptors for Agricultural Imagery

Analyzing all 63 articles from the literature, common descriptors were selected to characterize the scene in real-world agricultural applications and challenges with the scene for image processing. Semantically similar descriptors were kept in a single category, and the number of instances was counted when the authors used these descriptors to explain a scene. The low-level descriptors included color/contrast, texture, morphology, and illumination. The high-level descriptors were occlusion/overlap, background description, and complex/complicated. The count for the number of instances is provided in Figure 3. If an article described the minimal difference in color or the presence of low-contrast, a +1 count was added in the Color/Contrast descriptor. Similarly, if the similarity in texture between foreground and background was mentioned and varying lighting conditions were discussed, a +1 count was added to the Texture and Illumination descriptors, respectively. The Morphology descriptor was given a +1 count if the article discussed morphological features (shape, size, etc.). In the case of high-level descriptors, if the article expressed the interaction of foreground and background as complicated or complex to explain a scene, a +1 count was placed in the Complex/Complicated descriptor. When the article reported the types of objects or noise in the scene's background, a +1 count was given to the Background description category. Finally, if the objects in the foreground or background were partially or fully blocking the object of interest in the particular study, +1 count was added to the Occlusion/Overlap descriptor. The Texture descriptor was used 16 times, the Morphology descriptor was used 39 times, the Color/Contrast descriptor was employed 43 times, the Illumination descriptor 47 times, the Complex/Complicated descriptor 51 times, the Background descriptor 40 times, and the Occlusion/Overlap descriptor was used 42 times.

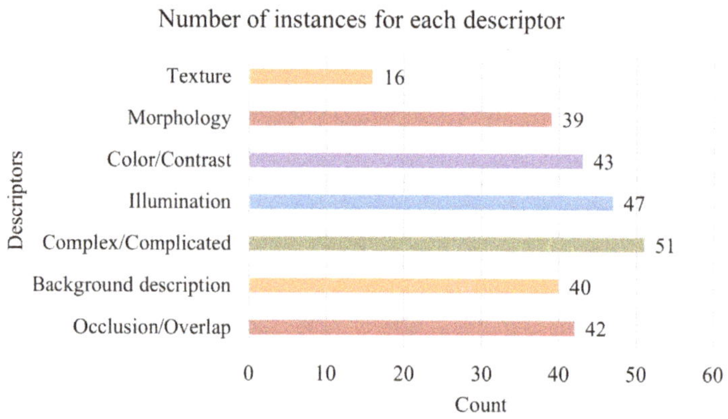

Figure 3. Number of instances for each descriptor for real-world agricultural scenes.

Except for the Texture descriptor, the count of other descriptors used to explain a scene present in real-world agricultural image datasets was substantially higher. There is evidence of common patterns in the terminologies that can be used to derive a more universal definition for real-world agricultural datasets.

3. Low-Contrast Complex Background as a Unified Term

Since 2018, there has been a steady increase in the number of studies conducted each year for real-world deep learning agricultural applications. Each study attempted to define the scene in the images for each agricultural dataset. Common phrases and descriptions were observed but were limited to a focused scene. There is a need for a universal definition and unified terms to describe the scenes in real-world agricultural datasets. Defining a universal definition will assist the scientific community in performing more focused studies for agricultural scenes and develop deep learning models that can easily be generalized to other agricultural applications specific to low-contrast complex background situations. The increase in the number of publications each year and common themes within these publications indicates the growing importance of developing deep learning methods for real-world agricultural applications. Deep learning methods for real-world low-contrast complex background agricultural applications should be a sub-discipline within computer vision and machine learning applications in agriculture. A unified definition will be useful to further solidify this research space.

The Merriam–Webster dictionary defines complex (adjective) as "hard to separate, analyze, or solve" and contrast (noun) as "the difference or degree of difference between things having similar or comparable nature". Based on the common descriptors used, the English language, and the literature reviewed, the low-contrast complex background can be used as a unified term and can be defined as follows:

"An image taken for an agricultural application can be said to contain low-contrast complex background if the object of interest has pixel value, saturation, and hue comparable to the entities present in the background, and/or the object of interest is either surrounded or occluded by other objects of interest, debris, biological materials, shadows, soil, and man-made materials of similar shape and sizes".

4. Materials and Methods

4.1. Object Detection Models

The process of detecting an object using deep learning models involves two steps. The first step is to locate the objects of interest in the image and the second step is to classify those objects into different classes [96]. Deep convolutional neural networks are known for their feature extraction capabilities from images; hence, the current standard object detection architectures are based upon deep convolutional neural networks [97,98]. The object detection architectures are divided into two categories. The first one is two-stage object detectors. In two-stage architectures the task of object localization and object classification is divided into two different networks within the architecture. The main advantage of two-stage detectors is high accuracy of detection and the major drawback is slow detection speed. Some of the two-stage object detectors are as follows: RCNN [99], SPPNet [100], Fast RCNN [101], Faster RCNN [102], Mask RCNN [103], etc. The other type of architecture is one-stage object detectors. One-stage detectors do not separate the tasks of object localization and classification but directly perform both of the tasks through a single network. The main advantage of one-stage or single-stage detectors is the high speed of detection and the main limitation of these architectures is that the accuracy is comparatively lower than for the two-stage detectors. Some of the common one-stage detectors are the YOLO series [29,104–106], SSD [107], RetinaNet [108], etc.

4.1.1. Two-Stage Detectors

RCNN [99] used a deep convolutional neural network (DCNN) to extract image features and SVM to classify regions. In the case of SPPNet [100], it again used DCNN to extract image features and maps regions to propose the feature maps. Furthermore, spatial pyramid pooling is used to input multi-scale images to the DCNN. Fast-RCNN [101] uses DCNN to extract image features and maps down-scaled region proposals using a Region-Of-Interest (ROI) pooling layer to the feature maps. Faster-RCNN [102] uses a Region Proposal Network (RPN) to replace the DCNN and then the RPN shares feature maps with the backbone network. Mask-RCNN [103] uses an ROI Align pooling layer instead of an ROI pooling layer improving the detection accuracy, and later combines training object detection and segmentation to improve accuracy in detection. The model architecture of the selected networks is shown in Figure 4.

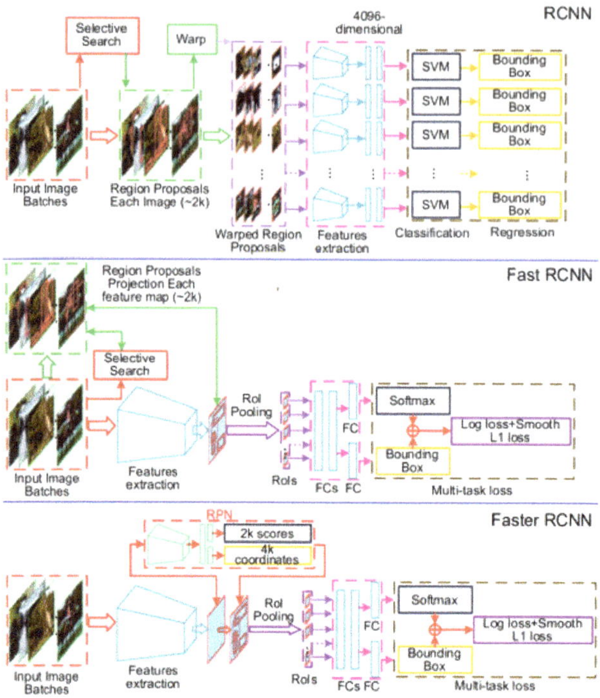

Figure 4. Model architecture of some of the 2-stage detectors [96].

4.1.2. One-Stage Detectors

YOLOv1 [104] is an end-to-end single neural network and can implement class probabilities and bounding-boxes regression directly from a full image. YOLOv2 [105] introduced a new backbone network (DarkNet19) and used the k-means clustering algorithm to generate anchor boxes. YOLOv3 [106] used multi-level feature fusion to improve the accuracy of multi-scale detections and introduced a new backbone network (DarkNet53). YOLOv4 [29] experimented with different combinations of model features, such as Weighted-Residual-Connections (WRCs), Cross Mini-Batch Normalization (CMBN), Self-adversarial Training, etc. to achieve state-of-the-art results. It also introduced mosaic data augmentation as a new method for image augmentation. SSD [107] presented a multi-layer detection mechanism and multi-scale anchor mechanisms at different neural layers. RetinaNet [108] used a feature pyramid network to extract features and developed a new focal loss for the model training. The model architecture of some of the networks is shown in Figures 5 and 6.

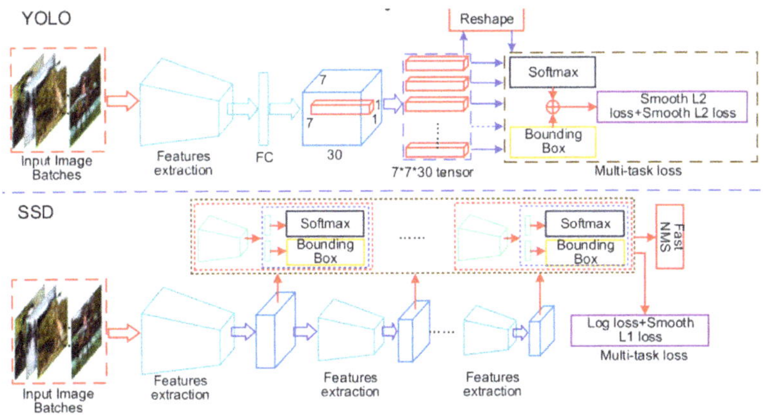

Figure 5. Model architecture of some of the 1-stage detectors [96].

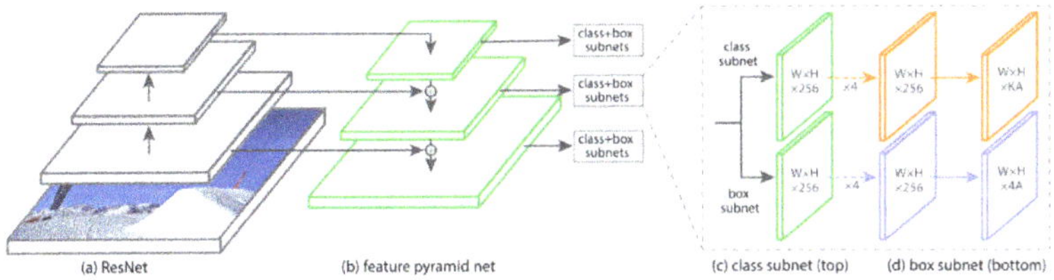

Figure 6. RetinaNet [108].

4.1.3. Backbone Networks

Backbone networks for object detection are networks which are used for feature extraction in state-of-the-art models. The main objective of backbone networks is to identify important features in images and learn their characteristics, such as color, shape, texture, etc. With more and more complex applications, there is a need to improve the model performance. To achieve this, the network architecture becomes more and more complex. In some other situations where limited memory and computing power is available, lightweight networks are proposed, which simplify the architecture without affecting the feature extraction capacity. Some of the backbone networks are discussed below.

VGGNet [109] added more layers to AlexNet [97], increasing the network size to 16–19 layers. This improved the feature extraction capabilities of the network. VGG16 and VGG19 are widely known model architectures. ResNet or Residual Network [110] addressed the problem of gradient dispersion and gradient explosion problems which arises due to increasing the network by adding more layers. The authors achieved this by adding a residual to the output of the previous layer or stack of layers before creating the input of the next layer. ResNet50 to ResNet 152 are widely used as backbone networks. DetNet [111] was proposed as a backbone for object detection to address some shortcomings of generic backbone networks. DetNet used a dilated convolutional network instead of down-sampling some of the last layers to ensure enhanced resolution and a receptive field, which helps in locating large and small objects.

The backbones discussed above are complex in nature to deepen the depth of the network. However, some methods were also developed which sought to reduce the number of parameters in the network. This was performed to address the problem of storage space and processing time to develop a network that can be deployed on a mobile device. InceptionV1/V2/V3 [112–114] used kernel decomposition to make a lightweight model. Xception improved over InceptionV3 with a depth-wise separable convolution [115]. MobileNet [116] improved on the structure of the depth-wise separable method. MobileNetV2 [117] used short-cut connections like ResNet [110] along with depth-wise separable convolutions to improve performance.

In this study, Faster-RCNN and RetinaNet were used for comparing networks as they were available in TensorFlow API pre-trained on Imagenet and both networks were able to use ResNet50, ResNet101, and ResNet152 as backbone networks.

4.2. Global Wheat Dataset

The Global Wheat Dataset was used for this study as a reference dataset [118]. The Global Wheat Dataset is composed of more than 6000 images of 1024 × 1024 pixels containing 300k+ unique wheat heads with corresponding bounding boxes. The dataset includes images from 11 different countries and 44 unique measurement sessions. The dataset is a collection of images taken between 2016 to 2019 by nine different institutes and contains genotypes from North America, Asia, Europe, and Australia. The row spacing varied from 12.5 cm to 30.5 cm. The dataset covers varying sowing density (186 to 450 seeds/m^2), soil types (loamy soil, silt-clay, etc.), and stage of growth (flowering to ripening). Cameras with different fields of view (7.1° to 45.5°) and focal length (7.7 mm to 60 mm) were employed to collect the imagery with the ground sampling distance varying from 0.2 mm/pixel to 0.56 mm/pixel. The dataset was chosen as it contains images with different lighting conditions and a range of hues with a complex background including occlusion of wheat heads. Selected example images from the Global Wheat Dataset are shown in Figure 7.

Figure 7. A few sample images from the Global Wheat Dataset.

4.3. System and Training Specifications

In this study, the experiments were conducted using the deep learning framework provided by TensorFlow API (version: 2.8.0) from Google. Python 3.8.9 version was used for developing the object detection model experimental setup. For training the different neural networks, a machine running the Windows 10 professional operating system was used. The models were trained on Intel i-7 11700K 3.6 GHz CPU, 128 GB ram, and NVIDIA RTX 3090 28 GB VRAM GPU.

The Global Wheat Dataset was divided into three parts: 3000+ images in the training set, 1400+ in the validation set, and 1700+ in the testing set. Each model was trained for 80,000 steps.

4.4. Image Augmentations

Image augmentations are used in deep learning models to address the issue of overfitting during training. Furthermore, image augmentations help when large datasets cannot be collected and labeled for training a deep learning model. Also, image augmentations are used to resolve the problems associated with class imbalance [119]. Image augmentations either warp the data or carry out oversampling to artificially inflate the training dataset. Augmentations dealing with oversampling create synthetic instances and add them to the training set, whereas data warping augmentations transform existing images while preserving their labels [120]. Data warping image augmentations can be further classified as photo-metric augmentations and geometric augmentations. Photo-metric augmentations alter the pixel values of the features in the image, including brightness, contrast, hue, etc., whereas geometric augmentations change the layout of the image, such as by cropping, flipping, rotating, etc.

Some of the image augmentation methods are as follows:

Flipping: In this augmentation, the images are flipped by either the horizontal or vertical axis. It is one of the easiest to implement.

Color Space: An image consists of a 3D matrix, height, width, and color channels. Each element in this matrix consists of a pixel value for all three color channels. These pixel values or channels can be manipulated in order to perform image augmentations. One method is isolation of a single color channel. Changing the brightness, contrast, or saturation randomly is another way to alter the color space.

Cropping: Randomly cropping images can be used as a method of image augmentation. A mixed height and width patch can be cropped from anywhere in the image.

Rotation: In this augmentation, the image is rotated either to the left or right between 1° and 359°. This augmentation helps in training the model to make the objects appear in different orientations.

Noise Injection: Noise injection involves injecting a matrix of random values drawn from a Gaussian distribution within the image. Adding noise to images can help neural networks learn more robust features.

Color Space Transformations: One way to perform color space augmentation is to loop through the images and decrease or increase the pixel values by a constant value. Another quick color space manipulation is to splice out individual RGB color matrices. A further transformation consists of restricting pixel values to a certain min or max value. Also, individual color channels can be distorted to create various color gradient images.

Kernel Filters: Sharpening and blurring images are very widely known kernel filter augmentations. These filters work by sliding an $n \times n$ matrix across an image with either a Gaussian blur filter, which will result in a blurrier image, or a high-contrast vertical or horizontal edge filter, which will result in a sharper image along the edges.

Mixing Images: Mixing images together by averaging their pixel values is another approach to data augmentation.

Random Erasing: Random erasing augmentation works by randomly selecting an $n \times m$ patch of an image and masking it with either minimum, maximum, mean pixel values, or random values. This technique was specifically designed to combat image recognition challenges due to occlusion. The authors of [119] demonstrated that not all the augmentations which were used in the study, including both geometric and photo-metric augmentations, were contributing equally towards improving a convolutional neural network's model performance. Ref. [29] also reported similar results, mostly focused on using geometric augmentation to improve model efficacy. Therefore, in this study, five photo-metric augmentations which transformed the color space were tested to quantify their effects on object detection model performance in low-contrast complex background applications. The augmentations were random brightness, random contrast, random saturation, random distort color, and random Red-Green-Blue (RGB) to gray-scale conversion. Random brightness was applied by randomly selecting a brightness value from a uniform distribution; random contrast was applied on an image by randomly selecting contrast values from a uniform distribution; random saturation was applied on an image by randomly selecting saturation values from a uniform distribution; random distort color was applied by randomly selecting a color channel and randomly changing the pixel values; and random RGB to gray-scale was applied by randomly selecting an image and converting it into a gray-scale image. Visual examples of these augmentations on the Global Wheat Dataset are shown in Figure 8.

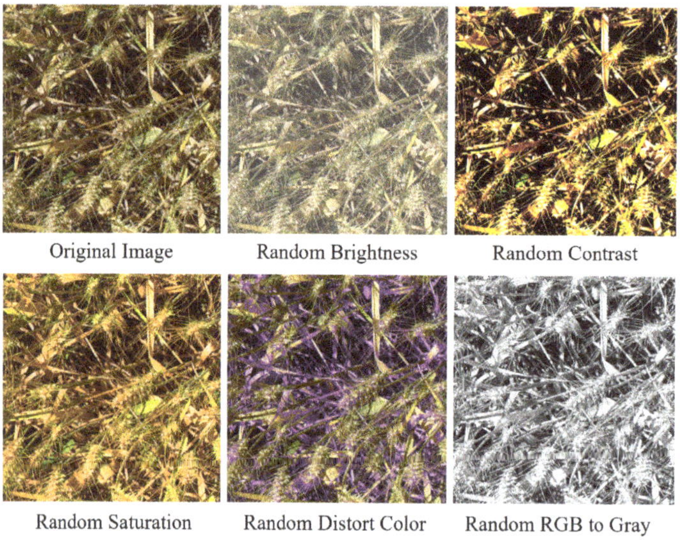

Figure 8. Visual examples of random brightness, random contrast, random saturation, random distort color, and random Red-Green-Blue (RGB) to gray-scale conversion on the Global Wheat Dataset.

4.5. *Performance Metrics*

In supervised machine learning, it is necessary to provide ground truth labels to the model, which assists the model to learn important features in the data. For object detection, the ground truth is defined as a bounding box that gives the location of the object of interest in the image. The model is provided as rectangular coordinates with respect to the image Cartesian plane, i.e., x-coordinate for the upper left corner of the bounding box, y-coordinate for the upper left corner of the bounding box, with the width of the bounding box in pixels, and the height of the bounding box in pixels. To evaluate the performance of the object detection models, the overlap of the detected bounding box provided by the model and the ground truth bounding box is calculated.

The Intersection-Over-Union (IOU) is the metric used to calculate the overlap of the predicted and ground truth bounding boxes (1). The IOU, also known as the Jaccard Index, is the ratio of the area of intersection of the two regions (boxes) in the numerator and the area of union of the two regions (boxes) in the denominator (Figure 9). If there is no overlap, but the boxes have common edges, the ratio will have a value of 0 as there is no intersection area. If there is a slight overlap, the ratio will achieve a value close to zero as the area of intersection is very small but the area of union is very large. As the overlap between the boxes increases, the area of intersection will increase and the area of union will decrease; therefore, the value of the ratio will approach 1. At full overlap, both the intersection and the union area are equal and the value of the ratio is 1.

$$IoU = \frac{area\ of\ intersection}{area\ of\ union} \tag{1}$$

Figure 9. Illustration showing the difference between a poor detection and a good detection while comparing the overlap of two regions. The green box is the ground truth bounding box and the red box is the predicted bounding box.

For a detection to be successful, the minimum percentage of overlap should be defined. This definition varies from project to project. In most cases, an IoU threshold of 0.5 (50% overlap) is considered a successful detection. Other common thresholds are 0.75 and 0.95 (75% and 95% overlap). By increasing the IoU threshold, the detected bounding box becomes highly accurate but it increases the difficulty and training time needed to detect a successful bounding box [121].

An object detection model provides multiple predicted bounding boxes for a single image. It is important to quantify how many predicted boxes were correct and how many out of all the ground truth boxes were detected. With every detection event, there will be some boxes which were predicted correctly depending on the set threshold (True Positive), there will be boxes that were not predicted correctly (False Positive), and finally, there will be some ground truth boxes which were never predicted (False Negative). These terms can be efficiently quantified as Precision and Recall. Precision can be defined as the proportion of the predicted positives that were actually correct (2) and Recall can be defined as the proportion of the actual positives that were predicted correctly (3).

$$Precision = \frac{True\ Positives}{True\ Positives + False\ Positives} \tag{2}$$

$$Recall = \frac{True\ Positives}{True\ Positives + False\ Negatives} \tag{3}$$

For all the classes present in a dataset, the precision and recall values can be calculated and can be plotted to create a Precision-Recall curve. The Precision Recall curve is a good way to evaluate the performance of an object detection model. An object detection model is considered to be performing well if the precision of the model remains high as the recall of the model is increasing. A model sometimes can have high precision (0 False Positives) or high recall (0 False Negatives). Precision-Recall curves usually start with a precision value—the precision starts decreasing as the value of recall increases (Figure 10).

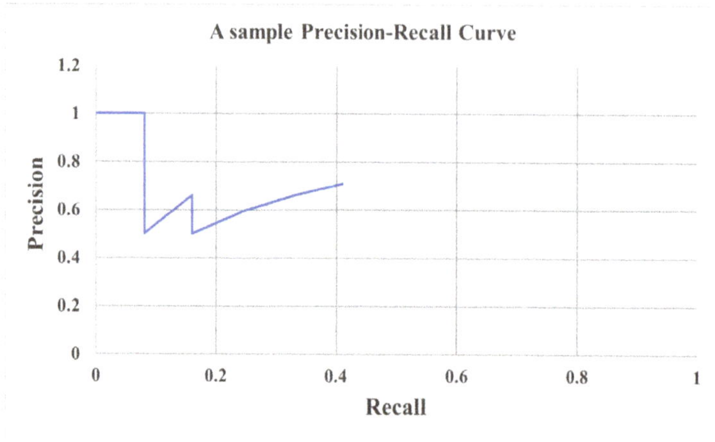

Figure 10. A sample precision-recall curve.

A drawback of Precision-Recall curves is that they cannot be compared from model to model. To address this, a quantitative value of Average Precision (AP) is used. The AP is the area under the curve of the Precision-Recall curve. AP is a numerical value which makes it easy to compare different models and can be defined as the average value of precision for all values of recall between 0 and 1 (4) [121]. Furthermore, if the APs of each class are averaged for all the classes, the resultant value is called the mean Average Precision (mAP).

$$AP = \frac{1}{n} \sum_{r \in [0, 0.1, \ldots, 1]} \rho_{interp(r)} \tag{4}$$

4.6. Experimental Design

In the first set of experiments, both of the selected object detection models, i.e., RetinaNet and Faster-RCNN, were trained on three different backbones. The backbones were as follows: ResNet50, ResNet101, and ResNet152. The models were compared on the COCO detection evaluation metrics, which include the average mAP at IOU = 0.5:0.05:0.95, the mAP at IOU = 0.50, the mAP at IOU = 0.75, the mAP for small objects: $area < 32^2$ pixels, the mAP for medium objects: $32^2 < area < 95^2$ pixels, and the mAP for large objects: $area > 96^2$ pixels.

In the second set of experiment, both RetinaNet and Faster-RCNN were trained using different color space augmentations for 80,000 steps. The backbone was fixed to ResNet101 for all the image augmentation experiments. The input image size was kept fixed at 1024 × 1024 pixels and no resizing was performed to avoid variations due to pixel interpolation. The color space augmentations used were random brightness, random contrast, random saturation, random distort color, and random Red-Green-Blue (RGB) to gray-scale conversion. Again, the models were compared on the COCO detection metrics.

Furthermore, all the models were compared on the total testing loss along with their overall mAP to quantify their effects on model fitting. Also, the percentage improvement of each model was compared either to a baseline or to the smallest model in the training.

In this study, six different mean Average Precision (mAP) measures from the COCO evaluation metrics were used to quantify the performance of the different object detection models. The measures are, namely, the average mAP at IOU = 0.5:0.05:0.95, the mAP at IOU = 0.50, the mAP at IOU = 0.75, the mAP for small objects: $area < 32^2$ pixels, the mAP for medium objects: $32^2 < area < 95^2$ pixels, and the mAP for large objects: $area > 96^2$ pixels. The resolution of the images used for this study was 1024 × 1024 pixels, meaning small objects occupy less than 0.1% and large objects occupy more than 0.9% area. In just the training set of the Global Wheat Dataset containing 3658 images, out of 163,644 individual bounding boxes, there were 2109 small objects, 134,299 medium objects, and 27,236 large objects. There were 1477 images in the testing set, containing 44,331 bounding boxes, out of which 1132 were small objects, 24,851 were medium objects, and 18,348 were large objects. The mAP values were calculated at the end of each training session on the validation dataset.

5. Results

In Table 2, the results from the first set of experiments are listed. In the case of RetinaNet, it was observed that the model with the smallest backbone, i.e., ResNet50, performed best compared to the other backbones for RetinaNet, with a mAP of 0.381. The other backbones had mAP values of 0.362 and 0.366 for ResNet101 and ResNet152, respectively. For the Faster-RCNN models, the model with the largest backbone performed best, i.e., ResNet152, with a mAP of 0.418. The backbones mAP values were 0.377 and 0.403 for ResNet50 and ResNet101, respectively.

Table 2. Comparison of 1-stage and 2-stage detectors with different backbones.

Model	Backbone	Input	mAP	mAP_{50}	mAP_{75}	mAP_{small}	mAP_{medium}	mAP_{large}
RetinaNet	ResNet50	1024 × 1024	0.381	0.782	0.320	0.022	0.308	0.489
RetinaNet	ResNet101	1024 × 1024	0.362	0.744	0.301	0.011	0.289	0.478
RetinaNet	ResNet152	1024 × 1024	0.366	0.746	0.313	0.009	0.288	0.488
Faster-RCNN	ResNet50	1024 × 1024	0.377	0.821	0.281	0.037	0.309	0.487
Faster-RCNN	ResNet101	1024 × 1024	0.403	0.838	0.325	0.050	0.336	0.503
Faster-RCNN	ResNet152	1024 × 1024	0.418	0.844	0.355	0.057	0.346	0.523

Table 3 presents the results for the second set of experiments, i.e., the image augmentation methods for the RetinaNet model with a backbone of ResNet101. The baseline model with no data augmentations involved had a mAP of 0.362. The first augmentation performed was random contrast. The resultant mAP was 0.351, meaning the model performed poorly compared to the baseline. The next augmentation was random brightness, with a mAP of 0.385, representing a reasonable improvement compared to the baseline. The augmentation after random brightness was random saturation which was performed similarly, producing a mAP value of 0.385. The random distort color was the best performing augmentation compared to the other methods tested, giving a mAP of 0.409. Finally, the random RGB to gray-scale performed worse than the other methods except random contrast, with a mAP of 0.375.

Table 3. Comparison of different image augmentations for RetinaNet.

Augmentation	Backbone	Input	mAP	mAP_{50}	mAP_{75}	mAP_{small}	mAP_{medium}	mAP_{large}
Baseline	ResNet101	1024 × 1024	0.362	0.744	0.301	0.011	0.289	0.478
Random Contrast	ResNet101	1024 × 1024	0.351	0.740	0.278	0.010	0.279	0.464
Random Brightness	ResNet101	1024 × 1024	0.385	0.781	0.328	0.019	0.311	0.494
Random Saturation	ResNet101	1024 × 1024	0.385	0.779	0.332	0.014	0.308	0.500
Distort Color	ResNet101	1024 × 1024	0.409	0.796	0.375	0.014	0.327	0.526
RGB to Gray	ResNet101	1024 × 1024	0.375	0.766	0.320	0.020	0.301	0.491

The results for the second set of experiments with the Faster-RCNN model with a backbone of ResNet101 are shown in Table 4. The baseline mAP for Faster-RCNN with no augmentations performed was 0.403. It was comparably higher compared to RetinaNet's baseline. This can be accounted for by Faster-RCNN being a two-stage detector, which is supposed to be more robust and accurate compared to RetinaNet, which is a single-stage detector. The first augmentation method tested was random contrast, providing a mAP of 0.406, which was a little greater than the baseline. A mAP of 0.414 was obtained when using the random brightness augmentation method during the training. This was a considerable increase compared to the baseline. The next augmentation was random saturation, giving a mAP of 0.412, which was slightly lower than the mAP for random brightness. The highest mAP, similar to the case with RetineNet, was provided by tthe random distort color augmentation method, with a value of 0.428. Lastly, the random RGB to gray-scale method gave a mAP of 0.418, which was comparable to the random brightness performance.

Table 4. Comparison of different image augmentations for Faster-RCNN.

Augmentation	Backbone	Input	mAP	mAP_{50}	mAP_{75}	mAP_{small}	mAP_{medium}	mAP_{large}
Baseline	ResNet101	1024 × 1024	0.403	0.838	0.325	0.050	0.336	0.503
Random Contrast	ResNet101	1024 × 1024	0.406	0.839	0.333	0.050	0.337	0.508
Random Brightness	ResNet101	1024 × 1024	0.414	0.847	0.344	0.055	0.343	0.515
Random Saturation	ResNet101	1024 × 1024	0.412	0.841	0.343	0.049	0.341	0.513
Distort Color	ResNet101	1024 × 1024	0.428	0.852	0.347	0.056	0.353	0.534
RGB to Gray	ResNet101	1024 × 1024	0.418	0.846	0.385	0.046	0.346	0.523

Furthermore, to better understand the performance for each image augmentation method, the testing loss was also compared along with the mAP for both the models. The methods which provided a higher mAP and a lower testing loss performed overall better compared to the other methods used to train the RetinaNet models. In Figure 11, random distort color is shown to have a testing loss of 1.305 with a mAP of 0.409, exhibiting the best performance for the RetinaNet model. Also, the lowest mAP value was produced by the random contrast trained model (0.362) and the maximum loss was produced by the baseline model (1.442). In the case of Faster-RCNN, a similar pattern was seen. The random distort color-trained model provided the least testing loss (1.503) and maximum mAP (0.428) when compared to the other Faster-RCNN-trained models in this experiment. For the Faster-RCNN models, the random contrast-trained model gave the highest training loss (1.616) and the baseline model had the lowest mAP (0.403) (Figure 12). In Table 5, the % improvement provided by each augmentation method in the case of RetinaNet is displayed. The best performing augmentation method, i.e., random distort color, gave a model improvement of 12.98% and the random contrast method decreased the model performance by −3.03%. The percentage model performance improvements using different image augmentation methods for Faster-RCNN are presented in Table 6. Similar to the RetinaNet case, the best performance improvement was obtained by using the random distort color method (6.2%). The least improvement with Faster-RCNN was produced by the random contrast method, i.e., 0.74%.

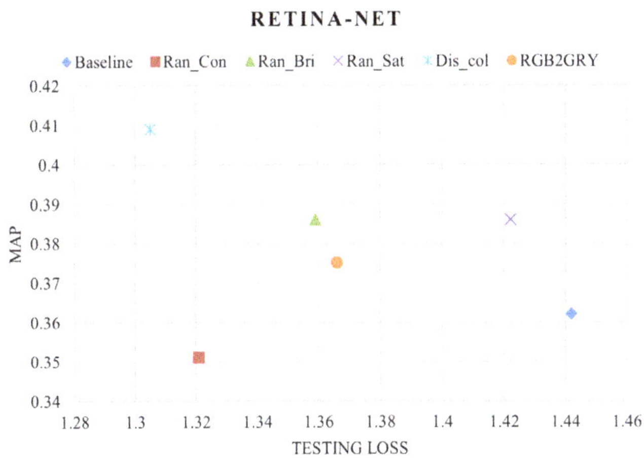

Figure 11. Comparison of mAP and testing loss for all five augmentations with RetinaNet.

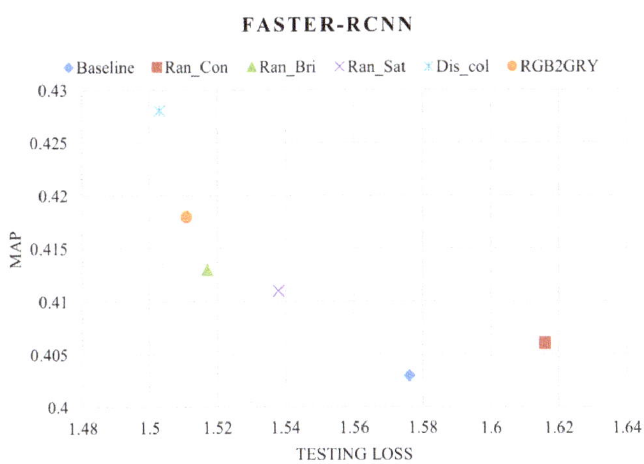

Figure 12. Comparison of mAP and testing loss for all five augmentations with Faster-RCNN.

Table 5. % Improvement using different image augmentations for RetinaNet.

Augmentation	*mAP*	% Improvement
Baseline	0.362	-
Random Contrast	0.351	−3.03%
Random Brightness	0.385	6.35%
Random Saturation	0.385	6.35%
Distort Color	0.409	12.98%
RGB to Gray	0.375	3.59%

Table 6. % Improvement using different image augmentations for Faster-RCNNt.

Augmentation	*mAP*	% Improvement
Baseline	0.403	-
Random Contrast	0.406	0.74%
Random Brightness	0.414	2.72%
Random Saturation	0.412	2.23%
Distort Color	0.428	6.20%
RGB to Gray	0.418	3.72%

6. Discussion

In both the cases with RetinaNet and Faster-RCNN, it was observed that the image augmentation method random contrast either decreased the model performance or did not improve the model performance significantly compared to the baseline trained model. This can be attributed to the random contrast method while augmenting pixel values sometimes creating features which behave as negative examples during model training. These negative examples cause the model to learn them as features, which are not always indicative that the object of interest is present there. The next step will be to check how the random contrast augmentation method affects the model performance of the other augmentation methods.

All the other four types of image augmentation methods, which are random brightness, random saturation, random distort color, and random RGB to gray-scale were used in combination with random contrast to train new models for both RetinaNet and Faster-RCNN. They were then compared to the models which were trained using only the first four aforementioned image augmentation methods. In Tables 7 and 8, the percentage changes in the model performance using all the other augmentation methods, in combination with random contrast for training both RetinaNet and Faster-RCNN, are presented. In both cases, random contrast did not provide any significant improvement except when used with random RGB to gray-scale for training RetinaNet. This may be because in combination with RGB to gray-scale, random contrast adds more difficult examples for the model to train on, making the model quite robust to unknown sets of images. Overall, random contrast, specifically when used with RetinaNet and Faster-RCNN to train on the agricultural dataset, did not provide any significant advantage. It either reduced or slightly increased the model performance when compared to the other image augmentation methods, which altered the pixel values to produce diversity in the dataset. In this study, a mid-size dataset was used. These findings might not apply to very small datasets as a small amount of examples are present for training. For small-scale datasets, further investigation is required to understand the effects of the data augmentations.

Table 7. % Change while using random contrast augmentation with other image augmentation methods for RetinaNet.

Augmentation	*mAP*	Augmentation Combinations	*mAP*	% Change
Random Brightness	0.385	Brightness-Contrast	0.389	1%
Random Saturation	0.385	Saturation-Contrast	0.376	−2%
Distort Color	0.409	Distort Color-Contrast	0.409	0%
RGB to Gray	0.375	RGB to Gray-Contrast	0.394	5%

Table 8. % Change while using random contrast augmentation with different image augmentation methods for Faster-RCNN.

Augmentation	*mAP*	Augmentation Combinations	*mAP*	% Change
Random Brightness	0.414	Brightness-Contrast	0.412	−0.5%
Random Saturation	0.412	Saturation-Contrast	0.413	0.2%
Distort Color	0.428	Distort Color-Contrast	0.426	−0.4%
RGB to Gray	0.418	RGB to Gray-Contrast	0.419	0.2%

7. Conclusions

Low-level and high-level descriptors can be used to explain a scene in an image. Low-level features relate to color, texture, etc., while high-level features deal with objects and the relative information of those objects. For any image-based dataset, a description of the scene in the images is required to explain the information the images provide. Real-world agricultural datasets can have various low-level and high-level features present in them. Due to this, instances were found in the literature where different descriptions were given to explain the same type of scene. Numerous studies were reviewed that attempted to define scenes in real-world agricultural datasets to address the variations found in dataset descriptions. Later, common themes and patterns were identified, and a definition was provided to explain scenes in real-world agricultural datasets as low-contrast complex backgrounds.

Furthermore, the effects of model size and photo-metric image augmentation methods on one-stage (RetinaNet) and two-stage (Faster-RCNN) object detection deep learning models was studied. There are countless examples in the literature where it is stated that larger models perform better. There was a need to study the effects of model size while training on low-contrast complex background agricultural datasets. It was observed that for one-stage detectors, smaller models performed better compared to larger models. The model sizes were varied using different backbones for the networks, including ResNet50, ResNet101, and ResNet152. In the case of two-stage detectors, the backbone size did not affect the model performance, while with the bigger model, the model performance improved. In the second set of experiments, different photo-metric-based image augmentation methods were compared to understand the affect they have on model performance. It is often suggested that in the case of limited data availability, image augmentations can be used to make the deep learning model more robust and generalized. In the literature, it was found that some augmentations might affect the model performance negatively. Comparing different augmentation methods helped in understanding their effects on one-stage and two-stage detectors when training on low-contrast complex background applications. It was observed that except for the random contrast image augmentation method, every other method significantly improved model performance. Even when random contrast was used with other augmentation methods, there was no significant improvement.

In future research, the study can be expanded further to include other single-stage and two-stage detectors, which may have backbones similar to the ones used in this study. For models with different backbone structures, a broader comparison can be made to evaluate the effects of different backbone model structures containing a similar number of parameters. In addition, other datasets related to agricultural operations should be explored to verify the patterns observed in this study. This study used row crop images as a dataset; future studies can use image datasets from other agricultural focus areas, such as horticulture, floriculture, specialty crops, etc. This will help expand knowledge of objection detection models for low-contrast complex background applications in agriculture.

Author Contributions: Conceptualization, H.S.S. and M.J.D.; methodology, H.S.S.; formal analysis, H.S.S.; investigation, H.S.S.; resources, M.J.D.; writing—original draft preparation, H.S.S.; writing—review and editing, M.J.D.; visualization, H.S.S.; supervision, M.J.D.; project administration, M.J.D. All authors have read and agreed to the published version of the manuscript.

Funding: This research received no external funding.

Institutional Review Board Statement: Not Applicable.

Informed Consent Statement: Not Applicable.

Data Availability Statement: The data presented in this study are available in Zenodo at https://zenodo.org/records/5092309. These data were derived from the following resources available in the public domain: [122]. The data was accessed on 1 June 2021.

Conflicts of Interest: The authors declare no conflicts of interest.

References

1. Zhang, N.; Wang, M.; Wang, N. Precision agriculture—A worldwide overview. *Comput. Electron. Agric.* **2002**, *36*, 113–132. [CrossRef]
2. Kayad, A.; Sozzi, M.; Gatto, S.; Whelan, B.; Sartori, L.; Marinello, F. Ten years of corn yield dynamics at field scale under digital agriculture solutions: A case study from North Italy. *Comput. Electron. Agric.* **2021**, *185*, 106126. [CrossRef]
3. De Alwis, S.; Hou, Z.; Zhang, Y.; Na, M.H.; Ofoghi, B.; Sajjanhar, A. A survey on smart farming data, applications and techniques. *Comput. Ind.* **2022**, *138*, 103624. [CrossRef]
4. Chamara, N.; Islam, M.D.; Bai, G.F.; Shi, Y.; Ge, Y. Ag-IoT for crop and environment monitoring: Past, present, and future. *Agric. Syst.* **2022**, *203*, 103497. [CrossRef]
5. Feng, L.; Chen, S.; Zhang, C.; Zhang, Y.; He, Y. A comprehensive review on recent applications of unmanned aerial vehicle remote sensing with various sensors for high-throughput plant phenotyping. *Comput. Electron. Agric.* **2021**, *182*, 106033. [CrossRef]
6. Ullo, S.L.; Sinha, G.R. Advances in IoT and smart sensors for remote sensing and agriculture applications. *Remote Sens.* **2021**, *13*, 2585. [CrossRef]
7. Cobbenhagen, A.; Antunes, D.J.; van de Molengraft, M.; Heemels, W. Opportunities for control engineering in arable precision agriculture. *Annu. Rev. Control* **2021**, *51*, 47–55. [CrossRef]
8. Sood, A.; Sharma, R.K.; Bhardwaj, A.K. Artificial intelligence research in agriculture: A review. *Online Inf. Rev.* **2021**, *46*, 1054–1075. [CrossRef]
9. Bari, B.S.; Islam, M.N.; Rashid, M.; Hasan, M.J.; Razman, M.A.M.; Musa, R.M.; Ab Nasir, A.F.; P.P. Abdul Majeed, A. A real-time approach of diagnosing rice leaf disease using deep learning-based faster R-CNN framework. *PeerJ Comput. Sci.* **2021**, *7*, e432. [CrossRef]
10. Costa, L.; Ampatzidis, Y.; Rohla, C.; Maness, N.; Cheary, B.; Zhang, L. Measuring pecan nut growth utilizing machine vision and deep learning for the better understanding of the fruit growth curve. *Comput. Electron. Agric.* **2021**, *181*, 105964. [CrossRef]

11. Fue, K.G.; Porter, W.M.; Barnes, E.M.; Rains, G.C. Ensemble Method of Deep Learning, Color Segmentation, and Image Transformation to Track, Localize, and Count Cotton Bolls Using a Moving Camera in Real-Time. *Trans. ASABE* **2021**, *64*, 341–352. [CrossRef]
12. Jabir, B.; Falih, N.; Rahmani, K. Accuracy and Efficiency Comparison of Object Detection Open-Source Models. *Int. J. Online Biomed. Eng. (iJOE)* **2021**, *17*, 165–184. [CrossRef]
13. Li, Z.; Li, Y.; Yang, Y.; Guo, R.; Yang, J.; Yue, J.; Wang, Y. A high-precision detection method of hydroponic lettuce seedlings status based on improved Faster RCNN. *Comput. Electron. Agric.* **2021**, *182*, 106054. [CrossRef]
14. Patrício, D.I.; Rieder, R. Computer vision and artificial intelligence in precision agriculture for grain crops: A systematic review. *Comput. Electron. Agric.* **2018**, *153*, 69–81. [CrossRef]
15. Wang, R.; Liu, L.; Xie, C.; Yang, P.; Li, R.; Zhou, M. Agripest: A large-scale domain-specific benchmark dataset for practical agricultural pest detection in the wild. *Sensors* **2021**, *21*, 1601. [CrossRef]
16. Weyler, J.; Milioto, A.; Falck, T.; Behley, J.; Stachniss, C. Joint Plant Instance Detection and Leaf Count Estimation for In-Field Plant Phenotyping. *IEEE Robot. Autom. Lett.* **2021**, *6*, 3599–3606. [CrossRef]
17. Yandun Narváez, F.; Gregorio, E.; Escolà, A.; Rosell-Polo, J.R.; Torres-Torriti, M.; Auat Cheein, F. Terrain classification using ToF sensors for the enhancement of agricultural machinery traversability. *J. Terramechanics* **2018**, *76*, 1–13. [CrossRef]
18. Li, W.; Wang, D.; Li, M.; Gao, Y.; Wu, J.; Yang, X. Field detection of tiny pests from sticky trap images using deep learning in agricultural greenhouse. *Comput. Electron. Agric.* **2021**, *183*, 106048. [CrossRef]
19. Bazame, H.C.; Molin, J.P.; Althoff, D.; Martello, M. Detection, classification, and mapping of coffee fruits during harvest with computer vision. *Comput. Electron. Agric.* **2021**, *183*, 106066. [CrossRef]
20. Chen, Y.; Wu, Z.; Zhao, B.; Fan, C.; Shi, S. Weed and corn seedling detection in field based on multi feature fusion and support vector machine. *Sensors* **2021**, *21*, 212. [CrossRef]
21. Li, Y.; Li, M.; Qi, J.; Zhou, D.; Zou, Z.; Liu, K. Detection of typical obstacles in orchards based on deep convolutional neural network. *Comput. Electron. Agric.* **2021**, *181*, 105932. [CrossRef]
22. Parvathi, S.; Tamil Selvi, S. Detection of maturity stages of coconuts in complex background using Faster R-CNN model. *Biosyst. Eng.* **2021**, *202*, 119–132. [CrossRef]
23. Gao, J.; Westergaard, J.C.; Sundmark, E.H.R.; Bagge, M.; Liljeroth, E.; Alexandersson, E. Automatic late blight lesion recognition and severity quantification based on field imagery of diverse potato genotypes by deep learning. *Knowl.-Based Syst.* **2021**, *214*, 106723. [CrossRef]
24. Anagnostis, A.; Tagarakis, A.C.; Asiminari, G.; Papageorgiou, E.; Kateris, D.; Moshou, D.; Bochtis, D. A deep learning approach for anthracnose infected trees classification in walnut orchards. *Comput. Electron. Agric.* **2021**, *182*, 105998. [CrossRef]
25. Gomez, A.S.; Aptoula, E.; Parsons, S.; Bosilj, P. Deep Regression Versus Detection for Counting in Robotic Phenotyping. *IEEE Robot. Autom. Lett.* **2021**, *6*, 2902–2907. [CrossRef]
26. Fu, L.; Duan, J.; Zou, X.; Lin, J.; Zhao, L.; Li, J.; Yang, Z. Fast and accurate detection of banana fruits in complex background orchards. *IEEE Access* **2020**, *8*, 196835–196846. [CrossRef]
27. Li, G.; Huang, X.; Ai, J.; Yi, Z.; Xie, W. Lemon-YOLO: An efficient object detection method for lemons in the natural environment. *IET Image Process.* **2021**, *15*, 1998–2009. [CrossRef]
28. Zhao, W.; Yamada, W.; Li, T.; Digman, M.; Runge, T. Augmenting crop detection for precision agriculture with deep visual transfer learning—A case study of bale detection. *Remote Sens.* **2021**, *13*, 23. [CrossRef]
29. Bochkovskiy, A.; Wang, C.Y.; Liao, H.Y.M. YOLOv4: Optimal Speed and Accuracy of Object Detection. *arXiv* **2020**, arXiv:2004.10934.
30. Sponchioni, G.; Vezzoni, M.; Bacchetti, A.; Pavesi, M.; Renga, F. The 4.0 revolution in agriculture: A multi-perspective definition. In Proceedings of the Summer School F. Turco-Industrial Systems Engineering, Brescia, Italy, 11–13 September 2019; pp. 143–149.
31. Schreefel, L.; Schulte, R.; De Boer, I.; Schrijver, A.P.; Van Zanten, H. Regenerative agriculture—The soil is the base. *Glob. Food Secur.* **2020**, *26*, 100404. [CrossRef]
32. Zhang, W.; Yang, G.; Lin, Y.; Ji, C.; Gupta, M.M. On Definition of Deep Learning. In Proceedings of the 2018 World Automation Congress (WAC), Stevenson, WA, USA, 3–6 June 2018; IEEE: Piscataway, NJ, USA, 2018; Volume 2018, pp. 1–5. [CrossRef]
33. Wang, D.; He, D. Fusion of Mask RCNN and attention mechanism for instance segmentation of apples under complex background. *Comput. Electron. Agric.* **2022**, *196*, 106864. [CrossRef]
34. Zhou, J.; Li, J.; Wang, C.; Wu, H.; Zhao, C.; Wang, Q. A vegetable disease recognition model for complex background based on region proposal and progressive learning. *Comput. Electron. Agric.* **2021**, *184*, 106101. [CrossRef]
35. Jiang, Y.; Li, C.; Paterson, A.H.; Robertson, J.S. DeepSeedling: Deep convolutional network and Kalman filter for plant seedling detection and counting in the field. *Plant Methods* **2019**, *15*, 141. [CrossRef] [PubMed]
36. Gao, L.; Lin, X. A method for accurately segmenting images of medicinal plant leaves with complex backgrounds. *Comput. Electron. Agric.* **2018**, *155*, 426–445. [CrossRef]

37. Albattah, W.; Javed, A.; Nawaz, M.; Masood, M.; Albahli, S. Artificial Intelligence-Based Drone System for Multiclass Plant Disease Detection Using an Improved Efficient Convolutional Neural Network. *Front. Plant Sci.* **2022**, *13*, 1313. [CrossRef]
38. Cheng, Z.; Qi, L.; Cheng, Y. Cherry Tree Crown Extraction from Natural Orchard Images with Complex Backgrounds. *Agriculture* **2021**, *11*, 431. [CrossRef]
39. Lin, G.; Tang, Y.; Zou, X.; Cheng, J.; Xiong, J. Fruit detection in natural environment using partial shape matching and probabilistic Hough transform. *Precis. Agric.* **2020**, *21*, 160–177. [CrossRef]
40. Yang, L.; Luo, J.; Wang, Z.; Chen, Y.; Wu, C. Research on recognition for cotton spider mites' damage level based on deep learning. *Int. J. Agric. Biol. Eng.* **2019**, *12*, 129–134. [CrossRef]
41. Sun, J.; He, X.; Ge, X.; Wu, X.; Shen, J.; Song, Y. Detection of key organs in tomato based on deep migration learning in a complex background. *Agriculture* **2018**, *8*, 196. [CrossRef]
42. Wang, N.; Qian, T.; Yang, J.; Li, L.; Zhang, Y.; Zheng, X.; Xu, Y.; Zhao, H.; Zhao, J. An Enhanced YOLOv5 Model for Greenhouse Cucumber Fruit Recognition Based on Color Space Features. *Agriculture* **2022**, *12*, 1556. [CrossRef]
43. Li, X.; Pan, J.; Xie, F.; Zeng, J.; Li, Q.; Huang, X.; Liu, D.; Wang, X. Fast and accurate green pepper detection in complex backgrounds via an improved Yolov4-tiny model. *Comput. Electron. Agric.* **2021**, *191*, 106503. [CrossRef]
44. Jiang, Y.; Li, C.; Xu, R.; Sun, S.; Robertson, J.S.; Paterson, A.H. DeepFlower: A deep learning-based approach to characterize flowering patterns of cotton plants in the field. *Plant Methods* **2020**, *16*, 156. [CrossRef]
45. Cao, Q.; Xu, L. Unsupervised greenhouse tomato plant segmentation based on self-adaptive iterative latent dirichlet allocation from surveillance camera. *Agronomy* **2019**, *9*, 91. [CrossRef]
46. Wang, X.; Wang, X.; Hu, C.; Dai, F.; Xing, J.; Wang, E.; Du, Z.; Wang, L.; Guo, W. Study on the Detection of Defoliation Effect of an Improved YOLOv5x Cotton. *Agriculture* **2022**, *12*, 1583. [CrossRef]
47. Liu, C.; Zhu, H.; Guo, W.; Han, X.; Chen, C.; Wu, H. EFDet: An efficient detection method for cucumber disease under natural complex environments. *Comput. Electron. Agric.* **2021**, *189*, 106378. [CrossRef]
48. Afonso, M.; Fonteijn, H.; Fiorentin, F.S.; Lensink, D.; Mooij, M.; Faber, N.; Polder, G.; Wehrens, R. Tomato fruit detection and counting in greenhouses using deep learning. *Front. Plant Sci.* **2020**, *11*, 571299. [CrossRef] [PubMed]
49. Moradi, F.; Javan, F.D.; Samadzadegan, F. Potential evaluation of visible-thermal UAV image fusion for individual tree detection based on convolutional neural network. *Int. J. Appl. Earth Obs. Geoinf.* **2022**, *113*, 103011. [CrossRef]
50. ur Rehman, Z.; Khan, M.A.; Ahmed, F.; Damaševičius, R.; Naqvi, S.R.; Nisar, W.; Javed, K. Recognizing apple leaf diseases using a novel parallel real-time processing framework based on MASK RCNN and transfer learning: An application for smart agriculture. *IET Image Process.* **2021**, *15*, 2157–2168. [CrossRef]
51. Osorio, K.; Puerto, A.; Pedraza, C.; Jamaica, D.; Rodríguez, L. A deep learning approach for weed detection in lettuce crops using multispectral images. *AgriEngineering* **2020**, *2*, 471–488. [CrossRef]
52. Ahmad, A.; Aggarwal, V.; Saraswat, D.; El Gamal, A.; Johal, G.S. GeoDLS: A Deep Learning-Based Corn Disease Tracking and Location System Using RTK Geolocated UAS Imagery. *Remote Sens.* **2022**, *14*, 4140. [CrossRef]
53. Wang, L.; Xiang, L.; Tang, L.; Jiang, H. A convolutional neural network-based method for corn stand counting in the field. *Sensors* **2021**, *21*, 507. [CrossRef] [PubMed]
54. Yang, Z.; Gao, S.; Xiao, F.; Li, G.; Ding, Y.; Guo, Q.; Paul, M.J.; Liu, Z. Leaf to panicle ratio (LPR): A new physiological trait indicative of source and sink relation in japonica rice based on deep learning. *Plant Methods* **2020**, *16*, 117. [CrossRef] [PubMed]
55. Maji, A.K.; Marwaha, S.; Kumar, S.; Arora, A.; Chinnusamy, V.; Islam, S. SlypNet: Spikelet-based yield prediction of wheat using advanced plant phenotyping and computer vision techniques. *Front. Plant Sci.* **2022**, *13*, 889853. [CrossRef]
56. Wang, Y.; Qin, Y.; Cui, J. Occlusion robust wheat ear counting algorithm based on deep learning. *Front. Plant Sci.* **2021**, *12*, 645899. [CrossRef]
57. Xu, X.; Li, H.; Yin, F.; Xi, L.; Qiao, H.; Ma, Z.; Shen, S.; Jiang, B.; Ma, X. Wheat ear counting using K-means clustering segmentation and convolutional neural network. *Plant Methods* **2020**, *16*, 106. [CrossRef] [PubMed]
58. Craze, H.A.; Pillay, N.; Joubert, F.; Berger, D.K. Deep Learning Diagnostics of Gray Leaf Spot in Maize under Mixed Disease Field Conditions. *Plants* **2022**, *11*, 1942. [CrossRef]
59. Wang, X.; Liu, J. Diseases detection of occlusion and overlapping tomato leaves based on deep learning. *Front. Plant Sci.* **2021**, *12*, 792244. [CrossRef]
60. Zhou, C.; Hu, J.; Xu, Z.; Yue, J.; Ye, H.; Yang, G. A monitoring system for the segmentation and grading of broccoli head based on deep learning and neural networks. *Front. Plant Sci.* **2020**, *11*, 402. [CrossRef]
61. Kim, T.; Kim, H.; Baik, K.; Choi, Y. Instance-Aware Plant Disease Detection by Utilizing Saliency Map and Self-Supervised Pre-Training. *Agriculture* **2022**, *12*, 1084. [CrossRef]
62. Afzaal, U.; Bhattarai, B.; Pandeya, Y.R.; Lee, J. An instance segmentation model for strawberry diseases based on mask R-CNN. *Sensors* **2021**, *21*, 6565. [CrossRef]
63. Chandra, A.L.; Desai, S.V.; Balasubramanian, V.N.; Ninomiya, S.; Guo, W. Active learning with point supervision for cost-effective panicle detection in cereal crops. *Plant Methods* **2020**, *16*, 34. [CrossRef]

64. Deng, J.; Lv, X.; Yang, L.; Zhao, B.; Zhou, C.; Yang, Z.; Jiang, J.; Ning, N.; Zhang, J.; Shi, J.; et al. Assessing Macro Disease Index of Wheat Stripe Rust Based on Segformer with Complex Background in the Field. *Sensors* **2022**, *22*, 5676. [CrossRef] [PubMed]
65. Albattah, W.; Nawaz, M.; Javed, A.; Masood, M.; Albahli, S. A novel deep learning method for detection and classification of plant diseases. *Complex Intell. Syst.* **2021**, *8*, 507–524. [CrossRef]
66. Xuan, G.; Gao, C.; Shao, Y.; Zhang, M.; Wang, Y.; Zhong, J.; Li, Q.; Peng, H. Apple detection in natural environment using deep learning algorithms. *IEEE Access* **2020**, *8*, 216772–216780. [CrossRef]
67. Jin, X.; Bagavathiannan, M.; Maity, A.; Chen, Y.; Yu, J. Deep learning for detecting herbicide weed control spectrum in turfgrass. *Plant Methods* **2022**, *18*, 94. [CrossRef]
68. Zheng, Z.; Xiong, J.; Lin, H.; Han, Y.; Sun, B.; Xie, Z.; Yang, Z.; Wang, C. A Method of Green Citrus Detection in Natural Environment using a Deep Convolutional Neural Network. *Front. Plant Sci.* **2021**, *7*, 705737. [CrossRef]
69. Han, Z.; Hu, W.; Peng, S.; Lin, H.; Zhang, J.; Zhou, J.; Wang, P.; Dian, Y. Detection of Standing Dead Trees after Pine Wilt Disease Outbreak with Airborne Remote Sensing Imagery by Multi-Scale Spatial Attention Deep Learning and Gaussian Kernel Approach. *Remote Sens.* **2022**, *14*, 3075. [CrossRef]
70. Kc, K.; Yin, Z.; Li, D.; Wu, Z. Impacts of background removal on convolutional neural networks for plant disease classification in-situ. *Agriculture* **2021**, *11*, 827. [CrossRef]
71. Zhang, Z.; Qiao, Y.; Guo, Y.; He, D. Deep learning based automatic grape downy mildew detection. *Front. Plant Sci.* **2022**, *13*, 872107. [CrossRef]
72. Shao, H.; Tang, R.; Lei, Y.; Mu, J.; Guan, Y.; Xiang, Y. Rice ear counting based on image segmentation and establishment of a dataset. *Plants* **2021**, *10*, 1625. [CrossRef]
73. Wang, C.; Wang, Y.; Liu, S.; Lin, G.; He, P.; Zhang, Z.; Zhou, Y. Study on pear flowers detection performance of YOLO-PEFL model trained with synthetic target images. *Front. Plant Sci.* **2022**, *13*, 911473. [CrossRef]
74. Miao, Y.; Huang, L.; Zhang, S. A Two-Step Phenotypic Parameter Measurement Strategy for Overlapped Grapes under Different Light Conditions. *Sensors* **2021**, *21*, 4532. [CrossRef] [PubMed]
75. Xiang, R.; Zhang, M.; Zhang, J. Recognition for Stems of Tomato Plants at Night Based on a Hybrid Joint Neural Network. *Agriculture* **2022**, *12*, 743. [CrossRef]
76. Zhu, J.; Cheng, M.; Wang, Q.; Yuan, H.; Cai, Z. Grape leaf black rot detection based on super-resolution image enhancement and deep learning. *Front. Plant Sci.* **2021**, *12*, 695749. [CrossRef] [PubMed]
77. Suo, J.; Zhan, J.; Zhou, G.; Chen, A.; Hu, Y.; Huang, W.; Cai, W.; Hu, Y.; Li, L. Casm-amfmnet: A network based on coordinate attention shuffle mechanism and asymmetric multi-scale fusion module for classification of grape leaf diseases. *Front. Plant Sci.* **2022**, *13*, 846767. [CrossRef]
78. Zhang, R.; Tian, Y.; Zhang, J.; Dai, S.; Hou, X.; Wang, J.; Guo, Q. Metric learning for image-based flower cultivars identification. *Plant Methods* **2021**, *17*, 65. [CrossRef] [PubMed]
79. Teimouri, N.; Jørgensen, R.N.; Green, O. Novel Assessment of Region-Based CNNs for Detecting Monocot/Dicot Weeds in Dense Field Environments. *Agronomy* **2022**, *12*, 1167. [CrossRef]
80. Magalhães, S.A.; Castro, L.; Moreira, G.; Dos Santos, F.N.; Cunha, M.; Dias, J.; Moreira, A.P. Evaluating the single-shot multibox detector and YOLO deep learning models for the detection of tomatoes in a greenhouse. *Sensors* **2021**, *21*, 3569. [CrossRef]
81. Yadav, A.; Thakur, U.; Saxena, R.; Pal, V.; Bhateja, V.; Lin, J.C.W. AFD-Net: Apple Foliar Disease multi classification using deep learning on plant pathology dataset. *Plant Soil* **2022**, *477*, 595–611. [CrossRef]
82. Wang, X.; Liu, J.; Zhu, X. Early real-time detection algorithm of tomato diseases and pests in the natural environment. *Plant Methods* **2021**, *17*, 1–17. [CrossRef]
83. Qian, X.; Zhang, C.; Chen, L.; Li, K. Deep learning-based identification of maize leaf diseases is improved by an attention mechanism: Self-Attention. *Front. Plant Sci.* **2022**, *13*, 864486. [CrossRef] [PubMed]
84. Wang, X.; Liu, J. Tomato anomalies detection in greenhouse scenarios based on YOLO-Dense. *Front. Plant Sci.* **2021**, *12*, 634103. [CrossRef] [PubMed]
85. Li, F.; Bai, J.; Zhang, M.; Zhang, R. Yield estimation of high-density cotton fields using low-altitude UAV imaging and deep learning. *Plant Methods* **2022**, *18*, 55. [CrossRef] [PubMed]
86. Tufail, M.; Iqbal, J.; Tiwana, M.I.; Alam, M.S.; Khan, Z.A.; Khan, M.T. Identification of tobacco crop based on machine learning for a precision agricultural sprayer. *IEEE Access* **2021**, *9*, 23814–23825. [CrossRef]
87. Buxbaum, N.; Lieth, J.H.; Earles, M. Non-destructive Plant Biomass Monitoring with High Spatio-Temporal Resolution via Proximal RGB-D Imagery and End-to-End Deep Learning. *Front. Plant Sci.* **2022**, *13*, 758818. [CrossRef]
88. Liu, C.; Zhao, C.; Wu, H.; Han, X.; Li, S. ADDLight: An Energy-Saving Adder Neural Network for Cucumber Disease Classification. *Agriculture* **2022**, *12*, 452. [CrossRef]
89. Yao, J.; Wang, Y.; Xiang, Y.; Yang, J.; Zhu, Y.; Li, X.; Li, S.; Zhang, J.; Gong, G. Two-Stage Detection Algorithm for Kiwifruit Leaf Diseases Based on Deep Learning. *Plants* **2022**, *11*, 768. [CrossRef]

90. Sozzi, M.; Cantalamessa, S.; Cogato, A.; Kayad, A.; Marinello, F. Automatic bunch detection in white grape varieties using YOLOv3, YOLOv4, and YOLOv5 deep learning algorithms. *Agronomy* **2022**, *12*, 319. [CrossRef]
91. Wang, C.; Zhou, J.; Zhang, Y.; Wu, H.; Zhao, C.; Teng, G.; Li, J. A Plant Disease Recognition Method Based on Fusion of Images and Graph Structure Text. *Front. Plant Sci.* **2022**, *12*, 731688. [CrossRef]
92. Zenkl, R.; Timofte, R.; Kirchgessner, N.; Roth, L.; Hund, A.; Van Gool, L.; Walter, A.; Aasen, H. Outdoor plant segmentation with deep learning for high-throughput field phenotyping on a diverse wheat dataset. *Front. Plant Sci.* **2021**, *12*, 774068. [CrossRef]
93. Abreu, F.F.; Rodrigues, L.H.A. Monitoring mini-tomatoes growth: A non-destructive machine vision-based alternative. *J. Agric. Eng.* **2022**, *53*, 3.
94. Wan, P.; Toudeshki, A.; Tan, H.; Ehsani, R. A methodology for fresh tomato maturity detection using computer vision. *Comput. Electron. Agric.* **2018**, *146*, 43–50. [CrossRef]
95. Buzzy, M.; Thesma, V.; Davoodi, M.; Mohammadpour Velni, J. Real-time plant leaf counting using deep object detection networks. *Sensors* **2020**, *20*, 6896. [CrossRef] [PubMed]
96. Xiao, Y.; Tian, Z.; Yu, J.; Zhang, Y.; Liu, S.; Du, S.; Lan, X. A review of object detection based on deep learning. *Multimed. Tools Appl.* **2020**, *79*, 23729–23791. [CrossRef]
97. Krizhevsky, A.; Sutskever, I.; Hinton, G.E. ImageNet classification with deep convolutional neural networks. *Commun. ACM* **2017**, *60*, 84–90. [CrossRef]
98. Lecun, Y.; Bengio, Y.; Hinton, G. Deep learning. *Nature* **2015**, *521*, 436–444. [CrossRef]
99. Girshick, R.; Donahue, J.; Darrell, T.; Malik, J. Rich feature hierarchies for accurate object detection and semantic segmentation. In Proceedings of the IEEE Conference on Computer Vision and Pattern Recognition, Columbus, OH, USA, 23–28 June 2014; pp. 580–587.
100. He, K.; Zhang, X.; Ren, S.; Sun, J. Spatial pyramid pooling in deep convolutional networks for visual recognition. *IEEE Trans. Pattern Anal. Mach. Intell.* **2015**, *37*, 1904–1916. [CrossRef]
101. Girshick, R. Fast r-cnn. In Proceedings of the IEEE International Conference on Computer Vision, Santiago, Chile, 7–13 December 2015; pp. 1440–1448.
102. Ren, S.; He, K.; Girshick, R.; Sun, J. Faster r-cnn: Towards real-time object detection with region proposal networks. *arXiv* **2015**, arXiv:1506.01497. [CrossRef]
103. He, K.; Gkioxari, G.; Dollár, P.; Girshick, R. Mask r-cnn. In Proceedings of the IEEE International Conference on Computer Vision, Venice, Italy, 22–29 October 2017; pp. 2961–2969.
104. Redmon, J.; Divvala, S.; Girshick, R.; Farhadi, A. You only look once: Unified, real-time object detection. In Proceedings of the IEEE Conference on Computer Vision and Pattern Recognition, Las Vegas, NV, USA, 27–30 June 2016; pp. 779–788.
105. Redmon, J.; Farhadi, A. YOLO9000: Better, faster, stronger. In Proceedings of the IEEE Conference on Computer Vision and Pattern Recognition, Honolulu, HI, USA, 21–26 July 2017; pp. 7263–7271.
106. Redmon, J.; Farhadi, A. Yolov3: An incremental improvement. *arXiv* **2018**, arXiv:1804.02767.
107. Liu, W.; Anguelov, D.; Erhan, D.; Szegedy, C.; Reed, S.; Fu, C.Y.; Berg, A.C. Ssd: Single shot multibox detector. In Proceedings of the European Conference on Computer Vision, Amsterdam, the Netherlands, 11–14 October 2016; Springer: Berlin/Heidelberg, Germany, 2016; pp. 21–37.
108. Lin, T.Y.; Goyal, P.; Girshick, R.; He, K.; Dollar, P. Focal Loss for Dense Object Detection. *arXiv* **2018**, arXiv:1708.02002.
109. Simonyan, K.; Zisserman, A. Very deep convolutional networks for large-scale image recognition. *arXiv* **2014**, arXiv:1409.1556.
110. He, K.; Zhang, X.; Ren, S.; Sun, J. Deep residual learning for image recognition. In Proceedings of the IEEE Conference on Computer Vision and Pattern Recognition, Las Vegas, NV, USA, 27–30 June 2016; pp. 770–778.
111. Li, Z.; Peng, C.; Yu, G.; Zhang, X.; Deng, Y.; Sun, J. Detnet: Design backbone for object detection. In Proceedings of the European Conference on Computer Vision (ECCV), Munich, Germany, 8–14 September 2018; pp. 334–350.
112. Ioffe, S.; Szegedy, C. Batch normalization: Accelerating deep network training by reducing internal covariate shift. In Proceedings of the International Conference on Machine Learning, PMLR, Lille, France, 7–9 July 2015; pp. 448–456.
113. Szegedy, C.; Liu, W.; Jia, Y.; Sermanet, P.; Reed, S.; Anguelov, D.; Erhan, D.; Vanhoucke, V.; Rabinovich, A. Going deeper with convolutions. In Proceedings of the IEEE Conference on Computer Vision and Pattern Recognition, Boston, MA, USA, 7–12 June 2015; pp. 1–9.
114. Szegedy, C.; Vanhoucke, V.; Ioffe, S.; Shlens, J.; Wojna, Z. Rethinking the inception architecture for computer vision. In Proceedings of the IEEE Conference on Computer Vision and Pattern Recognition, Las Vegas, NV, USA, 27–30 June 2016; pp. 2818–2826.
115. Chollet, F. Xception: Deep learning with depthwise separable convolutions. In Proceedings of the IEEE Conference on Computer Vision and Pattern Recognition, Honolulu, HI, USA, 21–26 July 2017; pp. 1251–1258.
116. Howard, A.G.; Zhu, M.; Chen, B.; Kalenichenko, D.; Wang, W.; Weyand, T.; Andreetto, M.; Adam, H. Mobilenets: Efficient convolutional neural networks for mobile vision applications. *arXiv* **2017**, arXiv:1704.04861.

117. Sandler, M.; Howard, A.; Zhu, M.; Zhmoginov, A.; Chen, L.C. Mobilenetv2: Inverted residuals and linear bottlenecks. In Proceedings of the IEEE Conference on Computer Vision and Pattern Recognition, Salt Lake City, UT, USA, 18–23 June 2018; pp. 4510–4520.
118. David, E.; Ogidi, F.; Guo, W.; Baret, F.; Stavness, I. Global Wheat Challenge 2020: Analysis of the competition design and winning models. *arXiv* **2021**, arXiv:2105.06182.
119. Taylor, L.; Nitschke, G. Improving deep learning with generic data augmentation. In Proceedings of the 2018 IEEE Symposium Series on Computational Intelligence (SSCI), Bangalore, India, 18–21 November 2018; IEEE: Piscataway, NJ, USA, 2018; pp. 1542–1547.
120. Shorten, C.; Khoshgoftaar, T.M. A survey on Image Data Augmentation for Deep Learning. *J. Big Data* **2019**, *6*, 60. [CrossRef]
121. Padilla, R.; Passos, W.L.; Dias, T.L.B.; Netto, S.L.; da Silva, E.A.B. A Comparative Analysis of Object Detection Metrics with a Companion Open-Source Toolkit. *Electronics* **2021**, *10*, 279. [CrossRef]
122. David, E.; Serouart, M.; Smith, D.; Madec, S.; Velumani, K.; Liu, S.; Wang, X.; Pinto, F.; Shafiee, S.; Tahir, I.S.A.; et al. Global Wheat Head Detection 2021: An Improved Dataset for Benchmarking Wheat Head Detection Methods. *Plant Phenomics* **2021**, *2021*, 9846158. [CrossRef]

AI

MDPI

Article

The Detection and Counting of Olive Tree Fruits Using Deep Learning Models in Tacna, Perú

Erbert Osco-Mamani [1,*], Oliver Santana-Carbajal [1], Israel Chaparro-Cruz [1], Daniel Ochoa-Donoso [2] and Sylvia Alcazar-Alay [3]

[1] Department of Informatics and Systems, Jorge Basadre Grohmann National University, Tacna 23003, Peru; osantanac@unjbg.edu.pe (O.S.-C.); ichaparroc@unjbg.edu.pe (I.C.-C.)

[2] Faculty of Electrical and Computer Engineering, ESPOL Polytechnic University, Escuela Superior Politécnica del Litoral, Guayaquil 0909022, Ecuador; dochoa@espol.edu.ec

[3] Department of Food Engineering, National Intercultural University of Quillabamba, Cusco 08741, Peru; sylvia.alcazar@uniq.edu.pe

* Correspondence: eosco@unjbg.edu.pe

Abstract: Predicting crop performance is key to decision making for farmers and business owners. Tacna is the main olive-producing region in Perú, with an annual yield of 6.4 t/ha, mainly of the Sevillana variety. Recently, olive production levels have fluctuated due to severe weather conditions and disease outbreaks. These climatic phenomena are expected to continue in the coming years. The objective of the study was to evaluate the performance of the model in natural and specific environments of the olive grove and counting olive fruits using CNNs from images. Among the models evaluated, YOLOv8m proved to be the most effective (94.960), followed by YOLOv8s, Faster R-CNN and RetinaNet. For the mAP50-95 metric, YOLOv8m was also the most effective (0.775). YOLOv8m achieved the best performance with an RMSE of 402.458 and a coefficient of determination R^2 of (0.944), indicating a high correlation with the actual fruit count. As part of this study, a novel olive fruit dataset was developed to capture the variability under different fruit conditions. Concluded that the predicting crop from images requires consideration of field imaging conditions, color tones, and the similarity between olives and leaves.

Keywords: fruit detection; fruit counting; deep learning; YOLOv8; Faster R-CNN; RetinaNet

1. Introduction

Olive production is an important agro-industrial activity for the economic and social development of the southern region of Perú, with 95% of the national production and an average yield of 6.4 t/ha [1]. The Tacna Region is characterized as a pioneer in olive production, with its economic benefits determined by the productivity and efficiency of profitability achieved in recent years. This has driven economic dynamism at the regional level, as evidenced by the production and agro-export of olives [1]. Tacna's local economy depends on having stable production levels. However, climate change has caused significant variability in annual olive production in recent years; between 2022 and 2023, Tacna olive production decreased by 27%, and the early forecast for 2024 indicated an even more drastic decrease in production, caused by the El Niño–Southern Oscillation [2,3].

The automatic estimation of olive crop yield is necessary to select production strategies to mitigate the effects of climate change [4] and can also reduce labor costs, improve the accuracy of yield forecasts, and support farmers in developing more rational harvesting and marketing strategies. Past works on quantitative crops have exploited computer vision techniques for oranges [5], lemons [6], apples [7], blueberries [8], plums [9], walnuts [10], tomatoes [11], and grape bunches [12]. Changing lighting conditions, occlusion, shadows, relative scale, and complex backgrounds negatively affect fruit detection accuracy [13–15].

Regarding olive detection, past works have applied feature engineering to compute image descriptors based on mathematical morphology and relied on local contrast thresholding for segmentation [16,17]. Clustering in RGB color space using k-nearest neighbors was used to detect the ripening stage of olive fruit batches using a real-time computer vision method [18]. Although fruit detection based on deep learning has shown good results, the accurate counting of olive fruits is a challenging problem. Detection and localization errors may occur due to artifacts, background clutter, lighting fluctuations, scale variations, and occlusion. Even for a trained eye, it may be difficult to distinguish an olive from the foliage [19].

In this paper, we present an initial attempt to detect olives using widely varying deep learning techniques. The main objective of the study was to evaluate the model's performance in natural and specific olive grove environments. To carry this out, we built a dataset of Sevillana olives and compared six deep learning models (YOLOv8m, YOLOv8s, Faster R-CNN 101, Faster R-CNN 50, RetinaNet 101, and RetinaNet 50) using RGB images captured in real production conditions. The suitability of each model was established by computing well-known object detection metrics.

In this study, a novel contribution is presented: (a) It is the first time that we have a dataset of images of harvested olive fruits in Tacna, Perú. (b) Although learning models have been used in previous research for detection tasks in agricultural environments, our work introduces an exhaustive evaluation of latest YOLOv8 models, applied specifically to the counting and detection of olive fruits of the Sevillian variety, which has not yet been explored in this domain. (c) Our approach not only uses YOLOv8 but also compares it with YOLOv8m, YOLOv8s, Faster R-CNN 101, Faster R-CNN 50, RetinaNet 101, and RetinaNet 50, evaluating key metrics such as average precision (map), training time, and average inference time. This comparative study provides information on the advantages and limitations of the models to provide a solid basis for selecting models based on the reader's needs. (d) Although there is previous research related to the detection of fruits of crops such as tomatoes, apples or citrus fruits, our work focuses on olive fruits (olives), which present unique challenges due to size (1.5 to 3.5 cm), branch density, and variability in colors and textures.

2. Literarature Review

In recent years, significant advancements have been made in the automated detection of fruits and disease-related lesions. These advancements can be divided into two types of approaches: the use of feature engineering and supervised or unsupervised classification. For instance, the Hough transform [20,21] and the k-means algorithm are used in [22,23]. Features such as color, shape, texture, and edges of a fruit [24,25] can be exploited for feature extraction. The ability to recognize fruits in complex scenes is the main disadvantage of this type of detection. Changes in lighting, clutter, and noise negatively affect detection rates.

A more recent approach relies on convolutional neural networks (CNNs). Deep-learning-based methods automatically extract image features from large sample datasets [25]. CNNs have been used for crop disease [26], weed [27], and fruit ripeness detection [28].

Deep-learning-based detectors may have one or two processing stages. Single-stage detectors are trained by optimizing both object classification and localization simultaneously. Two-stage algorithms generate a large number of potential detections, which are then tested to determine if they contain the target object [29]. R-CNN [30], Fast R-CNN [31], and Mask R-CNN [32] are examples of two-stage CNNs used in fruit detection. Two-stage detection algorithms report higher accuracy than single-stage algorithms to detect objects at different scales and with complex spatial relationships. However, they are computationally demanding, as they perform classification and refinement in separate steps.

In Fu et al. [33], a Fast R-CNN is used for kiwi recognition, which demonstrated good robustness against changes in lighting and leaf occlusion. Kim et al. [9] used RGBD images of plums for growth estimation, considering various environments, including the depth information of objects in an outdoor field, based on the R-CNN model. Mu et al. [11] used a model to detect ripe tomatoes, addressing issues such as fruit occlusion and variable lighting conditions, based on the R-CNN model. Zhou et al. [34] used Faster R-CNN to identify oranges at different maturity levels in a natural environment under various weather conditions. Jia et al. [35] used a detection and segmentation method for apples based on Mask-R-CNN to improve the detection of overlapping fruits.

The YOLO architecture Redmon et al. [36] has also been applied for the detection and counting of grape clusters using YOLOv3, YOLOv4, and YOLOv5 models Sozzi et al. [37]. Kuznetsova et al. [38] compared the YOLOv3, YOLOv3-Dense, Faster R-CNN, DaSNet-v2, and LedNet models for apple detection. Yang et al. [39] used YOLOv8 for tomato detection and compared it with SSD, Faster R-CNN, YOLOv4, YOLOv5, and YOLOv7. Ma et al. [40] proposed a lightweight detection algorithm based on YOLOv7-tiny using RMSE and MAE metrics to evaluate counting accuracy. In Wu et al. [10], YOLOv8 was used for the identification and counting of walnuts in UAVs images. Xiao et al. [41] trained two models, YOLOv8 and CenterNet, extracting visual features from images of apples and pears, focusing on detailed analysis of the skin to predict the fruit class.

3. Materials and Methods

3.1. Detection Architectures

The six models that were subjected to analysis are outlined below.

3.1.1. YOLOv8

YOLOv8 is a popular real-time object detection algorithm that has been improved over time, with its first version (YOLOv1) proposed in 2015 and the most recent, YOLOv8, introduced in 2023. YOLOv8 has 5 versions with different scales: YOLOv8n (nano), YOLOv8s (small), YOLOv8m (medium), YOLOv8l (large), and YOLOv8x (extra large). These versions share a common architectural approach but differ in the depth and width of their layers [42]. The versions YOLOv8s (depth: 0.33, width: 0.5) and YOLOv8m (depth: 0.67, width: 0.75) were chosen for use.

3.1.2. Faster R-CNN

Faster R-CNN is an object detection architecture that efficiently and accurately addresses the problem of generating region of interest (ROI) proposals. It consists of three main components: a shared convolutional network, a Region Proposal Network (RPN), and a detection module [43]. Faster R-CNN 50 (Backbone: ResNet50+FPN) [44] and Faster R-CNN 101 (Backbone: X101-FPN) [45] were chosen for use.

3.1.3. RetinaNet

RetinaNet is a single, unified network composed of a base network and two task-specific subnetworks. The base network is responsible for computing a convolutional feature map over the input image and is a pre-existing convolutional network. The first subnetwork performs convolutional object classification on the output of the base network, and the second subnetwork performs convolutional regression of bounding boxes.

Both subnetworks have a simple design specifically proposed for dense single-stage detection [46]. RetinaNet 50 (Backbone: ResNet-50) and RetinaNet 101 (Backbone: ResNet-101) were chosen for use [47].

Table 1 shows the number of parameters of the models, as well as the AP (average precision), traditionally called "mean average precision" (mAP) [48]. The "mAP50-95" value was obtained by the models when evaluated on the COCO dataset [42,45].

Table 1. Characteristics of deep learning models.

Model	Number of Parameters	COCO mAP50-95
YOLOv8s	11.2 M	44.9
YOLOv8m	25.9 M	50.2
Faster R-CNN 50	42 M	40.2
Faster R-CNN 101	105 M	43.0
RetinaNet 50	38 M	38.7
RetinaNet 101	57 M	40.4

YOLOv8 stands out for its speed and accuracy, making it ideal for real-time applications [5,36]. Faster R-CNN performs well in handling complex backgrounds and occlusion due to its robust design [43], while RetinaNet balances precision and efficiency by using focal loss, making it ideal for handling detection imbalances [46].

3.2. *Tacna Olive Dataset*

The proposed dataset was used for training and evaluating object detection models. Olive trees of the Sevillana variety were selected for the study, as shown in Figure 1. The tress images are located in the district of La Yarada, Los Palos, in the outskirts of Tacna (coordinates: 18°15′16.1″ S 70°26′30.7″ W).

(a) (b)

Figure 1. Image of the olive tree (**a**) and crop obtained from the image of the olive tree (**b**).

3.2.1. Data Acquisition

A Canon EOS Rebel T6i camera with a CMOS sensor of 22.3 mm × 14.9 mm and a resolution of 24.2 MP, a measuring tape to measure the distance between the olive tree and the camera (5 m approximately), and a tripod were used to record images at the same distance from each tree trunk. The lens aperture was set to f/4.5 so that the depth of field allowed well-focused images of both the inner and outer leaves of the tree while having a full tree in the scene. The image acquisition work was performed between 10:00 and 17:00 h. The camera gain was set to 100 and the aperture to f/4.5. A total of 8 photos per tree were captured by positioning the tripod around each tree at 45-degree steps. In total, 100 trees were photographed, mostly resulting in 815 RAW images in CR2 format. Photos were selected based on their focus quality, resulting in a sample of 503 images for analysis (Figure 1).

Variations in lighting and camera position are inherent in the outdoor agricultural environment (Yarada Los Palos) due to factors such as changing sunlight, shadows cast by branches, or different recording angles. These conditions reflect the actual environment in which our model was implemented and contribute to the ecological validity of the study. To reduce inconsistencies, images were taken in groups in the morning and afternoon to account for direct light, partial shade, and cloudy days and at eight different angles of the tree. This ensures that the model is exposed to different weather conditions (sunny, partly cloudy, and cloudy). This allowed us to have images that represent the variability of natural illumination, position with respect to the sun, and that projected by the sun. This variability will allow the models trained from this dataset to be robust to a variety of environmental conditions.

3.2.2. Data Generation

The olive trees were photographed, resulting in 815 images. These images were formatted to uncompressed .PNG format to reduce storage without lost of detail. Finally, 24 regions of each tree image were cropped to optimize data processing. The dataset consisted of 19,560 images of 1000 × 1000 pixels. A representative sample of 12,072 images was carefully selected for labeling, taking into account the variation in lighting based on the time of capture in a balanced manner, thus reducing the time spent on the labeling process.

3.2.3. Data Labeling

The labeling phase began with 480 randomly selected .png images of olive fruits, manually labeled by an expert using the software Label Studio v14.3. Olive fruits were delimited using "bounding boxes". An additional set of 576 images was subsequently labeled by the research team, with manual refinement by an expert to correct potential errors. This cyclic process of team labeling and expert labeling refinement was repeated over 5 iterations, during which the number of expert-labeled images gradually increased from 480 in the first iteration to 5064 in the final iteration. In each cycle, newly labeled data were used to improve the labeling process, enabling the efficient expert annotation of 10,728 images in the dataset.

The dataset was divided into 8549 (70.82%) images for training, 2179 (18.05%) images for validation, and 1344 (11.13%) images for testing, ensuring representativeness for each tree. Table 2 shows the information regarding the number of olive tree images, as well as the crops and pixels, after labeling 12,072 images. Ten training iterations of the studied models were performed to compare and generate the indicator reports for each epoch and model in a .csv file for subsequent analysis and evaluation.

Table 2. Sevillana variety olive tree dataset information

Number of Trees	Olive Tree Images	Image Crops	Image Size (px)	Image Crop Size (px)	Training	Validation	Test
62	503	12,072	6000 × 4000	1000 × 1000	8549	2179	1344

3.3. *Data Processing*

3.3.1. Training

The training of the learning models was carried out on a workstation with the following specifications: Intel® Xeon® Silver 4214 2.20 GHz processor, 64 GB DDR4 2933 366 MHz RAM, 1 TB SSD storage, NVIDIA A5000 24GB GPU. Operating system: Ubuntu

22.04 LTS, CUDA Ver. 12.1, OpenCV, Jupyter-Lab v4.0.6, and other libraries focused on implementing deep learning models for object detection.

During this stage, each of the selected models was trained 10 times, as explained below: YOLOv8s, YOLOv8m, Faster R-CNN 50, Faster R-CNN 101, RetinaNet 50 and RetinaNet 101 models.Training was set to 100 epochs, with a batch size of 4 and a learning rate of 0.001, using pre-trained weights from the COCO dataset.

Figure 2 shows the technical process and training developed in this work.

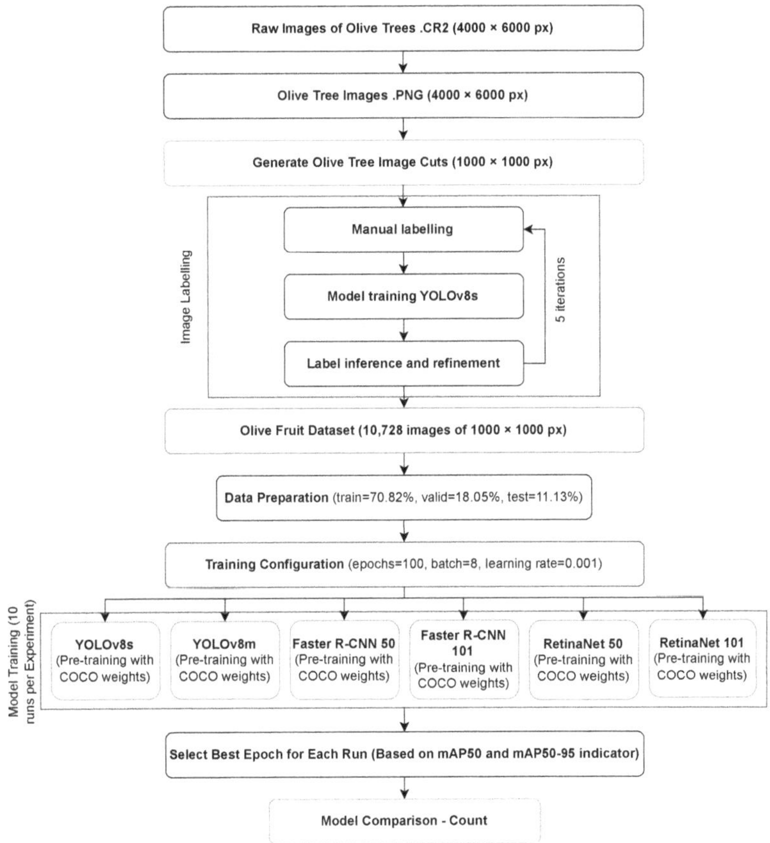

Figure 2. Technical process (from top to bottom): data acquisition, generation, and labeling; partitioning; training; and results.

3.3.2. Count

To carry out the counting measurement, all the olive fruits predicted by the models in the image crops were summed. Subsequently, the results from these crops were grouped by each complete image, and finally, 8 images were grouped for each tree, thus allowing a total fruit count for each tree to be obtained.

3.4. Evaluation Metrics

3.4.1. Intersection over Union (IoU)

According to Padilla et al. [49], the IoU measures the overlap area between the predicted bounding box and the actual bounding box, divided by the area of union between them. By comparing the IoU with a given threshold "t", we can classify a detection as correct or incorrect. If IoU $\geqslant$ t, the detection is considered correct. If IoU < t, the detection is considered incorrect. Its calculation is determined in Equation (1).

$$\text{IoU} = \frac{\text{area of overlap}}{\text{area of union}} \tag{1}$$

3.4.2. Mean Average Precision (mAP)

The mean average precision (mAP) is a metric used to assess the accuracy of object detectors across all classes in a given dataset [49].

$$\text{mAP} = \frac{1}{N}\sum_{i=1}^{N} \text{AP}_i \tag{2}$$

3.4.3. MAP50

The mAP50 value is calculated with a single Intersection over Union (IoU) value of 0.50 [48].

3.4.4. MAP50-95

The mAP50-95 value is calculated over multiple Intersection over Union (IoU) values. Specifically, 10 IoU thresholds are used, ranging from 0.50 to 0.95 in steps of 0.05 [48]. In order to estimate mAP50-95, it was necessary to take steps of 0.05, starting from an IoU threshold of 0.5 and stopping at 0.95. The average precision in this interval is the AP of a single class. By repeating this process for all classes and calculating the mean of all classes, it is possible to calculate the mAP50-95 [50].

3.4.5. Root Mean Square Error (RMSE)

The root mean square error (RMSE) has been used as a standard statistical metric to measure model performance in meteorology, air quality, and climate research studies [51]. The RMSE and the MAE are calculated for the dataset as

$$\text{RMSE} = \sqrt{\frac{1}{n}\sum_{i=1}^{n}(\hat{y}_i - y_i)^2} \tag{3}$$

4. Results

4.1. Model Results for the mAP50 Indicator

In Figure 3, the mAP50 value is shown on the validation set for the models over the 100 epochs of training. The best results from the runs are highlighted, and the highest mAP50 values for each model are marked with stars in the respective colors indicated in the figure legend.

It can be observed that among the six models studied, the YOLOv8m model shows the best performance with an mAP50 value of 0.9496 at epoch 77. The YOLOv8s model reached a maximum mAP50 value of 0.94685 at epoch 78. The Faster R-CNN 101 model obtained a maximum mAP50 value of 0.723431 at epoch 97, while the Faster R-CNN 50 model achieved a maximum mAP50 value of 0.722168 at epoch 99. The RetinaNet 101 model reached a maximum mAP50 value of 0.693666 at epoch 97, and finally, the RetinaNet 50 model achieved a maximum mAP50 value of 0.683174 at epoch 99.

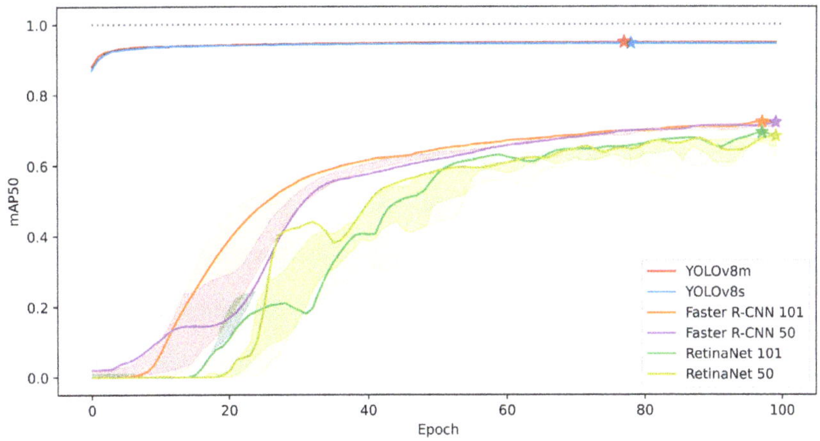

Figure 3. Mean average precision (mAP50) of the deep learning models.

4.2. Model Results for the mAP50-95 Indicator

Figure 4 shows the mAP50-95 value on the validation set for the models over the 100 epochs of training. The best results from the runs are highlighted, and the highest mAP50-95 values for each model are marked with stars in the respective colors indicated in the figure legend.

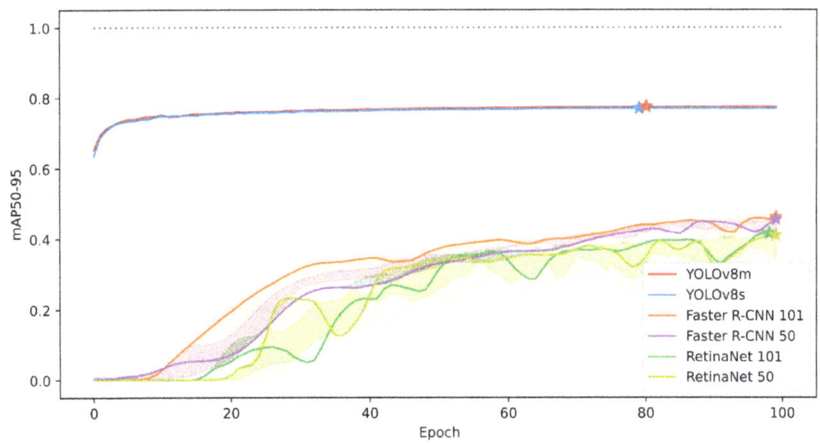

Figure 4. Mean average precision (mAP50-95) of the deep learning models.

It can be observed that the YOLOv8m model achieved the best performance with an mAP50-95 value of 0.77533 at epoch 80. The YOLOv8s model reached a maximum mAP50-95 value of 0.7707 at epoch 79. The Faster R-CNN 101 model obtained a maximum mAP50-95 value of 0.462735 at epoch 99, while the Faster R-CNN 50 model achieved a maximum mAP50-95 value of 0.455714 at epoch 99. The RetinaNet 101 model reached a maximum mAP50-95 value of 0.416039 at epoch 98, and finally, the RetinaNet 50 model achieved a maximum mAP50-95 value of 0.410199 at epoch 99.

4.3. Model Results for the mAP50 Indicator on the Box Plot

In Figure 5, the box plot shows the distribution of performance in terms of mAP50 for the RetinaNet, Faster R-CNN, and YOLOv8 models, with the dispersion represented by the Interquartile Range (IQR). The YOLOv8m and YOLOv8s models stand out for their high performance and minimal dispersion, reflected in IQRs of 0.00014 and 0.000165, indicating great consistency in their predictions. On the other hand, the Faster R-CNN 101 and 50 models, with IQRs of 0.00142625 and 0.006719, show greater variability in their results. The RetinaNet 101 and 50 models present the lowest performance and a broader dispersion, with IQRs of 0.0099085 and 0.004745.

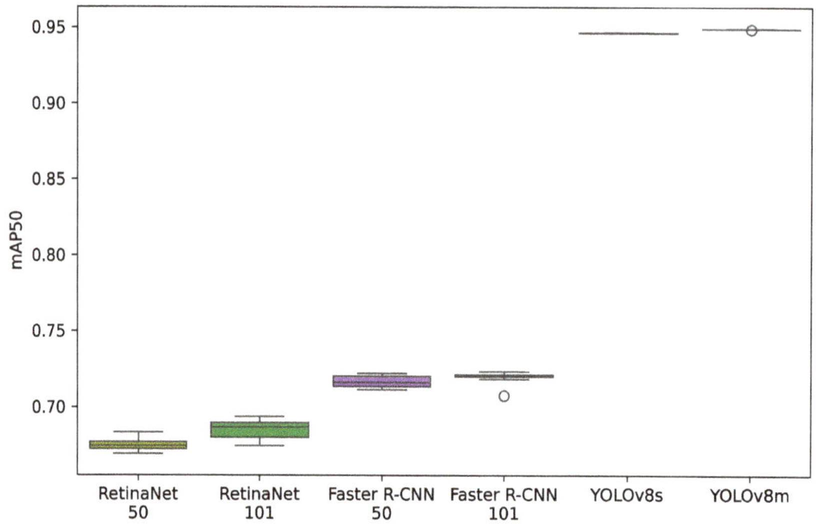

Figure 5. Box plot of mean average precision (mAP50) of the deep learning models.

4.4. Model Results for the mAP50-95 Indicator on the Box Plot

In Figure 6, the box plot presents the distribution of performance in terms of mAP50-95 for the RetinaNet, Faster R-CNN, and YOLOv8 models, with dispersion measured by the Interquartile Range (IQR).

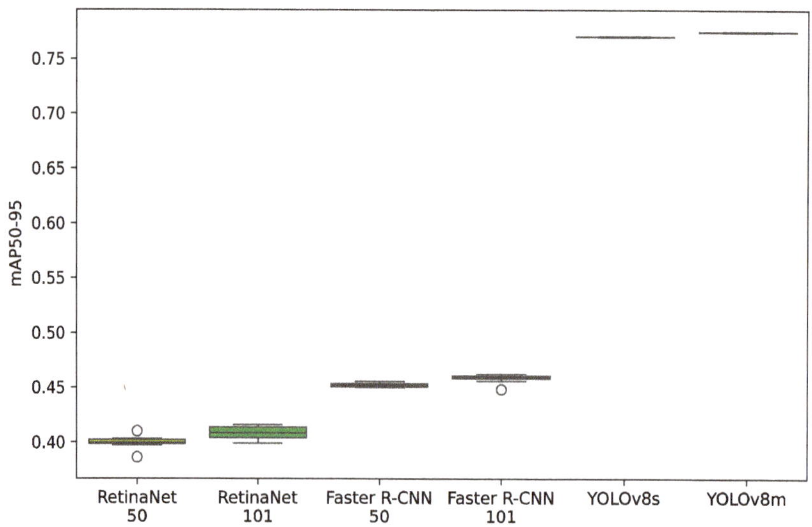

Figure 6. Box plot of mean average precision (mAP50-95) of the deep learning models.

The YOLOv8m and YOLOv8s models stand out once again, showing the highest results and the lowest dispersion, with IQRs of 0.00026 and 0.0005025, respectively, reflecting high consistency in their performance. The Faster R-CNN 101 and 50 models show lower results compared to YOLOv8 and greater variability, with IQRs of 0.0028417 and 0.00269775, respectively. The RetinaNet 101 and 50 models present the lowest medians and the highest dispersion among the evaluated models, with IQRs of 0.0098295 for RetinaNet 101 and 0.003743 for RetinaNet 50. These results confirm the superiority of YOLOv8 not only in terms of accuracy but also in consistency, outperforming the Faster R-CNN and RetinaNet models in the mAP50-95 metric.

4.5. Model Comparison

Table 3 shows the results of the YOLO, Faster, and RetinaNet models, for both the mAP50 and mAP50-95 metrics.

Table 3. Comparison of the performance of models YOLOv8, Faster R-CNN, and RetinaNet on the validation and test set.

Set	Metric	Config	YOLOv8m	YOLOv8s	Faster R-CNN 101	Faster R-CNN 50	RetinaNet 101	RetinaNet 50
Validation Set	mAP50	Min ↑	0.94908	0.94642	0.70764	0.71145	0.67431	0.66907
		Max ↑	0.94960	0.94685	0.72343	0.72217	0.69367	0.68317
		Median ↑	0.94943	0.94656	0.72044	0.71628	0.68663	0.67432
		Mean ↑	0.94941	0.94660	0.71950	0.71662	0.68499	0.67460
		IQR ↓	0.00014	0.00017	0.00143	0.00672	0.00991	0.00475
	mAP50-95	Min ↑	0.77443	0.77001	0.44847	0.45016	0.39926	0.38647
		Max ↑	0.77533	0.77070	0.46274	0.45571	0.41604	0.41020
		Median ↑	0.77485	0.77030	0.45975	0.45217	0.40858	0.39963
		Mean ↑	0.77486	0.77035	0.45874	0.45238	0.40840	0.39980
		IQR ↓	0.00026	0.00050	0.00284	0.00270	0.00983	0.00374
Test Set	mAP50	Min ↑	0.98052	0.97302	0.74823	0.70900	0.65702	0.67012
		Max ↑	0.98273	0.97396	0.77196	0.74358	0.74416	0.71648
		Median ↑	0.98127	0.97321	0.75698	0.73403	0.72824	0.71083
		Mean ↑	0.98142	0.97328	0.75822	0.73230	0.72033	0.70649
		IQR ↓	0.00027	0.00027	0.00764	0.01577	0.00845	0.00843
	mAP50-95	Min ↑	0.95183	0.89984	0.47001	0.46071	0.35718	0.23806
		Max ↑	0.96060	0.90068	0.51599	0.49783	0.44312	0.42318
		Median ↑	0.95319	0.90021	0.49278	0.48431	0.41572	0.37883
		Mean ↑	0.95365	0.90023	0.49370	0.48224	0.40890	0.37242
		IQR ↓	0.00171	0.00029	0.01553	0.00940	0.04628	0.05260

The results in Table 3 demonstrate that the YOLOv8m model not only achieves the best performance in the validation set with an mAP50 of 0.9496 and an mAP50-95 of 0.77533, but also outperforms the other models in the test set. Specifically, YOLOv8m achieves the highest values for mAP50 (0.98273) and mAP50-95 (0.96060) across all configurations. In contrast, YOLOv8s shows slightly lower performance, while Faster R-CNN 101, Faster R-CNN 50, RetinaNet 101, and RetinaNet 50 exhibit significantly reduced mAP values. The interquartile range (IQR) for YOLOv8m remains the lowest across both he validation and test sets, indicating a more stable and consistent performance. These results reinforce the efficiency and accuracy of the YOLOv8m model for olive fruit detection tasks compared to the other evaluated models.

In Table 4, the training time and average inference time for each analyzed model are presented, highlighting the efficiency of YOLOv8s compared to the other models. The study emphasizes that YOLOv8s not only delivers outstanding performance in fruit identification and counting but is also highly efficient in terms of speed, even though the task to be solved does not require real-time inference.

Table 4. Training and inference time

Model	Average Training Time (Hrs)	Average Inference Time (ms)
YOLOv8m	8.640	13.22
YOLOv8s	5.014	07.18
Faster R-CNN 101	4.896	87.04
Faster R-CNN 50	2.221	43.60
RetinaNet 101	2.640	52.29
RetinaNet 50	2.200	42.51

The results in Table 4 indicate that YOLOv8m takes longer to train due to its complexity, but with YOLOv8s, it manages to reduce training to approximately 5 h, which justifies its use in the detection and counting of olive fruits. The processing of the models utilized YOLOv8m and YOLOv8s, with YOLOv8m being more accurate (Table 3) and YOLOv8s being faster in inference (Table 4). However, for field applications, such as drones or mobile systems, lighter models like YOLOv8n are recommended. These are designed for hardware with low-performance GPUs due to constraints related to energy, cost, and portability.

Table 5 presents the RMSE results of the number of predictions from the models (using a confidence threshold of 0.5) applied to the 447 images, which were grouped into batches of 8 images per tree, generating 55 results corresponding to 55 trees. The results of the models were compared with the sum of the manual labels. The R^2 value for YOLOv8m, which is 0.944, indicates the precision and quality of the model's predictions, reflecting a high correlation between the predictions and the actual values.

Table 5. RMSE and R^2 of deep learning models

Model	RMSE	R^2
YOLOv8m	402.46	0.94
YOLOv8s	452.98	0.93
Faster R-CNN 101	1758.15	−0.06
Faster R-CNN 50	1435.63	0.29
RetinaNet 101	3417.34	−3.02
RetinaNet 50	1942.45	−0.30

Figure 7 shows the relationship between the number of labeled fruits on olive trees and the predictions with a confidence value of 0.5 from the six deep learning models. Each graph indicates how the predictions of each model fit the real values through the blue reference line (perfect prediction), showing a scatter pattern where the YOLOv8m and YOLOv8s models demonstrate a closer fit to the reference line and come closer to the actual number of fruits labeled per tree.

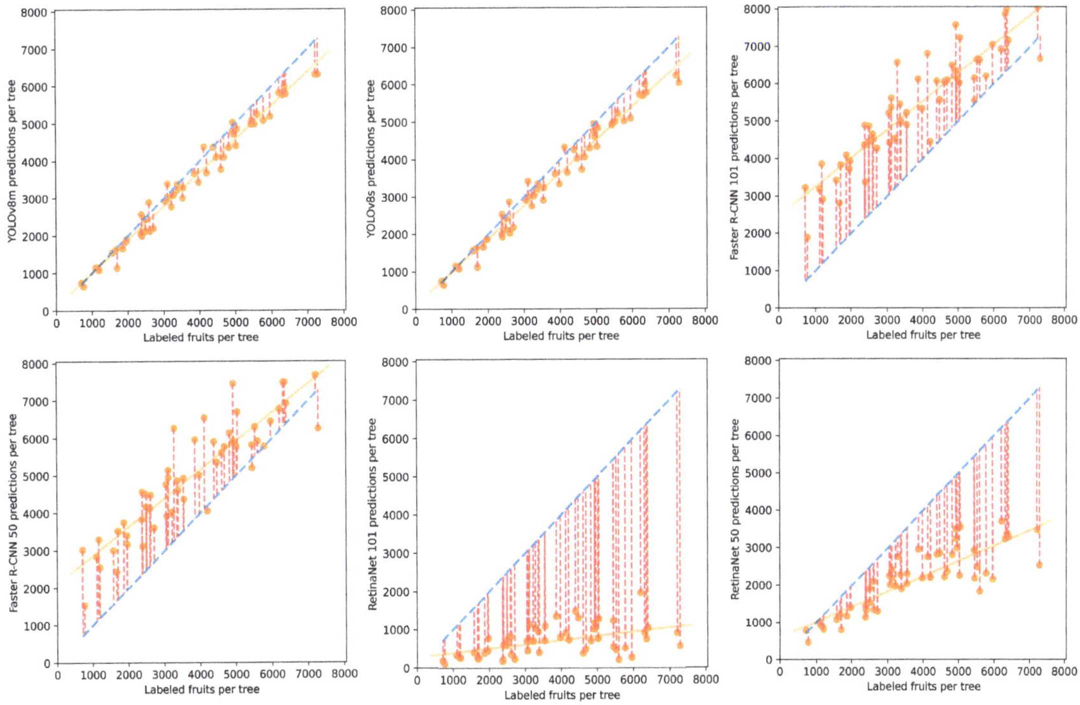

Figure 7. The relationship of the actual number of olive fruits vs. the predictions of six deep learning models.

As can be seen from the results in Table 5, the YOLOv8m model shows the lowest RMSE value for confidence thresholds (0.4, 0.5, 0.6), presenting the smallest prediction error. Regarding the average inference time, the YOLOv8s model shows the shortest time of 7.183 ms.

The results described above show the effectiveness of the models studied. Additionally, we have performed a qualitative analysis of the most important errors made by the model by categorizing the probable causes into (a) multiple fruits, (b) lighting conditions, (c) differentiation between green olives and olive leaves, and (d) fruits located at the edges of the image. Representative visual samples are shown in Figure 8.

(a) Multiple fruits (b) Lighting conditions (c) Differentiation between green olives and olive leaves (d) Fruits on the edges of the image

Figure 8. Representativevisual examples of errors where the model failed.

The results of the normality test indicate a normal distribution of the data. In Table 6, the results of the ANOVA statistical test for the mAP50 and mAP50-95 metrics across the models are presented. The analysis shows that the performance differences in terms of mAP50 and mAP50-95 between some models are statistically significant ($p < 0.05$), validating the superiority of YOLOv8 under specific conditions. Table 6 shows the results of the ANOVA statistical test for the mAP 50 indicator of the models.

Table 6. Statistical test results of ANOVA (mAP50 and mAP50-95).

		Sum of Squares	gl	Root Mean Square	F	Sig.
mAP50	Between groups	0.842	5	0.168	10,640.937	0.000
	Within groups	0.001	54	0.000		
	Total	0.843	59			
mAP50-95	Between groups	1.594	5	0.319	19,747.337	0.000
	Within groups	0.001	54	0.000		
	Total	1.595	59			

Figure 9 shows a cropped image from the validation set and the inference with a confidence threshold of 0.25 from the YOLOv8m model.

Figure 10 shows the result of applying the best model, YOLOv8m, to a full image of an olive tree using the sliced inference technique with SAHI.

Figure 9. Inferenceof the best YOLOv8m model on a crop image.

Figure 10. Inference of the best YOLOv8m model on a tree image.

5. Discussion

5.1. mAP50 Indicator

The data obtained for the mAP50 indicator (Figure 3) show that the best model was YOLOv8m (94.960), followed by YOLOv8s (94.685), Faster R-CNN 101 (72.343), Faster R-CNN 50 (72.217), RetinaNet 101 (69.367), and RetinaNet 50 (68.317) in the detection of Sevillana olive varieties. These results contrast with those obtained by Zhu et al. [28], who studied the detection of four olive varieties (Frantoio, Ezhi8, Leccino, and Picholine), achieving mAP50 indicators for YOLOv3 (90.87, 92.95, 91.24, and 93.67) and Faster R-CNN (93.56, 94.51, 93.05, and 94.81). In their research, the best performance was achieved by Faster R-CNN, with an mAP50 of 94.81 in the Picholine variety, and they also proposed the Olive-EfficientDet model. However, it is important to consider that the YOLO version used by Zhu et al. [28] is earlier than the one used in this study.

Studies by Yang et al. [39] for the detection of tomato fruits compared mAP50 measurements by applying YOLOv8 (91.9), YOLOv7 (91.6), YOLOv5 (91.2), and Faster R-CNN, which performed lower at (80.8). It is evident that the YOLOv8 model achieved better results than Faster R-CNN, similar to the performance of the model demonstrated in the present research, as well as showing that the YOLO model improves across its different versions. Additionally, in the work of Wu et al. [10], various YOLO versions were compared for walnut detection, and the mAP50 values for YOLOv5, YOLOv7, and YOLOv8 ranged from 95.9 to 96.9. Similarly, Lin et al. [5] in their citrus fruit research obtained mAP50 values using YOLOv8s (77.6) and Faster R-CNN (77.3), where both models showed similar performance with no significant differences in their results.

Regarding the results of the research conducted by Kim et al. [9] for plum detection, mAP50 values can be observed in the Faster R-CNN 101 models (73.63) and RetinaNet 101 (70.34), where Faster R-CNN obtained better results than those shown in our work (72.217), possibly due to the characteristics of the fruits studied.

5.2. mAP50-95 Indicator

The results obtained in this study show that for the mAP50-95 indicator, the best model was YOLOv8m (77.533), followed by YOLOv8s (77.070), Faster R-CNN 101 (46.274), Faster R-CNN 50 (45.571), RetinaNet 101 (41.604), and RetinaNet 50 (41.020). Similarly, other investigations such as Aljaafreh et al. [52], conducted for the detection of olive fruits from digital videos using deep neural networks, reported mAP50-95 values with YOLO models in two different versions, YOLOv5x (77.08) and YOLOv5s (74.13), obtaining comparable results. On the other hand, the research by Lin et al. [5] focused on detecting citrus fruits, achieving mAP50-95 values for the YOLOv8s models (57.0) and Faster R-CNN (47.2). In contrast to the results of this study on olive fruits, the YOLOv8s model shows a lower mAP50-95 index, while Faster R-CNN presents similar values.

5.3. Olive Fruit Count

The study by Bellocchio et al. [19], focusing on counting fruits (olives, almonds, and apples), developed an architecture that learns to count without specific task labels (object bounding boxes). The results show that this approach achieves performance values for the unsupervised model with RMSE of 2.44 ± 0.06, and for the supervised model, RMSE values of 2.03±0.07, which are analogous to the results of this study.

For the olive counting, the number of fruits detected by the models was compared with the number of manually labeled olives in each image set of a tree (see Table 5). In the inferences, a confidence threshold of 0.5 was used; the YOLOv8m model demonstrated the best performance in olive counting, achieving the lowest RMSE values (402.46) and the highest R^2 values (0.94).

According to the results presented in Table 3, it is evident that YOLOv8m offers better performance compared to YOLOv8s, which can be attributed to its deeper architecture [42].

From the results obtained in out study, it is evident that the YOLOv8 model stands out for its balance between speed and precision, being ideal for real-time applications, as evidenced in our results in Tables 3 and 4; YOLOv8 outperforms Faster R-CNN and RetinaNet due to its optimized architecture (FPN and PAN), which enhances the detection of small objects with speed and efficiency [5,42]. YOLOv8m stands out for its balance between accuracy and speed, making it ideal for detecting olive fruits under field conditions and the best choice for this study.

In future work, testing lightweight architectures such as YOLOv8n is planned to evaluate their performance compared to the larger models trained in this study. Nonetheless, it was observed that the YOLOv8s model offers better average inference times, as evidenced in Table 4. This is due to the fact that YOLOv8s, with its fewer layers and parameters, provides a balance between accuracy and speed.

6. Conclusions

The results demonstrated that the YOLOv8 models outperformed Faster-RCNN and RetinaNet in metrics such as mAP50, mAP50-95, RMSE, and R^2, for both fruit detection and olive fruit counting. YOLOv8m achieved the best results in mAP50 (94.960) and mAP50-95 (77.533). Additionally, for counting, YOLOv8m also stood out with the lowest RMSE values (402.46) and the highest R^2 (0.94). These results indicate that YOLOv8 not only excels in terms of accuracy (mAP50) but also exhibits lower variability, making it a more reliable and consistent option for detection compared to the other evaluated models.

It is concluded that the YOLOv8 model is the most suitable image processing model for detecting Sevillana-type olive fruits according to the results of this research. However, it is important to consider that factors such as camera positioning, shot distance, lighting, shadows, the type of device used, and other uncontrolled variables can affect the effectiveness of the deep learning models evaluated. All of these aspects should be taken into account to optimize performance in future applications.

The new dataset could be used in future work to address the limitations of trained models, along with the deployment of models on edge devices. However, it is worth mentioning that there are inexpensive devices on the market that support real-time YOLOv8s model inference, although this is not necessary for counting olives.

Author Contributions: Conceptualization, E.O.-M., O.S.-C., I.C.-C., D.O.-D. and S.A.-A.; methodology, E.O.-M. and S.A.-A.; formal analysis, E.O.-M.; project administration, E.O.-M.; supervision, E.O.-M.; writing—original draft, E.O.-M.; validation, O.S.-C.; visualization, O.S.-C.; software, I.C.-C. and D.O.-D.; data curation, I.C.-C. and D.O.-D.; writing—review & editing, S.A.-A. All authors have read and agreed to the published version of the manuscript.

Funding: This research was funded by resources from Canon, Sobrecanon, and mining royalties at the National University Jorge Basadre Grohmann of Tacna.

Institutional Review Board Statement: The study was conducted in accordance and approved by the Institutional Review Board (or Ethics Committee) of Universidad Nacional Jorge Basadre Grohmann, Tacna, Perú (protocol code 2023-035-CEIUNJBG).

Informed Consent Statement: Not applicable.

Data Availability Statement: The data presented in this study are available upon request from the corresponding author and can be accessed via the following link: "https://kaggle.com/datasets/1f0a4b51c3d5b2e73459b1b384585edde0e08d0ddb9874099e22b51e8a93a914 (accessed on 27 December 2024)".

Conflicts of Interest: The authors declare no conflict of interest. The funders had no role in the design of the study; in the collection, analyses, or interpretation of data; in the writing of the manuscript; or in the decision to publish the results.

References

1. Casanova Núñez Melgar, D.P. *Guía Técnica del Cultivo de Olivo en la Región Tacna*; Instituto Nacional de Innovación Agraria: Lima, Peru, 2022.
2. Dawson, D. El Niño Diezma la Cosecha de Aceituna Peruana. Available online: https://es.oliveoiltimes.com/production/el-nino-decimates-peruvian-olive-harvest/127929 (accessed on 13 October 2024).
3. de Agricultura Tacna, D.R. Región Tacna: Serie Histórica Producción Agropecuaria 2014–2023. Available online: https://www.agritacna.gob.pe/direcciones/estadistica-agraria (accessed on 17 December 2024).
4. de Freitas Cunha, R.L.; Silva, B. Estimating crop yields with remote sensing and deep learning. In Proceedings of the 2020 IEEE Latin American GRSS & ISPRS Remote Sensing Conference (LAGIRS), Santiago, Chile, 22–26 March 2020; pp. 273–278. [CrossRef]
5. Lin, Y.; Huang, Z.; Liang, Y.; Liu, Y.; Jiang, W. AG-YOLO: A Rapid Citrus Fruit Detection Algorithm with Global Context Fusion. *Agriculture* **2024**, *14*, 114. [CrossRef]
6. Li, G.; Huang, X.; Ai, J.; Yi, Z.; Xie, W. Lemon-YOLO: An efficient object detection method for lemons in the natural environment. *IET Image Process.* **2021**, *15*, 1998–2009. [CrossRef]
7. Hu, J.; Fan, C.; Wang, Z.; Ruan, J.; Wu, S. Fruit detection and counting in apple orchards based on improved Yolov7 and multi-Object tracking methods. *Sensors* **2023**, *23*, 5903. [CrossRef] [PubMed]
8. Yang, W.; Ma, X.; An, H. Blueberry Ripeness Detection Model Based on Enhanced Detail Feature and Content-Aware Reassembly. *Agronomy* **2023**, *13*, 1613. [CrossRef]
9. Kim, E.; Hong, S.J.; Kim, S.Y.; Lee, C.H.; Kim, S.; Kim, H.J.; Kim, G. CNN-based object detection and growth estimation of plum fruit (Prunus mume) using RGB and depth imaging techniques. *Sci. Rep.* **2022**, *12*, 20796. [CrossRef] [PubMed]
10. Wu, M.; Yun, L.; Xue, C.; Chen, Z.; Xia, Y. Walnut Recognition Method for UAV Remote Sensing Images. *Agriculture* **2024**, *14*, 646. [CrossRef]
11. Mu, Y.; Chen, T.S.; Ninomiya, S.; Guo, W. Intact detection of highly occluded immature tomatoes on plants using deep learning techniques. *Sensors* **2020**, *20*, 2984. [CrossRef]
12. Peng, J.; Ouyang, C.; Peng, H.; Hu, W.; Wang, Y.; Jiang, P. MultiFuseYOLO: Redefining Wine Grape Variety Recognition through Multisource Information Fusion. *Sensors* **2024**, *24*, 2953. [CrossRef]
13. Tian, Y.; Yang, G.; Wang, Z.; Wang, H.; Li, E.; Liang, Z. Apple detection during different growth stages in orchards using the improved YOLO-V3 model. *Comput. Electron. Agric.* **2019**, *157*, 417–426. [CrossRef]
14. Ulutas, E.G.; ALTIN, C. Kiwi Fruit Detection with Deep Learning Methods. *J. Adv. Nat. Sci. Eng. Res.* **2023**, *7*, 39–45. [CrossRef]
15. Kang, H.; Chen, C. Fruit detection, segmentation and 3D visualisation of environments in apple orchards. *Comput. Electron. Agric.* **2020**, *171*, 105302. [CrossRef]

16. Ponce, J.M.; Aquino, A.; Millán, B.; Andújar, J.M. Olive-fruit mass and size estimation using image analysis and feature modeling. *Sensors* **2018**, *18*, 2930. [CrossRef] [PubMed]
17. Ponce, J.M.; Aquino, A.; Millan, B.; Andújar, J.M. Automatic Counting and Individual Size and Mass Estimation of Olive-Fruits Through Computer Vision Techniques. *IEEE Access* **2019**, *7*, 59451–59465. [CrossRef]
18. Ortenzi, L.; Figorilli, S.; Costa, C.; Pallottino, F.; Violino, S.; Pagano, M.; Imperi, G.; Manganiello, R.; Lanza, B.; Antonucci, F. A machine vision rapid method to determine the ripeness degree of olive lots. *Sensors* **2021**, *21*, 2940. [CrossRef]
19. Bellocchio, E.; Ciarfuglia, T.A.; Costante, G.; Valigi, P. Weakly supervised fruit counting for yield estimation using spatial consistency. *IEEE Robot. Autom. Lett.* **2019**, *4*, 2348–2355. [CrossRef]
20. Murillo-Bracamontes, E.A.; Martinez-Rosas, M.E.; Miranda-Velasco, M.M.; Martinez-Reyes, H.L.; Martinez-Sandoval, J.R.; Cervantes-de Avila, H. Implementation of Hough transform for fruit image segmentation. *Procedia Eng.* **2012**, *35*, 230–239. [CrossRef]
21. Lin, G.; Tang, Y.; Zou, X.; Cheng, J.; Xiong, J. Fruit detection in natural environment using partial shape matching and probabilistic Hough transform. *Precis. Agric.* **2020**, *21*, 160–177. [CrossRef]
22. Sidehabi, S.W.; Suyuti, A.; Areni, I.S.; Nurtanio, I. Classification on passion fruit's ripeness using K-means clustering and artificial neural network. In Proceedings of the 2018 International Conference on Information and Communications Technology (ICOIACT), Yogyakarta, Indonesia, 6–7 March 2018; pp. 304–309. [CrossRef]
23. Yu, Y.; Velastin, S.A.; Yin, F. Automatic grading of apples based on multi-features and weighted K-means clustering algorithm. *Inf. Process. Agric.* **2020**, *7*, 556–565. [CrossRef]
24. Bhargava, A.; Bansal, A. Fruits and vegetables quality evaluation using computer vision: A review. *J. King Saud-Univ.-Comput. Inf. Sci.* **2021**, *33*, 243–257. [CrossRef]
25. Vibhute, A.; Bodhe, S.K.; et al. Applications of image processing in agriculture: A survey. *Int. J. Comput. Appl.* **2012**, *52*, 34–40. [CrossRef]
26. Jafar, A.; Bibi, N.; Naqvi, R.A.; Sadeghi-Niaraki, A.; Jeong, D. Revolutionizing agriculture with artificial intelligence: Plant disease detection methods, applications, and their limitations. *Front. Plant Sci.* **2024**, *15*, 1356260. [CrossRef] [PubMed]
27. Mesías-Ruiz, G.A.; Peña, J.M.; de Castro, A.I.; Borra-Serrano, I.; Dorado, J. Cognitive Computing Advancements: Improving Precision Crop Protection through UAV Imagery for Targeted Weed Monitoring. *Remote Sens.* **2024**, *16*, 3026. [CrossRef]
28. Zhu, X.; Chen, F.; Zhang, X.; Zheng, Y.; Peng, X.; Chen, C. Detection the maturity of multi-cultivar olive fruit in orchard environments based on Olive-EfficientDet. *Sci. Hortic.* **2024**, *324*, 112607. [CrossRef]
29. Wang, W.; Shi, Y.; Liu, W.; Che, Z. An Unstructured Orchard Grape Detection Method Utilizing YOLOv5s. *Agriculture* **2024**, *14*, 262. [CrossRef]
30. Girshick, R.; Donahue, J.; Darrell, T.; Malik, J. Rich feature hierarchies for accurate object detection and semantic segmentation. In Proceedings of the IEEE Conference on Computer Vision and Pattern Recognition, Columbus, OH, USA, 23–28 June 2014; pp. 580–587. [CrossRef]
31. Girshick, R. Fast R-CNN. In Proceedings of the 2015 IEEE International Conference on Computer Vision (ICCV), Santiago, Chile, 7–13 December 2015; pp. 1440–1448. [CrossRef]
32. He, K.; Gkioxari, G.; Dollár, P.; Girshick, R. Mask r-cnn. In Proceedings of the IEEE International Conference on Computer Vision, Venice, Italy, 22–29 October 2017; pp. 2961–2969. [CrossRef]
33. Fu, L.; Feng, Y.; Majeed, Y.; Zhang, X.; Zhang, J.; Karkee, M.; Zhang, Q. Kiwifruit detection in field images using Faster R-CNN with ZFNet. *IFAC-PapersOnLine* **2018**, *51*, 45–50. [CrossRef]
34. Zhou, B.; Wu, K.; Chen, M. Detection of Gannan Navel Orange Ripeness in Natural Environment Based on YOLOv5-NMM. *Agronomy* **2024**, *14*, 910. [CrossRef]
35. Jia, W.; Tian, Y.; Luo, R.; Zhang, Z.; Lian, J.; Zheng, Y. Detection and segmentation of overlapped fruits based on optimized mask R-CNN application in apple harvesting robot. *Comput. Electron. Agric.* **2020**, *172*, 105380. [CrossRef]
36. Redmon, J.; Divvala, S.; Girshick, R.; Farhadi, A. You Only Look Once: Unified, Real-Time Object Detection. In Proceedings of the 2016 IEEE Conference on Computer Vision and Pattern Recognition (CVPR), Las Vegas, NV, USA, 27–30 June 2016; pp. 779–788. [CrossRef]
37. Sozzi, M.; Cantalamessa, S.; Cogato, A.; Kayad, A.; Marinello, F. Automatic bunch detection in white grape varieties using YOLOv3, YOLOv4, and YOLOv5 deep learning algorithms. *Agronomy* **2022**, *12*, 319. [CrossRef]
38. Kuznetsova, A.; Maleva, T.; Soloviev, V. Using YOLOv3 algorithm with pre-and post-processing for apple detection in fruit-harvesting robot. *Agronomy* **2020**, *10*, 1016. [CrossRef]
39. Yang, G.; Wang, J.; Nie, Z.; Yang, H.; Yu, S. A lightweight YOLOv8 tomato detection algorithm combining feature enhancement and attention. *Agronomy* **2023**, *13*, 1824. [CrossRef]
40. Ma, L.; Zhao, L.; Wang, Z.; Zhang, J.; Chen, G. Detection and counting of small target apples under complicated environments by using improved YOLOv7-tiny. *Agronomy* **2023**, *13*, 1419. [CrossRef]

41. Xiao, B.; Nguyen, M.; Yan, W.Q. Fruit ripeness identification using YOLOv8 model. *Multimed. Tools Appl.* **2024**, *83*, 28039–28056. [CrossRef]
42. Jocher, G.; Qiu, J.; Chaurasia, A. Detectron2. Available online: https://github.com/ultralytics/ultralytics (accessed on 13 October 2024).
43. Ren, S.; He, K.; Girshick, R.; Sun, J. Faster R-CNN: Towards Real-Time Object Detection with Region Proposal Networks. *IEEE Trans. Pattern Anal. Mach. Intell.* **2017**, *39*, 1137–1149. [CrossRef] [PubMed]
44. Chincholi, F.; Koestler, H. Detectron2 for lesion detection in diabetic retinopathy. *Algorithms* **2023**, *16*, 147. [CrossRef]
45. Wu, Y.; Kirillov, A.; Massa, F.; Lo, W.Y.; Girshick, R. Detectron2. Available online: https://github.com/facebookresearch/detectron2 (accessed on 14 October 2024).
46. Lin, T.Y.; Goyal, P.; Girshick, R.; He, K.; Dollár, P. Focal Loss for Dense Object Detection. *IEEE Trans. Pattern Anal. Mach. Intell.* **2020**, *42*, 318–327. [CrossRef] [PubMed]
47. Santos, A.; Marcato Junior, J.; de Andrade Silva, J.; Pereira, R.; Matos, D.; Menezes, G.; Higa, L.; Eltner, A.; Ramos, A.P.; Osco, L.; et al. Storm-drain and manhole detection using the retinanet method. *Sensors* **2020**, *20*, 4450. [CrossRef] [PubMed]
48. COCOdataset. Detection Evaluation. Available online: https://cocodataset.org/#detection-eval (accessed on 13 October 2024).
49. Padilla, R.; Netto, S.L.; Da Silva, E.A. A survey on performance metrics for object-detection algorithms. In Proceedings of the 2020 International Conference on Systems, Signals and Image Processing (IWSSIP), Niteroi, Brazil, 1–3 July 2020; pp. 237–242. [CrossRef]
50. Reis, D.; Kupec, J.; Hong, J.; Daoudi, A. Real-time flying object detection with YOLOv8. *arXiv preprint* **2023**, arXiv:2305.09972.
51. Chai, T.; Draxler, R.R. Root mean square error (RMSE) or mean absolute error (MAE)?–Arguments against avoiding RMSE in the literature. *Geosci. Model Dev.* **2014**, *7*, 1247–1250. [CrossRef]
52. Aljaafreh, A.; Elzagzoug, E.Y.; Abukhait, J.; Soliman, A.H.; Alja'Afreh, S.S.; Sivanathan, A.; Hughes, J. A Real-Time Olive Fruit Detection for Harvesting Robot Based on YOLO Algorithms. *Acta Technol. Agric.* **2023**, *26*, 121–132. [CrossRef]

AI MDPI

Article

Dimensionality Reduction Statistical Models for Soil Attribute Prediction Based on Raw Spectral Data

Marcelo Chan Fu Wei [1,*], Ricardo Canal Filho [1], Tiago Rodrigues Tavares [1,2], José Paulo Molin [1] and Afrânio Márcio Corrêa Vieira [3,4]

1 Laboratory of Precision Agriculture (LAP), Department of Biosystems Engineering, "Luiz de Queiroz" College of Agriculture (ESALQ), University of São Paulo (USP), Piracicaba 13418900, Brazil
2 Center for Nuclear Energy in Agriculture (CENA), University of São Paulo (USP), Piracicaba 13416000, Brazil
3 Department of Statistics, Federal University of Sao Carlos (UFSCar), São Carlos 13565905, Brazil
4 Statistics and Agricultural Experimentation Graduate Program, "Luiz de Queiroz" College of Agriculture (ESALQ), University of São Paulo (USP), Piracicaba 13418900, Brazil
* Correspondence: marcelochan@usp.br

Abstract: To obtain a better performance when modeling soil spectral data for attribute prediction, researchers frequently resort to data pretreatment, aiming to reduce noise and highlight the spectral features. Even with the awareness of the existence of dimensionality reduction statistical approaches that can cope with data sparse dimensionality, few studies have explored its applicability in soil sensing. Therefore, this study's objective was to assess the predictive performance of two dimensionality reduction statistical models that are not widespread in the proximal soil sensing community: principal components regression (PCR) and least absolute shrinkage and selection operator (lasso). Here, these two approaches were compared with multiple linear regressions (MLR). All of the modelling strategies were applied without employing pretreatment techniques for soil attribute determination using X-ray fluorescence spectroscopy (XRF) and visible and near-infrared diffuse reflectance spectroscopy (Vis-NIR) data. In addition, the achieved results were compared against the ones reported in the literature that applied pretreatment techniques. The study was carried out with 102 soil samples from two distinct fields. Predictive models were developed for nine chemical and physical soil attributes, using lasso, PCR and MLR. Both Vis-NIR and XRF raw spectral data presented a great performance for soil attribute prediction when modelled with PCR and the lasso method. In general, similar results were found comparing the root mean squared error (RMSE) and coefficient of determination (R^2) from the literature that applied pretreatment techniques and this study. For example, considering base saturation (V%), for Vis-NIR combined with PCR, in this study, RMSE and R^2 values of 10.60 and 0.79 were found compared with 10.38 and 0.80, respectively, in the literature. In addition, looking at potassium (K), XRF associated with lasso yielded an RMSE value of 0.60 and R^2 of 0.92, and in the literature, RMSE and R^2 of 0.53 and 0.95, respectively, were found. The major discrepancy was observed for phosphorus (P) and organic matter (OM) prediction applying PCR in the XRF data, which showed R^2 of 0.33 (for P) and 0.52 (for OM) without using pretreatment techniques in this study, and R^2 of 0.01 (for P) and 0.74 (for OM) when using preprocessing techniques in the literature. These results indicate that data pretreatment can be disposable for predicting some soil attributes when using Vis-NIR and XRF raw data modeled with dimensionality reduction statistical models. Despite this, there is no consensus on the best way to calibrate data, as this seems to be attribute and area specific.

Keywords: multivariate analysis; visible near-infrared; X-ray fluorescence

1. Introduction

High spatial density monitoring of soil attributes is a crucial step to build soil maps that can guide better management decisions in crop fields. Fine-scale monitoring of soil properties is important because if this is neglected, the maps produced are inefficient and unreliable [1], as increasing the spatial density of the soil information directly implies an understanding of the spatial variability of these attributes [2].

The main issues with the traditional soil sampling and analysis methods are cost and time consumption, which become barriers for farmers to increase the density of soil data acquisition [3,4]. Nevertheless, the traditional laboratory analysis consumes reagents that the community aims to diminish, as agriculture production moves towards new global sustainability guidelines. In this scenario, alternative or complementary techniques have slowed down the efforts of soil scientists. Over the last decades, aiming to increase the amount of data related to soil, researchers began to study different soil sensing techniques and their applicability in agriculture [5]. These approaches allow for the rapid acquisition of soil data directly in the field. However, to convert sensor data into agronomic information, predictive models need to be calibrated.

Among the sensing techniques that have shown potential for soil research are the visible and near-infrared diffuse reflectance spectroscopy (Vis-NIR) and X-ray fluorescence spectroscopy (XRF). The core idea for the application of both is to use different spectral data, obtained quickly and without the use of reagents, in order to predict agronomic attributes [6].

The current approach will not replace laboratory analysis, but will allow for augmenting the number of observations, as it uses sensors' output and soil laboratorial analysis to build specific calibrations, transforming spectral data into predictions of physical and chemical soil properties. Therefore, the importance of machine learning (ML) techniques to leverage this novel analysis method is settled. Once ML calibrations are built, soil spectral data can be acquired, allowing for predicting soil attributes in a sustainable, time saving, and cost-effective way [7].

In addition, another highlight of the use of ML calibrations for this task is to create artificial intelligence systems to predict soil attributes. This can reduce human interference in the determination of the physical and chemical soil properties, aiming to simplify and standardize the sample acquisition, processing, and analysis steps [8].

The Vis-NIR spectra of the soil can be related with contents of clay, organic matter (OM), organic carbon (OC), and moisture, and its applications in the literature are observed mainly using laboratory spectral acquisition [9,10]. In the last few years, researchers have developed prediction systems with spectral Vis-NIR acquisition directly in the field, using embedded sensors in agricultural machinery [11–13]. Conversely, XRF is based on the induction of fluorescence in a soil sample through its excitation with an incident X-ray source, as well as the subsequent measurement of specific photons emitted after this process [14]. This technique can accurately measure the total content of some soil elements (Fe, Al, Si, Ca, and K), and can be used to generate indirect calibration models to predict other soil parameters (e.g., cation exchange capacity (CEC), potential of hydrogen (pH), etc.) [15,16].

After data acquisition from the sensors, samples must be taken for chemical and physical analysis in the laboratory to obtain the reference values for each sample in order to fit the predictive models. This last step is usually conducted after applying several combinations of spectra pretreatment methods (e.g., normalization, smoothing, derivative algorithms, and stepwise procedures) aiming to highlight specific features, reduce spectral noise, and select variables to reduce covariates [17,18]. There is not a unique formula to perform this step, nor a method or a set of methods, that have been found to be the best pretreatment sequence. This means that the performance of the prediction will rely on the ability of the researcher who calibrates the models. On the other hand, there are also studies that have applied statistical methods to predict attributes based on spectral data without using pretreatment methods, that have argued that the application of pretreatments can reduce the sensors' predictive performance [19].

In fact, the spectra, whether from Vis-NIR or XRF, have a large number of variables (e.g., $n > 300$), and some of them will deliver a low contribution to prediction models, which, in some cases, can hinder its performance. Nevertheless, it is necessary to consider the cost of data processing (time consumption), and especially to evaluate its trade-off with performance gain. This is particularly advantageous, because there are methods that can handle the high dimensionality of spectral data without pretreatment or excluding part of the raw data. If a machine learning method can maintain the performance of prediction, using less steps of calibration, or even standardizing the procedure, it can be useful for the soil spectroscopy community. For this purpose, the modeling techniques of principal component regression (PCR) and least absolute shrinkage and selection operator (lasso) regression can be highlighted. These techniques are not widespread in the proximal soil sensing community, and, if successfully applied for predicting the soil attributes, may be an alternative to avoid the spectral preprocessing step. To the best of our knowledge, the performance of these techniques has not yet been evaluated for the prediction of fertility attributes using data from XRF and Vis-NIR sensors, which is the motivation for the execution of this study.

PCR is a regression method that combines principal components analysis with least squares regression [20]. It is a relatively simple, but very useful method [21]. On the other hand, lasso allows for the identification of relevant and irrelevant predictor variables, assigning different weights to each of them [22], a process known as shrinkage or regularization. In the context of proximal soil sensing, although it is easier to find studies applying PCR [18,23] than lasso [24], both methods have not been explored much in the literature and are hardly cited. In this sense, partial least squares (PLS) is often cited as the best method to fit predictive models using soil spectroscopy data, outperforming other models such as artificial neural network (ANN) [25], random forest (RF) [26], or multiple linear regression (MLR) [27]. Therefore, PLS is the most documented technique in the literature [28,29].

Thus, the goal of the present study was to evaluate the predictive performance of the dimensionality reduction statistical models of PCR, lasso, and MLR for soil attribute determination using XRF and Vis-NIR publicly available data without pretreatment, and comparing these results against the ones reported in the literature that applied pretreatment methods.

2. Materials and Methods

2.1. Soil Samples

The dataset used consists of 102 soil samples from the soil database of the Laboratory of Precision Agriculture (LAP) from Luiz de Queiroz College of Agriculture, University of São Paulo [30]. The samples were collected from 0–20 cm depth in two fields under active agricultural production. Field 1 (22°41′57″ S and 47°38′33″ W) is located in the municipality of Piracicaba, State of São Paulo, where 58 samples were collected. Field 2 (14°06′05″ S and 57°45′58″ W) is located in Campo Novo do Parecis, State of Mato Grosso, where the remaining 44 samples were collected. Both fields have considerable textural dissimilarity, as observed from their classification—Lixisol with a clayey texture and Ferralsol with a sandy loam to sandy clay loam texture, respectively. The samples were stored after being air-dried and sieved at 2 mm.

2.2. Physical and Chemical Attributes Analyses

The reference analysis of the clay content, OM, CEC, pH, base saturation (V%), extractable phosphorus (P), extractable potassium (K), extractable calcium (Ca), and extractable magnesium (Mg) were determined in a commercial laboratory for the soil fertility analyses. The laboratorial procedure followed the methodology described by Van Raij et al. [31]. Extractable nutrients (P, K, Ca, and Mg) were determined using ion exchange resin extraction. The OM content was determined via oxidation with a potassium dichromate solution, and the pH was determined via the calcium chloride solution. The texture was determined using the Bouyoucos hydrometer method in a dispersing solution for the clay content.

2.3. Spectral Data Acquisition

The soil spectral data were acquired for all samples using commercial Vis-NIR (351 spectral variables) and XRF (1458 spectral variables) equipment. The Vis-NIR analysis was performed using Veris Vis-NIR spectrometer (Veris Technologies, Salina, KS, USA), which collected the spectra from 343 to 2222 nm, with a spectral resolution of around 5 nm. All of the acquisitions were performed under laboratory conditions, placing the sample at a circular sapphire window located in the bottom portion of a shank module. This spectrometer self-calibrated before each spectra acquisition by collecting a dark reference measurement and a known internal reference material measurement. Spectral regions at 343–432 and 2153–2222 nm were removed due to the high presence of noise, resulting in a raw spectra from 437 to 2149 nm. For the XRF spectra acquisition, the portable XRF equipment Tracer III-SD (Bruker AXS, Madison, WI, USA) was used. The equipment was configured using the instrumental conditions suggested by Tavares et al. [32]. Samples were scanned in triplicate and then averaged for further analysis.

2.4. Data Modeling

For all of the three methods, data from the two sensors were individually used. All of the processes were conducted in RStudio [33]. Raw spectral data were used as the input variables to fit the prediction models of the clay, OM, CEC, pH, V%, P, K, Ca, and Mg. The prediction models were fitted by applying three regression models: MLR, PCR, and lasso regression.

The MLR method is a regression analysis that has one target related to more than one feature, where the target can be estimated by Equation (1) [34].

$$\mathrm{Y} = \mathrm{X}\beta + e \tag{1}$$

where Y is a $(n \times 1)$ target vector, X is a $(n \times p)$ features matrix (predictor variables), β is a $p \times 1$ vector of unknown coefficients, and e is a $n \times 1$ random vector of errors.

The PCR method occurs in three steps: (a) perform principal component analysis on the observed data matrix to obtain principal components; (b) apply linear regression to obtain the vector of estimated regression coefficients; and (c) use PCA loadings (eigenvectors) to obtain the PCR estimator (β), as shown in Equation (2).

$$\widehat{\beta_k} = V_k \widehat{\delta_k} \tag{2}$$

where $\widehat{\beta}$ is the PCR estimator, k belongs to $\{1, \ldots, p\}$, p is the number of covariates, V is the orthonormal set of eigenvectors, and $\widehat{\delta}$ is the vector of estimated regression coefficients.

Lasso is a regression with an l1-norm penalty aiming to find β = {βj}, which minimizes Equation (3) [35].

$$\sum_{i=1}^{N} \left(y_i - \sum_j x_{ij}\beta_j\right)^2 + \lambda \Sigma_{j=1}^{p} \left|\beta_j\right| \tag{3}$$

where x_{ij} is the standardized features, y_i is the centered target values, and $i = 1, 2, \ldots, \mathrm{N}$ and $j = 1, 2, \ldots, p$.

Data were split into 75% as the training set and 25% as the validation set. Therefore, 76 samples were used for training and 26 for validation. The data split was randomly selected using the seed (666). Although this is a random procedure to avoid bias, the seed allows for repeatability after the process is finished. Both PCR and lasso were performed using the R language library caret [36], setting the method as pcr() for PCR and glmnet() for lasso.

In PCR, the optimal number of principal components was defined using the tuning control function trainControl with the cross-validation (cv) as 10 and the tuneLength (number of principal components tested) equal to 30 in the training dataset. The elbow rule [37] was used to minimize the root mean squared error (RMSE) of cross-validation and to maximize the variance explained, aiming, when possible, to obtain at least 70% of the explained variance. Although this rule will not always choose the model that present the highest coefficient of determination (R^2) and lowest RMSE, it follows the principle of parsimony in multivariate calibration, assuming that of the two models with meaningful predictions, the one with fewer parameters is preferred [38]. The scale within train function was set to TRUE as it standardized each variable before the generation of the principal components. Then, the model was tuned and applied to the validation dataset, calculating the RMSE and R^2 of the prediction.

For the lasso method, standardize within the function train was set to TRUE for data scaling, which removes the effect of features that present different unit/magnitude. The best alpha (α) and lambda (λ) values were extracted from the fitted model, which was applied to the validation dataset and obtained the RMSE and R^2 of prediction.

MLR method was applied using the function lm. The model was fitted on the training dataset and then applied to the validation dataset, obtaining the respective RMSE and R^2 for each attribute predicted.

3. Results and Discussion

For the PCR prediction, both Vis-NIR and XRF soil attributes prediction presented R^2 values from 0.52 to 0.85, being P the only exception (Table 1). The R^2 for P was 0.03 and 0.33 for Vis-NIR and XRF, respectively, indicating that P was better predicted, for the data used in this study, using PCR prediction and XRF sensor technique. P prediction yielded poor parameters in both studies because it has no direct spectral response in Vis-NIR [9] nor emission lines for XRF technique [32]. The accurate prediction of P, sometimes observed in literature [39,40], lies into the covariation with other soil properties of the studied area, allowing indirect calibrations [41].

Table 1. Results from the principal component regression (PCR) fitting model applied to the soil physical-chemical attributes studied.

Sensor	Target Variable	Number of Components Used	Variance Explained (%)	Validation		Tavares et al. [42]	
				RMSE	R^2	RMSE	R^2
Vis-NIR	Clay	2	84.91	42.80	0.80	27.32	0.93
	OM	7	72.29	2.95	0.72	2.10	0.86
	CEC	10	71.39	18.91	0.56	18.66	0.51
	pH	19	70.47	0.28	0.58	0.34	0.19
	V%	2	75.99	10.60	0.79	10.38	0.80
	P	3	11.87	13.76	0.03	12.05	0.07
	K	4	66.29	0.90	0.82	1.20	0.74
	Ca	2	69.61	12.80	0.63	10.98	0.68
	Mg	10	71.05	9.34	0.59	8.85	0.52
XRF	Clay	1	83.00	45.63	0.78	29.40	0.92
	OM	19	66.28	3.89	0.52	3.01	0.74
	CEC	4	72.46	15.37	0.70	10.19	0.88
	pH	13	53.80	0.24	0.68	0.33	0.34
	V%	1	78.87	9.68	0.83	5.60	0.95
	P	15	41.26	11.39	0.33	13.27	0.01
	K	2	67.47	0.82	0.85	0.53	0.95
	Ca	1	72.06	11.99	0.68	4.09	0.96
	Mg	4	71.54	8.13	0.68	4.28	0.89

Vis-NIR: visible and near-infrared diffuse reflectance spectroscopy; XRF: X-ray fluorescence spectroscopy; Clay (g kg^{-1}); OM: organic matter content (g kg^{-1}); CEC: cation exchange capacity ($mmol_c$ kg^{-1}); pH: potential of hydrogen; V%: base saturation ($mmol_c$ kg^{-1}); P: phosphorus ($mmol_c$ kg^{-1}); K: potassium ($mmol_c$ kg^{-1}); Ca: calcium ($mmol_c$ kg^{-1}); Mg: magnesium ($mmol_c$ kg^{-1}); RMSE: root mean squared error; R^2: coefficient of determination.

The application of PCR on XRF raw data used less components when compared with the Vis-NIR raw data, except for OM, pH, and P. It is well known that clay and organic matter have a direct spectral response on Vis-NIR [41]. Therefore, the model can maximize the explained variance with a low number of principal components (PC), as it will identify well established spectral regions that are modified due to the present amount of these attributes [9]. In this context, other soil attributes that present a strong linear correlation with clay and OM (i.e., Vis-NIR primary response attributes) can be indirectly predicted [41]. Once this covariation exists, the model will tend to identify the same spectral regions used for the primary response attributes. However, the portion of the variance that are not correlated with primary response attributes will be randomly assigned by the model. Hence, the weaker the linear correlation, the greater the number of PCs the model will need to increase the explained variance.

The indirect calibrations built with PCR (i.e., CEC, pH, V%, K, Ca, and Mg) presented a similar or slightly improved performance in comparison with the results in Tavares et al. [42], although this applied the elbow rule. In this sense, the pH prediction of this study was highlighted due to the R^2 of 0.58 (Vis-NIR) and 0.68 (XRF) obtained by PCR. These results reinforced the capacity of dimensionality reduction models for identifying important pH-related spectral regions as those reported in the literature by wavebands of around 2200 nm of O–H and N–H bonds, for OM direct responses that can also be indirectly related to pH, as described by Chang et al. [43] and Li et al. [44].

The different calibration strategies used in Tavares et al. [42] achieved slightly better performances than the results reached by the present study. For example, the R^2 for clay and OM obtained by the authors were 0.93 and 0.86, respectively, compared with 0.80 and 0.72 achieved by our study. Despite this difference, the method used in this study was justified by the principle of parsimony applied to the multivariate calibration [38], reducing the calibration parameters and providing satisfactory predictions, even diminishing the performance.

Comparing the lasso predictions obtained in this study with the results reported in Tavares et al. [42] (Table 2), a similar pattern as that reported for PCR predictions was noted. Despite the sensor used, clay and OM had their prediction performance reduced, but still presented a satisfactory prediction. The other calibrations presented a similar performance, except for the CEC, Ca, and Mg prediction from the XRF data, for which the RMSE from lasso increased by over 50% when compared with the results from the above-mentioned authors.

Table 2. Results from least absolute shrinkage and selection operator (lasso) regression fitting models applied to the soil physical-chemical attributes studied.

Sensor	Target Variable	α	λ	Validation		Tavares et al. [42]	
				RMSE	R^2	RMSE	R^2
Vis-NIR	Clay	0.10	5.140	38.03	0.84	27.32	0.93
	OM	0.10	0.218	2.95	0.71	2.10	0.86
	CEC	0.10	1.149	21.27	0.44	18.66	0.51
	pH	0.55	0.044	0.35	0.33	0.34	0.19
	V%	0.10	1.140	10.98	0.78	10.38	0.80
	P	1.00	3.220	13.30	0.06	12.05	0.07
	K	0.10	0.387	0.94	0.81	1.20	0.74
	Ca	0.55	0.929	12.95	0.62	10.98	0.68
	Mg	0.10	0.557	10.72	0.45	8.85	0.52
XRF	Clay	0.10	5.554	42.23	0.81	29.40	0.92
	OM	0.10	2.513	3.58	0.59	3.01	0.74
	CEC	0.55	4.329	16.45	0.68	10.19	0.88
	pH	0.10	0.148	0.24	0.70	0.33	0.34
	V%	0.10	4.051	4.63	0.96	5.60	0.95
	P	1.00	4.432	12.42	0.34	13.27	0.01
	K	0.10	0.145	0.60	0.92	0.53	0.95
	Ca	1.00	1.110	7.91	0.86	4.09	0.96
	Mg	0.10	0.660	6.63	0.81	4.28	0.89

Vis-NIR: visible and near-infrared diffuse reflectance spectroscopy; XRF: X-ray fluorescence spectroscopy; Clay (g kg^{-1}); OM: organic matter content (g kg^{-1}); CEC: cation exchange capacity ($mmol_c$ kg^{-1}); pH: potential of hydrogen; V%: base saturation ($mmol_c$ kg^{-1}); P: phosphorus ($mmol_c$ kg^{-1}); K: potassium ($mmol_c$ kg^{-1}); Ca: calcium ($mmol_c$ kg^{-1}); Mg: magnesium ($mmol_c$ kg^{-1}); RMSE: root mean squared error; R^2: coefficient of determination.

Considering Vis-NIR, the P predictions were not accurately predicted in this study (RMSE and R^2 of 13.30 mmol$_c$ kg^{-1} and 0.06, respectively) nor in that of Tavares et al. [42] (RMSE and R^2 of 12.05 mmol$_c$ kg^{-1} and 0.07, respectively). Both Vis-NIR and XRF predictions of V% presented low RMSE (Vis-NIR using lasso achieved 10.98 mmol$_c$ kg^{-1} and Tavares et al. [42] reached 10.38 mmol$_c$ kg^{-1}; XRF using lasso obtained 4.63 mmol$_c$ kg^{-1} and Tavares et al. [42] achieved 5.60 mmol$_c$ kg^{-1}) and high R^2 (Vis-NIR using lasso achieved 0.78 and Tavares et al. [42] reached 0.80; XRF using lasso obtained 0.96 and Tavares et al. [42] achieved 0.95) values. This comparison highlights that instead of using pretreatment techniques, which can be time consuming, dimensionality reduction statistical models are capable of coping with soil spectral data for the successful predictions of the fertility attributes.

In a study of moisture sensibility of soil attributes prediction using Vis-NIR spectra [45], the authors applied the average pretreatment (averaging values of the spectra within a given interval to reduce dimensionality) before fitting partial least squares regression (PLSR), random forest (RF) regression, artificial neural network (ANN), and support vector machine (SVM) models. As a result, no major discrepancy in prediction among the different models tested was found. This corroborates the difficulty in defining whether pretreatment methods can regularly cope with a singular statistical model, but not with others, as no pattern was observed in the results. This emphasizes the complexity to settle a standard procedure to build ML calibrations for the soil spectral data using pretreatment techniques.

Another study used Vis-NIR spectra to predict the soil attributes, which aimed to define the lime requirement doses [46]. The pretreatments involved smoothing and resampling for all of the data, and then the authors tested the maximum normalization, multiplicative scatter correction, and standard normal variate to build PLSR predictive models. The best results found for RMSE were 0.33 for pH. Compared with the results reported in this study, applying lasso (RMSE values of 0.35 for pH) and PCR (RMSE values of 0.28 for pH), no major differences were observed, even compared with the results in Tavares et al. [42], which presented RMSE values of 0.34.

More recently, in a study using Vis-NIR spectra, testing several pretreatment sequences, the authors used the strategy of choosing a specific combination for each individual soil property, aiming to predict using PLSR [39]. Comparing the results of the prediction from the models that contained only laboratory measured spectra, the pH presented R^2 of 0.45 and RMSE of 0.56. The R^2 values for K, Ca, and Mg were 0.13, 0.48, and 0.25, respectively. Thus, we can infer that the results in this study, without pretreatment and using alternative statistical modelling presented a higher R^2, indicating that dimensionality reduction statistical models can be used to predict soil attributes, with the advantage of not requiring spectral preprocessing.

Another study evaluated alternative statistical approaches (RF regression, and SVM with radial and linear kernel) to predict the soil physical attributes using both Vis-NIR and XRF spectra, separately and in tandem [47]. The authors tested the datasets with and without pretreatment. Analyzing the results, just as observed in the present study, there was also no major difference between the predictions, and no pattern was found. The main factors of variation appeared to be the technique used (Vis-NIR, XRF, or the combination of both) and the soil sampling depth.

These comparisons corroborate and incite the question of whether the use of pretreatment methods is decisive or not for soil attribute prediction using spectral data, as there are few methods that automatically apply pretreatment techniques and fit predictive models, such as the 'all-possibilities' approach (APA) of Kopačková et al. [48]. The APA method is an algorithm that automatically fits predictive models (e.g., ANN and PLSR) using several pretreatment methods, including averaging, centering, smoothing, standardization, normalization, and transformations, among others. Furthermore, for the application of this algorithm, it is necessary to have a high computational processing power. In their study, the PARACUDA® computing engine was used, which the authors defined as "an extremely computer power-consuming method and thus it runs on a grid based supercomputer with many processing cores for rapid analysis". Without automatic methods, the majority of studies that apply pretreatment techniques on soil spectral data rely on the expertise of the user to calibrate the predictive models. Therefore, the use of alternative approaches is an opportunity to standardize the procedure, and to reduce the time and cost consumption.

Depending on the purpose of the study, pretreatment is indeed needed for some applications, e.g., detailed chemometrics analysis, where the visualization of the spectra range is important for a given attribute. On the other hand, looking at only predictive models based on spectral data, there may be an opportunity to explore the application of predictive models based on raw soil spectral data, as observed in a study that applied deep learning on raw spectral data to predict fresh fruit attributes [19].

Further in this subject, Velliangiri and Alagumuthukrishnan [49] described that using dimensionality reduction models, such as PCR and lasso, can aid in the removal of noisy, redundant, and irrelevant data, which are similar characteristics to the goals of pretreatment methods. This corroborates the methods tested in this study and also explains the results that were found by the models calibrated using PCR and lasso

The results for the MLR fitted models (Table 3) indicate that the use of raw spectral data to calibrate MLR models for soil attribute prediction is not feasible independently of the source of the data tested (Vis-NIR or XRF). This is because MLR is a method that handles high-dimensional data poorly in comparison with PCR and lasso. The assumptions underlying the dimensionality reduction statistical models include the correlations among the predictors, the noise to signal, and model errors [49]. This is often more realistic than the MLR assumptions of independent and error free predictors, allowing these models to handle high-dimensional data better than ordinary MLR [27].

Table 3. The results from the multiple linear regression fitting models applied to the soil physical-chemical attributes studied.

Sensor	Target Variable	Validation		Tavares et al. [42]	
		RMSE	R^2	RMSE	R^2
Vis-NIR	Clay	158.60	0.32	27.32	0.93
	OM	38.98	0.01	2.10	0.86
	CEC	86.61	0.10	18.66	0.51
	pH	1.37	0.01	0.34	0.19
	V%	81.66	0.15	10.38	0.80
	P	70.10	0.03	12.05	0.07
	K	9.54	0.22	1.20	0.74
	Ca	83.37	0.07	10.98	0.68
	Mg	35.72	0.09	8.85	0.52
XRF	Clay	448.21	0.02	29.40	0.92
	OM	33.79	0.13	3.01	0.74
	CEC	444.65	0.01	10.19	0.88
	pH	10.20	0.01	0.33	0.34
	V%	482.51	0.00	5.60	0.95
	P	51.07	0.09	13.27	0.01
	K	23.35	0.01	0.53	0.95
	Ca	394.45	0.00	4.09	0.96
	Mg	314.98	0.03	4.28	0.89

Vis-NIR: visible and near-infrared diffuse reflectance spectroscopy; XRF: X-ray fluorescence spectroscopy; Clay (g kg^{-1}); OM: organic matter content (g kg^{-1}); CEC: cation exchange capacity ($mmol_c$ kg^{-1}); pH: potential of hydrogen; V%: base saturation ($mmol_c$ kg^{-1}); P: Phosphorus ($mmol_c$ kg^{-1}); K: potassium ($mmol_c$ kg^{-1}); Ca: calcium ($mmol_c$ kg^{-1}); Mg: magnesium ($mmol_c$ kg^{-1}); RMSE: root mean squared error; R^2: coefficient of determination.

Note that pretreatment followed by MLR improved the prediction accuracy, as shown in the results from Tavares et al. [42], highlighting that MLR can be applied to predict soil attributes based on spectral data. However, the application of pretreatment techniques is necessary.

Despite the close values of the quality indicators of the dimensionality reduction statistical models and the models using pretreatment methods, it was not possible to observe an optimal method that is unanimously the best for all attributes and sensors. It is important to highlight that the aim of this study was not to indicate the best set (sensor and method) to predict soil physical-chemical attributes, but to show the potential of applying unusual methods that can be suitable to fit predictive models based on the soil raw spectral data.

Spectra pretreatment relies on the ability of the person who calibrates the model, to choose the combination of techniques that will extract the best prediction. The proposed method in this study is advantageous as it presents the potential for standardizing the model calibration step, thus diminishing human interference.

The sensors used in the laboratory to obtain spectral data are being adapted to be used directly in the field, with Vis-NIR being the only that has already been tested. The main use in the laboratory justifies the appliance of pretreatment techniques before fitting the predictive model. However, looking at the advances in agricultural machines towards proximal soil sensing, strategies to predict soil attributes online in a standard manner, and reducing the time and processing cost needed as much as possible, can be a helpful way to leverage soil sensing techniques with a high spatial resolution [3]. Therefore, the advances in this research area can further decide whether methods using less processing steps become more interesting than the classical approaches. The comparison of the results found in this study was restricted to laboratory measured spectra, as there was a disadvantage in the quality of online acquired spectra when compared with the laboratory acquisition [41].

Future works should evaluate different statistical approaches, including those conducted in this study, comparing the results from models built with and without pretreatment techniques for online and laboratory spectra acquisition. If possible, they should be applied in a dataset that presents a large variability and number of samples. In addition, they should be postulated not only by metrics that assess mean values, such as RMSE, but also by the correlation among predicted values, so as to observe the discrepancy among the different modelling methods.

4. Conclusions

The application of principal component regression and the lasso method overperformed the multiple linear regression approach. The R^2 values found in this study applying PCR and the lasso method ranged from 0.33 to 0.96 and 0.03 to 0.84 for XRF and Vis-NIR sensor data, respectively. The MLR approach did not present satisfactory results for any of the evaluated attributes ($R^2 \leq 0.32$), indicating that it cannot be applied on raw spectral data. When comparing the results achieved in the present study using the tested dimensionality reduction statistical models with the studies in the literature that applied data pretreatment methods, we noticed that both strategies presented similarly satisfactory results. Hence, this suggests the possibility of predicting soil attributes without applying pretreatment methods, making the data processing faster.

Author Contributions: Conceptualization, M.C.F.W. and R.C.F.; methodology, M.C.F.W., R.C.F. and A.M.C.V.; formal analysis, M.C.F.W., R.C.F. and A.M.C.V.; writing—review and editing, M.C.F.W., R.C.F., T.R.T., J.P.M. and A.M.C.V. All authors have read and agreed to the published version of the manuscript.

Funding: Soil fertility tests were funded by CNPq, "Edital de Chamada Universal", grant number 458180/2014-9. Tiago R. Tavares was funded by the São Paulo Research Foundation (FAPESP), Grant No. 2020/16670-9.

Institutional Review Board Statement: Not applicable.

Informed Consent Statement: Not applicable.

Data Availability Statement: Data is available at: Ref. [30].

Conflicts of Interest: The authors declare no conflict of interest.

References

1. Wollenhaupt, N.C.; Wolkowski, R.P.; Clayton, M.K. Mapping Soil Test Phosphorus and Potassium for Variable-Rate Fertilizer Application. *J. Prod. Agric.* **1994**, *7*, 441–448. [CrossRef]
2. Cherubin, M.R.; Santi, A.L.; Eitelwein, M.T.; Amado, T.J.C.; Simon, D.H.; Damian, J.M. Dimensão da malha amostral para caracterização da variabilidade espacial de fósforo e potássio em Latossolo Vermelho. *Pesqui. Agropecuária Bras.* **2015**, *50*, 168–177. (In Portuguese) [CrossRef]
3. Viscarra Rossel, R.A.; Adamchuk, V.I.; Sudduth, K.A.; McKenzie, N.J.; Lobsey, C. Proximal soil sensing: An effective approach for soil measurements in space and time. In *Advances in Agronomy*; Sparks, D.L., Ed.; Academic Press: London, UK, 2011; pp. 243–291.
4. Wang, D.; Chakraborty, S.; Weindorf, D.C.; Li, B.; Sharma, A.; Paul, S.; Ali, N. Synthesized use of VisNIR DRS and PXRF for soil characterization: Total carbon and total nitrogen. *Geoderma* **2015**, *243–244*, 157–167. [CrossRef]
5. Molin, J.P.; Tavares, T.R. Sensor Systems for Mapping Soil Fertility Attributes: Challenges, Advances, and Perspectives in Brazilian Tropical Soils. *Eng. Agrícola* **2019**, *39*, 126–147. [CrossRef]
6. Kuang, B.; Mahmood, H.S.; Quraishi, M.Z.; Hoogmoed, W.B.; Mouazen, A.M.; van Henten, E.J. Sensing soil properties in the laboratory, in situ, and on-line: A review. In *Advances in Agronomy*; Sparks, D.L., Ed.; Academic Press: London, UK, 2012; pp. 155–223.
7. Sharma, R.; Kamble, S.S.; Gunasekaran, A.; Kumar, V.; Kumar, A. A systematic literature review on machine learning applications for sustainable agriculture supply chain performance. *Comput. Oper. Res.* **2020**, *119*, 104926. [CrossRef]
8. Rossi, P.; Mangiavacchi, P.L.; Monarca, D.; Cecchini, M. Smart Machinery and Devices for Reducing Risks from Human-Machine Interference in Agriculture: A Review. In *Safety, Health and Welfare in Agriculture and Agro-food Systems*; Biocca, M., Cavallo, E., Cecchine, M., Failla, S., Romano, E., Eds.; Springer: Cham, Switzerland, 2022; pp. 195–204.
9. Nocita, M.; Stevens, A.; van Wesemael, B.; Aitkenhead, M.; Bachmann, M.; Barthès, B.; Dor, E.B.; Brown, D.J.; Clairotte, M.; Csorba, A.; et al. Soil spectroscopy: An alternative to wet chemistry for soil monitoring. In *Advances in Agronomy*; Sparks, D.L., Ed.; Academic Press: London, UK, 2015; pp. 139–159.
10. Pasquini, C. Near infrared spectroscopy: A mature analytical technique with new perspectives–A review. *Anal. Chim. Acta* **2018**, *1026*, 8–36. [CrossRef]
11. Bönecke, E.; Meyer, S.; Vogel, S.; Schröter, I.; Gebbers, R.; Kling, C.; Kramer, E.; Lück, K.; Nagel, A.; Philipp, G.; et al. Guidelines for precise lime management based on high-resolution soil pH, texture and SOM maps generated from proximal soil sensing data. *Precis. Agric.* **2020**, *22*, 493–523. [CrossRef]
12. Vogel, S.; Bönecke, E.; Kling, C.; Kramer, E.; Lück, K.; Philipp, G.; Rühlmann, J.; Schröter, I.; Gebbers, R. Direct prediction of site-specific lime requirement of arable fields using the base neutralizing capacity and a multi-sensor platform for on-the-go soil mapping. *Precis. Agric.* **2022**, *23*, 127–149. [CrossRef]
13. Munnaf, M.A.; Haesaert, G.; Van Meirvenne, M.; Mouazen, A.M. Multi-sensors data fusion approach for site-specific seeding of consumption and seed potato production. *Precis. Agric.* **2021**, *22*, 1890–1917. [CrossRef]
14. Kalnicky, D.J.; Singhvi, R. Field portable XRF analysis of environmental samples. *J. Hazard. Mater.* **2001**, *83*, 93–122. [CrossRef]
15. Nawar, S.; Delbecque, N.; Declercq, Y.; De Smedt, P.; Finke, P.; Verdoodt, A.; Mouazen, A.M. Can spectral analyses improve measurement of key soil fertility parameters with X-ray fluorescence spectrometry? *Geoderma* **2019**, *350*, 29–39. [CrossRef]
16. Tavares, T.R.; Molin, J.P.; Javadi, S.H.; De Carvalho, H.W.P.; Mouazen, A.M. Combined Use of Vis-NIR and XRF Sensors for Tropical Soil Fertility Analysis: Assessing Different Data Fusion Approaches. *Sensors* **2021**, *21*, 148. [CrossRef] [PubMed]
17. Munnaf, M.A.; Guerrero, A.; Nawar, S.; Haesaert, G.; Van Meirvenne, M.; Mouazen, A.M. A combined data mining approach for on-line prediction of key soil quality indicators by Vis-NIR spectroscopy. *Soil Tillage Res.* **2021**, *205*, 104808. [CrossRef]
18. Javadi, S.H.; Munnaf, M.A.; Mouazen, A.M. Fusion of Vis-NIR and XRF spectra for estimation of key soil attributes. *Geoderma* **2021**, *385*, 114851. [CrossRef]
19. Mishra, P.; Rutledge, D.N.; Roger, J.-M.; Wali, K.; Khan, H.A. Chemometric pre-processing can negatively affect the performance of near-infrared spectroscopy models for fruit quality prediction. *Talanta* **2021**, *229*, 122303. [CrossRef] [PubMed]
20. Jolliffe, I. *Principal Component Analysis (pp. 1094–1096). Springer Berlin Heidelberg. RESUME SELİN DEĞİRMECİ Marmara University, Goztepe Campus ProQuest Number: ProQuest). Copyright of the Dissertation Is Held by the Author. All Rights Reserved, 28243034*; Springer: New York, NY, USA, 2011.
21. Agarwal, A.; Shah, D.; Shen, D.; Song, D. On robustness of principal component regression. *J. Am. Stat. Assoc.* **2021**, *10*, 1–34. [CrossRef]
22. Lee, J.H.; Shi, Z.; Gao, Z. On LASSO for predictive regression. *J. Econ.* **2021**, *229*, 322–349. [CrossRef]
23. Pudełko, A.; Chodak, M. Estimation of total nitrogen and organic carbon contents in mine soils with NIR reflectance spectroscopy and various chemometric methods. *Geoderma* **2020**, *368*, 114306. [CrossRef]
24. Bricklemyer, R.S.; Brown, D.J.; Turk, P.J.; Clegg, S. Comparing vis-NIRS, LIBS, and combined vis-NIRS-LIBS for intact soil core soil carbon measurement. *Soil Sci. Soc. Am. J.* **2018**, *82*, 1482–1496. [CrossRef]
25. Kuang, B.; Tekin, Y.; Mouazen, A.M. Comparison between artificial neural network and partial least squares for on-line visible and near infrared spectroscopy measurement of soil organic carbon, pH and clay content. *Soil Tillage Res.* **2015**, *146*, 243–252. [CrossRef]

26. Knox, N.; Grunwald, S.; McDowell, M.; Bruland, G.; Myers, D.; Harris, W. Modelling soil carbon fractions with visible near-infrared (VNIR) and mid-infrared (MIR) spectroscopy. *Geoderma* **2015**, *239–240*, 229–239. [CrossRef]
27. Wold, S.; Sjöström, M.; Eriksson, L. PLS-regression: A basic tool of chemometrics. *Chemom. Intell. Lab. Syst.* **2001**, *58*, 109–130. [CrossRef]
28. Bellon-Maurel, V.; McBratney, A. Near-infrared (NIR) and mid-infrared (MIR) spectroscopic techniques for assessing the amount of carbon stock in soils–Critical review and research perspectives. *Soil Biol. Biochem.* **2011**, *43*, 1398–1410. [CrossRef]
29. Rossel, R.V.; Behrens, T. Using data mining to model and interpret soil diffuse reflectance spectra. *Geoderma* **2010**, *158*, 46–54. [CrossRef]
30. Tavares, T.R.; Molin, J.P.; Nunes, L.C.; Alves, E.E.N.; Krug, F.J.; de Carvalho, H.W.P. Spectral data of tropical soils using dry-chemistry techniques (Vis-NIR, XRF, and LIBS): A dataset for soil fertility prediction. *Data Brief* **2022**, *41*, 108004. [CrossRef]
31. Van Raij, B.; Andrade, J.C.; Cantarela, H.; Quaggio, J.A. *Análise Química Para Avaliação de Solos Tropicais*; IAC: Campinas, Brazil, 2001; 285p. (In Portuguese)
32. Tavares, T.R.; Molin, J.P.; Nunes, L.C.; Alves, E.E.N.; Melquiades, F.L.; de Carvalho, H.W.P.; Mouazen, A.M. Effect of X-Ray Tube Configuration on Measurement of Key Soil Fertility Attributes with XRF. *Remote Sens.* **2020**, *12*, 963. [CrossRef]
33. R Core Team. *R: A Language and Environment for Statistical Computing*; R Foundation for Statistical Computing: Vienna, Austria, 2018.
34. Olive, D.J. *Linear Regression*, 1st ed.; Springer: Cham, Switzerland, 2017; Multiple linear regression; pp. 17–83.
35. Tibshirani, R. Regression Shrinkage and Selection via the Lasso: A retrospective. *J. R. Stat. Soc.* **2011**, *73*, 267–288. [CrossRef]
36. Kuhn, M. Building Predictive Models in R Using the caret Package. *J. Stat. Softw.* **2008**, *28*, 1548–7660. [CrossRef]
37. Tracy, T.; Fu, Y.; Roy, I.; Jonas, E.; Glendenning, P. *High Performance Computing*; Kunkel, J., Balaji, P., Dongarra, J., Eds.; Springer: Cham, Switzerland, 2015; pp. 200–218.
38. Seasholtz, M.B.; Kowalski, B. The parsimony principle applied to multivariate calibration. *Anal. Chim. Acta* **1993**, *277*, 165–177. [CrossRef]
39. Abdul Munnaf, M.; Nawar, S.; Mouazen, A.M. Estimation of Secondary Soil Properties by Fusion of Laboratory and On-Line Measured Vis–NIR Spectra. *Remote. Sens.* **2019**, *11*, 2819. [CrossRef]
40. Mouazen, A.M.; Kuang, B. On-line visible and near infrared spectroscopy for in-field phosphorous management. *Soil Tillage Res.* **2016**, *155*, 471–477. [CrossRef]
41. Stenberg, B.; Viscarra-Rossel, R.A.; Mouazen, A.M.; Wetterlind, J. Visible and near infrared spectroscopy in soil science. In *Advances in Agronomy*; Sparks, D.L., Ed.; Academic Press: London, UK, 2010; pp. 163–215.
42. Tavares, T.; Molin, J.; Nunes, L.; Wei, M.; Krug, F.; de Carvalho, H.; Mouazen, A. Multi-Sensor Approach for Tropical Soil Fertility Analysis: Comparison of Individual and Combined Performance of VNIR, XRF, and LIBS Spectroscopies. *Agronomy* **2021**, *11*, 1028. [CrossRef]
43. Chang, C.-W.; Laird, D.A.; Mausbach, M.J.; Hurburgh, C.R., Jr. Near-Infrared Reflectance Spectroscopy-Principal Components Regression Analyses of Soil Properties. *Soil Sci. Soc. Am. J.* **2001**, *65*, 480–490. [CrossRef]
44. Li, S.; Ji, W.; Chen, S.; Peng, J.; Zhou, Y.; Shi, Z. Potential of VIS-NIR-SWIR Spectroscopy from the Chinese Soil Spectral Library for Assessment of Nitrogen Fertilization Rates in the Paddy-Rice Region, China. *Remote Sens.* **2015**, *7*, 7029–7043. [CrossRef]
45. Wijewardane, N.K.; Ge, Y.; Morgan, C.L.S. Moisture insensitive prediction of soil properties from Vis-NIR reflectance spectra based on external parameter orthogonalization. *Geoderma* **2016**, *267*, 92–101. [CrossRef]
46. Franceschini, M.; Demattê, J.; Kooistra, L.; Bartholomeus, H.; Rizzo, R.; Fongaro, C.; Molin, J. Effects of external factors on soil reflectance measured on-the-go and assessment of potential spectral correction through orthogonalisation and standardisation procedures. *Soil Tillage Res.* **2018**, *177*, 19–36. [CrossRef]
47. Benedet, L.; Faria, W.M.; Silva, S.H.G.; Mancini, M.; Demattê, J.A.M.; Guilherme, L.R.G.; Curi, N. Soil texture prediction using portable X-ray fluorescence spectrometry and visible near-infrared diffuse reflectance spectroscopy. *Geoderma* **2020**, *376*, 114553. [CrossRef]
48. Kopačková, V.; Ben-Dor, E.; Carmon, N.; Notesco, G. Modelling Diverse Soil Attributes with Visible to Longwave Infrared Spectroscopy Using PLSR Employed by an Automatic Modelling Engine. *Remote Sens.* **2017**, *9*, 134. [CrossRef]
49. Velliangiri, S.; Alagumuthukrishnan, S.; Joseph, S.I.T. A Review of Dimensionality Reduction Techniques for Efficient Computation. *Procedia Comput. Sci.* **2019**, *165*, 104–111. [CrossRef]

AI MDPI

Article

A General Machine Learning Model for Assessing Fruit Quality Using Deep Image Features

Ioannis D. Apostolopoulos [1,*], Mpesi Tzani [2] and Sokratis I. Aznaouridis [3]

1 Department of Medical Physics, School of Medicine, University of Patras, 26504 Rio, Greece
2 Department of Electrical and Computer Technology Engineering, University of Patras, 26504 Rio, Greece
3 Department of Computer Engineering and Informatics, University of Patras, 26504 Rio, Greece
* Correspondence: ece7216@upnet.gr

Abstract: Fruit quality is a critical factor in the produce industry, affecting producers, distributors, consumers, and the economy. High-quality fruits are more appealing, nutritious, and safe, boosting consumer satisfaction and revenue for producers. Artificial intelligence can aid in assessing the quality of fruit using images. This paper presents a general machine learning model for assessing fruit quality using deep image features. This model leverages the learning capabilities of the recent successful networks for image classification called vision transformers (ViT). The ViT model is built and trained with a combination of various fruit datasets and taught to distinguish between good and rotten fruit images based on their visual appearance and not predefined quality attributes. The general model demonstrated impressive results in accurately identifying the quality of various fruits, such as apples (with a 99.50% accuracy), cucumbers (99%), grapes (100%), kakis (99.50%), oranges (99.50%), papayas (98%), peaches (98%), tomatoes (99.50%), and watermelons (98%). However, it showed slightly lower performance in identifying guavas (97%), lemons (97%), limes (97.50%), mangoes (97.50%), pears (97%), and pomegranates (97%).

Keywords: fruit quality; machine learning; deep learning

1. Introduction

Fruit quality refers to a fruit's overall characteristics that determine its desirability, nutritional content, and safety for consumption [1]. It is determined by the fruit's appearance, flavour, texture, nutritional value, and safety [2]. For several reasons, high fruit quality is crucial for the industry, consumers, and the economy.

High-quality fruits benefit growers and sellers economically, promote healthy eating habits, reduce healthcare costs, positively impact the environment, ensure food safety, and promote international trade [3]. Promoting high fruit quality requires using sustainable farming practices, implementing food safety regulations, and promoting healthy eating habits [3]. For the industry, fruit quality is critical for market competitiveness and profitability. The produce industry is highly competitive, and consumers are more discerning than ever, demanding high-quality fruits that meet their flavour, appearance, and nutrition expectations. Furthermore, the reputation of producers and distributors depends on the quality of their products [3]. Consumers who are satisfied with the quality of fruits are more likely to become repeat customers and recommend the products to others, which can help to build a strong brand image and increase sales [3].

In addition, fruit quality is critical for food safety [1]. Poor-quality fruits are more prone to contamination by pathogens and spoilage microorganisms, leading to foodborne illness outbreaks and damaging the industry's reputation. For people, fruit quality is crucial because it determines the taste, nutritional value, and safety of their consumed fruits [1]. High-quality fruits are more nutritious, flavourful, and appealing, making them more likely to be consumed and incorporated into a healthy diet. Furthermore, high-quality fruits are less likely to contain harmful contaminants or spoilage microorganisms, reducing the risk of foodborne illness and promoting public health.

Fruit quality impacts the entire supply chain, from producers to distributors to retailers. High-quality fruits are less likely to spoil during transportation and storage, reducing waste and increasing profits for all parties involved. Furthermore, high-quality fruits are more likely to be sold at premium prices, increasing the value of the entire supply chain.

Several factors determine fruit quality, including variety, growing conditions, harvesting practices, transportation, and storage [1]. For example, the timing of the harvest can have a significant impact. Harvesting fruits too early can result in poor taste, texture, and aroma; harvesting fruits too late can lead to overripening, loss of nutrients, and spoilage. Growing conditions such as soil quality, irrigation, and pest management can also impact fruit quality. Fruits grown in nutrient-rich soil, with proper irrigation and pest management practices, are more likely to be of higher quality than those grown in poor soil conditions or with inadequate pest control measures. Transportation and storage conditions are also crucial for maintaining fruit quality. Fruits must be transported and stored at optimal temperatures and humidity levels to prevent spoilage, maintain freshness, and preserve nutritional value.

Artificial intelligence (AI) can aid in assessing the quality of the fruit using images [4–7]. AI-based technologies such as computer vision and machine learning (ML) algorithms can analyse the visual characteristics of the fruit and provide an objective quality assessment [8,9]. The AI algorithms can be trained using a large dataset of images [10] of different fruits with varying quality. They can learn to identify the specific features that indicate the quality of the fruit [11,12].

This study is the first to introduce the concept of a general ML model for visually assessing the fruit quality of various types of fruits. While our research focuses on this specific application, it is important to acknowledge that the field of machine learning has witnessed the development of general models for various other applications as well, such as low-cost sensor calibration [13], small molecule substrates of enzyme prediction [14], and topology optimization [15].

We considered the development of a vision transformer (ViT) network [16], a type of neural network architecture designed for image classification tasks that use the transformer architecture, introduced initially for natural language processing. In ViT, an image is first divided into fixed-size patches. These patches are then flattened and linearly projected into a lower dimensional space, creating a sequence of embeddings. These embeddings are then fed into a multi-head self-attention mechanism, which allows the network to learn to attend to essential patches and relationships between patches.

The self-attention mechanism [17] is followed by a feedforward neural network, which processes the attended embeddings and outputs class probabilities. ViT also includes additional techniques, such as layer normalisation, residual connections, and token embedding, which help improve the network's performance. ViT allows for effective self-attention mechanisms in image classification tasks, providing a promising alternative to traditional convolutional neural networks (CNNs) [18].

A collection of fruit-quality datasets of various fruit types, such as apples, cucumbers, grapes, kakis, oranges, papayas, peaches, tomatoes, watermelons, guavas, lemons, limes, mangoes, pears, and pomegranates, served to train the general model and inspect its performance against fruit-dedicated trained models.

The contributions of this study can be summarised as follows:

- We present a general ML model for determining the quality of various fruit based on their visual appearance;
- This general model performs better or equal to dedicated per-fruit models;
- Comparisons with the State-of-the-Art architectures reveal the superiority of ViTs in fruit quality assessment.

2. Related Work

Recent studies have reported remarkable success in visually estimating fruit quality.

Rodríguez et al. [19] focused on identifying plum varieties during early maturity stages, a difficult task even for experts. The authors proposed a two-step approach where images are first processed to isolate the plum. Then, a deep convolutional neural network is used to determine its variety. The results demonstrate high accuracy, ranging from 91 to 97%.

In [20], the authors proposed a CNN to help with sorting by detecting defects in mangosteen. Indonesia has identified mangosteen as a fruit with significant export potential, but not all are defect free. Quality assurance for export is performed manually by sorting experts, which can lead to inconsistent and inaccurate results due to human error. The suggested method achieved a classification accuracy of 97% in defect recognition.

During the growth process of apple fruit crops, there are instances where biological damage occurs on the surface or inside of the fruit. These lesions are typically caused by external factors such as the incorrect application of fertilisers, pest infestations, or changes in meteorological conditions such as temperature, sunlight, and humidity. Wenxue et al. [21] employed a CNN for real-time recognition of apple skin lesions captured by infrared video sensors, capable of intelligent, unattended alerting for disease pests. Experimental results show that the proposed method achieves a high accuracy and recall rate of up to 97.5% and 98.5%, respectively.

In [22], the authors proposed an automated method to distinguish between naturally and artificially ripened bananas using spectral and RGB data. They used a neural network on RGB data and achieved an accuracy of up to 90%. They used spectral data classifiers such as random forest, multilayer perceptron, and feedforward neural networks. They achieved accuracies of up to 98.74% and 89.49%, respectively. These findings could help ensure the safety of banana consumption by identifying artificially ripened bananas, which can harm human health.

In [23], hyperspectral reflectance imaging (400~1000 nm) was used to evaluate and classify three common types of peach diseases by analysing spectral and imaging information. Principal component analysis was used to reduce the high dimensionality of the hyperspectral images, and 54 imaging features were extracted from each sample. The proposed model had 82.5%, 92.5%, and 100% accuracy for slightly decayed, moderately decayed, and severely decayed samples, respectively.

Ref. [24] proposed developing a deep learning-based model called Fruit-CNN for recognising fruits and assessing their quality. The dataset used in this study includes twelve categories of six different fruits based on their quality. It comprises 12,000 images in real-world situations with varying backgrounds. The proposed model outperformed other State-of-the-Art models, achieving an accuracy of 99.6% on a test set of previously unseen images. In [25], the authors utilised a CNN to create an efficient fruit classification model. The model was trained using the Fruits 360 dataset, which consists of 131 varieties of fruits and vegetables. This study focused on three specific fruits, divided into the following three categories based on quality: good, raw, and damaged. The model was developed using Keras and trained for 50 epochs, achieving an accuracy rate of 95%. In [11], the authors used two banana fruit datasets to train and assess their presented model. The original dataset contains 2100 images categorised into ripe, unripe, and over-ripe, with 700 images in each category. This study employed a handcrafted CNN for the classification. The CNN model achieved an accuracy of 98.25% and 81.96% regarding the two datasets.

In [26], the authors developed a model to identify rotting fruits from input images. This study used three types of fruits: apples, bananas, and oranges. The features of the fruit images were collected using the MobileNetV2 [27] architecture. The model's performance was evaluated on a Kaggle dataset, and it achieved a validation accuracy of 99.61%. In [28], the authors proposed two approaches for classifying the maturity status of papaya: machine learning (ML) and transfer learning. The experiment used 300 papaya fruit images, with 100 images for each maturity level. The ML approach a utilised local binary pattern, histogram of directed gradients, grey level co-occurrence matrix, and classification approaches including k-nearest neighbours, support vector machine, and naive Bayes. In contrast, transfer learning utilised seven pre-trained models, including VGG-19 [29]. Both methods achieved 100% accuracy, with the ML method achieving this in 0.0995 s of training time and the transfer learning method achieving 100% accuracy.

Most related works have focused on building fruit-specific models. Subsequently, they utilised datasets containing fruits from a single variety. There is a need for general fruit quality prediction models, which are transferrable from industry to industry and are trained using large-scale datasets. Moreover, recent advances in deep learning models can be benchmarked for fruit quality assessment to investigate their performance.

3. Materials and Methods

3.1. Deep Learning Framework

We propose a ViT model for the classification task. The current section describes the fundamental concepts of the ViT model and the parameters of the proposed model.

3.1.1. Convolutional Neural Networks (CNNs)

CNNs are a class of neural networks designed explicitly for image-processing tasks [30,31]. CNNs use convolutional and pooling layers to extract features from an input image. Convolutional layers work by convolving a set of learnable filters (kernels) over the input image to produce feature maps [18]. The filters are designed to detect specific patterns in the image, such as edges or corners.

Pooling layers are used to downsample the feature maps produced by convolutional layers, reducing their size while retaining the most critical information. The most common type of pooling layer is max pooling, which takes the maximum value from each subregion of the feature map.

CNNs have succeeded highly in image classification tasks, achieving State-of-the-Art performance on benchmark datasets such as ImageNet. However, they are limited in their ability to capture global relationships between different parts of an image.

3.1.2. Transformers

Transformers are a type of neural network architecture initially developed for natural language processing tasks, such as machine translation and text summarisation. Transformers use a self-attention mechanism [32] to capture relationships between different parts of an input sequence [33].

The self-attention mechanism works by computing a weighted sum of the input sequence, where the weights are taught based on the importance of each element to the other elements in the sequence. This allows the model to focus on relevant parts of the input sequence while ignoring irrelevant parts.

Transformers have been highly successful in natural language processing tasks, achieving State-of-the-Art performance on benchmark datasets such as GLUE and SuperGLUE.

3.1.3. ViT Model

ViTs are a type of deep learning model that combines the power of CNNs with the attention mechanism of transformers to process images. This hybrid architecture is highly effective for image classification tasks, as it allows the model to focus on relevant parts of an image while capturing spatial relationships between them.

ViTs use the following two main components: CNNs and transformer networks. The CNNs are used for feature extraction from the images, while transformer networks are used for attention mechanisms. CNNs are particularly good at capturing local image features such as edges and corners. In contrast, transformer networks can capture the global structure of images by attending to relevant regions. By combining the two, visual transformer CNNs can capture local and global features, improving performance.

The ViT of this study divides the input image into a grid of 33 smaller patches, similar to how image segmentation works [16]. Each patch is flattened and passed through convolutional layers to extract features. The transformer network then processes these features, which attends to the most relevant features and aggregates them to generate a representation of the image. This representation is then passed through a series of fully connected layers to classify the image.

The proposed ViT model in Figure 1 consists of multiple layers of self-attention and feedforward networks. The self-attention mechanism allows the network to attend to different input parts and weight them based on relevance. The feedforward network generates a new representation of the input, which is then used in the next self-attention layer.

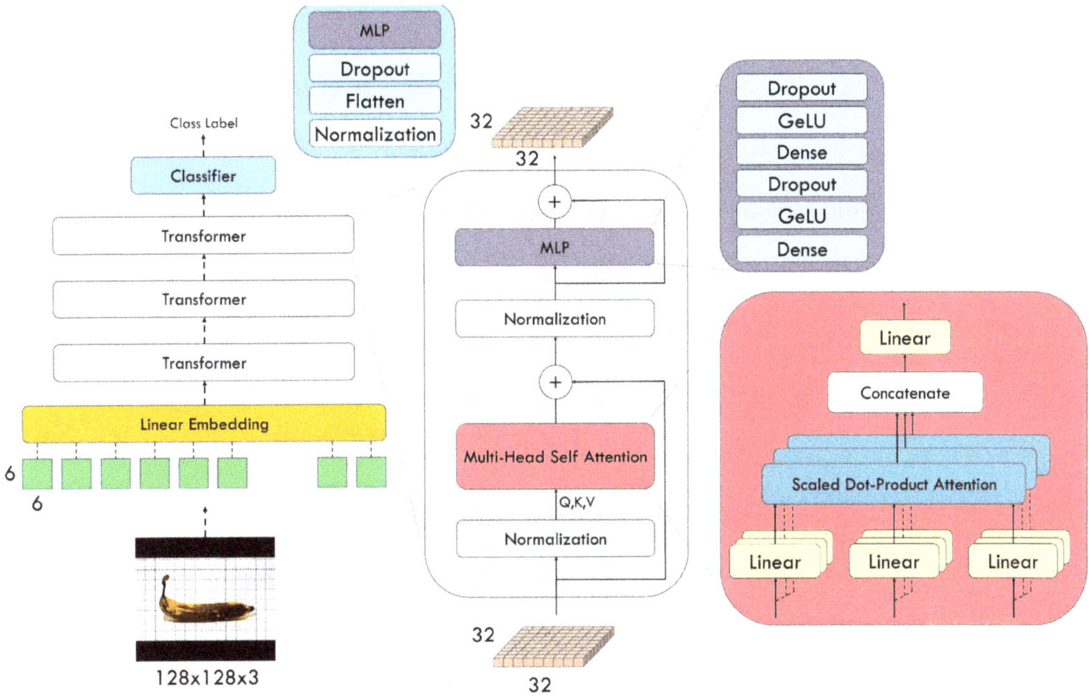

Figure 1. The proposed vision transformer network.

The model takes input images of size (200,200,3) and returns a prediction of one of the two classes. The model's architecture consists of a series of transformer blocks, each with a multi-head attention layer and a multilayer perceptron (MLP) layer. The input images are divided into patches and fed into the transformer blocks. The model is trained using the sparse categorical cross entropy loss function and the AdamW optimiser.

The model first processes the input images by dividing them into smaller patches. Each patch is then encoded using a patch encoder layer, which applies a dense layer and an embedding layer. The encoded patches are then passed through a series of transformer blocks. Each block applies a layer of multi-head attention followed by an MLP. The multi-head attention layer allows the model to attend to different image parts. In contrast, the MLP layer applies non-linear transformations to the encoded patches.

After the final transformer block, the encoded patches are flattened and fed into an MLP that produces the final classification. The MLP applies two dense layers with 500 and 250 units to the encoded patches. The output of the MLP is then passed through a dense layer with two units and a Softmax activation function to produce the final prediction.

The model is trained using the sparse categorical cross-entropy loss function, which compares the predicted class probabilities to the actual class labels. The AdamW optimiser optimises the model, which applies weight decay to the model parameters. The model is evaluated using the sparse categorical accuracy metric, which measures the proportion of correctly classified examples.

3.2. Datasets

3.2.1. Sources

We used various sources for collecting fruit images classified between quality-related categories. We used the extracted image collection to develop this study's large-scale dataset. The image sources comprise the following:

- FruitNet: Indian fruits dataset with quality: https://www.kaggle.com/datasets/shashwatwork/fruitnet-indian-fruits-dataset-with-quality (accessed on 2 February 2023);
- FruitQ dataset: https://www.kaggle.com/datasets/sholzz/fruitq-dataset (accessed on 2 February 2023);
- Lemon quality dataset: https://www.kaggle.com/datasets/yusufemir/lemon-quality-dataset (accessed on 2 February 2023);
- Mango varieties classification and grading: https://www.kaggle.com/datasets/saurabhshahane/mango-varieties-classification (accessed on 2 February 2023).

3.2.2. Characteristics and Preprocessing

The datasets mentioned above were processed to create this study's dataset. The analysis identified 16 fruit types.

We have followed the steps described below to create the dataset:

Step 1. Download all files from each source.
Step 2. Create the initial list of examined fruit types.
Step 3. For each dataset, validate the availability of each fruit in the list.
Step 4. For each dataset, exclude corrupted and low-resolution images.
Step 5. Create a large-scale dataset that contains all available fruit types.
Step 6. Exclude fruits that are not labelled.
Step 7. Define the two classes: good quality (GQ) and bad quality (BQ).
Step 8. Exclude fruit types that include less than 50 images per class.

Table 1 presents the image distribution between the classes of the final dataset, the total number of images per fruit, the initial image format, and image size.

Table 1. Per-fruit characteristics of this study's dataset.

Datasets	Number of Images Representing Good Quality Fruit	Number of Images Representing Bad Quality Fruit	Total	Format	Image Size (Height, Width)
Apple	1149	1141	2290	PNG	(192, 256)
Banana	1292	1520	2812	PNG	(720, 1280)
Cucumber	250	461	711	PNG	(720, 1280)
Grape	227	482	709	PNG	(720, 1280)
Guava	1152	1129	2281	JPEG	(256, 256)
Kaki	545	566	1111	PNG	(720, 1280)
Lemon	1125	951	2076	PNG	(300, 300)
Lime	1094	1085	2179	JPEG	(192, 256)
Mango	200	200	400	JPEG	(424, 752)
Orange	1216	1159	2375	PNG	(256, 256)
Papaya	130	663	793	PNG	(720, 1280)
Peach	425	720	1145	PNG	(720, 1280)
Pear	504	593	1097	JPEG	(720, 1280)
Pomegranate	5940	1187	7127	JPEG	(256, 256)
Tomato	600	1255	1855	PNG	(720, 1280)
Watermelon	51	203	254	PNG	(720, 1280)
Total (UD dataset)	15,900	13,315	29,215	-	-

Apart from the 16 separate datasets, which have been organised to represent one fruit each, we created an ultimate dataset of all fruit types for training the general model. This dataset will henceforth be addressed as the Union dataset (UD).

We also collected 200 images per fruit that serve the purpose of the external evaluation dataset. The characteristics of this dataset are presented in Table 2.

Table 2. Per-fruit characteristics of this study's external evaluation dataset.

External Dataset	Number of Images Representing Good Quality Fruit	Number of Images Representing Bad Quality Fruit	Total	Format	Image Size (Height, Width)
Apple	100	100	200	JPEG	(192, 256)
Banana	100	100	200	JPEG	(720, 1280)
Cucumber	100	100	200	JPEG	(256, 256)
Grape	100	100	200	PNG	(256, 256)
Guava	100	100	200	JPEG	(256, 256)
Kaki	100	100	200	PNG	(720, 1280)
Lemon	100	100	200	PNG	(300, 300)
Lime	100	100	200	JPEG	(192, 256)
Mango	100	100	200	JPEG	(424, 752)
Orange	100	100	200	JPEG	(256, 256)
Papaya	100	100	200	PNG	(256, 256)
Peach	100	100	200	JPEG	(256, 256)
Pear	100	100	200	JPEG	(720, 1280)
Pomegranate	100	100	200	JPEG	(256, 256)
Tomato	100	100	200	PNG	(256, 256)
Watermelon	100	100	200	JPEG	(720, 1280)

Figure 2 illustrates the data collection and preprocessing steps for creating the datasets of this study.

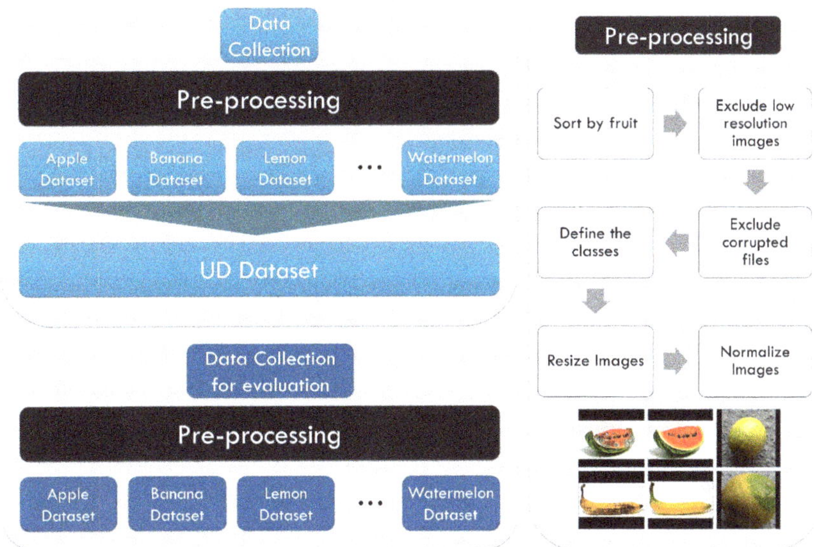

Figure 2. Data collection and processing procedure. The top-left box describes the process of creating the UD dataset. The lower-left box presents the creating of the external evaluation datasets. Both datasets share the same pre-processing steps, visualized in the right box.

We emphasize that, in this study, we exclusively assessed the quality of fruits based on their visual appearance. We did not consider other features, such as taste, texture, nutritional content, or internal characteristics such as ripeness, which are undoubtedly critical factors in determining overall fruit quality. This limitation is important to acknowledge, as it implies that our quality assessment is solely based on external attributes such as colour, shape, size, and visual defects. While visual appearance can provide valuable insights into fruit quality, it is not a comprehensive measure.

Dataset preprocessing includes sorting the images by fruit, excluding low-resolution and corrupted images, grouping the images into classes, resizing the images to fit in a black background with a 200×200-pixel canvas, and normalisation.

CNNs require input images to have a consistent size. Resizing ensures that all input images have the same dimensions, which is essential for the network to process them effectively. This standardization simplifies the architecture and reduces the need for complex resizing operations within the network. Resized images are computationally more efficient to process. Large variations in image sizes can increase the computational load on the network, slowing down training and inference. Resizing images to a uniform size reduces this computational burden.

Normalizing pixel values to a common range (e.g., [0, 1] or [−1, 1]) helps to stabilize and accelerate the training process. It ensures that the network's weights are updated uniformly, preventing saturation of activation functions. Normalization also helps mitigate the effects of the differences in lighting and contrast across images, making the network more robust to variations in input data. Normalizing inputs helps maintain a consistent scale of gradients across layers during backpropagation. This can prevent vanishing or exploding gradients, which are common issues in deep networks, and enable more stable and faster convergence during training. Normalization can act as a form of regularization by reducing the likelihood of overfitting. It imposes constraints on the network's weights and activations, making the model more resistant to noise in the training data.

Data augmentation is a crucial strategy to artificially increase the effective size of the training dataset and improve model generalization. The following methods were applied:

- Width shift: We randomly shifted the image horizontally, changing the position of the fruit within the frame. This helps the model learn to recognize the same fruit from different viewpoints.
- Height shift: similar to width shift, we randomly shifted the image vertically to introduce variations in the fruit's vertical position within the frame.
- Rotation: We applied random rotations to the images to simulate different orientations of the fruits. This helps the model become more invariant to rotation.
- Gaussian noise: we added Gaussian noise to the images to simulate variations in lighting conditions and improved the model's robustness to noise.
- Sheer: sheer transformations were applied to deform the image, introducing slight distortions that mimic real-world deformations in fruit appearance.

3.3. Experiment Design

Figure 3 illustrates the methodology of this research study. We designed the experiments as follows:

a. Build a ViT network and perform a 10-fold cross-validation procedure using the UD dataset.
b. Evaluate the model's per-fruit performance in detecting rotten- and good-quality fruits.
c. Build ViT models for each fruit and perform a 10-fold cross-validation procedure using data from the specific fruit.
d. Evaluate the models' performance in detecting rotten- and good-quality fruits.

Figure 3 illustrates the methodology of the present research study.

In evaluating a classification model's performance, several key metrics are commonly used, such as accuracy, precision, recall, and the F1 score. Accuracy measures the proportion of correctly classified instances, providing a general overview of a model's correctness. Precision, conversely, gauges the model's ability to correctly identify positive instances among those it predicted as positive, focusing on minimizing false positives. Recall, also known as sensitivity, assesses the model's capability to identify all positive instances among the actual positives, concentrating on minimizing false negatives. The F1 score, which harmonizes precision and recall, offers a balanced metric that considers false positives and false negatives, making it particularly useful when class imbalance is present in the data. These evaluation criteria collectively provide a comprehensive assessment of a model's performance, aiding in informed decision-making and model refinement.

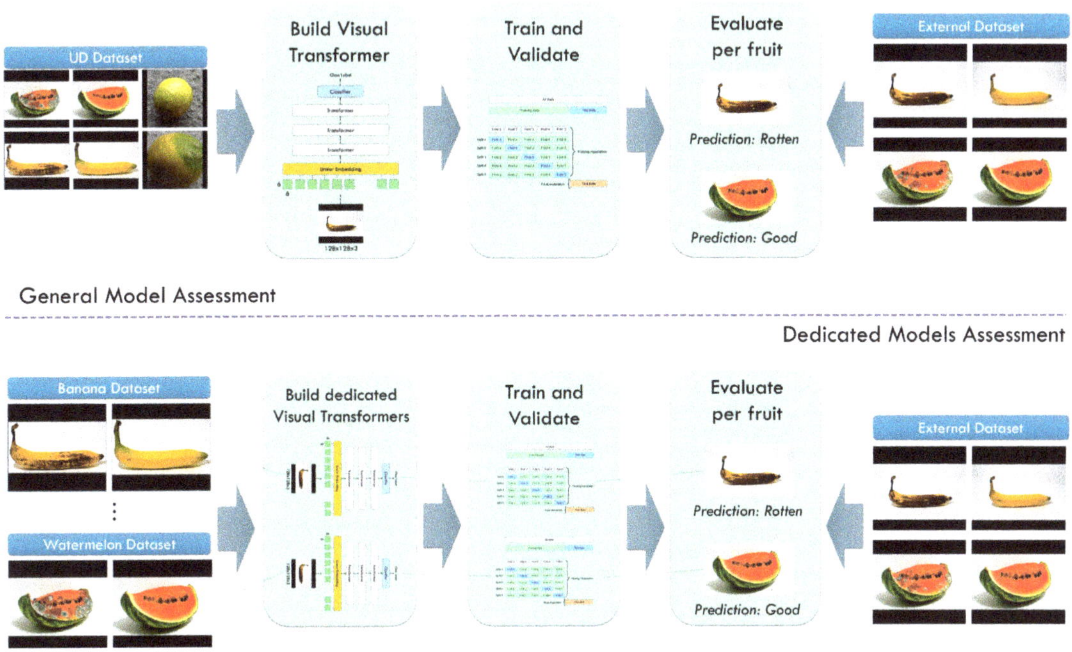

Figure 3. Research methodology.

4. Results

4.1. General Model

In this section, we present the classification results of the general model, which was trained using the large-scale UD dataset.

4.1.1. Training and Validation Performance

Under the 10-fold cross-validation procedure, the general model achieves an accuracy of 0.9794. The latter is computed regardless of the fruit under examination. The model obtains a 0.9886 precision, 0.9733 recall, and 0.9809 F1 score (Table 3).

Table 3. Results of the general model under a 10-fold cross-validation procedure. UD refers to the training dataset.

Training Data	Testing Data	Accuracy	Precision	Recall	F1
UD	UD	0.9794	0.9886	0.9733	0.9809

The above scores represent the aggregated scores derived from each iteration over the ten-fold procedure. The model performs excellently in identifying the general condition of any fruit of the dataset. It yields 178 false-good predictions and 424 false-rotten predictions. Correct predictions include 15,476 true-good cases and 13,137 true-rotten cases.

4.1.2. External Per-Fruit Evaluation

The general model has been evaluated using the external datasets of various fruit types. The reader shall recall that each external dataset includes 100 good and 100 rotten fruit representations. Table 4 presents the results.

Table 4. Results of the general model when testing with external data. The testing fruit column refers to the type of fruits used for testing the model. The latter images originate from the test dataset.

Training Data	Testing Fruit	Accuracy	Precision	Recall	F1
UD	Apple	0.9950	1.0000	0.9900	0.9950
UD	Banana	0.9800	0.9615	1.0000	0.9804
UD	Cucumber	0.9900	0.9804	1.0000	0.9901
UD	Grape	1.0000	1.0000	1.0000	1.0000
UD	Guava	0.9700	0.9796	0.9600	0.9697
UD	Kaki	0.9950	0.9901	1.0000	0.9950
UD	Lemon	0.9700	0.9608	0.9800	0.9703
UD	Lime	0.9750	0.9798	0.9700	0.9749
UD	Mango	0.9750	0.9897	0.9600	0.9746
UD	Orange	0.9950	0.9901	1.0000	0.9950
UD	Papaya	0.9800	0.9898	0.9700	0.9798
UD	Peach	0.9800	0.9706	0.9900	0.9802
UD	Pear	0.9700	0.9796	0.9600	0.9697
UD	Pomegranate	0.9700	0.9796	0.9600	0.9697
UD	Tomato	0.9950	0.9901	1.0000	0.9950
UD	Watermelon	0.9800	0.9706	0.9900	0.9802

The general model shows remarkable performance in identifying the quality of apples (accuracy of 0.9950), cucumbers (accuracy of 0.99), grapes (accuracy of 1.00), kakis (accuracy of 0.9950), oranges (accuracy of 0.9950), papayas (accuracy of 0.98), peaches (accuracy of 0.98), tomatoes (accuracy of 0.9950), and watermelons (accuracy of 0.98).

Slight worse performance was recorded concerning guavas (accuracy of 0.9700), lemons (accuracy of 0.9700), limes (accuracy of 0.9750), mangoes (accuracy of 0.9750), pears (accuracy of 0.9700), and pomegranates (accuracy of 0.9700).

It is worth noticing that the general model achieved equal or higher classification scores in the external datasets than the scores from the Union dataset (UD) which contains the training data. This phenomenon is strong evidence of the generalisation capabilities of the model.

4.2. Dedicated Models

In this section, we present the results of the dedicated models. Each model is trained to distinguish between good and rotten images of a specific fruit. Subsequently, each model can operate using images of a single fruit variety.

4.2.1. Training and Validation Performance

Table 5 summarises the 10-fold cross-validation results of the dedicated models.

All models obtain high-performance metrics except for the grape and papaya models.

Table 5. Results of dedicated models under a 10-fold cross-validation procedure. The testing fruit column refers to the type of fruits used for testing the model. The latter images originate from the test dataset.

Training Data (UD)	Testing Fruit	Accuracy	Precision	Recall	F1
Apple	Apple	0.9948	0.9974	0.9922	0.9948
Banana	Banana	0.9904	0.9854	0.9938	0.9896
Cucumber	Cucumber	0.9887	0.9764	0.9920	0.9841
Grape	Grape	0.9661	0.9511	0.9427	0.9469
Guava	Guava	0.9965	0.9974	0.9957	0.9965
Kaki	Kaki	0.9928	0.9873	0.9982	0.9927
Lemon	Lemon	0.9981	1.0000	0.9964	0.9982
Lime	Lime	0.9991	0.9982	1.0000	0.9991
Mango	Mango	0.9625	0.9793	0.9450	0.9618
Orange	Orange	0.9971	0.9984	0.9959	0.9971
Papaya	Papaya	0.9546	0.7831	1.0000	0.8784
Peach	Peach	0.9965	0.9953	0.9953	0.9953
Pear	Pear	0.9909	0.9940	0.9861	0.9900
Pomegranate	Pomegranate	0.9964	0.9975	0.9981	0.9978
Tomato	Tomato	0.9957	0.9933	0.9933	0.9933
Watermelon	Watermelon	0.9055	0.6800	1.0000	0.8095

4.2.2. External Per-Fruit Evaluation

Table 6 summarises the classification metrics of each dedicated model when predicting the classes of the external dataset.

Table 6. Results of dedicated models. The testing fruit column refers to the type of fruits used for testing the model. The latter images originate from the test dataset.

Training Data	Testing Fruit	Accuracy	Precision	Recall	F1
Apple	Apple	0.9950	1.0000	0.9900	0.9950
Banana	Banana	0.9950	0.9901	1.0000	0.9950
Cucumber	Cucumber	0.9850	0.9899	0.9800	0.9849
Grape	Grape	0.9900	0.9900	0.9900	0.9900
Guava	Guava	0.9850	0.9709	1.0000	0.9852
Kaki	Kaki	0.9900	1.0000	0.9800	0.9899
Lemon	Lemon	0.9950	1.0000	0.9900	0.9950
Lime	Lime	0.9800	0.9898	0.9700	0.9798
Mango	Mango	0.9500	0.9412	0.9600	0.9505
Orange	Orange	0.9950	1.0000	0.9900	0.9950
Papaya	Papaya	0.9500	0.9688	0.9300	0.9490
Peach	Peach	0.9800	0.9706	0.9900	0.9802
Pear	Pear	0.9650	0.9697	0.9600	0.9648
Pomegranate	Pomegranate	0.9950	0.9901	1.0000	0.9950
Tomato	Tomato	0.9800	0.9800	0.9800	0.9800
Watermelon	Watermelon	0.9550	0.9505	0.9600	0.9552

The dedicated models perform remarkably for apples (accuracy of 0.9950), bananas (accuracy of 0.9950), cucumbers (accuracy of 0.9850), grapes (accuracy of 0.99), kakis (accuracy of 0.99), lemons (accuracy of 0.9950), oranges (accuracy of 0.9950), and pomegranates (accuracy of 0.9950).

A slight decrease in accuracy is observed for the limes (accuracy of 0.98), peaches (accuracy of 0.98), and tomatoes (accuracy of 0.98).

The dedicated models show suboptimal results in classifying mangos (accuracy of 0.95), papayas (accuracy of 0.95), pears (accuracy of 0.9650), and watermelons (accuracy of 0.9550).

We compared the results of the general model and the dedicated models (Table 7).

Table 7. Comparison between dedicated models and the general model in per-fruit accuracy measured using the external test set.

Fruit	Dedicated Model	General Model
Apple	0.9950	0.9950
Banana	0.9950	0.9800
Cucumber	0.9850	0.9900
Grape	0.9900	1.0000
Guava	0.9850	0.9700
Kaki	0.9900	0.9950
Lemon	0.9950	0.9700
Lime	0.9800	0.9750
Mango	0.9500	0.9750
Orange	0.9950	0.9950
Papaya	0.9500	0.9800
Peach	0.9800	0.9800
Pear	0.9650	0.9700
Pomegranate	0.9950	0.9700
Tomato	0.9800	0.9950
Watermelon	0.9550	0.9800

The general model is more effective than the dedicated models for predicting the quality of cucumbers, grapes, kakis, mangos, papayas, pears, tomatoes, and watermelons.

It yields equal classification accuracy in apples, oranges, and peaches. Subsequently, the dedicated models are better when built for bananas, guavas, lemons, limes, and pomegranates. Of the sixteen fruit types, the dedicated models performed better only in five of them (Table 7, Figure 4).

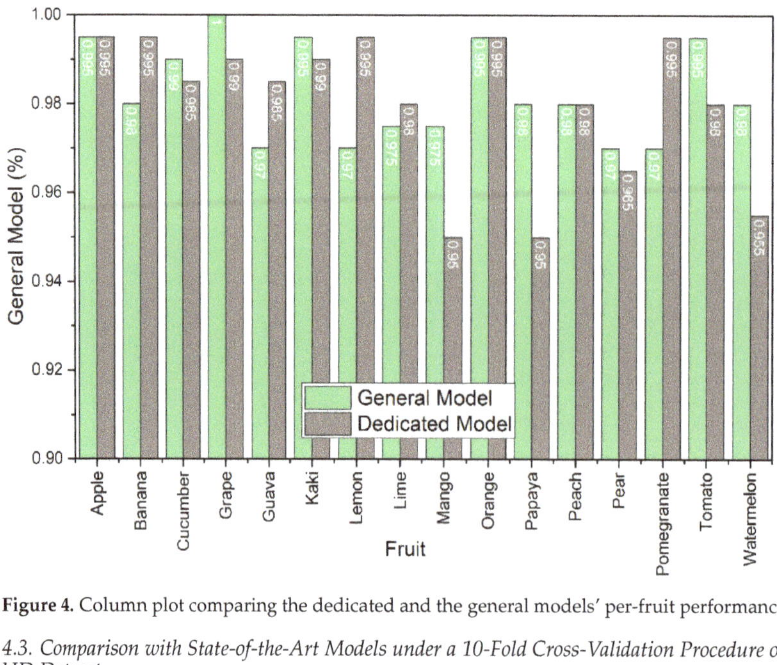

Figure 4. Column plot comparing the dedicated and the general models' per-fruit performance.

4.3. Comparison with State-of-the-Art Models under a 10-Fold Cross-Validation Procedure on the UD Dataset

We oppose the proposed general model (ViT) to various State-of-the-Art networks implemented using the Keras Python library. Each network was trained and evaluated under the same conditions. Table 8 presents the obtained performance metrics.

Table 8. UD dataset classification of various State-of-the-Art CNN-based networks under a 10-fold cross-validation procedure.

Model	Accuracy	Precision	Recall	F1
Xception [34]	0.9524	0.9726	0.9390	0.9555
VGG16 [29]	0.9446	0.9647	0.9323	0.9482
VGG19 [29]	0.9671	0.9875	0.9516	0.9693
ResNet152 [35]	0.9785	0.9887	0.9716	0.9800
ResNet152V2 [35]	0.9606	0.9861	0.9409	0.9630
InceptionV3 [36]	0.9539	0.9711	0.9433	0.9570
InceptionResNetV2 [36]	0.9641	0.9796	0.9539	0.9666
MobileNet [27]	0.9536	0.9820	0.9319	0.9563
MobileNetV2 [27]	0.9624	0.9805	0.9499	0.9649
DenseNet169 [37]	0.9631	0.9669	0.9652	0.9660
DenseNet201 [37]	0.9598	0.9736	0.9519	0.9627
NASNetMobile [38]	0.9547	0.9819	0.9340	0.9574
EfficientNetB6 [39]	0.9660	0.9718	0.9655	0.9686
EfficientNetB7 [39]	0.9705	0.9842	0.9611	0.9725
EfficientNetV2B3 [39]	0.9591	0.9716	0.9526	0.9620
ConvNeXtLarge [40]	0.9732	0.9870	0.9634	0.9750
ConvNeXtXLarge [40]	0.9486	0.9651	0.9396	0.9522
Swin Transformer [41]	0.9632	0.9874	0.9445	0.9654
Perceiver Network [42]	0.9643	0.9711	0.9631	0.9671
Involutional Neural Network [43]	0.9635	0.9725	0.9601	0.9663
ConvMixer [16,44,45]	0.9591	0.9715	0.9529	0.9621
BigTransfer [46]	0.9574	0.9659	0.9555	0.9606
EANet [47]	0.9732	0.9874	0.9630	0.9750
FNet [33]	0.9690	0.9709	0.9722	0.9716
gMLP [48]	0.9597	0.9818	0.9435	0.9623
MLP-Mixer [46]	0.9564	0.9656	0.9539	0.9597
Attention VGG19 [49]	0.9644	0.9852	0.9489	0.9667
Vision Transformer (this study)	0.9794	0.9886	0.9733	0.9809

The top networks exhibiting equivalent performance include ResNet152 [35], ConvNeXtLarge [40], and EANet [47]. Figure 5 provides a visual comparison regarding the recorder accuracy of each model. The ViT model of this study is slightly better than the rest. Further and extensive fine tuning of other models may reveal that other models can perform equally well. However, the latter is beyond the scope of this paper.

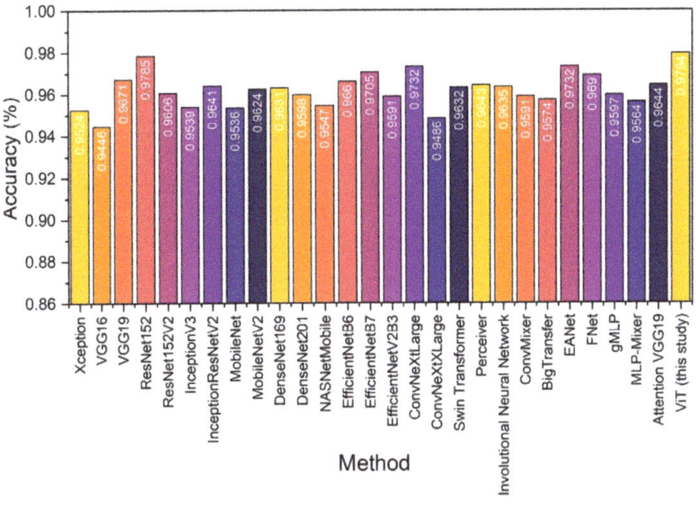

Figure 5. UD dataset classification performance comparison between various State-of-the-Art CNN-based networks under a 10-fold cross-validation procedure.

4.4. Comparison with Classic Machine Learning Models

We also oppose the proposed general model (ViT) to various classic machine learning models implemented with the aid of the scikit-learn Python library. Each network was trained and evaluated under the same conditions. To prepare the images for such networks, the initial image was flattened to form a one-dimensional vector that is mandatory for processing by these classifiers.

For the random forest algorithm, we set the number of trees (n_estimators) to 1000; the maximum depth of each tree (max_depth) to 25; the minimum samples required to split a node (min_samples_split) to 2; the minimum samples required at a leaf node (min_samples_leaf) to 1; and the maximum number of features considered for each split (max_features) to 'sqrt' (square root of the total number of features).

For the XGBoost algorithm, we chose a learning rate (learning_rate) of 0.1; 1000 boosting rounds (n_estimators); a maximum tree depth (max_depth) of 15; a minimum sum of instance weight in a child (min_child_weight) of 1; a subsample fraction (subsample) of 0.8; and a fraction of features used for each tree (colsample_bytree) of 0.8.

For the k-nearest neighbours (KNN) algorithm, we set the number of neighbours (k) to 11. We used the default Euclidean distance metric for neighbour selection.

Regarding the support vector machine (SVM) algorithm, we applied a linear kernel, set the regularization parameter (C) to 1.0, and used class weights balanced according to the input data. For the naive Bayes method, we applied Laplace smoothing (alpha) with a value of 1.0. We used standard text preprocessing and feature extraction techniques.

In our neural network, we defined 6 hidden layers with 128 neurons each; used ReLU activation functions; a batch size of 32; 50 training epochs' a learning rate of 0.001; and applied dropout regularization with a rate of 0.2. We used the Adam optimizer and categorical cross-entropy loss for a classification task.

Table 9 presents the obtained performance metrics.

Table 9. UD dataset classification of various State-of-the-Art ML networks under a 10-fold cross-validation procedure.

Model	Accuracy	Precision	Recall	F1
random forest	0.9343	0.9693	0.9081	0.9377
XGBoost	0.9343	0.9635	0.9140	0.9381
K-Nearest Neighbours	0.9213	0.9767	0.8764	0.9238
Support Vector Machine	0.9159	0.9773	0.8655	0.9180
Naive Bayes	0.8733	0.9615	0.7991	0.8728
Neural Network	0.8732	0.9752	0.7870	0.8710
Vision Transformer (this study)	0.9794	0.9886	0.9733	0.9809

The tree-based algorithms (random forest and XGBoost 2.0.0.) outperform the rest, yielding an accuracy of 0.9343. Still, the classical ML algorithms exhibit lower evaluation metrics compared to the vision transformer network.

4.5. Comparison with the Literature

We collected recent literature employing either dedicated models and examining a single fruit or general models applied to various fruit representations. Table 10 compares the general model of this study and models suggested by related works.

Table 10. Comparison with the literature.

Fruit	Study	Objective	Method(s)	Accuracy
Plum	[19]	Determination of plum maturity from images	Deep CNN	91–97%
Mangosteen	[20]	Quality assurance in mangosteen export	Deep CNN	97%
Apple	[21]	Apple lesions identification	Deep CNN	97.5%
Banana	[22]	Differentiation between naturally and artificially ripened bananas	Neural Network	98.74%
Peach	[23]	Peach disease identification	Deep Belief Network	82.5–100%
Multiple (6)	[24]	Quality Assessment	Deep CNN	99.6%
Multiple (3)	[25]	Quality Assessment	Deep CNN	95%
Banana	[11]	Quality Assessment	Deep CNN	81.75–98.25%
Multiple (3)	[26]	Quality Assessment	Deep CNN	99.61%
Papaya	[28]	Quality Assessment	Deep CNN	100%
Pomegranate	[50]	Quality Assessment	Recurrent Neural Network	95%
Grapes	[51]	Quality Assessment	Artificial Neural Network	87.8%
Mango	[52]	Quality Assessment	SVM	98.6%
Apple	[52]	Quality Assessment	Deep CNN	98.6%

The comparison reveals that the suggested general and dedicated models are consistent with the literature and may exhibit better performance regarding specific fruit types. More precisely, most studies report an accuracy between 97% and 99% in determining the quality of the fruits. The general model of this study reports per-fruit accuracies that vary between 97% and 100%.

The comparisons also verify that the general model is often better than the dedicated models.

5. Discussion

The quality of fruits is essential in determining their market value and consumer satisfaction. High-quality fruits are visually appealing, flavourful, and nutritionally dense. However, assessing fruit quality can be laborious and time-consuming, especially when performed manually. This is where deep learning technology can be applied to automate and optimise the process of fruit quality assessment. By processing a large dataset of fruit images, deep learning algorithms can be trained to recognise specific patterns and features indicative of fruit quality. For instance, a deep learning model can be trained to identify specific colouration, texture, and shape characteristics that indicate freshness, ripeness, or maturity in a fruit. Deep learning can be used to assess the quality of fruits at different stages of production, from the farm to the market. Farmers can use deep learning algorithms to assess the quality of their products in real-time, allowing them to make informed decisions on when to harvest or transport their fruits.

Additionally, food retailers can use deep learning technology to sort and grade fruits based on their quality, reducing waste, and ensuring consistent product quality for consumers. Furthermore, deep learning can also be applied to preserve fruit quality during storage and transportation. By detecting and removing low-quality fruits before shipping, deep learning algorithms can reduce the chances of damage or spoilage during transportation, ensuring that consumers receive only high-quality fruits.

This research study presented a general ML model based on vision transformers for estimating fruit quality based on photographs. We proposed a general model that can be trained with multiple fruits and predict the quality of any fruit variety that participated in the training set. This general model was superior to dedicated models, in which training was performed using a single fruit variety. According to the results, a generalised model predicts the quality of cucumbers, grapes, kakis, mangos, papayas, pears, tomatoes, and watermelons more efficiently than dedicated models. However, the classification accuracy of the generalised and dedicated models is similar for apples, oranges, and peaches.

On the other hand, the dedicated models perform better for bananas, guavas, lemons, limes, and pomegranates. Only five of the sixteen fruits analysed showed improved results when using dedicated models.

This suggests that while a generalised model may provide satisfactory results for most fruits, dedicated models tailored to specific fruits can significantly enhance the accuracy of the predictions, particularly for fruits with unique characteristics or qualities that are difficult to generalise.

To summarize, we presented a machine learning model based on ViT networks capable of assessing the quality of various fruits based solely on their visual appearance, eliminating the need for fruit-specific models. Our general model showcases performance that either equals or surpasses dedicated, fruit-specific models, simplifying the process while maintaining or enhancing accuracy. Through rigorous comparisons with State-of-the-Art techniques, our research establishes vision transformers (ViTs) as the superior choice for fruit quality assessment, setting a new benchmark in computer vision for agriculture and quality control. This study has some limitations. Firstly, fruit quality can be evaluated based on several factors, including appearance, flavour, texture, and nutritional content. While the appearance of the fruit can be an indicator of quality, it is not always reliable.

In some cases, the appearance of the fruit can provide some clues about its quality. For example, ripe fruit should have a bright and uniform colour, be free of bruises or blemishes, and have a firm and smooth texture. However, some exceptions exist to these guidelines, such as bananas, which develop brown spots as they ripen but are still perfectly edible. Other factors affecting fruit quality, such as flavour and nutritional content, cannot be assessed based on appearance alone. For example, some fruits may look perfectly fine but lack flavour or be low in certain nutrients. While some fruit characteristics such as colour, shape, and texture can be visually evaluated. Other vital factors such as flavour, aroma, and nutritional content cannot be assessed visually. Moreover, the visual appearance of the fruit can be influenced by various factors, such as lighting, the angle of the camera, and post-harvest treatments, which can affect the quality assessment. The latter can be considered a limitation of this study.

Integrating machine learning models into existing fruit sorting and grading systems may improve efficiency and accuracy but also open the door to a holistic approach that combines image and non-image characteristics for more comprehensive fruit quality assessments. This synergy between different data sources maximizes the potential for optimizing fruit grading processes across various agricultural contexts.

Adapting machine learning models to account for variations in fruit quality stemming from diverse factors such as climate, soil, and growing conditions is crucial for ensuring the robustness and applicability of these models in real-world agricultural settings. One approach involves incorporating these environmental variables as features in the training dataset. By including climate data (e.g., temperature, humidity, and precipitation); soil characteristics (e.g., pH levels and nutrient content); and growing conditions (e.g., irrigation methods and pesticide usage), the existing model can learn to recognize patterns and correlations between these variables and fruit quality. This enables the model to make more nuanced and context-aware quality assessments. Regular updates of these environmental data help the model adapt to changing conditions over time.

Secondly, while studying 16 fruit types provides valuable insights, it is essential to note that this sample size may not represent all fruit types. To fully assess the effectiveness of generalised versus dedicated models for predicting fruit quality, a more comprehensive and diverse dataset should be used.

Including a broader range of fruit varieties in future studies can help to identify patterns and trends across different types of fruit and further establish the efficacy of generalised and dedicated models. Additionally, expanding the sample size can provide more accurate and robust results, allowing for greater confidence in the findings and a better understanding of the strengths and limitations of these modelling approaches.

The integration of machine learning into fruit quality assessment raises important ethical considerations. Privacy and consent are paramount, demanding robust data anonymization and comprehensive consent procedures. Transparency and fairness are crucial. Biases inherited from data must be addressed with fairness-aware algorithms, ongoing monitoring, and clear model explanations. Environmental responsibility is key, as machine learning can impact resource consumption. Ethical practices involve optimizing algorithms for sustainability. Labour displacement concerns the call for plans to retrain and reskill affected workers. Finally, ensuring equitable access to these technologies, especially for small-scale farmers, is vital. Initiatives for technology transfer and knowledge sharing promote fairness and broad benefits.

6. Conclusions

AI-based technologies can potentially revolutionise the fruit industry by providing objective and efficient quality assessment. This study introduced a general machine learning model based on vision transformers to assess fruit quality from images. The model outperformed dedicated models trained on single fruit types, except for apples, oranges, and peaches, where both had similar accuracy. Dedicated models were better for specific fruits such as bananas and pomegranates. Overall, a generalised model worked well for most fruit types. However, dedicated models could improve the accuracy for fruit types with unique features. Fruit quality depends on multiple factors, including appearance, flavour, and nutrition. Appearance can be misleading and affected by various factors. This study has limitations in this regard. Finally, while the 16 fruit types used in this study provide a valid starting point, future research should include a more diverse and extensive range of fruit types to better evaluate the effectiveness of generalised and dedicated models in predicting fruit quality.

Author Contributions: Conceptualization, I.D.A. and M.T.; data curation, I.D.A.; investigation, M.T.; methodology, I.D.A.; resources, I.D.A. and M.T.; software, I.D.A. and S.I.A.; validation, I.D.A.; visualization, I.D.A.; writing—original draft preparation, I.D.A.; writing—review and editing, M.T. and S.I.A. All authors have read and agreed to the published version of the manuscript.

Funding: This research received no external funding.

Institutional Review Board Statement: Not applicable.

Informed Consent Statement: Not applicable.

Data Availability Statement: The datasets of the study are open-access.

Conflicts of Interest: The authors declare no conflict of interest.

References

1. Prasad, K.; Jacob, S.; Siddiqui, M.W. Fruit maturity, harvesting, and quality standards. In *Preharvest Modulation of Postharvest Fruit and Vegetable Quality*; Elsevier: Amsterdam, The Netherlands, 2018; pp. 41–69. [CrossRef]
2. Pathmanaban, P.; Gnanavel, B.; Anandan, S.S. Recent application of imaging techniques for fruit quality assessment. *Trends Food Sci. Technol.* **2019**, *94*, 32–42. [CrossRef]
3. Mowat, A.; Collins, R. Consumer behaviour and fruit quality: Supply chain management in an emerging industry. *Supply Chain Manag. Int. J.* **2000**, *5*, 45–54. [CrossRef]
4. Zhou, L.; Zhang, C.; Liu, F.; Qiu, Z.; He, Y. Application of Deep Learning in Food: A Review. *Compr. Rev. Food Sci. Food Saf.* **2019**, *18*, 1793–1811. [CrossRef] [PubMed]
5. Melesse, T.Y.; Bollo, M.; Di Pasquale, V.; Centro, F.; Riemma, S. Machine Learning-Based Digital Twin for Monitoring Fruit Quality Evolution. *Procedia Comput. Sci.* **2022**, *200*, 13–20. [CrossRef]
6. Hemamalini, V.; Rajarajeswari, S.; Nachiyappan, S.; Sambath, M.; Devi, T.; Singh, B.K.; Raghuvanshi, A. Food Quality Inspection and Grading Using Efficient Image Segmentation and Machine Learning-Based System. *J. Food Qual.* **2022**, *2022*, 5262294. [CrossRef]
7. Han, J.; Li, T.; He, Y.; Gao, Q. Using Machine Learning Approaches for Food Quality Detection. *Math. Probl. Eng.* **2022**, *2022*, 6852022. [CrossRef]
8. Dhiman, B.; Kumar, Y.; Kumar, M. Fruit quality evaluation using machine learning techniques: Review, motivation and future perspectives. *Multimedia Tools Appl.* **2022**, *81*, 16255–16277. [CrossRef]

9. Bhargava, A.; Bansal, A.; Goyal, V. Machine Learning–Based Detection and Sorting of Multiple Vegetables and Fruits. *Food Anal. Methods* **2022**, *15*, 228–242. [CrossRef]
10. Cheng, S.; Jin, Y.; Harrison, S.P.; Quilodrán-Casas, C.; Prentice, I.C.; Guo, Y.-K.; Arcucci, R. Parameter Flexible Wildfire Prediction Using Machine Learning Techniques: Forward and Inverse Modelling. *Remote Sens.* **2022**, *14*, 3228. [CrossRef]
11. Aherwadi, N.; Mittal, U.; Singla, J.; Jhanjhi, N.Z.; Yassine, A.; Hossain, M.S. Prediction of Fruit Maturity, Quality, and Its Life Using Deep Learning Algorithms. *Electronics* **2022**, *11*, 4100. [CrossRef]
12. Liu, Y.; Pu, H.; Sun, D.-W. Efficient extraction of deep image features using convolutional neural network (CNN) for applications in detecting and analysing complex food matrices. *Trends Food Sci. Technol.* **2021**, *113*, 193–204. [CrossRef]
13. Zimmerman, N.; Presto, A.A.; Kumar, S.P.N.; Gu, J.; Hauryliuk, A.; Robinson, E.S.; Robinson, A.L.; Subramanian, R. A machine learning calibration model using random forests to improve sensor performance for lower-cost air quality monitoring. *Atmospheric Meas. Tech.* **2018**, *11*, 291–313. [CrossRef]
14. Kroll, A.; Ranjan, S.; Engqvist, M.K.M.; Lercher, M.J. A general model to predict small molecule substrates of enzymes based on machine and deep learning. *Nat. Commun.* **2023**, *14*, 2787. [CrossRef]
15. Chi, H.; Zhang, Y.; Tang, T.L.E.; Mirabella, L.; Dalloro, L.; Song, L.; Paulino, G.H. Universal machine learning for topology optimization. *Comput. Methods Appl. Mech. Eng.* **2020**, *375*, 112739. [CrossRef]
16. d'Ascoli, S.; Touvron, H.; Leavitt, M.; Morcos, A.; Biroli, G.; Sagun, L. Convit: Improving vision transformers with soft convolutional inductive biases. *arXiv* **2021**, arXiv:2103.10697. [CrossRef]
17. Tian, C.; Xu, Y.; Li, Z.; Zuo, W.; Fei, L.; Liu, H. Attention-guided CNN for image denoising. *Neural Netw.* **2020**, *124*, 117–129. [CrossRef]
18. LeCun, Y.; Kavukcuoglu, K.; Farabet, C. Convolutional networks and applications in vision. In Proceedings of the 2010 IEEE International Symposium on Circuits and Systems, Paris, France, 30 May–2 June 2010; IEEE: Piscataway, NJ, USA, 2010; pp. 253–256.
19. Rodríguez, F.J.; García, A.; Pardo, P.J.; Chávez, F.; Luque-Baena, R.M. Study and classification of plum varieties using image analysis and deep learning techniques. *Prog. Artif. Intell.* **2017**, *7*, 119–127. [CrossRef]
20. Azizah, L.M.; Umayah, S.F.; Riyadi, S.; Damarjati, C.; Utama, N.A. Deep learning implementation using convolutional neural network in mangosteen surface defect detection. In Proceedings of the 2017 7th IEEE International Conference on Control System, Computing and Engineering (ICCSCE), Penang, Malaysia, 24–26 November 2017; pp. 242–246. [CrossRef]
21. Tan, W.; Zhao, C.; Wu, H. Intelligent alerting for fruit-melon lesion image based on momentum deep learning. *Multimedia Tools Appl.* **2016**, *75*, 16741–16761. [CrossRef]
22. Mithun, B.S.; Sujit, S.; Karan, B.; Arijit, C.; Shalini, M.; Kavya, G.; Brojeshwar, B.; and Sanjay, K. Non-destructive method to detect artificially ripened banana using hyperspectral sensing and RGB imaging. In Proceedings of the SPIE 10665, Sensing for Agriculture and Food Quality and Safety X, 106650T, Orlando, FL, USA, 15 May 2018. [CrossRef]
23. Sun, Y.; Wei, K.; Liu, Q.; Pan, L.; Tu, K. Classification and Discrimination of Different Fungal Diseases of Three Infection Levels on Peaches Using Hyperspectral Reflectance Imaging Analysis. *Sensors* **2018**, *18*, 1295. [CrossRef]
24. Kumar, A.; Joshi, R.C.; Dutta, M.K.; Jonak, M.; Burget, R. Fruit-CNN: An Efficient Deep learning-based Fruit Classification and Quality Assessment for Precision Agriculture. In Proceedings of the 2021 13th International Congress on Ultra Modern Telecommunications and Control Systems and Workshops (ICUMT), Brno, Czech Republic, 25–27 October 2021; pp. 60–65. [CrossRef]
25. Bobde, S.; Jaiswal, S.; Kulkarni, P.; Patil, O.; Khode, P.; Jha, R. Fruit Quality Recognition using Deep Learning Algorithm. In Proceedings of the 2021 International Conference on Smart Generation Computing, Communication and Networking (SMART GENCON), Pune, India, 29–30 October 2021; pp. 1–5. [CrossRef]
26. Chakraborty, S.; Shamrat, F.J.M.; Billah, M.; Jubair, A.; Alauddin; Ranjan, R. Implementation of Deep Learning Methods to Identify Rotten Fruits. In Proceedings of the 2021 5th International Conference on Trends in Electronics and Informatics (ICOEI), Tirunelveli, India, 3–5 June 2021; pp. 1207–1212.
27. Howard, A.G.; Zhu, M.; Chen, B.; Kalenichenko, D.; Wang, W.; Weyand, T.; Andreetto, M.; Adam, H. Mobilenets: Efficient convolutional neural networks for mobile vision applications. *arXiv* **2017**, arXiv:1704.04861.
28. Behera, S.K.; Rath, A.K.; Sethy, P.K. Maturity status classification of papaya fruits based on machine learning and transfer learning approach. *Inf. Process Agric.* **2021**, *8*, 244–250. [CrossRef]
29. Simonyan, K.; Zisserman, A. Very Deep Convolutional Networks for Large-Scale Image Recognition. *arXiv* **2015**, arXiv:14091556.
30. Goodfellow, I.; Bengio, Y.; Courville, A.; Bengio, Y. *Deep Learning*; MIT Press: Cambridge, MA, USA, 2016; Volume 1.
31. Sinha, R.S.; Hwang, S.-H. Comparison of CNN Applications for RSSI-Based Fingerprint Indoor Localization. *Electronics* **2019**, *8*, 989. [CrossRef]
32. Mou, L.; Zhu, X.X. Learning to Pay Attention on Spectral Domain: A Spectral Attention Module-Based Convolutional Network for Hyperspectral Image Classification. *IEEE Trans. Geosci. Remote Sens.* **2020**, *58*, 110–122. [CrossRef]
33. Lee-Thorp, J.; Ainslie, J.; Eckstein, I.; Ontanon, S. FNet: Mixing Tokens with Fourier Transforms. *arXiv* **2022**, arXiv:2105.03824. [CrossRef]
34. Chollet, F. Xception: Deep learning with depthwise separable convolutions. In Proceedings of the 2017 IEEE Conference on Computer Vision and Pattern Recognition (CVPR), Honolulu, HI, USA, 21–26 July 2017.

35. He, K.; Zhang, X.; Ren, S.; Sun, J. Deep residual learning for image recognition. In Proceedings of the IEEE Computer Society Conference on Computer Vision and Pattern Recognition (CVPR), Las Vegas, NV, USA, 27–30 June 2016; pp. 770–778. [CrossRef]
36. Szegedy, C.; Liu, W.; Jia, Y.; Sermanet, P.; Reed, S.; Anguelov, D.; Erhan, D.; Vanhoucke, V.; Rabinovich, A.; Liu, W.; et al. Going deeper with convolutions. In Proceedings of the 2015 IEEE Conference on Computer Vision and Pattern Recognition (CVPR), Boston, MA, USA, 7–12 June 2015; pp. 1–9. [CrossRef]
37. Huang, G.; Liu, Z.; van der Maaten, L.; Weinberger, K.Q. Densely Connected Convolutional Networks. *arXiv* **2018**, arXiv:160806993.
38. Zoph, B.; Vasudevan, V.; Shlens, J.; Le, Q.V. Learning Transferable Architectures for Scalable Image Recognition. *arXiv* **2018**, arXiv:170707012.
39. Zhang, L.; Shen, H.; Luo, Y.; Cao, X.; Pan, L.; Wang, T.; Feng, Q. Efficient CNN Architecture Design Guided by Visualization. *arXiv* **2022**, arXiv:2207.10318. [CrossRef]
40. Liu, Z.; Mao, H.; Wu, C.-Y.; Feichtenhofer, C.; Darrell, T.; Xie, S. A ConvNet for the 2020s. *arXiv* **2022**, arXiv:2201.03545. [CrossRef]
41. Liu, Z.; Lin, Y.; Cao, Y.; Hu, H.; Wei, Y.; Zhang, Z.; Lin, S.; Guo, B. Swin Transformer: Hierarchical Vision Transformer using Shifted Windows. *arXiv* **2021**, arXiv:2103.14030. [CrossRef]
42. Jaegle, A.; Gimeno, F.; Brock, A.; Zisserman, A.; Vinyals, O.; Carreira, J. Perceiver: General Perception with Iterative Attention. *arXiv* **2021**, arXiv:2103.03206. [CrossRef]
43. Li, D.; Hu, J.; Wang, C.; Li, X.; She, Q.; Zhu, L.; Zhang, T.; Chen, Q. Involution: Inverting the Inherence of Convolution for Visual Recognition. *arXiv* **2021**, arXiv:2103.06255. [CrossRef]
44. Dai, Z.; Liu, H.; Le, Q.V.; Tan, M. CoAtNet: Marrying Convolution and Attention for All Data Sizes. *arXiv* **2021**, arXiv:2106.04803. [CrossRef]
45. Hassani, A.; Walton, S.; Shah, N.; Abuduweili, A.; Li, J.; Shi, H. Escaping the Big Data Paradigm with Compact Transformers. *arXiv* **2021**, arXiv:2104.05704. [CrossRef]
46. Tolstikhin, I.; Houlsby, N.; Kolesnikov, A.; Beyer, L.; Zhai, X.; Unterthiner, T.; Yung, J.; Steiner, A.; Keysers, D.; Uszkoreit, J.; et al. MLP-Mixer: An all-MLP Architecture for Vision. *arXiv* **2021**, arXiv:2105.01601. [CrossRef]
47. Guo, M.-H.; Liu, Z.-N.; Mu, T.-J.; Hu, S.-M. Beyond Self-attention: External Attention using Two Linear Layers for Visual Tasks. *arXiv* **2021**, arXiv:2105.02358. [CrossRef]
48. Liu, H.; Dai, Z.; So, D.R.; Le, Q.V. Pay Attention to MLPs. *arXiv* **2021**, arXiv:2105.08050.
49. Apostolopoulos, I.D.; Aznaouridis, S.; Tzani, M. An Attention-Based Deep Convolutional Neural Network for Brain Tumor and Disorder Classification and Grading in Magnetic Resonance Imaging. *Information* **2023**, *14*, 174. [CrossRef]
50. Mureşan, H.; Oltean, M. Fruit recognition from images using deep learning. *Acta Univ. Sapientiae Inform.* **2018**, *10*, 26–42. [CrossRef]
51. Pujari, J.D.; Yakkundimath, R.; Byadgi, A.S. Recognition and classification of Produce affected by identically looking Powdery Mildew disease. *Acta Technol. Agric.* **2014**, *17*, 29–34. [CrossRef]
52. Khan, M.A.; Akram, T.; Sharif, M.; Awais, M.; Javed, K.; Ali, H.; Saba, T. CCDF: Automatic system for segmentation and recognition of fruit crops diseases based on correlation coefficient and deep CNN features. *Comput. Electron. Agric.* **2018**, *155*, 220–236. [CrossRef]

AI MDPI

Article

Comparison of Various Nitrogen and Water Dual Stress Effects for Predicting Relative Water Content and Nitrogen Content in Maize Plants through Hyperspectral Imaging

Hideki Maki [1,2,3], Valerie Lynch [4], Dongdong Ma [1], Mitchell R. Tuinstra [4], Masanori Yamasaki [2,†] and Jian Jin [1,*]

1 Department of Agricultural and Biological Engineering, Purdue University, West Lafayette, IN 47907, USA; makih3@sc.sumitomo-chem.co.jp (H.M.)
2 Food Resources Education and Research Center, Graduate School of Agricultural Science, Kobe University, Kasai 675-2103, Japan; yamasakimn@agr.niigata-u.ac.jp
3 Health and Crop Sciences Research Laboratory, Sumitomo Chemical Co., Ltd., Takarazuka 665-8555, Japan
4 Department of Agronomy, Purdue University, West Lafayette, IN 47907, USA; mtuinstr@purdue.edu (M.R.T.)
* Correspondence: jinjian@purdue.edu; Tel.: +1-765-494-1182
† Present address: Graduate School of Science and Technology, Niigata University, Niigata 950-2181, Japan.

Abstract: Water and nitrogen (N) are major factors in plant growth and agricultural production. However, these are often confounded and produce overlapping symptoms of plant stress. The objective of this study is to verify whether the different levels of N treatment influence water status prediction and vice versa with hyperspectral modeling. We cultivated 108 maize plants in a greenhouse under three-level N treatments in combination with three-level water treatments. Hyperspectral images were collected from those plants, then Relative Water Content (RWC), as well as N content, was measured as ground truth. A Partial Least Squares (PLS) regression analysis was used to build prediction models for RWC and N content. Then, their accuracy and robustness were compared according to the different N treatment datasets and different water treatment datasets, respectively. The results demonstrated that the PLS prediction for RWC using hyperspectral data was impacted by N stress difference (Ratio of Performance to Deviation; RPD from 0.87 to 2.27). Furthermore, the dataset with water and N dual stresses improved model accuracy and robustness (RPD from 1.69 to 2.64). Conversely, the PLS prediction for N content was found to be robust against water stress difference (RPD from 2.33 to 3.06). In conclusion, we suggest that water and N dual treatments can be helpful in building models with wide applicability and high accuracy for evaluating plant water status such as RWC.

Keywords: nitrogen treatment; drought stress; hyperspectral camera; plant phenotyping; partial least squares regression

1. Introduction

Water and nitrogen (N) are major factors in plant growth and agricultural production. Water is necessary for normal plant growth and development. Drought is one of the major constraints limiting crop production [1] and many actual cases have been reported [2–4]. Drought stress reduces leaf size, stem extension, and root proliferation, disturbs plant water relations, and reduces water-use efficiency [5]. Drought stress also reduces photosynthesis in a variety of ways. For example, drought progressively decreases CO_2 assimilation rates due to reduced stomatal conductance [6]. Moreover, it induces a reduction in the contents and activities of photosynthetic carbon reduction cycle enzymes, including the key enzyme, ribulose-1,5-bisphosphate carboxylase/oxygenase, Rubisco [6].

N is one of the most important nutrients for a plant. Organically bound N is used to build all amino acids, amines, ureides, peptides, and proteins in plants [7]. Much of the plant response to N deficiency is centered on photosynthetic elements. One study showed that N deficiency leads to a breakdown of proteins in wheat chloroplasts, peroxisomes, and cytosol [8]. Chloroplasts contain N involved in photosynthesis as soluble protein, dominated by the enzyme Rubisco, and protein in the thylakoid membranes [9]. Another study determined that N deficiency decreased the leaf area, chlorophyll content, and photosynthetic rate, ultimately resulting in lower dry matter accumulation of sorghum plants [10]. The decreased photosynthetic rate due to N deficiency was mainly associated with the decreases in stomatal conductance and intercellular CO_2 concentration [10].

For the reasons described above, water and N condition monitoring in plants play an important role for plant science, breeding, and agricultural production purposes. There are a variety of ways to analyze the water and N status of plants [11,12]. However, most of these analyses are slow, labor-intensive, and destructive to the plant. Remote sensing of plant water and N status could solve these problems. Particularly, hyperspectral imaging is one of the promising tools to evaluate plant traits of interest. Compared to traditional RGB (Red, Green, and Blue) images, hyperspectral images are able to collect much more information from hundreds of narrow spectral bands. Water and N content prediction models can be built using a bunch of spectral information. Researchers have used hyperspectral imaging to evaluate the water content [13–17] or the N content [18–21] of various plant species. Several previous works have combined water and nutrient stresses in the experiment, but only demonstrated that hyperspectral data could be used to predict water content and nutrient content, respectively [22–25].

Water and N stresses are confounded, and the application of one stress affects the other stress. Previous biological studies have determined that water and N stresses are involved in each other's expression of plant phenotypes. One of the large contributors to a decrease in growth under drought stress is N deficiency [26]. In one drought stress study, the plant N content decreased by almost 4% [27]. In other studies, drought reduced plant growth by reducing N uptake, transport, and distribution [28]. Within the plant, as a consequence of the link between nutrient mass flow and transpiration, nutrient availability, particularly that of nitrate, partially regulates plant water flux [29]. However, as described above, there are not enough previous hyperspectral imaging studies of combined water and N stresses, and there is no direct discussion about the impact of dual stresses to predict water and N content in plants.

Our main contributions and novelty in this study are as follows:

- In order to investigate how the popular imaging models confuse between different types of plant stresses, we conducted hyperspectral imaging for maize plants under three levels of N treatments inter-leaved with three levels of water treatments.
- Partial Least Squares (PLS) models to predict Relative Water Content (RWC) and N content were developed. We compared their accuracy and robustness according to the different N and water treatment datasets.
- RWC prediction using hyperspectral data was severely impacted by N stress difference. Conversely, N content prediction was robust against water stress difference.
- The new models developed from the inter-leaved assay significantly helped to relieve the impacts on RWC prediction.

This paper is organized as follows: Following the introduction, Section 2 briefly reviews previous works related to water and N prediction models using hyperspectral imaging. Section 3 describes the materials and methods and Section 4 provides the experimental results. Section 5 discusses experimental results and the limitation of this study, and finally, Section 6 concludes with some remarks and future works.

2. Related Work

Water and N condition monitoring can be performed in a variety of ways. Plant physiological studies commonly measure water and N content using plant tissue as a sample [11,12]. Most of these analyses are time-consuming, labor-intensive, and inherently destructive to the plant, making it impossible to measure the same plant multiple times during its growth period. For example, RWC is one of the most commonly used measurements to evaluate water status in plants [12].

Remote sensing of plant water and N status could solve these problems. Vegetation indices are often used to evaluate them in the field. Normalized Difference Vegetation Index (NDVI) is one of the most common vegetation indices and is calculated with reflectance in red and infrared bands or is measured by a spectroradiometer. Yousfi et al. [30] used NDVI measured by a spectroradiometer to evaluate water status in the turfgrass field. Colovic et al. [31] also used NDVI and other vegetation indices calculated with reflectance data from a hyperspectral image to assess water and N status. Using the vegetation index is an easy and quick, especially for field measurements, but difficult way to predict the actual content of water and N.

Image-based plant phenotyping offers a high-throughput and non-destructive method that can significantly reduce time and labor costs. It is also a repeatable way to measure the same plant sample. A hyperspectral camera is gradually becoming more common in the plant research area. Traditional RGB images capture data on light intensities in three spectral bands, whereas hyperspectral images are able to collect much more information from hundreds of narrow spectral bands. Using a bunch of spectral information from hyperspectral imaging, models to predict the water and N contents of the plant can be built. In previous studies, hyperspectral imaging has successfully evaluated the water content such as in maize [13], wheat [14], soybean [15], rice [16], and grapevine [17], and N content evaluation has been also reported in sugarcane [18], wheat [19], cucumber [20], and oilseed rape [21].

Several prior works have combined water and nutrient stresses in their experiments. Pandey et al. [22] combined different levels of water and nutrient stress in maize and soybean. Then, water content and nutrient content (N, phosphorus, potassium, sulfur, and so on) were predicted respectively. Similar studies of combined water and N stresses only demonstrated that hyperspectral data could be used to predict water content and N content, respectively, in wheat [23], spinach [24], and sorghum and maize [25].

However, there are not enough previous hyperspectral imaging studies of combined water and N stresses, and there is no direct discussion about how the popular imaging models confuse between different types of plant stresses. This work fills the gap in previous works by conducting hyperspectral imaging for maize plants under three levels of N treatments inter-leaved with three levels of water treatments to investigate the impact of dual stresses to predict water and N content in plants.

3. Materials and Methods

3.1. Hyperspectral Imaging System

The imaging system consisted of an L-shaped cabinet with top-view and side-view hyperspectral cameras and lighting mounted inside. The cameras were 5.5 megapixels sCMOS visible near-infrared hyperspectral sensors (MSV-500, Middleton Spectral Vision, Middleton, WI, USA). The spectral resolution of the sensor was approximately 0.64 nm, and there were 969 bands in the spectral range 400–1019 nm. Inside the imaging tower, there were eight 500 W halogen lamps to provide lighting: four around the top-view camera and four around the side-view camera.

3.2. Experimental Design

Plant growth and data collection were conducted in the Purdue University Lily greenhouse from January to February 2017. A total of 108 maize (*Zea mays*, B73xMo17 hybrid) plants were grown in 5.7 L pots. B73xMo17 is an F1 hybrid. The Dr. Tuinstra Lab produced seeds of B73xMo17 by crossing the public inbred lines B73 and Mo17 at Purdue University. Seeds were used without any pretreatment. Pots were filled with a 50/50 mixture of sand and Turface AthleticsTM MVP® (PROFILE Products LLC, Buffalo Grove, IL, USA). All maize plants were initially given one round 24-8-16 commercial fertilizer (Miracle-Gro® Water Soluble All Purpose Plant Food, The Scotts Miracle-Gro Company, Marysville, OH,

USA). The greenhouse was set to 23–29 °C and supplemental lighting was on 12 h a day. Each pot had an irrigation line that delivered 500 mL water to each plant every morning.

Maize plants were grown under 3 different N treatments by 3 different water treatments and therefore 9 different treatment groups (Table 1). Treatments were assigned to pots following a randomized complete block design with 12 replications.

Table 1. N and water treatment combinations.

N Treatment	Water Treatment	Number of Plants		
		Training	Test	Total
Low N	Watered	10	2	12
	Low Drought	10	2	12
	Drought	10	2	12
Medium N	Watered	10	2	12
	Low Drought	10	2	12
	Drought	10	2	12
High N	Watered	10	2	12
	Low Drought	10	2	12
	Drought	10	2	12
(Total)		90	18	108

N treatments were started from the fourth week after sowing (Figure 1b). Plants were in the vegetative growth stage three or four at this point in time. Plants were fertilized weekly with 250 mL of 3 different modified Hoagland's solutions, which contain different levels of N (0, 3, and 20 mM ammonium nitrate). The original Hoagland solution was developed in 1938 [32] and revised in 1950 [33]. The solutions used in this experiment were modified from the original recipe in order to provide all macronutrients and micronutrients with three different levels of N.

(a) (b) (c)

Figure 1. The growth of maize plants on 2nd, 4th, and 7th week after sowing: (**a**) 2nd week, (**b**) 4th week, (**c**) 7th week.

While the N treatments were applied over the entirety of plant growth, water treatments were applied for a short period of time. At the vegetative growth stage seven, the plants assigned to receive low drought and drought treatments were withheld from watering for 3 or 4 days before data collection day. The other plants assigned to the watered treatment were kept watered every morning.

3.3. Image Acquisition and Sample Collection

Data collection was performed at the vegetative growth stage seven on 16 February 2017 (Figure 1c). All the plants were placed inside the imaging tower and scanned one by one from top-view. The distance from the camera to the plant was 2.3 m. To calibrate the acquired images, a flat polyvinyl chloride (PVC) panel was scanned as the white reference. The white reference calibration was used to minimize the effect of the uneven intensity of the lighting source in different bands [34,35].

After image acquisition, a small section of the top collared leaf was taken to measure relative water content (RWC). Then, the rest of the top-collared leaf was collected for N content analysis. The collected RWC and N content data were used as the ground truth data in a later hyperspectral image-based prediction model.

3.4. Relative Water Content Measurements

RWC was measured to estimate the water status of the leaf samples as described by Turner [12]. In this study, a piece of the top collared leaf sampled for RWC measurement was approximately 2.5 cm × 5.0 cm and weighed to obtain the fresh weight (FW). The samples were immediately submerged in deionized water overnight to ensure the tissue was fully turgid and weighed to obtain the turgid weight (TW). Finally, all samples were fully dried in a 60 °C dryer and weighed to obtain the dried weight (DW). These values were used to determine the RWC using the equation below.

$$RWC\ (\%) = [(FW - DW)/(TW - DW)] \times 100 \tag{1}$$

3.5. Nitrogen Content Measurements

The harvested top-collared leaf materials were dried in a 60 °C dryer for a week. Dried leaf materials were ground and stored at room temperature, and subsequently, N concentration of these samples was analyzed using a Thermo Scientific FlashEA® 1112 Nitrogen and Carbon Analyzer for Soils, Sediments, and Filters (Thermo Fisher Scientific, Waltham, MA, USA) based on the flash dynamic combustion method [25].

3.6. Image Processing

Figure 2 shows whole image processing steps. To process the hyperspectral images, the raw images were first calibrated with a white reference. The calibrated hyperspectral images were processed using a segmentation procedure with simple convolution methodology that was the same as used in our previous study. [34]. A vector of sequential integers from −20 to 20 was multiplied by the reflectance intensity vector from the red-edge region (680–720 nm). By choosing threshold 7 as the boundary between plant tissue and the background, the plant was successfully segmented out (Figure 3).

The hyperspectral imaging produced a reflectance data point at every pixel of every plant at every wavelength. After segmentation, the pixels representing plant tissues were extracted from the hyperspectral images. Then, the reflectance spectrum of each plant pixel was calculated from those extracted plant images. The average of all the pixels was calculated for every plant at every wavelength and this average was used to represent the plant at a specific wavelength. The spectral resolution was approximately 0.64 nm, and there were 969 bands in the spectral range 400–1019 nm. We used 777 wavebands (from 450 to 962 nm) instead of 969 bands (from 400 to 1019 nm) due to the low signal-to-noise ratio in the two tails of the spectra. Matlab R2016a (MATLAB 9.0) software (The MathWorks Inc., Natick, MA, USA) was used to develop image processing algorithms.

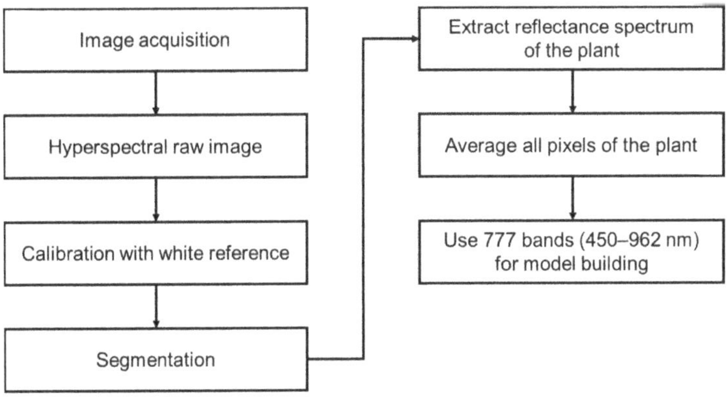

Figure 2. The diagram of whole image processing steps.

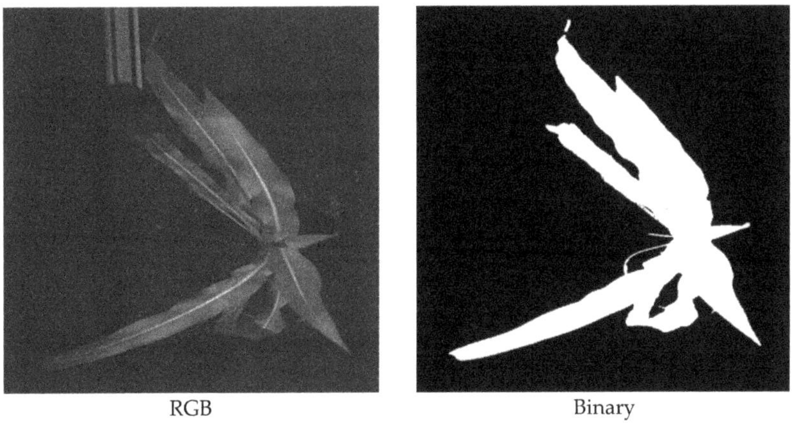

Figure 3. The top view images of maize in the hyperspectral imaging tower and the segmented binary images.

3.7. Data Analysis

PLS is a widely used predictive method for regression and has been applied in previous studies to analyze hyperspectral images for estimating plant water status [13,22,36] and N content [19,37]. In this study, PLS was used to model and predict RWC and N content. The final dataset contained 108 maize samples reflectance at every wavelength along with RWC and N content ground truth collections.

PLS modeling was carried out in Matlab R2016a software and PLS Toolbox (Eigenvector Research Inc., Manson, WA, USA). The preprocessing steps were first introduced and applied to spectra features to remove scaling effects and so on. All 108 samples were separated into two groups as shown in Table 1: a training dataset with 90 samples for model building and calibration, and a test dataset with 18 samples for validating purposes. Leave-one-out cross-validation was used for model calibration. The number of components was determined by choosing those that gave the first local minimum in Root Mean Squared Error of Cross Validation ($RMSE_{CV}$).

To determine how N treatment influences RWC prediction by hyperspectral imaging, three PLS models were built using three different sub-datasets. Each sub-dataset represented an individual N treatment and was referred to as Low-N, Medium-N, and High-N. Each sub-dataset consisted of 30 samples with one N level but varying water stress levels (Watered, Low drought, and Drought) from 90 samples of training dataset. Then, these models were applied to the test dataset including all N and water level combinations, to predict RWC. In addition, another three models were developed with another three sub-datasets named R1, R2, and R3. A total of 30 samples were chosen for each sub-dataset by randomly dividing all 90 samples of the whole training dataset into 3 groups. These models were also applied to the same test dataset to predict RWC.

Similar approaches were used to investigate the water treatment effect on N content prediction. Three PLS models for N content prediction were built using three different sub-datasets. Each sub-dataset represented individual water treatment and was referred to as Watered, Low drought, or Drought. Each sub-dataset consisted of 30 samples, which have one water level with varying N stress levels (Low-N, Medium-N, and High-N) from 90 samples of training dataset. Then, these models were applied to the test dataset including all water and N level combinations, to predict N content. Then, another 3 prediction models for N content were developed with R1, R2, and R3 sub-dataset. These models were also applied to the same test dataset to predict N content.

To assess the model performance, the following three indicators were calculated for both the training and testing datasets: Coefficient of Determination (R2) between measured and predicted values; Root Mean Square Error (RMSE), defined as the Equation (2) in which y_i are measured values and $\hat{y}_i$ are predicted values; and Ratio of Performance to Deviation (RPD), calculated as the ratio of the Standard Deviation (SD) of the measured values of the samples to the RMSE using Equation (3).

$$\text{RMSE} = \sqrt{\frac{1}{\text{N}} \times \sum(\text{y}_\text{i} - \hat{\text{y}}_\text{i})^2} \tag{2}$$

$$\text{RPD} = \text{SD}/\text{RMSE} \tag{3}$$

3.8. Statistical Analysis

A two-way analysis of variance (ANOVA) was carried out to determine the main effect of each N and water treatment as well as their interaction effect on RWC ground truth and on N content ground truth, respectively. These ground truth data were checked for normality before the ANOVA. Data were analyzed in R statistical environment [38].

4. Results

4.1. RWC and N Content Ground Truth

Figure 4 shows the range of RWC levels measured in maize leaves under all different N and water treatment combinations. The RWC ranged from 54.3% to 98.9% of the leaf section. In each N treatment, watered plants had consistently high RWCs close to 100% while both the low drought and drought treatment plants had RWCs in the clearly wider range toward the lower side. There was a significantly clear water treatment effect as well as an N treatment effect shown in the ANOVA, but no significant water × N interaction effect was found for RWC ground truth (Table 2).

Figure 5 illustrates the range of N content levels measured in maize leaves under all different N and water treatment combinations. The N content ranged from 1.11% to 2.96% of the leaf section. In each water treatment, plants grown with higher N treatment tended to have higher N content in the leaves. As in the case with RWC, the ANOVA confirmed that N content ground truth was significantly affected by N treatment as well as water treatment, but no significant water × N interaction effect was found for N content ground truth (Table 2).

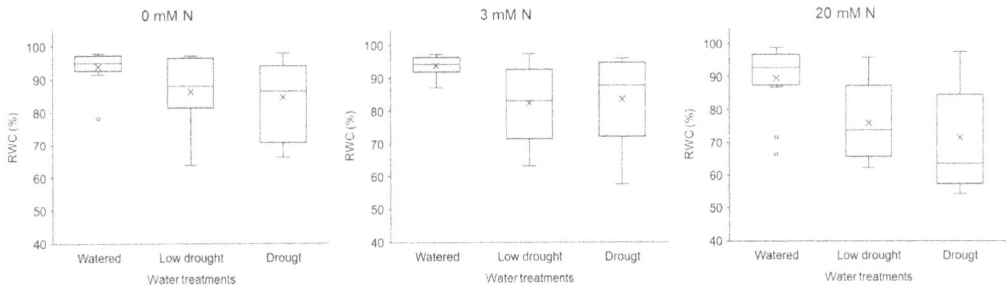

Figure 4. Boxplots of RWC in maize plant leaves. X indicates mean values and circle indicates outliers in the boxplots.

Table 2. Two-way ANOVA results of the effect of N and water treatments on RWC and N content.

Factor	Df	ANOVA on RWC			ANOVA on N Content		
		F	*p*		*F*	*p*	
Nitrogen	2	6.864	0.002	**	98.785	<0.001	***
Water	2	13.044	<0.001	***	5.136	0.008	**
Nitrogen × Water	4	0.634	0.640		1.013	0.405	

Note: Significant levels are indicated with ** $p < 0.01$, *** $p < 0.001$.

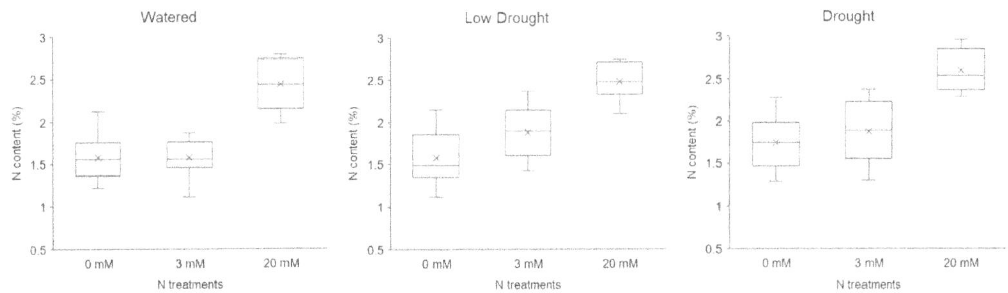

Figure 5. Boxplots of N content in maize plant leaves. X indicates mean values in the boxplots.

4.2. PLS Modeling for RWC Prediction

N and water stresses are often confounded and produce overlapping symptoms of plant stress. To determine how N treatment influences RWC prediction by hyperspectral imaging, three PLS models were built using three different N treatment sub-datasets. Figure 6 shows the relationships between measured and predicted RWC. The predicted RWC from each model was plotted against the reference RWC. Table 3 provides detailed statistical results from the different N-level training dataset approach. For comparison purposes, another three models (R1, R2, and R3) using randomly selected datasets were applied. Then, the predicted value was plotted against the reference value in Figure 5, and the statistical results are shown in Table 3.

4.3. PLS Modeling for N Content Prediction

Similar approaches were tested to investigate the water treatment effect on N content prediction. Three PLS models for N content prediction were built using three different water treatment sub-datasets. Figure 7 shows the relationships between the measured and predicted N content. The predicted N content from each model was plotted against the reference N content. Table 4 provides detailed statistical results from different water-level training dataset approaches. In addition, another three prediction models (R1, R2, and R3) using randomly selected datasets were also applied. Then, the predicted value was plotted against the reference value in Figure 6, and the statistical results are shown in Table 4.

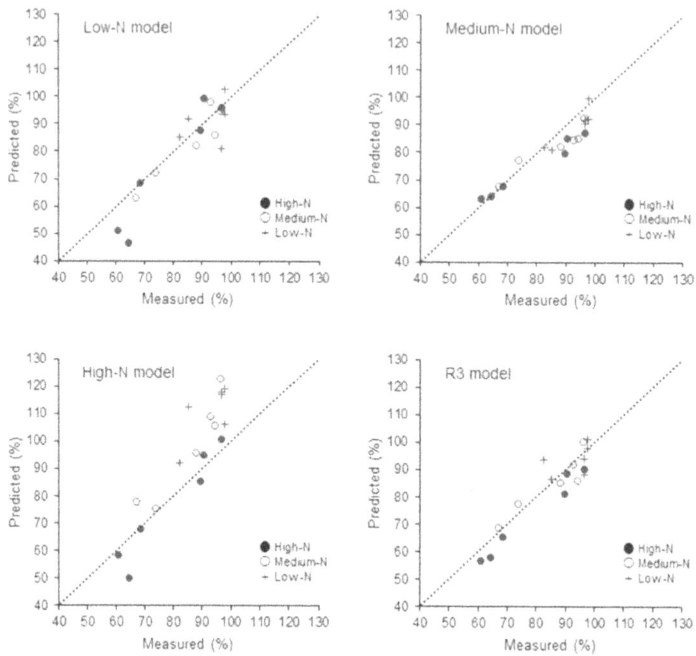

Figure 6. Scatterplots of measured RWC vs. predicted RWC for the test set (n = 18).

Table 3. Results of predicting RWC of maize plants using hyperspectral imaging with different datasets.

Sub-Dataset			Training (n = 30)			Test (n = 18)		
Approach	Label	PLS Comp	R^2	$RMSE_{CV}$	RPD	R^2	$RMSE_V$	RPD
Individual water treatment	Watered	8	0.698	6.01	1.80	0.831	7.35	1.70
	Low drought	4	0.632	6.96	1.63	0.911	5.50	2.27
	Drought	5	0.849	5.73	2.56	0.832	14.44	0.87
Randomly selected	R1	7	0.757	7.36	2.00	0.934	4.73	2.64
	R2	4	0.623	7.10	1.61	0.723	7.38	1.69
	R3	4	0.728	6.27	1.90	0.854	5.27	2.37

Table 4. Results of predicting N content of maize plants using hyperspectral imaging with different datasets.

Sub-Dataset			Training (n = 30)			Test (n = 18)		
Approach	Label	PLS Comp	R^2	$RMSE_{CV}$	RPD	R^2	$RMSE_V$	RPD
Individual water treatment	Watered	1	0.799	0.204	2.23	0.896	0.170	3.06
	Low drought	3	0.809	0.193	2.28	0.842	0.210	2.47
	Drought	5	0.820	0.204	2.34	0.835	0.223	2.33
Randomly selected	R1	4	0.772	0.232	2.08	0.853	0.212	2.45
	R2	1	0.800	0.170	2.24	0.896	0.172	3.02
	R3	3	0.854	0.197	2.59	0.863	0.206	2.52

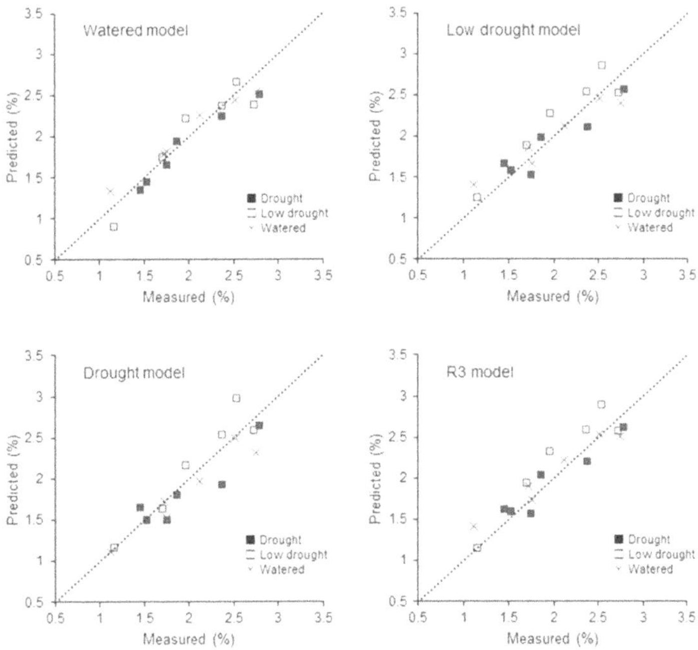

Figure 7. Scatterplots of measured N content vs. predicted N content for the test set (n = 18).

5. Discussion

5.1. Effects of N and Water Treatments on RWC and N Content Ground Truth

There was a significantly clear water treatment effect as well as an N treatment effect shown in the ANOVA, but no significant water × N interaction effect was found for RWC ground truth (Table 2). This means that both water and N treatment certainly affected RWC ground truth, but each treatment worked independently and there was no positive or negative synergistic effect. As in the case with RWC, the ANOVA confirmed that N content ground truth was significantly affected by N treatment as well as water treatment, but no significant water × N interaction effect was found for N content ground truth (Table 2). This means both N and water treatment certainly affected N content ground truth, but each treatment worked independently and there was no positive or negative synergistic effect. These results indicated that the design of this experiment was successful and created a large variation in plant RWC and N content across the water and N treatments. This also provided a reliable base to examine both the N treatment effect for RWC prediction and the water treatment effect for N content prediction with hyperspectral images.

5.2. N Treatment Effect on PLS Modeling for RWC Prediction

N and water stresses are often confounded and produce overlapping symptoms of plant stress. To determine how N treatment influences RWC prediction by hyperspectral imaging, three PLS models were built using three different N treatment sub-datasets.

Figure 6 shows the relationships between measured and predicted RWC. In the case of RWC prediction with the High-N base model, many of the lower N-level plants (crosses and open circles) were plotted above the 1:1 line. This means lower N levels cause the predictions to be higher than the reference measurements. The Low-N base model displayed an opposite trend. Some of the higher N-level plants (open and filled circles) tended to be plotted below the 1:1 line. This means that higher N levels cause the predictions to be lower than the reference measurements. The Medium-N base model seems to have no bias.

Table 3 provided detailed statistical results from different N-level training dataset approaches. RMSE is a good measure of how accurately the model predicted the response, and it is the most important criterion for fit if the main purpose of the model was prediction. Among the three individual N treatment approaches, the accuracy of RWC prediction fluctuated depending on the N treatment level. The Medium-N model showed the highest accuracy (lowest RMSE) and the High-N-based model showed the lowest accuracy (highest RMSE). RPD also showed a similar trend. RPD is widely used as a criterion to evaluate the usefulness of prediction models in analytical chemistry and chemometric modeling [39,40]. Ge et al. [13] and Pandey et al. [22] used RPD to assess their hyperspectral models to predict the water content as well as the nutrient concentration of plants. There were three categories: (A). good models RPD > 2.0; (B). fair models RPD = 1.4~2.0; and (C). non-reliable models RPD < 1.4 [39]. Following this category standard, only the Medium-N model belonged to good models. Conversely, the High-N model indicated non-reliable model performance. These results demonstrated that RWC prediction models using hyperspectral data are potentially impacted by N stress. We expected that using the dataset with water and N dual stresses can help to improve the accuracy and robustness of the model.

To confirm this idea, another three models (R1, R2, and R3) using randomly selected datasets were applied. The R3 scatterplot in Figure 5 showed less bias. Samples from Low-N, Medium-N, and High-N were plotted uniformly around the 1:1 line. $RMSE_V$ and RPD in Table 3 also looked improved. In conclusion, the dataset with water and N dual stresses can help to build an accurate and robust PLS model for RWC prediction by using hyperspectral data.

5.3. Water Treatment Effect on PLS Modeling for N Content Prediction N

Correspondingly, three PLS models for N content prediction were built using three different water treatment sub-datasets to investigate the water treatment effect on N content prediction.

Figure 7 shows the relationships between measured and predicted N content. As contrasted with the RWC prediction in Figure 6, the scatter of crosses (samples from Watered), open squares (samples from Low drought), and filled squares (samples from Drought) around the 1:1 line were quite consistent. There was no overestimation or underestimation of the difference in water treatment. Table 4 provides detailed statistical results from different water-level training dataset approaches. The results of RMSE and RPD for N content prediction also showed steady performance regardless of the water level difference in model building.

In addition, another three prediction models (R1, R2, and R3) using randomly selected datasets were also applied. Figure 7 and Table 4 show that there was no significant improvement in N content prediction by using the dataset with water and N dual stresses. We expected the water and N stress phenotypes to overlap each other because N deficiency as a result of drought is the largest contributor to the decrease in growth under drought stress [26]. However, in contrast to the N treatment impact on RWC prediction, the PLS prediction for N content was found to be robust against the water status difference in plants.

5.4. The Limitation of This Study

Though the importance of growing plants under both water and N stresses has been suggested, further study is still needed to improve the model applicability and accuracy of plant water status estimation. For example, the two types of stresses may show significantly different color distribution patterns between the different leaf areas such as the base, edge, tip, mid-rib, secondary veins, and mesophyll. Analysis of these color distribution patterns may provide more accurate models to clearly differentiate between N and water stresses. Purdue University's LeafSpec hyperspectral leaf imager [41] is a new sensor technology providing high-quality single-leaf hyperspectral images with 0.5 mm resolution, which enables such spatial stress distribution analyses. Preliminary studies with this 2021 Davidson Prize winner technology have already shown advantages in accurate plant stress detection [42].

6. Conclusions

In this study, we investigated how N stress influences water status predictions, as well as how water stress affects N status predictions through hyperspectral imaging. We conducted hyperspectral imaging for maize plants under three-level N treatments inter-leaved with three-level water treatments. Then, PLS models to predict RWC and N content were developed and their accuracy and robustness were compared according to the different N treatment datasets and different water treatment datasets, respectively. The results demonstrated that the PLS prediction for RWC using hyperspectral data was potentially impacted by the N stress difference. Furthermore, the dataset with water and N dual stresses improved model accuracy and robustness. Conversely, the PLS prediction for N content was found to be robust against the water stress difference. In conclusion, we suggested that water and N dual treatments can be helpful in building models with wide applicability and high accuracy for evaluating plant water status, such as RWC.

In the future, other limitation factors should be included to further improve the model's applicability and accuracy. We conducted similar hyperspectral plant phenotyping research previously in terms of multi-species [25] and different imaging angles [34]. These findings should be combined to develop a broader solution. We also plan to include our new sensor technology, the LeafSpec hyperspectral imager [41], to detect single-leaf high-quality stress distribution. This approach could provide high-quality predictions of plant conditions.

Author Contributions: Conceptualization, H.M. and J.J.; methodology, H.M., V.L., D.M., M.R.T. and J.J.; software, H.M. and D.M.; validation, H.M.; formal analysis, H.M.; investigation, H.M. and V.L.; resources, M.R.T. and J.J.; data curation, H.M. and D.M.; writing—original draft preparation, H.M.; writing—review and editing, H.M., M.Y. and J.J.; visualization, H.M. and D.M.; supervision, M.R.T., M.Y. and J.J.; project administration, J.J.; funding acquisition, J.J. All authors have read and agreed to the published version of the manuscript.

Funding: This research was funded by Sumitomo Chemical, grant number 16121941.

Institutional Review Board Statement: Not applicable.

Informed Consent Statement: Not applicable.

Data Availability Statement: The data presented in this study are available on request from the corresponding author. The data are not publicly available due to confidentiality.

Acknowledgments: We are grateful to Sylvie Brouder for help in analyzing the nitrogen content, and Michael Mickelbart for help in analyzing the relative water content.

Conflicts of Interest: The authors declare no conflict of interest.

References

1. Farooq, M.; Hussain, M.; Wahid, A.; Siddique, K.H.M. Drought Stress in Plants: An Overview. In *Plant Responses to Drought Stress*; Springer: Berlin/Heidelberg, Germany, 2012; pp. 1–33.
2. Jamieson, P.D.; Martin, R.J.; Francis, G.S. Drought influences on grain yield of barley, wheat, and maize. *New Zealand J. Crop. Hortic. Sci.* **1995**, *23*, 55–66. [CrossRef]
3. Shao, H.B.; Chu, L.Y.; Jaleel, C.A.; Manivannan, P.; Panneerselvam, R.; Shao, M.A. Understanding Water Deficit Stress-Induced Changes in the Basic Metabolism of Higher Plants-Biotechnologically and Sustainably Improving Agri-culture and the Ecoenvironment in Arid Regions of the Globe. *Crit. Rev. Biotechnol.* **2009**, *29*, 131–151. [CrossRef] [PubMed]
4. Harrison, M.T.; Tardieu, F.; Dong, Z.; Messina, C.D.; Hammer, G.L. Characterizing drought stress and trait influence on maize yield under current and future conditions. *Glob. Chang. Biol.* **2014**, *20*, 867–878. [CrossRef] [PubMed]
5. Farooq, M.; Wahid, A.; Kobayashi, N.; Fujita, D.; Basra, S.M.A. Plant drought stress: Effects, mechanisms and management. *Agron. Sustain. Dev.* **2009**, *29*, 185–212. [CrossRef]
6. Reddy, A.R.; Chaitanya, K.V.; Vivekanandan, M. Drought-induced responses of photosynthesis and antioxidant metabolism in higher plants. *J. Plant Physiol.* **2004**, *161*, 1189–1202. [CrossRef]

7. Marschner, H. Functions of Mineral Nutrients; Macronutrients. In *Mineral Nutrition of Higher Plants*; Academic Press: Cambridge, MA, USA, 1995; pp. 229–312.
8. Crafts-Brandner, S.J.; Hölzer, R.; Feller, U. Influence of nitrogen deficiency on senescence and the amounts of RNA and proteins in wheat leaves. *Physiol. Plant.* **1998**, *102*, 192–200. [CrossRef]
9. Evans, J.R. Photosynthesis and nitrogen relationships in leaves of C3 plants. *Oecologia* **1989**, *78*, 9–19. [CrossRef] [PubMed]
10. Zhao, D.; Reddy, K.R.; Kakani, V.G.; Reddy, V.R. Nitrogen deficiency effects on plant growth, leaf photosynthesis, and hyperspectral reflectance properties of sorghum. *Eur. J. Agron.* **2005**, *22*, 391–403. [CrossRef]
11. Muñoz-Huerta, R.F.; Guevara-Gonzalez, R.G.; Contreras-Medina, L.M.; Torres-Pacheco, I.; Prado-Olivarez, J.; Ocampo-Velazquez, R.V. A Review of Methods for Sensing the Nitrogen Status in Plants: Advantages, Disadvantages and Recent Advances. *Sensors* **2013**, *13*, 10823–10843. [CrossRef]
12. Turner, N.C. Techniques and experimental approaches for the measurement of plant water status. *Plant Soil* **1981**, *58*, 339–366. [CrossRef]
13. Ge, Y.; Bai, G.; Stoerger, V.; Schnable, J.C. Temporal dynamics of maize plant growth, water use, and leaf water content using automated high throughput RGB and hyperspectral imaging. *Comput. Electron. Agric.* **2016**, *127*, 625–632. [CrossRef]
14. Jin, X.; Xu, X.; Song, X.; Li, Z.; Wang, J.; Guo, W. Estimation of Leaf Water Content in Winter Wheat Using Grey Relational Analysis–Partial Least Squares Modeling with Hyperspectral Data. *Agron. J.* **2013**, *105*, 1385–1392. [CrossRef]
15. Kovar, M.; Brestic, M.; Sytar, O.; Barek, V.; Hauptvogel, P.; Zivcak, M. Evaluation of Hyperspectral Reflectance Parameters to Assess the Leaf Water Content in Soybean. *Water* **2019**, *11*, 443. [CrossRef]
16. Krishna, G.; Sahoo, R.N.; Singh, P.; Bajpai, V.; Patra, H.; Kumar, S.; Dandapani, R.; Gupta, V.K.; Viswanathan, C.; Ahmad, T.; et al. Comparison of various modelling approaches for water deficit stress monitoring in rice crop through hyperspectral remote sensing. *Agric. Water Manag.* **2019**, *213*, 231–244. [CrossRef]
17. Pôças, I.; Gonçalves, J.; Costa, P.M.; Gonçalves, I.; Pereira, L.S.; Cunha, M. Hyperspectral-based predictive modelling of grapevine water status in the Portuguese Douro wine region. *Int. J. Appl. Earth Obs. Geoinform.* **2017**, *58*, 177–190. [CrossRef]
18. Abdel-Rahman, E.M.; Ahmed, F.; Ismail, R. Random forest regression and spectral band selection for estimating sugarcane leaf nitrogen concentration using EO-1 Hyperion hyperspectral data. *Int. J. Remote Sens.* **2013**, *34*, 712–728. [CrossRef]
19. Li, F.; Mistele, B.; Hu, Y.; Chen, X.; Schmidhalter, U. Reflectance estimation of canopy nitrogen content in winter wheat using optimised hyperspectral spectral indices and partial least squares regression. *Eur. J. Agron.* **2014**, *52*, 198–209. [CrossRef]
20. Sabzi, S.; Pourdarbani, R.; Rohban, M.H.; García-Mateos, G.; Arribas, J.I. Estimation of nitrogen content in cucumber plant (*Cucumis sativus* L.) leaves using hyperspectral imaging data with neural network and partial least squares regressions. *Chemom. Intell. Lab. Syst.* **2021**, *217*, 104404. [CrossRef]
21. Yu, X.; Lu, H.; Liu, Q. Deep-learning-based regression model and hyperspectral imaging for rapid detection of nitrogen concentration in oilseed rape (*Brassica napus* L.) leaf. *Chemom. Intell. Lab. Syst.* **2018**, *172*, 188–193. [CrossRef]
22. Pandey, P.; Ge, Y.; Stoerger, V.; Schnable, J.C. High Throughput In vivo Analysis of Plant Leaf Chemical Properties Using Hyperspectral Imaging. *Front. Plant Sci.* **2017**, *8*, 1348. [CrossRef]
23. Bruning, B.; Liu, H.; Brien, C.; Berger, B.; Lewis, M.; Garnett, T. The Development of Hyperspectral Distribution Maps to Predict the Content and Distribution of Nitrogen and Water in Wheat (*Triticum aestivum*). *Front. Plant Sci.* **2019**, *10*, 1380. [CrossRef] [PubMed]
24. Corti, M.; Gallina, P.M.; Cavalli, D.; Cabassi, G. Hyperspectral imaging of spinach canopy under combined water and nitrogen stress to estimate biomass, water, and nitrogen content. *Biosyst. Eng.* **2017**, *158*, 38–50. [CrossRef]
25. Lin, M.-Y.; Lynch, V.; Ma, D.; Maki, H.; Jin, J.; Tuinstra, M. Multi-Species Prediction of Physiological Traits with Hyperspectral Modeling. *Plants* **2022**, *11*, 676. [CrossRef] [PubMed]
26. Heckathorn, S.A.; DeLucia, E.H.; Zielinski, R.E. The contribution of drought-related decreases in foliar nitrogen concentration to decreases in photosynthetic capacity during and after drought in prairie grasses. *Physiol. Plant.* **1997**, *101*, 173–182. [CrossRef]
27. He, M.; Dijkstra, F.A. Drought effect on plant nitrogen and phosphorus: A metaanalysis. *New Phytol.* **2014**, *204*, 924–931. [CrossRef]
28. Rouphael, Y.; Cardarelli, M.; Schwarz, D.; Franken, P.; Colla, G. Effects of Drought on Nutrient Uptake and Assimilation in Vegetable Crops. In *Plant Responses to Drought Stress*; Aroca, R., Ed.; Springer: Berlin/Heidelberg, Germany, 2012; pp. 171–195. [CrossRef]
29. Cramer, M.D.; Hawkins, H.-J.; Verboom, G.A. The importance of nutritional regulation of plant water flux. *Oecologia* **2009**, *161*, 15–24. [CrossRef]
30. Yousfi, S.; Marín, J.; Parra, L.; Lloret, J.; Mauri, P.V. Remote sensing devices as key methods in the advanced turfgrass phenotyping under different water regimes. *Agric. Water Manag.* **2022**, *266*, 107581. [CrossRef]
31. Colovic, M.; Yu, K.; Todorovic, M.; Cantore, V.; Hamze, M.; Albrizio, R.; Stellacci, A.M. Hyperspectral Vegetation Indices to Assess Water and Nitrogen Status of Sweet Maize Crop. *Agronomy* **2022**, *12*, 2181. [CrossRef]
32. Hoagland, D.R.; Arnon, D.I. Growing plants without soil by the water-culture method. *Circ. Calif. Agric. Exp. Stn.* **1938**, 1–16.
33. Hoagland, D.R.; Arnon, D.I. The water-culture method for growing plants without soil. *Circ. Calif. Agric. Exp. Stn.* **1950**, *347*, 1–32.
34. Zhang, L.; Maki, H.; Ma, D.; Sánchez-Gallego, J.A.; Mickelbart, M.V.; Wang, L.; Rehman, T.U.; Jin, J. Optimized angles of the swing hyperspectral imaging system for single corn plant. *Comput. Electron. Agric.* **2019**, *156*, 349–359. [CrossRef]

35. Zhao, Y.-R.; Yu, K.-Q.; He, Y. Hyperspectral Imaging Coupled with Random Frog and Calibration Models for Assessment of Total Soluble Solids in Mulberries. *J. Anal. Methods Chem.* **2015**, *2015*, 343782. [CrossRef] [PubMed]
36. Cotrozzi, L.; Couture, J.J.; Cavender-Bares, J.; Kingdon, C.C.; Fallon, B.; Pilz, G.; Pellegrini, E.; Nali, C.; Townsend, P.A. Using foliar spectral properties to assess the effects of drought on plant water potential. *Tree Physiol.* **2017**, *37*, 1582–1591. [CrossRef] [PubMed]
37. Nigon, T.J.; Mulla, D.J.; Rosen, C.J.; Cohen, Y.; Alchanatis, V.; Knight, J.; Rud, R. Hyperspectral aerial imagery for detecting nitrogen stress in two potato cultivars. *Comput. Electron. Agric.* **2015**, *112*, 36–46. [CrossRef]
38. R Core Team R: A Language and Environment for Statistical Computing. R Foundation for Statistical Computing, Vienna, Austria. 2018. Available online: https://www.R-project.org (accessed on 23 May 2023).
39. Chang, C.-W.; Laird, D.A.; Mausbach, M.J.; Hurburgh, C.R., Jr. Near-Infrared Reflectance Spectroscopy-Principal Components Regression Analyses of Soil Properties. *Soil Sci. Soc. Am. J.* **2001**, *65*, 480–490. [CrossRef]
40. Fearn, T. Assessing Calibrations: SEP, RPD, RER and R2. *NIR News* **2002**, *13*, 12–14. [CrossRef]
41. Wang, L.; Jin, J.; Song, Z.; Wang, J.; Zhang, L.; Rehman, T.U.; Ma, D.; Carpenter, N.R.; Tuinstra, M.R. LeafSpec: An accurate and portable hyperspectral corn leaf imager. *Comput. Electron. Agric.* **2020**, *169*, 105209. [CrossRef]
42. Ma, D.; Wang, L.; Zhang, L.; Song, Z.; Rehman, T.U.; Jin, J. Stress Distribution Analysis on Hyperspectral Corn Leaf Images for Improved Phenotyping Quality. *Sensors* **2020**, *20*, 3659. [CrossRef]

AI

MDPI

Article

Comparative Analysis of Machine Learning Techniques Using RGB Imaging for Nitrogen Stress Detection in Maize

Sumaira Ghazal [1], Namratha Kommineni [1] and Arslan Munir [2,*]

[1] Department of Computer Science, Kansas State University, Manhattan, KS 66506, USA; sghazal@ksu.edu (S.G.); komminen@ksu.edu (N.K.)

[2] Department of Electrical Engineering and Computer Science, Florida Atlantic University, Boca Raton, FL 33431, USA

* Correspondence: arslanm@fau.edu

Abstract: Proper nitrogen management in crops is crucial to ensure optimal growth and yield maximization. While hyperspectral imagery is often used for nitrogen status estimation in crops, it is not feasible for real-time applications due to the complexity and high cost associated with it. Much of the research utilizing RGB data for detecting nitrogen stress in plants relies on datasets obtained under laboratory settings, which limits its usability in practical applications. This study focuses on identifying nitrogen deficiency in maize crops using RGB imaging data from a publicly available dataset obtained under field conditions. We have proposed a custom-built vision transformer model for the classification of maize into three stress classes. Additionally, we have analyzed the performance of convolutional neural network models, including ResNet50, EfficientNetB0, InceptionV3, and DenseNet121, for nitrogen stress estimation. Our approach involves transfer learning with fine-tuning, adding layers tailored to our specific application. Our detailed analysis shows that while vision transformer models generalize well, they converge prematurely with a higher loss value, indicating the need for further optimization. In contrast, the fine-tuned CNN models classify the crop into stressed, non-stressed, and semi-stressed classes with higher accuracy, achieving a maximum accuracy of 97% with EfficientNetB0 as the base model. This makes our fine-tuned EfficientNetB0 model a suitable candidate for practical applications in nitrogen stress detection.

Keywords: computer vision; transfer learning; convolutional neural networks; vision transformers; nitrogen stress detection; maize

1. Introduction

Nitrogen is an important nutrient for the growth and development of plants. Naturally occurring nitrogen compounds absorbed by plants are not sufficiently available. Hence, farmers apply nitrogen fertilizer to crops to enhance nitrogen absorption and increase crop production. According to the study by Ritchie [1], 115 million tonnes of nitrogen fertilizer are annually applied to crops but only 35% is utilized by the crops. The excessive amount causes nitrogen pollution including the acidification of soil, water contamination, and nitrous oxide emissions that contribute to global warming. China is the largest contributor to this pollution, causing one third of the total pollution, followed by India with 18%, and USA with 11% of the total.

Nitrogen use efficiency (NUE) measures how much of the applied nitrogen is utilized by crops as opposed to being lost to the environment [2]. The overapplication of nitrogen fertilizer can reduce NUE. By improving farming practices using the latest technology advancements, we can work towards increasing NUE, which will increase crop yield with reduced fertilizer application. Nitrogen fertilization is necessary for better yield and crop production, but the excessive use of fertilizers decreases NUE, increases overall fertilization costs without any improvement in crop yield, and has an undesirable impact on the environment. Therefore, it is necessary to prevent the overuse of nitrogen fertilizer by the targeted application of fertilizer and its rate adjustment.

Nitrogen is critical for chlorophyll formation in plant leaves, which produces the green color of vegetation. Nitrogen-deficient plants lack sufficient chlorophyll content, so the early signs of nitrogen deficiency appear in mature leaves as slight discoloration to a lighter green color in early stages, turning leaves to yellow later and further causing the premature fading of older leaves. Apart from this, thin and weak vegetation, slow growth of stem, and a purple hue are also signs of nitrogen deficiency. In later stages, it can cause premature fading and necrosis, resulting in early plant death and crop loss. Different crops show different symptoms of nitrogen deficiency. In maize, visual symptoms include a V-shaped yellowing pattern on the leaves, smaller or poorly filled maize ears, glossy maize kernels indicating low protein content, and premature browning of stover in the late season [3]. These visual symptoms can be used to identify nitrogen deficiency in maize crop images using state-of-the-art computer vision techniques that are non-destructive and can be easily automated for faster response.

In this study, we have developed a custom-built vision transformer (ViT) model specifically designed for the classification of nitrogen stress levels in corn crops using RGB (red, green, and blue) image data. We also utilize the transfer learning approach to fine-tune some widely used classifiers for our application. Our approach focuses on leveraging the power of deep learning and vision transformers to accurately classify images into three distinct levels of nitrogen fertilization: no nitrogen applied, 75 kg nitrogen applied, and 136 kg nitrogen applied. Our contributions can be summarized as follows:

- We have designed and implemented a vision transformer model tailored for the agricultural domain, specifically for nitrogen stress detection in corn crops.
- Our developed vision and CNN models exclusively use RGB image data, making these models accessible and practical for widespread agricultural applications.
- We have classified the nitrogen stress levels into three categories, providing detailed insights into the nitrogen status of the crops. For this purpose, we have used a publicly available dataset collected under field conditions.
- We have utilized a transfer learning-based approach to fine-tune widely used vision transformer and CNN models for nitrogen stress level classification. This involved adapting pre-trained models to our specific application, enhancing their performance.
- We have performed a detailed comparative analysis using two different image resolutions (100×100 and 224×224) to evaluate the impact of image resolution on classifier accuracy and identify the best-performing classifier.

Our research focuses on nitrogen-level classification from RGB images of maize crop, which is a foundational step towards achieving the goal of developing a precision agriculture system that can accurately determine the nitrogen fertilizer requirements for specific areas within a maize crop field. By detecting different nitrogen levels in crops, detailed nitrogen deficiency maps can be generated, which can then be used to divide the field into management zones, each with specific fertilizer requirements. With the precise identification of nitrogen-deficient areas, farmers can apply fertilizers only where needed, thereby increasing NUE and reducing waste.

2. Related Works

Traditional nitrogen status analysis in plants is performed using destructive chemical testing; therefore, the research community has been focused on exploring visual analysis techniques based on optical properties and machine learning models to approach the problem of nitrogen stress detection in crops. Hyperspectral imaging is a widely used method for stress detection in crops. Being a non-destructive method, it can capture detailed spectral information across various narrow spectral bands. Nitrogen affects the overall health of plants, which is reflected in the spectral properties of their leaves. Hyperspectral imaging is sensitive to these changes related to nitrogen deficiency, and therefore, many researchers have utilized hyperspectral imaging and vegetation indices associated with spectral properties of leaves for nitrogen stress detection. Authors in [4–7] used hyperspectral remote sensing to extract nitrogen information in cucumber, hemp, potato and tea leaves. Wang et al. [8] used hyperspectral imagery data to determine nitrogen deficiency and its relation with crop yield in maize crop. The authors in [9] determined nitrogen status in maize for variable rate fertilizer application using spectral vegetation indices. Wu et al. [10] used a combination of spectral and texture-based features for estimating maize nitrogen content.

Even though hyperspectral imaging sensors have improved with time, due to the large volume of data requiring exhaustive processing and specialized expertise as well as high cost, hyperspectral imaging is not very feasible for real-time processing as compared to RGB imaging [11]. Multispectral imaging, on the other hand, is relatively feasible for real-time applications due to the lower data volume and lesser computational cost. Burns et al. [12] gave an overview of commonly used vegetation indices for assessing plant nitrogen stress. They used multispectral data to study the relationship between various vegetation indices with nitrogen stress and fertilizer rate application in maize. According to their study, chlorophyll index green (CI_{green}), the green normalized difference vegetation index (GNDVI), and the red edge normalized difference vegetation index (RENDVI) are the best vegetation indices for the detection of nitrogen deficiency in maize. Zheng et al. [13] used a five-band multispectral camera to obtain images of winter wheat using a drone during five growth stages of winter wheat. They compared various parametric and non-parametric modeling algorithms for nitrogen content determination, where parametric modeling is based on 19 vegetation indices. The best vegetation index for nitrogen content detection was reported to be the modified renormalized difference vegetation index (RDVI) with a coefficient of determination (R^2) of 0.73 and root mean squared error (RMSE) of 0.38. For non-parametric models, random forest performed best with R^2 of 0.79 and RMSE of 0.33. The authors in [14] used a combination of hyperspectral and multispectral data for nitrogen content estimation in tea plants and reported an R^2 and RMSE of 0.9186 and 0.0560, respectively.

In contrast to the multispectral and hyperspectral imaging techniques, many recent studies have explored the use of advanced machine learning algorithms, particularly convolutional neural networks (CNNs), applied to RGB imaging data for nitrogen stress detection in crops. This approach utilizes the valuable information provided by RGB images, which, when combined with the deep learning capabilities of CNNs, has shown promising results in accurately identifying nitrogen stress indicators in plants. Moreover, RGB images are more amenable for real-time processing compared to other approaches. Azimi, Kaur, and Gandhi [15] proposed a CNN for nitrogen stress indication in sorghum plant shoot images. The dataset they used for their study comprises RGB images of three classes: a healthy plants class representing 100% nitrogen treatment available, a semi-stressed class with 50% nitrogen treatment, and a severely stressed class with 10% nitrogen treatment available to the plants. They used a publicly available dataset obtained under laboratory conditions, consisting of images of plant shoots captured against a white background. They proposed a 23-layered CNN model with 5 convolutional layers, respectively, each followed by batch normalization and ReLU activation, where the first four convolutional layers are followed by max pooling layers. The fifth convolutional layer is followed by fully connected, softmax, and classification output layers. They reported a maximum accuracy of 84%.

Zermas et al. [16] proposed an annotation assistant tool, which identifies plant areas demonstrating nitrogen deficiency. The tool provides a recommendation to the annotator to create a training set that is then used to train a deep learning model. The images in their work were collected through a drone. The annotator tool uses k-means clustering and support vector machine (SVM) models in various steps to divide pixels in green (healthy), yellow (stressed), and brown (soil) pixels. This step generates a training set with bounding boxes representing nitrogen-stressed areas in images. This set is then used to train a Faster RCNN architecture with a ResNet50 feature extractor for detection. They reported a mean average precision of 82.3% and an intersection over union percentage of 50%. Zhang et al. [17] used RGB (red, green, and blue) imaging from a drone for nitrogen stress detection in rapeseed leaves during seedling stage. They used a U-Net model for purple rapeseed leaf segmentation and to establish a relationship between leaf purple area and nitrogen deficit. They reported that the purple area increases as the nitrogen application level decreases, suggesting a greater stress on the plants due to lower nitrogen levels. They determined the relationship between the purple area and nitrogen content to be negative exponential with an R^2 of 0.858, indicating a strong correlation between these factors.

Haider et al. [18] also proposed a vision-based analysis technique for nitrogen content estimation in spinach leaves. They used a specially designed board with white background and two reference colors that are represented by green and yellow circles.The first step of their methodology involves using Otsu's thresholding and the bounding box method to extract the region of interest, which includes the reference circles and the leaf. The leaf and reference circles are then distinguished from each other using the width-to-height ratio of their corresponding bounding boxes. The next step of their methodology uses the Gaussian mixture model technique for background segmentation of the leaf. Their methodology then computes the mean color value of the reference circles and leaf. The decision is made on the basis of distance of leaf mean color from the mean color of the reference circles. They reported an R^2 of 0.92 and RMSE of 0.0845.

3. Materials and Methods

Figure 1 depicts the flowchart of our methodology. The dataset utilized in this research is presented by Salaić et al. [19] and is publicly available through the Mendeley data repository [20]. The dataset is composed of 1200 images of maize canopy annotated by an agricultural expert. The images were collected in July 2023 over a 5-day period encompassing the flowering stage of maize. They were captured randomly along the rows between 7:30 a.m. and 11:00 a.m., with each image sized at 2400 × 1600 pixels. The images are categorized into three distinct classes, each representing a different nitrogen fertilization level. The classes include N0 (no fertilization), N75 (75 kg of nitrogen fertilizer applied), and NFull (136 kg of fertilizer applied), with 400 images per class. The images are captured using a digital single-lens reflex (DSLR) camera positioned at a 45° angle relative to the plots.

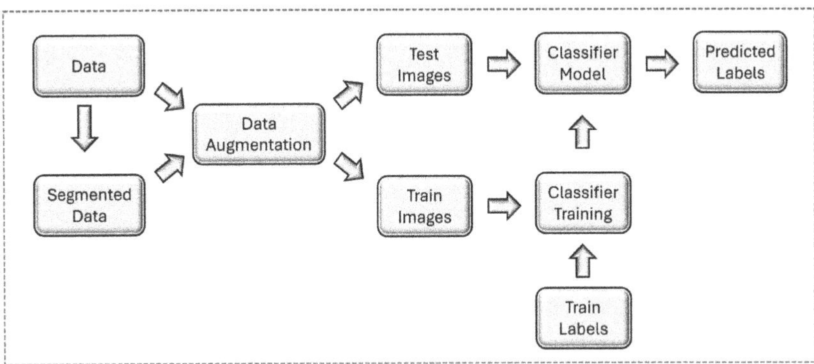

Figure 1. Flowchart of proposed methodology for nitrogen stress detection.

3.1. Pre-Processing

Pre-processing is a critical step in preparing data for classifier training. This process involves several key steps, including resizing the images to a uniform resolution. We performed experimentation using two image resolutions: 100 × 100 pixels and 224 × 224 pixels. Additionally, segmentation is performed to isolate the plant region. To achieve this, a threshold value of 30 is applied to the inverted blue image channel in the RGB color space to segment the sky. For removing the soil part, thresholding on the hue image is applied in the HSV color space. Two threshold values are selected based on the lighting and shadowing in the images. Images with higher exposure to sunlight are segmented using a threshold of 75, while images with lower exposure use a threshold of 40. Following thresholding, morphological closing is applied to refine the segmentation mask using a cross-shaped 5×5 kernel. All the values were determined empirically after extensive experimentation. Figure 2 shows sample images from the dataset before and after segmentation.

(a) (b) (c) (d) (e) (f)

Figure 2. Samples of maize images: (**a**) sample image from class N0 with no fertilization, (**b**) sample image from class N75 with 75 kg of fertilizer applied, (**c**) sample image from class NFull with 136 kg of fertilizer applied, (**d**–**f**) same images as (**a**–**c**) after segmentation.

3.2. Data Augmentation

Data augmentation is an important technique in expanding small datasets. Machine learning models require a large number of training samples to learn from, which are challenging to obtain in real-world settings. Data augmentation works by generating new data from the original data by introducing small variations, thereby artificially creating a bigger dataset to enhance the model's generalization and performance. To increase the robustness of our models and prevent overfitting, we apply various transformations including rotations, shifts, shear, zooms, and flips to augment the input images. We apply rotation and flip transformations to both the original and segmented images, increasing the total number of images in our dataset from 1200 to 4800. We use a random flip set to horizontal and apply a random rotation range of 20%. During the training process, we use the ImageDataGenerator class in TensorFlow to apply shift, shear, and zoom transformations. Width and height shift values are set to 15% of the image width and height, respectively. The shear angle is set to 15 degrees, and the zoom range is set to 15%. In this way, we are able to generate a diverse set of examples for our deep learning models to learn from. This is crucial to generate an extensive and varied dataset that could train robust models for the precise classification of images based on nitrogen deficiency levels.

3.3. Classifier Selection

We use supervised machine learning to classify three levels of nitrogen fertilization (i.e., N0, N75, and NFull). Our study focuses on comparing the performance of ViTs and CNNs in detecting nitrogen deficiency in maize images. ViT is a revolutionary deep learning model introduced by Dosovitskiy et al. [21] that extends the transformer architecture from natural language processing to image classification tasks. While traditional CNNs use a hierarchical approach to process images, ViTs use a sequential approach, treating images as a sequence of small tokens in a similar manner to words being processed in natural language processing. This enables ViTs to capture long-range dependencies more effectively in images, resulting in strong performance in image classification tasks. While CNNs have long been the dominant architecture for computer vision, delivering high effectiveness across various applications, ViTs have demonstrated remarkable performance in numerous image classification tasks, often matching or surpassing the accuracy of CNNs. Our research aims to examine the performance of vision transformers, particularly in identifying nitrogen stress in plants, by comparing their effectiveness against state-of-the-art neural networks.

4. Implementation

We use the Tensorflow library with Keras API to develop our framework. We compare the performance of state-of-the-art ViTs with widely used CNNs. We conduct two sets of experiments. In the first set, we implement a ViT model built from scratch and trained entirely on our dataset. In the second set, we implement a transfer learning approach, where we select various pre-trained deep learning architectures as the base models and fine-tune them on our dataset. The base models include a pre-trained vision transformer model, as well as four widely known CNN models: EfficientNetB0 [22], DenseNet121 [23], InceptionV3 [24], and ResNet50 [25]. The details for both sets of experiments are presented below.

4.1. Vision Transformer Custom-Built Model

In the first set of our experiments, we develop a custom vision transformer model tailored to our specific application. The vision transformer model is constructed in several key steps. The first step is to augment the input images and divide into patches. For this model, we perform experimentation using two image resolutions: image size of 100×100 with a patch size of 10 and image size of 224×224 with a patch size of 16. These patches are then fed into a dense layer, which learns positional embeddings for each patch. This step is crucial for the model to understand the spatial relationships between different parts of the image. The output of the dense layer is then passed into the transformer network. The transformer network consists of several blocks. The block diagram showing the transformer structure is given in Figure 3. The transformer block contains a normalization layer followed by a multihead attention layer. The output of this attention layer is added to the original input and then fed into the next part of the block, which is composed of another normalization layer followed by a multilayer perceptron (MLP) that processes the intermediate features. The MLP in the transformer block comprises two hidden layers with 128 and 64 neurons, respectively. The output of this MLP is then passed to the MLP head in the output block for further processing. In our model, the MLP head unit has two hidden layers with 2048 and 1024 neurons, respectively. Finally, the output of the MLP head is passed to a dense layer that computes the class probabilities, determining the predicted class for each input image.

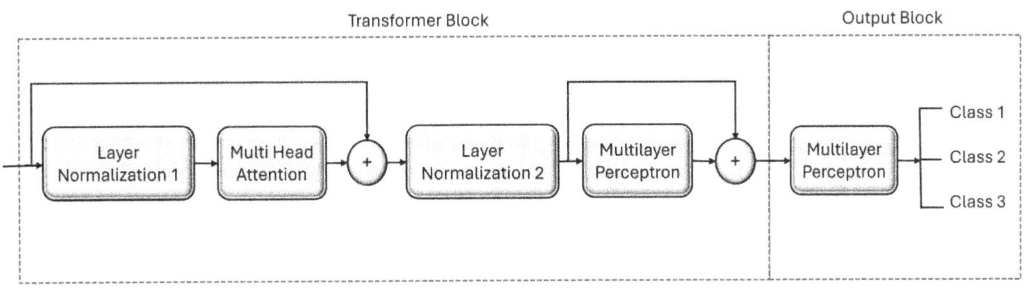

Figure 3. Architecture of our custom-built vision transformer model to classify three levels of nitrogen fertilization (i.e., N0, N75, and NFull).

4.2. Transfer Learning and Fine-Tuned Models

In the second set of our experiments, we employ a transfer learning-based approach. In this set of experiments, we further compare a pre-trained vision transformer model with selected CNN models.

4.2.1. Fine-Tuned Vision Transformer Model

We select ViT-B/16 as the base model for fine-tuning on our dataset, which comprises 12 layers and 12 head units. This model configuration is known for its ability to capture intricate patterns in image data, making it a suitable choice for our classification task. Since this ViT base model accepts an image resolution of 224 × 224, we use only this resolution for training and testing in this part of the experiments. Following the base model, we incorporate a batch normalization layer, succeeded by an intermediate dense layer with 64 units and L2 regularization. Finally, the output layer, consisting of three units with a softmax activation function, completes the model. Figure 4 shows the detailed architecture of the model.

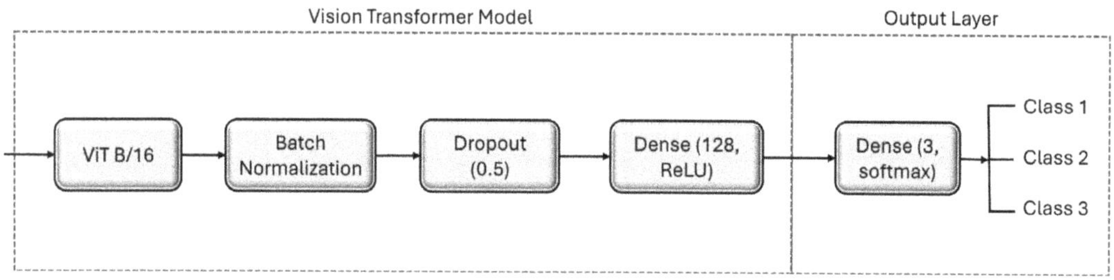

Figure 4. Architecture of the fine-tuned vision transformer model to classify three levels of nitrogen fertilization (i.e., N0, N75, and NFull).

4.2.2. Fine-Tuned Neural Network Models

The neural network architectures chosen as the base model for the second set of experiments are EfficientNetB0, DenseNet121, InceptionV3, and ResNet50. Similar to the custom-built ViT model, we perform experimentation using two image resolutions of 100 × 100 and 224 × 224. The base model's output is fed into a global average pooling 2D layer, followed by a dropout layer with a 50% dropout rate to prevent overfitting. Subsequently, an intermediate dense layer with 128 units, ReLU activation, and L2 regularization is employed. Finally, the output layer consists of three units with a softmax activation function. Figure 5 presents the details of this architecture.

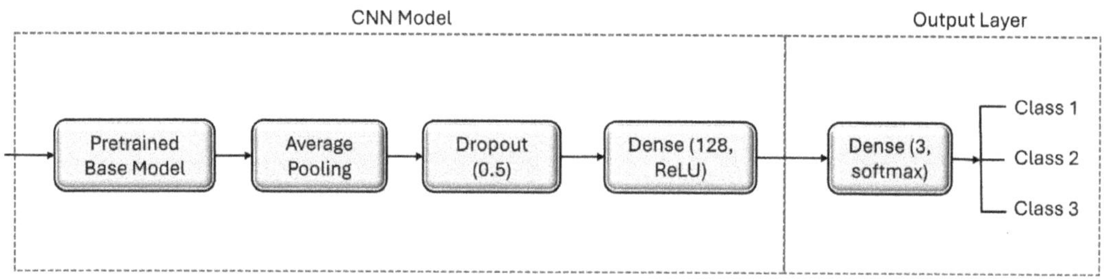

Figure 5. Architecture of fine-tuned CNN-based model to classify three levels of nitrogen fertilization (i.e., N0, N75, and NFull).

4.3. Training Hyperparameters

We incorporate the Adam optimizer in our models with an initial learning rate of 0.001 and sparse categorical cross-entropy as the loss function. We implement several training callbacks to enhance model performance and prevent overfitting. Early stopping is employed to halt training if there is no improvement in performance for 10 consecutive epochs, thus promoting generalization. The Model Checkpoint callback saves the best-performing model based on validation accuracy, ensuring that the model with the highest predictive capability is retained. The ReduceLROnPlateau callback reduces the learning rate by a factor of 0.2 if the validation loss fails to improve for 5 epochs, with a lower bound set at 0.00001, allowing for fine tuning of the model training. Additionally, a Learning Rate Scheduler adjusts the learning rate based on the epoch number to control the speed of model convergence, with an exponential decay scheduler employed in this instance. Each model has been trained for 50 epochs, and the results are detailed in the subsequent section.

5. Results and Discussion

For our experiments, we split the dataset into train and test sets, with 90% of the data allocated to the train set and 10% to the test set. We conduct the experiments using three-fold cross-validation for each classifier, with the validation ratio set to 20%. The best model configuration is selected based on the cross-validation results and evaluated on the test set. Accuracy, precision and recall are selected as the performance parameters. Additionally, training and validation loss curves are utilized to assess the performance of the classifiers.

Accuracy is defined as the ratio of correctly predicted instances to the total number of instances evaluated. Precision is defined as the ratio of true positive predictions to the total number of positive predictions made by the model. Recall is the ratio of true positive predictions to the total number of actual positive samples in the class. The training loss measures the error between the model's predicted output and the actual target during training. Training aims at minimizing training loss through optimization techniques such as gradient descent. Decreasing training loss indicates improved fitting of the model to the training data. Validation loss, on the other hand, measures the error on a separate validation dataset that the model has not encountered during training, providing insight into its performance on unseen data. Validation loss evaluates the model's generalization capability, with lower validation loss indicating effective generalization and higher validation loss suggesting potential overfitting.

5.1. *Custom ViT Results*

Figure 6 shows the training and validation loss curves for the custom ViT model for both image resolutions. For both cases, the training loss continues to decrease gradually, reaching a steady value by the end of the training. This gradual decrease suggests that the model continues to learn at a slower pace. The validation loss decreases at a relatively faster pace until epoch 20, after which it decreases very slowly and smoothly until the last epoch. This behavior indicates that while the model continues to improve its generalization, the rate of improvement becomes minimal after epoch 20. Overall, the model reaches a convergence point around epoch 30, where both training and validation losses stabilize, indicating effective learning and parameter stabilization.

From Figure 6, it can be observed that even though the final training loss for the image resolution 224×224 is much lower (0.3) compared to the image resolution 100×100 (0.6), the validation loss is the same for both the cases (0.6). This indicates that the model with a larger image size might be overfitting to the training data. However, the test accuracy with the large-image resolution is higher (76%) as compared to the small-image resolution (70%). The improved test accuracy with the larger image size suggests that the model benefits from the additional information provided by larger images which contain more detail and spatial information than smaller images, making them the preferred choice for vision transformer classifier.

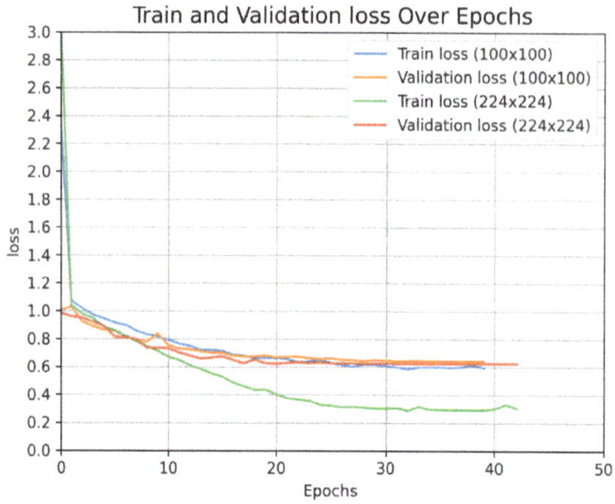

Figure 6. Custom-built vision transformer model results for train and validation losses for two image resolutions (100 × 100 and 224 × 224) for nitrogen fertilization-level prediction.

5.2. *Transfer Learning and Fine-Tuned Models Results*

5.2.1. Fine-Tuned ViT Results

Figure 7 shows the graphs of training and validation losses over 50 epochs for our fine-tuned model with the pre-trained ViT base. The graph demonstrates that the training loss gradually decreases, indicating that the model is effectively learning from the data. Initially, there are fluctuations in the validation loss, suggesting that the model is adjusting to the data and fine-tuning its parameters to find an optimal configuration. After epoch 25, the loss curve smooths out and remains relatively constant. This stability and similarity between the training and validation losses suggest good generalization. The curves indicate that by 50 epochs, the model has likely reached convergence, with its weights stabilized, and further training does not result in significant changes to the loss values. The pre-trained ViT model attains the highest training accuracy of 70% for the best configuration, and the best test accuracy of 64%.

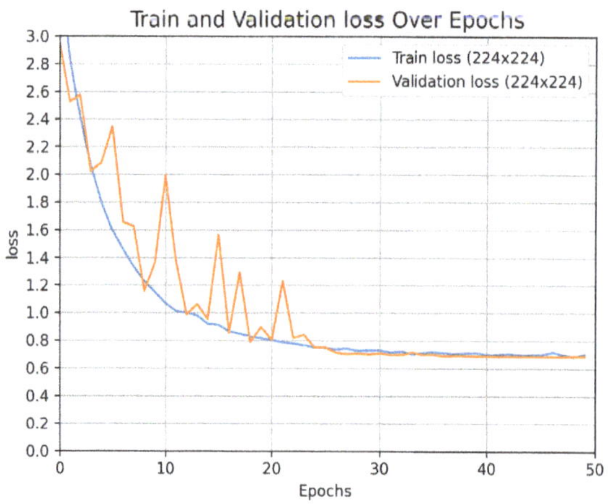

Figure 7. Fine-tuned vision transformer model results for train and validation losses for image resolution 224 × 224 for nitrogen fertilization-level prediction.

Both ViT-based models exhibit similar performance, with the custom-built model showing better performance. Both models achieve their best loss values within 30 epochs, although the pre-trained model displays more fluctuations at the beginning of training. The stability of the loss curves and the consistent values of both training and validation losses after 30 epochs suggest that the models have likely reached a point of convergence. Further training is unlikely to lead to significant improvements, as it can be inferred from the training curve that the models have already learned fully from the data given the current architecture and hyperparameters.

5.2.2. Fine-Tuned CNN Models Results

Table 1 gives a comparison of train and test accuracies for all CNN-based models for both image resolutions (100 × 100 and 224 × 224). It has been observed that all base models perform better with a higher image resolution of 224 × 224 as compared to a lower resolution of 100 × 100. Among all the models, the largest difference in test accuracy is observed with InceptionV3, which achieves a test accuracy of 74% for smaller images and 91% for larger images. ResNet50 shows slightly better performance, with 78% test accuracy for smaller images and 84.5% for larger images. The DenseNet121-based model performs reasonably well, with 90% and 93% test accuracy for small and large images, respectively. The best performance is obtained with the EfficientNetB0 base model, achieving a test accuracy of 92% for smaller images (100 × 100), and 97% for larger images (224 × 224), making it the most suitable model for our application.

Table 1. Comparison of test accuracies of fine-tuned CNN classifier models for two image resolutions (100 × 100 and 224 × 224) for nitrogen fertilization-level prediction.

CNN-Based Model	Train Accuracy (%)		Test Accuracy (%)	
	100 × 100	224 × 224	100 × 100	224 × 224
EfficientNetB0	99	99.7	92	97
DenseNet121	98	98	90	93
InceptionV3	76	99	74	91
ResNet50	84	90	78	84.5

Figure 8 gives the training and validation loss curves for the fine-tuned EfficientNetB0 model for both image resolutions. A gradually decreasing and smooth training curve in both cases indicates that the model's optimization process is stable and effective. The fluctuations in the validation loss early in the training are due to the model making adjustments to the weights and learning the data. Steady values of training and validation losses indicate that the model has converged. Training loss values are similar for both resolutions and become steady after 30 epochs. However, the validation loss for the small-image resolution is higher (0.23) compared to the large-image resolution (0.08). The significant difference in validation loss suggests that higher resolution images provide more detailed information, enabling the model to make more accurate predictions. The test accuracy values are 92% for small images and 97% for large images, indicating better performance with higher image resolution.

The training curves of both ViT and CNN models indicate that CNN-based fine-tuned models outperform ViT models for detecting nitrogen stress in maize crop images. While ViT models demonstrate good generalization, they tend to converge prematurely, ceasing to learn once a stable loss curve is achieved. Despite their generalization ability, ViT models exhibit higher loss values, indicating potential for improvement. Table 2 gives a comparison of train and test accuracies for the best CNN and vision transformer models for the image size 224 × 224. It can be seen that the fine-tuned CNN with EfficientNetB0 as the base model achieves higher accuracy as compared to ViT models.

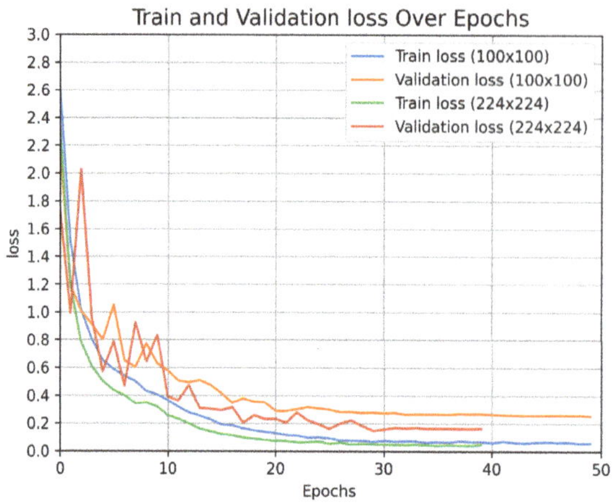

Figure 8. Fine-tuned CNN best model results for train and validation losses for two image resolutions for nitrogen fertilization-level prediction.

Table 2. Comparison of train and test accuracies of vision transformer models and CNN best model for image resolution 224×224 for nitrogen fertilization-level prediction.

Architecture	Train Accuracy (%)	Test Accuracy (%)
ViT (Custom)	87	76
Fine-tuned ViT (VIT-b/16)	71	69
CNN (EfficientNetB0)	99.7	97

The individual class results (i.e., fertilization-levels N0, N75, and NFull) are presented in Figure 9. Figure 9 only presents the results for the CNN model with EfficientNetB0 as the base model, as it outperforms other CNN models (as depicted in Table 1). For class N0, although the recall value for the CNN model (EfficientNetB0) is higher compared to the ViT models, it remains the lowest when compared to the other two classes, which have recall values of 0.99 and 1. This indicates that the CNN fine-tuned model performs better for N75 and NFull than for N0. However, the recall value for class N0 for the CNN model is still the best performing among the three classifiers for the N0 class. The fine-tuned ViT model shows both precision and recall values of less than 50% for the N0 class, indicating poor performance in identifying N0 samples. However, the fine-tuned ViT model demonstrates better recall for the N75 class as compared to the other classes, with the highest value of 0.81. The custom ViT model has moderate recall and precision values for the N0 class, suggesting there is room for improvement as compared to the other classes, where it performs relatively better. For N75, the custom ViT shows the best performance with a recall of 0.9 and precision of 0.82, both of which are higher than the values for the other two classes (N0 and NFull). This indicates that the custom ViT is particularly effective at classifying the N75 class compared to the other classes. Overall, the custom ViT outperforms the fine-tuned ViT for all the three classes; however, the fine-tuned CNN model with EfficientNetB0 as the base outperforms both ViTs for all the classes.

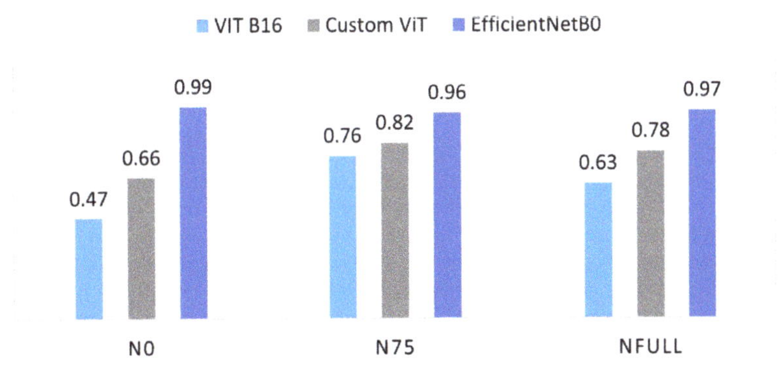

(**a**) Precision

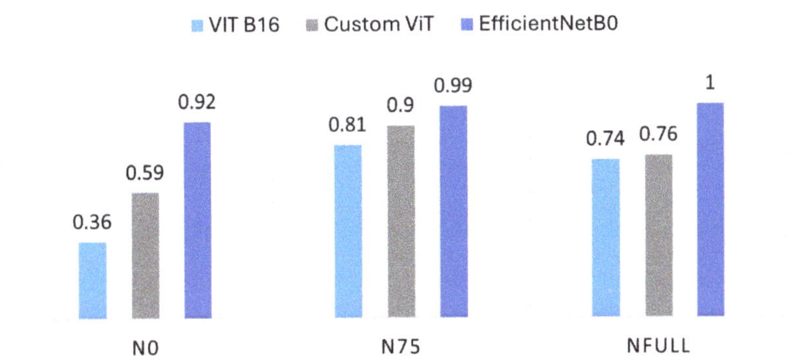

(**b**) Recall

Figure 9. Classification results for individual fertilization-level classes for image resolution 224 × 224.

6. Conclusions

The current research uses computer vision-based analysis techniques and deep learning models to assess the nitrogen status of maize crops. In this study, we have employed two distinct approaches to develop models for nitrogen fertilization-level classification task. The first approach involves constructing a vision transformer (ViT) model from scratch. This custom-built model has been designed and trained entirely on our dataset, allowing us to tailor the architecture and training process specifically to our needs. The second approach utilizes transfer learning, leveraging a pre-trained ViT model and four widely implemented CNN models. These pre-trained models have been fine-tuned on our dataset, allowing us to benefit from the rich feature representations learned from large-scale datasets like ImageNet. By adding custom layers on top of the pre-trained base and fine-tuning the models, we have enhanced the models' effectiveness by combining the generalization capabilities of the pre-trained models with the specific characteristics of our dataset. Additionally, we have tested two image resolutions, 100 × 100 and 224 × 224, to observe the relationship between image resolution and model performance. Our findings indicate that the performance of all the models is significantly affected by image size, with larger images (224 × 224) yielding better results as compared to smaller images (100 × 100). Furthermore, through a comparative analysis of fine-tuned and custom-built ViT models against fine-tuned convolutional neural networks (CNNs), we observe that CNN models outperform ViT models at both image resolutions. Specifically, the CNN model with EfficientNetB0 as the base successfully classifies the crops into stressed, non-stressed, and semi-stressed classes, achieving a best test accuracy of 97%. Even though the final training and validation losses for vision transformer models are consistent, the overall loss values are large. These results indicate that while vision transformers slowly learn and generalize well to the data, to reduce loss values and increase accuracy, further optimization is needed, which might be achieved with a larger number of data samples.

The current research on nitrogen deficiency detection using RGB images is a critical step towards the long-term goal of developing a real-time precision agriculture system for targeted nitrogen fertilizer application. Our research focuses on using only RGB images as compared to hyperspectral imagery, which is costly and less suitable for real-time implementation. The superior performance of EfficientNetB0 suggests its potential for real-time decision-making in precision agriculture applications. Vision transformers require a large number of data samples and significantly more training time and parameters, suggesting their somewhat lower suitability for typical agricultural datasets (particularly for nitrogen stress identification), which are usually smaller. After a detailed comparison and analysis, it is concluded that CNNs are more practical for real-time agricultural applications, and the vision transformers need further refinement for agricultural datasets.

Author Contributions: Conceptualization, A.M.; methodology, S.G., N.K. and A.M.; software, S.G. and N.K.; validation, S.G. and N.K.; formal analysis, S.G. and N.K.; investigation, A.M.; resources, A.M.; data curation, S.G. and N.K.; writing—original draft preparation, S.G. and N.K.; writing—review and editing, A.M.; visualization, S.G. and N.K.; supervision, A.M.; project administration, A.M.; funding acquisition, A.M. All authors have read and agreed to the published version of the manuscript.

Funding: This work was supported in part by the United States Department of Agriculture (USDA) National Institute of Food and Agriculture (NIFA), Award Number 2023-67021-40614. Any opinions, findings, and conclusions or recommendations expressed in this material are those of the author(s) and do not necessarily reflect the views of the USDA NIFA.

Institutional Review Board Statement: Not Applicable.

Informed Consent Statement: Not Applicable.

Data Availability Statement: Dataset used in this research is available publicly at https://data.mendeley.com/datasets/g7xnn2bm4g/1 accessed on 12 July 2024.

Conflicts of Interest: The authors declare no conflicts of interest.

Abbreviations

RGB	Red, green, blue
NUE	Nitrogen use efficiency
CI_{green}	Chlorophyll index green
GNDVI	Green normalized difference vegetation index
RENDVI	Red edge normalized difference vegetation index
RDVI	Renormalized difference vegetation index
R^2	Coefficient of determination
RMSE	Root mean squared error
ReLU	Rectified linear unit
CNN	Convolutional neural network
SVM	Support vector machine
RCNN	Region-based convolutional neural networks
DSLR	Digital single-lens reflex
ViT	Vision transformer
MLP	Multilayer perceptron

References

1. Ritchie, H. Excess Fertilizer Use: Which Countries Cause Environmental Damage by Overapplying Fertilizers? 2021. Available online: https://ourworldindata.org/excess-fertilizer (accessed on 12 July 2024).
2. Sainju, U.M.; Ghimire, R.; Pradhan, G.P. Nitrogen Fertilization I: Impact on Crop, Soil, and Environment. In *Nitrogen Fixation*; Rigobelo, E.C., Serra, A.P., Eds.; IntechOpen: Rijeka, Croatia, 2019; Chapter 5. [CrossRef]
3. Nitrogen Deficiency in Crops: How to Detect & Fix It. 2021. Available online: https://eos.com/blog/nitrogen-deficiency/ (accessed on 31 May 2024).
4. Shi, J.-Y.; Zou, X.-B.; Zhao, J.-W.; Wang, K.-L.; Chen, Z.-W.; Huang, X.-W.; Zhang, D.-T.; Holmes, M. Nondestructive diagnostics of nitrogen deficiency by cucumber leaf chlorophyll distribution map based on near infrared hyperspectral imaging. *Sci. Hortic.* **2012**, *138*, 190–197. [CrossRef]
5. Sanaeifar, A.; Yang, C.; Min, A.; Jones, C.R.; Michaels, T.E.; Krueger, Q.J.; Barnes, R.; Velte, T.J. Noninvasive Early Detection of Nutrient Deficiencies in Greenhouse-Grown Industrial Hemp Using Hyperspectral Imaging. *Remote Sens.* **2024**, *16*, 187. [CrossRef]
6. Liu, N.; Townsend, P.A.; Naber, M.R.; Bethke, P.C.; Hills, W.B.; Wang, Y. Hyperspectral imagery to monitor crop nutrient status within and across growing seasons. *Remote Sens. Environ.* **2021**, *255*, 112303. [CrossRef]
7. Yamashita, H.; Sonobe, R.; Hirono, Y.; Morita, A.; Ikka, T. Dissection of hyperspectral reflectance to estimate nitrogen and chlorophyll contents in tea leaves based on machine learning algorithms. *Sci. Rep.* **2020**, *10*, 17360. [CrossRef]
8. Wang, S.; Guan, K.; Wang, Z.; Ainsworth, E.A.; Zheng, T.; Townsend, P.A.; Liu, N.; Nafziger, E.; Masters, M.D.; Li, K.; et al. Airborne hyperspectral imaging of nitrogen deficiency on crop traits and yield of maize by machine learning and radiative transfer modeling. *Int. J. Appl. Earth Obs. Geoinf.* **2021**, *105*, 102617. [CrossRef]
9. Cilia, C.; Panigada, C.; Rossini, M.; Meroni, M.; Busetto, L.; Amaducci, S.; Boschetti, M.; Picchi, V.; Colombo, R. Nitrogen Status Assessment for Variable Rate Fertilization in Maize through Hyperspectral Imagery. *Remote Sens.* **2014**, *6*, 6549–6565. [CrossRef]
10. Wu, L.; Gong, Y.; Bai, X.; Wang, W.; Wang, Z. Nondestructive Determination of Leaf Nitrogen Content in Corn by Hyperspectral Imaging Using Spectral and Texture Fusion. *Appl. Sci.* **2023**, *13*, 1910. [CrossRef]
11. Ram, B.G.; Oduor, P.; Igathinathane, C.; Howatt, K.; Sun, X. A systematic review of hyperspectral imaging in precision agriculture: Analysis of its current state and future prospects. *Comput. Electron. Agric.* **2024**, *222*, 109037. [CrossRef]
12. Burns, B.W.; Green, V.S.; Hashem, A.A.; Massey, J.H.; Shew, A.M.; Adviento-Borbe, M.A.A.; Milad, M. Determining nitrogen deficiencies for maize using various remote sensing indices. *Precis. Agric.* **2022**, *23*, 791–811. [CrossRef]
13. Zheng, H.; Li, W.; Jiang, J.; Liu, Y.; Cheng, T.; Tian, Y.; Zhu, Y.; Cao, W.; Zhang, Y.; Yao, X. A Comparative Assessment of Different Modeling Algorithms for Estimating Leaf Nitrogen Content in Winter Wheat Using Multispectral Images from an Unmanned Aerial Vehicle. *Remote Sens.* **2018**, *10*, 2026. [CrossRef]
14. Cao, Q.; Yang, G.; Duan, D.; Chen, L.; Wang, F.; Xu, B.; Zhao, C.; Niu, F. Combining multispectral and hyperspectral data to estimate nitrogen status of tea plants (Camellia sinensis (L.) O. Kuntze) under field conditions. *Comput. Electron. Agric.* **2022**, *198*, 107084. [CrossRef]
15. Azimi, S.; Kaur, T.; Gandhi, T.K. A deep learning approach to measure stress level in plants due to Nitrogen deficiency. *Measurement* **2021**, *173*, 108650. [CrossRef]
16. Zermas, D.; Nelson, H.J.; Stanitsas, P.; Morellas, V.; Mulla, D.J.; Papanikolopoulos, N. A Methodology for the Detection of Nitrogen Deficiency in Corn Fields Using High-Resolution RGB Imagery. *IEEE Trans. Autom. Sci. Eng.* **2021**, *18*, 1879–1891. [CrossRef]
17. Zhang, J.; Xie, T.; Yang, C.; Song, H.; Jiang, Z.; Zhou, G.; Zhang, D.; Feng, H.; Xie, J. Segmenting Purple Rapeseed Leaves in the Field from UAV RGB Imagery Using Deep Learning as an Auxiliary Means for Nitrogen Stress Detection. *Remote Sens.* **2020**, *12*, 1403. [CrossRef]
18. Haider, T.; Farid, M.S.; Mahmood, R.; Ilyas, A.; Khan, M.H.; Haider, S.T.A.; Chaudhry, M.H.; Gul, M. A Computer-Vision-Based Approach for Nitrogen Content Estimation in Plant Leaves. *Agriculture* **2021**, *11*, 766. [CrossRef]
19. Salaić, M.; Novoselnik, F.; Žarko, I.P.; Galić, V. Nitrogen deficiency in maize: Annotated image classification dataset. *Data Brief* **2023**, *50*, 109625. [CrossRef] [PubMed]
20. Galic, V.; Podnar Žarko, I.; Novoselnik, F.; Salaic, M. Nitrogen deficiency in maize: Annotated image classification dataset. *Mendeley Data* **2023**. [CrossRef]
21. Dosovitskiy, A.; Beyer, L.; Kolesnikov, A.; Weissenborn, D.; Zhai, X.; Unterthiner, T.; Dehghani, M.; Minderer, M.; Heigold, G.; Gelly, S.; et al. An Image is Worth 16 × 16 Words: Transformers for Image Recognition at Scale. In Proceedings of the IEEE/CVF Conference on Computer Vision and Pattern Recognition, New Orleans, LA, USA, 18–24 June 2021; pp. 3156–3164.
22. Tan, M.; Le, Q.V. EfficientNet: Rethinking Model Scaling for Convolutional Neural Networks. *arXiv* **2019**, arXiv:1905.11946.
23. Huang, G.; Liu, Z.; van der Maaten, L.; Weinberger, K.Q. Densely Connected Convolutional Networks. *arXiv* **2018**, arXiv:1608.06993.

24. Szegedy, C.; Vanhoucke, V.; Ioffe, S.; Shlens, J.; Wojna, Z. Rethinking the Inception Architecture for Computer Vision. *arXiv* **2015**, arXiv:1512.00567.
25. He, K.; Zhang, X.; Ren, S.; Sun, J. Identity Mappings in Deep Residual Networks. In Proceedings of the Computer Vision–ECCV 2016, Amsterdam, The Netherlands, 11–14 October 2016; Leibe, B., Matas, J., Sebe, N., Welling, M., Eds.; Springer: Cham, Switzerland, 2016; pp. 630–645.

AI MDPI

Article

Enhancing *Tuta absoluta* Detection on Tomato Plants: Ensemble Techniques and Deep Learning

Nikolaos Giakoumoglou [1], Eleftheria-Maria Pechlivani [1,*], Nikolaos Frangakis [2] and Dimitrios Tzovaras [1]

[1] Centre for Research and Technology Hellas, Information Technologies Institute, 57001 Thermi, Greece; ngiakoumoglou@iti.gr (N.G.); dimitrios.tzovaras@iti.gr (D.T.)
[2] iKnowHow S.A., 15451 Athens, Greece; nfrangakis@iknowhow.com
* Correspondence: riapechl@iti.gr; Tel.: +30-231-125-7751

Abstract: Early detection and efficient management practices to control *Tuta absoluta* (Meyrick) infestation is crucial for safeguarding tomato production yield and minimizing economic losses. This study investigates the detection of *T. absoluta* infestation on tomato plants using object detection models combined with ensemble techniques. Additionally, this study highlights the importance of utilizing a dataset captured in real settings in open-field and greenhouse environments to address the complexity of real-life challenges in object detection of plant health scenarios. The effectiveness of deep-learning-based models, including Faster R-CNN and RetinaNet, was evaluated in terms of detecting *T. absoluta* damage. The initial model evaluations revealed diminishing performance levels across various model configurations, including different backbones and heads. To enhance detection predictions and improve mean Average Precision (mAP) scores, ensemble techniques were applied such as Non-Maximum Suppression (NMS), Soft Non-Maximum Suppression (Soft NMS), Non-Maximum Weighted (NMW), and Weighted Boxes Fusion (WBF). The outcomes shown that the WBF technique significantly improved the mAP scores, resulting in a 20% improvement from 0.58 (max mAP from individual models) to 0.70. The results of this study contribute to the field of agricultural pest detection by emphasizing the potential of deep learning and ensemble techniques in improving the accuracy and reliability of object detection models.

Keywords: deep learning; ensembles; Faster R-CNN; object detection; RetinaNet; tomato plants; *Tuta absoluta*; weighted boxes fusion

1. Introduction

The tomato plant (*Solanum lycopersicum* L.) plays a significant role in global agriculture, serving as a dietary staple and contributing significantly to the world's food systems, with an annual production exceeding 160 million tonnes globally [1]. Additionally, the tomato industry holds economic significance in various market segments such as canned goods, sauces, beverages, and cosmetics, with approximately a quarter of its production dedicated to processing, making it an essential commodity [2].

However, this thriving industry is facing a formidable adversary in the form of the tomato leaf miner, scientific name *Tuta absoluta* (Meyrick) (Lepidoptera: Gelechiidae) and henceforth referred to as *T. absoluta* [3]. This pest, native to South America, has managed to proliferate far beyond its region of origin, causing considerable damage to tomato crops worldwide. The invasive pest was first detected in Europe in late 2006 [4] and has spread to North Africa and the Middle East at an exceptional speed. The severity of this infestation is such that it can cause production losses ranging from 80% to 100% [5]. This leads to a ripple effect of economic distress, affecting not just the agricultural sector but also the livelihoods of millions who depend on it [6].

The widespread presence of *T. absoluta* not only threatens the worldwide tomato supply but also discourages growers due to the rising costs of its integrated pest management.

Despite the availability of various pest management methods, such as chemical pesticides, pheromone traps, and the cultivation of resistant tomato varieties [7], effective control of *T. absoluta* spread remains elusive. Therefore, for the surveillance of *T. absoluta*, there is a need for digital-based technologies that leverage deep learning techniques, enabling fast, accurate, and non-invasive early detection of this pest [8].

Recently, the successful application of deep learning models in many fields of computer vision for early pest and disease detection in crops using convolutional neural network (CNN) has presented a unique opportunity for enabling efficient and rapid detection of *T. absoluta* infestation at early stages [9].

In this study, four ensemble techniques were benchmarked to evaluate their performance in improving the detections of *T. absoluta* infestations on tomato plants, as generated by deep learning object detection models. A dataset captured under real-life conditions in open fields and greenhouses was utilized, presenting various challenges in detecting *T. absoluta* damage. The outputs from multiple configurations of the employed Faster R-CNN and RetinaNet deep learning models, combined with ensemble techniques such as Non-Maximum Suppression (NMS), Soft Non-Maximum Suppression (Soft NMS), Non-Maximum Weighted (NMW), and Weighted Boxes Fusion (WBF) exhibited improved performance. In particular, the fusion of detection results using WBF, considering the detection scores, effectively improved the mean Average Precision (mAP) metric by 20%, elevating it from 0.58 (the highest observed mAP from individual deep learning models) to 0.70. This study demonstrates the superiority of the WBF ensemble technique in addressing the challenges associated with *T. absoluta* digital detection on tomato plants, highlighting its significance in enhancing the accuracy and reliability of deep learning models for agricultural applications. The innovation in this study lies in the combination of established object detection models with ensemble techniques, which reveals an improvement in *T. absoluta* detection accuracy under real-world agricultural settings. This approach not only establishes a robust method for pest detection but also extends the applicability of deep learning models in precision agriculture, aligning with the broader goal of integrated pest management.

The rest of the paper is structured as follows. Section 2 presents the related work in the field of tomato disease and pest identification and *T. absoluta* detection. Section 3 delineates the methodology employed in this study, encompassing the dataset, the object detection models, the ensemble techniques utilized, and the evaluation metrics adopted for detecting the impacts of *T. absoluta* on tomato plants. Section 4 exhibits the results. The study concludes in Section 5.

2. Related Work

In this section, the application of deep learning techniques for the identification of tomato diseases and pests, as well as the detection of *T. absoluta* effects, is reviewed. Significant progress has been made in agricultural applications by employing deep learning methods. In particular, the identification of diverse plant diseases and pests has yielded promising and impressive outcomes, as evidenced by recent studies [10–12].

2.1. *Tomato Disease and Pest Identification*

Tomato disease and pest identification have been actively investigated in the field of deep learning. Nevertheless, it is crucial to acknowledge that the nature of the tasks addressed in these investigations differs, with certain studies emphasizing classification while others prioritize object detection. It is worth noting that, in real-world applications, classification tends to present limitations and poses a comparatively simpler challenge compared to detection.

Wang et al. (2019) [13] used Faster R-CNN and Mask R-CNN in tomato disease detection. The models achieved a mAP of 88.53% and 99.64% for Faster R-CNN and Mask R-CNN, respectively. Wang et al. (2021) [14] used a YOLOv3-tiny network on a self-built tomato disease and pest dataset and dealt with light shadow, branch occlusion, and overlapping leaves conditions. Yet, it should be noted that the objects depicted in the images were considerably prominent in size, which significantly facilitated the detection task by providing distinct and easily identifiable visual features. Fuentes et al. (2017) [15] used a deep-learning-based approach to detect diseases and pests in tomato plants using images captured in-place by camera devices with various resolutions. Using Faster R-CNN, R-FCN, and SSD, combined with various backbones such as VGG and ResNets, the authors demonstrated the performance of these models on images with diseases and pests.

Other studies primarily focused on the classification of tomato diseases rather than the detection of specific regions. Zaborowicz et al. (2017) [16] explored the application of computer image analysis and artificial neural modeling for assessing the quality of greenhouse tomatoes. Zaborowicz et al. (2013) [17] aimed to devise a neural network classification model for the automatic evaluation of greenhouse tomato quality, leveraging computer image analysis and artificial neural networks. Brahimi et al. (2017) [18] presented deep models, including AlexNet and GoogleNet, which were trained using a large dataset of 14,828 images to identify nine tomato diseases. Ferentinos (2018) [19] employed several deep models, namely AlexNet, GoogLeNet, and VGG, to recognize 58 diseases from a dataset consisting of 87,848 leaf images of various plants. Liang et al. (2019) [20] proposed a multitasking system based on the ResNet-50 architecture capable of diagnosing diseases, recognizing plant species, and estimating the severity of diseases. Rangarajan et al. (2018) [21] used two pre-trained deep learning models, VGG-16 and AlexNet, to classify six tomato diseases. Trivedi et al. (2021) [22] used CNNs to effectively define and classify tomato diseases with 98.49% accuracy.

In the field of tomato disease and pest identification, there is a distinguishable difference between classification and detection tasks, despite both utilizing deep learning models. Detection-focused studies, such as those by Wang et al. (2019) [13] and Fuentes et al. (2017) [15], are aimed at identifying and localizing specific affected areas within images, dealing with complexities like varied lighting and occlusions. Conversely, classification-centric research, exemplified by works by Brahimi et al. (2017) [18] and Ferentinos (2018) [19], is focused on categorizing images based on the type of disease or pest present, without localizing the affected areas. These differences underline the varied approaches and complexities within the broader scope of automated identification of plant diseases and pests.

2.2. *Tuta Absoluta Identification*

Tuta absoluta identification has been a subject of interest in the field of deep learning. However, while several studies have focused on the classification of *T. absoluta* effects, limited research has addressed the specific task of *T. absoluta* detection using object detection techniques. Notably, detection poses a more complex challenge compared to classification and carries greater relevance in real-world applications.

Among the studies that have explored *T. absoluta* detection, Loyani et al. (2021) [23] inoculated *T. absoluta* on randomly selected tomato plants and created a dataset of 5235 tomato images. The authors used semantic and instance segmentation models based on U-Net and Mask R-CNN to segment the effects of *T. absoluta* on tomato leaf images at the pixel level using field data. The Mask R-CNN achieved a mAP of 85.67%, while the U-Net model achieved an Intersection over Union (IoU) of 78.60% and Dice coefficient of 82.86%. Nonetheless, the large size of the depicted objects facilitated the detection task with distinct visual features. Georgantopoulos et al. (2023) [24] presented a dataset of multispectral images of tomato plants at various stages of infection with *Tuta absoluta* and Leveillula taurica. The authors applied a Faster-RCNN object detector for the localization and classification of lesions, achieving a 90% F1-score on binary classification, and 20.2% mAP on detection.

However, existing research on *T. absoluta* primarily focuses on its classification or the classification of the damage it causes. Rubanga et al. (2020) [25] determined the severity status of *T. absoluta*'s damage on tomato plants at an early stage of tomato growth. The authors pre-trained four CNN architectures to classify healthy and infested tomato leaves collected from real field experiments with an average accuracy of 87.2%. Mkonyi et al. (2020) [26] leveraged CNN network architectures for the identification of *T. absoluta* in the tomato plant. The approach reinforces the classification of a dataset that consists of 2145 images with an accuracy of 91.9% on the tested images.

Research on *Tuta absoluta* identification via deep learning exhibits a division between classification and detection. Loyani et al. (2021) [23] and Georgantopoulos et al. (2023) [24] delve into object detection, focusing on identifying and localizing the impacts of *T. absoluta* on tomato plants, a task accentuated by its complexity and applicative relevance. Conversely, works by Rubanga et al. (2020) [25] and Mkonyi et al. (2020) [26] are anchored in classification, assessing the severity and identifying the presence of *T. absoluta* damage without specific localization. Despite the shared foundation in deep learning, a clear distinction emerges from the depth of analysis and practical implications each approach entails.

3. Methodology

The methodology section outlines the dataset utilized, details the object detection models implemented, and describes the evaluation metrics adopted to assess the effectiveness of these models in the detection of *T. absoluta* infestation on tomato plants. The model is tailored to run on a mobile platform aboard a fully autonomous mobile robot [27]. The collected data is processed within a robust decision support system [28] to ensure precise detection and evaluation of the infestations.

3.1. Dataset

The dataset used in this study consists of color images collected under real-life conditions in both fields and greenhouses depicting *T. absoluta* damage on tomato plants. These images were obtained by the EDEN Library (https://edenlibrary.ai (accessed on 1 March 2023)) in collaboration with its partners, ensuring the clear distinction of *T. absoluta* damage on tomato plant cultivation while preserving background information. The collection process involved farmers using mobile phones and other devices to capture images from various angles and under different lighting conditions. This collection style is consistent with typical precision farming and robotic tasks, and it provides a suitable basis for training deep learning models to detect *T. absoluta* infestations under diverse scenarios in real settings. The dataset's robustness is enhanced by the intentional inclusion of images captured under non-optimal illumination conditions and with variable resolution and color balance. Any images that did not meet the high standards set by experts were excluded from the dataset.

The dataset comprises a total of 659 images, with 396 reserved for training and 263 designated for validation. The average size of the training images is $(1964 \pm 1068) \times (1863 \pm 802)$ with a median of 1504×1504 (ranging from 960×720 to 4640×4000). The validation images have a mean size of $(1772 \pm 801) \times (1769 \pm 725)$ and a median of 1504×1504 (ranging from 1200×1504 to 4640×4000).

Each of the 659 images has been annotated with bounding boxes that enclose areas of *T. absoluta* damage. These annotations were executed by expert agronomists from the EDEN Library. All annotations underwent manual review and adjustment to ensure their consistency and accuracy. The training set contains 5443 annotations while the validation set includes 3267, totaling 8710 annotations. This equates to an average of 13.83 annotations per image across the one class. The annotation area for the training set has a mean size of $14{,}508 \pm 31{,}806$ pixels and a median of 5764, while the validation set's annotation area has a mean size of $12{,}234 \pm 24{,}857$ pixels and a median of 5304. Therefore, it can be deduced that the mean object, as indicated by the bounding boxes, captures approximately 0.39% (as far as training set is concerned). For a visual reference, an example image depicting *T. absoluta* damage on tomato leaves, annotated with bounding boxes, is presented in Figure 1. The aforementioned details highlight the inherent difficulty associated with addressing the object detection task within this dataset.

Figure 1. Sample featuring bounding box annotations that point out *T. absoluta* damage (class 0) on tomato leaves under various conditions and illuminations. Six different images are displayed, each containing annotations highlighting the damage. The images have been resized to a resolution of 1024×1024 pixels.

3.2. Object Detection Models

Two popular object detection models, Faster R-CNN [29] and RetinaNet [30], were utilized in this study to detect *T. absoluta* damage in tomato crops.

Faster R-CNN is a pivotal model in the domain of object detection that efficiently handles the task of object localization and classification in images. Its architecture is orchestrated in a manner that first employs a deep convolutional network as a backbone to extract a rich set of feature maps from the input image. Following the backbone, a Region Proposal Network (RPN) operates as the neck, generating a set of object proposals with associated objectness scores, significantly speeding up the detection process while maintaining a high level of accuracy. These proposals are then forwarded to the head of the model, where two sibling layers perform classification and bounding box regression to accurately identify and localize objects within the image.

For the Faster R-CNN models, configurations were established using ResNet-50 and ResNet-101 backbones [31] with Feature Pyramid Network (FPN), Convolutional Layer 4 (C4), and Dilated Convolutional Layer 5 (DC5) heads. The FPN head, paired with a ResNet+FPN backbone, was used with standard convolutional (conv) and fully connected (FC) heads for box prediction due to its optimal speed/accuracy tradeoff [32]. The C4 head employed a ResNet conv4 backbone with a conv5 head, following the original baseline configuration in the Faster R-CNN paper [29]. Lastly, the DC5 head utilized a ResNet conv5 backbone with dilations in conv5, as well as standard conv and FC heads for box prediction.

Meanwhile, RetinaNet is recognized for its effectiveness in detecting objects across a wide range of scales. Its architecture also comprises a backbone for feature extraction, an FPN neck for multi-scale feature aggregation, and a head with two sub-heads for classification and bounding box regression. Notably, RetinaNet introduces a novel loss function, the Focal Loss, designed to focus training on hard negative examples and thereby mitigate the foreground–background class imbalance.

RetinaNet configurations were set up with ResNet-50 and ResNet-101 backbones [31], complemented with an FPN head. The choice of these configurations was informed by their demonstrated efficiency in various object detection tasks, thus enabling a comprehensive evaluation of their performance in the detection of *T. absoluta* infestations.

The employment of different backbones and heads in these models serves to explore the speed–accuracy trade-off and the level of detail captured in the object detection task. For instance, a ResNet-101 backbone, with its deeper architecture, is expected to capture more intricate features compared to a ResNet-50 backbone, albeit at a higher computational cost. Similarly, the choice of head (FPN, C4, or DC5) influences the model's capability to handle varying object scales and complexities, which is crucial for accurately detecting *T. absoluta* damage in diverse real-world agricultural scenarios.

3.3. Ensemble Techniques

The process of combining predictions from the multiple object detection models was achieved using several ensemble techniques, namely Non-Maximum Suppression (NMS), Soft Non-Maximum Suppression (Soft NMS) [33], Non-Maximum Weighted (NMW) [34], and Weighted Boxes Fusion (WBF) [35].

NMS is a fundamental algorithm in object detection that removes duplicate detections by keeping only the bounding box with the highest confidence score and suppressing others that have a high IoU overlap with it. Soft NMS, an improvement over NMS, instead of discarding overlapping detections entirely, decreases their confidence scores based on the degree of overlap [33]. The NMW method uses a weighted combination of the overlapping detections to create a more accurate bounding box [34]. The weights are calculated based on the confidence scores of the detections, resulting in a bounding box that better represents the detections. The WBF method uses the detection score to obtain a weighted box [35]. This allows for a fused result that reflects all detection results according to the score, unlike existing methods that use only one result with the highest score when multiple models detect the same object.

All ensemble techniques were applied post the training of the models; hence, no additional training was required. The ensemble process was executed by aggregating the prediction outputs of the individual trained models. This strategy not only conserves computational resources by avoiding the need for further training but also enhances the robustness and accuracy of the detection process, making it a pragmatic approach for real-world agricultural applications [36].

3.4. Object Detection and Ensemble Workflow

Figure 2 illustrates the workflow employed for object detection and ensemble aggregation in this study. The process initiates with an input image which is fed into multiple object detection models. Each of these models processes the image independently to identify and localize *T. absoluta* infestations, producing images annotated with bounding boxes around detected instances of damage. The individual outputs from these models are then aggregated using an ensemble technique that merges the detection results, leveraging the strengths of each model to enhance accuracy and reliability. This ensemble approach effectively combines the diverse detection capabilities of the employed models, ensuring a more comprehensive representation of *T. absoluta* damage instances within the input image. The final output of this workflow is an image with aggregated bounding boxes around detected instances of *T. absoluta* damage, representing a consensus among the models employed.

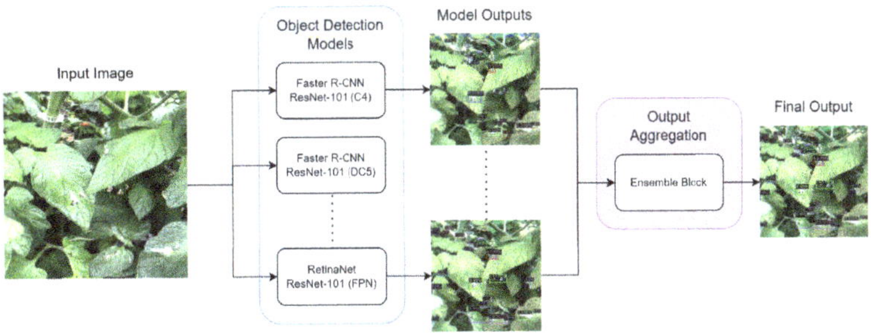

Figure 2. Object detection and ensemble workflow: the diagram delineates the process from input image, through individual model processing (Faster R-CNN, RetinaNet), to ensemble aggregation, producing the final output with aggregated bounding boxes around detected *T. absoluta* damage instances.

3.5. Object Detection Model Training

The object detection models were trained using a specific set of hyperparameters. Initially, each backbone was pre-trained on the COCO 2017 dataset [37]. The models were then trained on an image input size of 1024 × 1024 for a total of 20,000 iterations with a batch size of 2. The Stochastic Gradient Descent (SGD) optimization algorithm was used for model training. The learning rate was set at 0.001, with a momentum of 0.9 and a weight decay of 1×10^{-5}. In order to facilitate better learning, the learning rate was reduced by 0.1 at both the 80% and 90% completion stages of the total iterations. This strategy is known to help models converge more effectively towards the end of the training process. To improve model robustness and ensure that the models could generalize to new data, several data augmentation techniques were applied during training. These included vertical flipping, horizontal flipping, rotation, and Contrast Limited Adaptive Histogram Equalization (CLAHE) [38].

All models were trained using the Detectron2 framework [39] in PyTorch [40]. The training was executed on a PC equipped with an Intel Xeon CPU, an NVIDIA Tesla K80 GPU with 16GB of VRAM, and 12 GB of RAM.

3.6. Evaluation Metrics

The evaluation of the object detection models in this study was performed using the COCO detection metrics [37]. These metrics have been widely adopted in the field for their comprehensive evaluation of both localization and classification aspects of object detection. The primary metric of interest in this study was mAP_{50}, which represents the mean Average Precision (mAP) at an Intersection over Union (IoU) threshold of 0.5 across all classes (in this paper, the terms mAP and mAP_{50} are used interchangeably, indicating the same concept).

4. Results and Discussion

4.1. Initial Model Evaluations

The initial performance evaluation, as summarized in Table 1, revealed varying levels of effectiveness among the object detection models, taking both accuracy and inference time (in seconds per image) into consideration. An mAP_{50} score of 0.56 was achieved by the Faster R-CNN model with a ResNet-101 backbone and C4 head, with an inference time of 0.139 s per image. Its DC5 and FPN variants completed inferences in 0.086 and 0.051 s per image, respectively, and yielded mAP_{50} scores of 0.58 and 0.56.

Faster R-CNN with a ResNet-50 backbone and C4 head reached an mAP_{50} score of 0.58, taking 0.103 s per image for inference. The DC5 and FPN heads on the same backbone completed their inferences in 0.069 and 0.038 s per image, respectively, with mAP_{50} scores of 0.58 and 0.55. Notably, the Faster R-CNN model with the ResNet-50 backbone and FPN head boasted the fastest inference time among all evaluated configurations.

With a ResNeXt-101 backbone and FPN head, the Faster R-CNN model displayed an mAP_{50} score of 0.56 and took 0.098 s per image for inference. RetinaNet models equipped with ResNet-101 and ResNet-50 backbones, both with FPN heads, showed mAP_{50} scores of 0.57. Their inference times were 0.054 and 0.041 s per image, respectively.

Table 1. Initial performance evaluation (mAP_{50} and inference time in seconds per image) of object detection models across various model configurations, including different backbones and heads.

Model	Backbone	Head	mAP_{50}	Inference Time
Faster R-CNN	ResNet-101	C4	0.56	0.139
Faster R-CNN	ResNet-101	DC5	0.58	0.086
Faster R-CNN	ResNet-101	FPN	0.56	0.051
Faster R-CNN	ResNet-50	C4	0.58	0.103
Faster R-CNN	ResNet-50	DC5	0.58	0.069
Faster R-CNN	ResNet-50	FPN	0.55	0.038
Faster R-CNN	ResNeXt-101	FPN	0.56	0.098
RetinaNet	ResNet-101	FPN	0.57	0.054
RetinaNet	ResNet-50	FPN	0.57	0.041

4.2. Qualitative Results

Detecting *T. absoluta* damage on tomato plants is a challenging task due to the inherent imperfections of individual detection models. When multiple object detection models are employed, it is common for certain instances of damage to be detected by one model while being missed by others (Figure 3). Such inconsistencies in detection results can significantly impact the overall accuracy and reliability of the detection process. To address this limitation, ensemble techniques offer a viable solution for improving *T. absoluta* detection. By combining the outputs of multiple models, ensemble methods can effectively encompass a wider range of *T. absoluta* damage instances, capitalizing on the strengths of different models and compensating for their respective shortcomings.

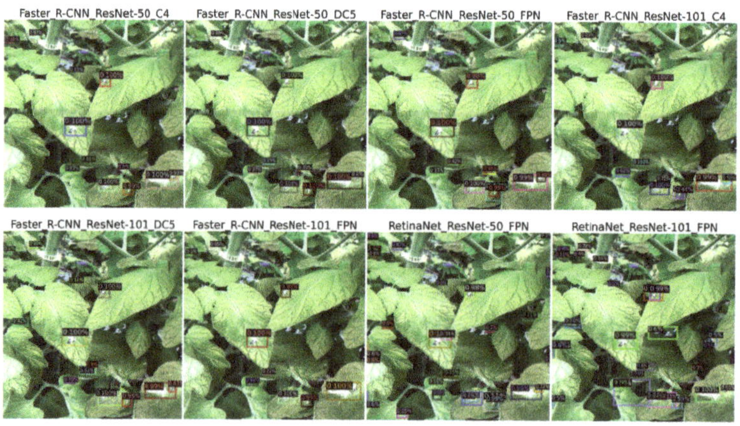

Figure 3. Example of *T. absoluta* damage instances identified by various models. The image represents the top left image of Figure 1, demonstrating varying bounding box predictions by each model. Each bounding box is annotated with its respective class, and class 0 signifies damage attributed to *T. absoluta*, accompanied by the predicted probability for each identified object.

Individual models may fail to accurately detect *T. absoluta* infestations due to various factors. The limitations often arise from the model's architecture, training data, and the inherent variability in real-world agricultural imagery [36]. Furthermore, the model's ability to generalize might be hindered if the training data is not diverse enough to capture the range of conditions under which *T. absoluta* damage occurs [41]. Furthermore, the hyperparameters and the architecture of the model might not be optimal for the specific task of detecting *T. absoluta* infestations on tomato plants [36].

4.3. Ensemble Results

In an effort to improve the object detection performance, four ensemble techniques were benchmarked to assess their effectiveness in enhancing the detection of *T. absoluta* infestations on tomato plants. Different configurations of models were ensembled using NMS, Soft NMS, NMW, and WBF. NMS was configured with an IoU threshold of 0.1, a sigma of 0.01, and a box-skipping threshold of 0.01. For Soft NMS, only the box-skipping threshold of 0.01 was applied. NMW utilized an IoU threshold of 0.4 and a box-skipping threshold of 0.01. Lastly, the WBF method was implemented with an IoU threshold of 0.5 and a threshold of 0.01 for skipping a box.

The ensemble results, as depicted in Table 2, demonstrated the impact of different ensemble configurations on the mAP_{50} scores and the associated inference time. The total inference time for each ensemble configuration is calculated as $T_{\text{total}} = \sum_{i=1}^{N} T_i + T_{\text{ensemble}}$, where T_{total} is the total inference time, T_i is the inference time of the ith model, T_{ensemble} is the fixed additional latency due to the ensemble technique, and N is the total number of models in the ensemble.

A detailed examination of the ensemble techniques uncovered the specific average times associated with each. The NMS method recorded an average inference time of 1.9046×10^{-5} s. The Soft NMS recorded an average time of 4.2064×10^{-5} s. The NMW method emerged as faster with an average time of 3.5040×10^{-5} s. Impressively, the WBF method was the fastest, clocking an average time of 1.0002×10^{-5} s. However, these ensemble techniques contribute a slight increase to the total inference time of the processing.

The ensemble of all models resulted in mAP_{50} scores of 0.70, 0.66, 0.63, and 0.61 for WBF, NMW, Soft NMS, and NMS, respectively. When ensembling only Faster R-CNN models, the scores were slightly lower: 0.69 for WBF, 0.66 for NMW, 0.63 for Soft NMS, and 0.61 for NMS. Ensembling the RetinaNet models, the scores were 0.64, 0.63, 0.60, and 0.59 for WBF, NMW, Soft NMS, and NMS, respectively. For ensembles of Faster R-CNN models with a ResNet-101 backbone, the scores ranged from 0.64 for WBF to 0.59 for NMS. When only the Faster R-CNN models with a ResNet-50 backbone were ensembled, the scores increased to 0.67 for WBF and 0.60 for NMS. The ensemble of all Faster R-CNN models with a C4 head resulted in scores of from 0.63 for WBF to 0.59 for NMS. For the DC5 head, the scores were similar, ranging from 0.63 for WBF to 0.60 for NMS. Lastly, the ensemble of all Faster R-CNN models with an FPN head produced the lowest scores, ranging from 0.61 for WBF to 0.56 for NMS.

The success of ensemble models in this study is clearly illustrated in Figure 4, which demonstrates the superior performance of ensemble models in terms of mAP_{50} scores compared to individual models, despite the inference time. For a balance between accuracy and inference time, the ensemble of RetinaNet models lies on the top left of Figure 4, showcasing a favorable trade-off. However, for enhanced accuracy, ensembles involving all models or just Faster R-CNN models are preferred as they achieve higher mAP_{50} scores. In particular, the WBF ensemble stands out in achieving a higher mAP_{50} score, showcasing the advantage of aggregating detection results from different models.

These ensemble techniques, namely NMS, Soft NMS, NMW, and WBF, effectively merge the detection results from different models, leading to a more robust and accurate detection of *T. absoluta* damage. The aggregated knowledge from different models provides a broader understanding of the data, which is crucial for handling the variability and complexities inherent in real-world agricultural imagery [42]. Ensemble models often outperform single models as they benefit from the diversity among the models in the ensemble [43]. The diversity can arise from differences in the model architectures, training data, or the learning algorithms, which contribute to a more comprehensive and reliable detection performance [36]. The fusion of detection results from diverse models, facilitated by ensemble techniques, forms a more complete representation of the data, thus enhancing the model's ability to accurately detect *T. absoluta* infestations across varying conditions encountered in agricultural settings.

Table 2. Ensemble results (mAP_{50}) using different configurations of models and ensemble techniques.

Ensemble Configuration	WBF	NMW	Soft NMS	NMS
All models	0.70	0.66	0.63	0.61
All Faster R-CNN	0.69	0.66	0.63	0.61
All RetinaNet	0.64	0.63	0.60	0.59
All Faster R-CNN ResNet-101	0.64	0.62	0.60	0.59
All Faster R-CNN ResNet-50	0.67	0.64	0.61	0.60
All Faster R-CNN C4	0.63	0.61	0.60	0.59
All Faster R-CNN DC5	0.63	0.62	0.60	0.60
All Faster R-CNN FPN	0.61	0.60	0.58	0.56

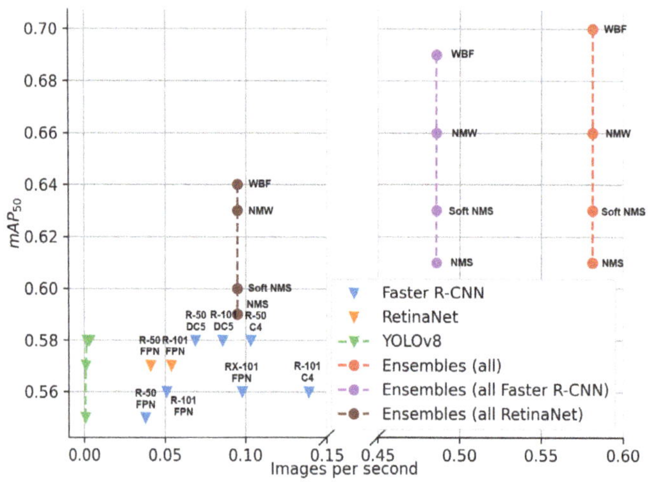

Figure 4. Comparison of mAP_{50} scores versus inference time (images per second) among individual and ensemble models. Higher mAP_{50} values indicate better model performance in accurately detecting *T. absoluta* damage.

4.4. Comparison with State-of-the-Art

The results generated in this study are significant, particularly when compared with those obtained by Loyani et al. (2021) [23] and Georgantopoulos et al. (2023) [24]. Loyani et al. managed to achieve a high mAP_{50} of 0.8567 using Mask R-CNN for detecting *T. absoluta* effects, which was facilitated by the large size of the depicted objects in their dataset. However, even though this study tackles a more challenging detection scenario, the ensemble methods, especially the WBF, managed to achieve an mAP_{50} score of 0.70, showcasing the robustness and effectiveness of the employed ensemble techniques.

In addition, Georgantopoulos et al. leveraged multispectral images, which provide richer spectral information compared to color images, yet their Faster-RCNN model only achieved a 0.202 mAP_{50} on detection. This stark contrast in performance, compared to the 0.70 mAP_{50} score achieved with the ensemble of all models in this study, underscores the significance of the ensemble techniques deployed. The refined configurations in the ensemble methods, such as combining models with different backbones and heads, demonstrated an ability to adjust to specific attributes of the data, thereby optimizing detection performance.

4.5. Benchmark Analysis with YOLOv8

For a more comprehensive benchmark analysis, additional state-of-the-art models were examined alongside the proposed ensemble techniques employed with Faster R-CNN and RetinaNet. Specifically, various versions of YOLOv8 [44], namely YOLOv8n, YOLOv8s, YOLOv8m, YOLOv8l, and YOLOv8x, were trained and evaluated. Every variant varied in network depth, denoted by a growing number of layers, and network width, demonstrated by an increasing number of filters in these layers. Each model was trained for 100 epochs, with a patience of 15, a batch size of 8 (4 in the case of YOLOv8x), and an image size of 1024×1024. The SGD optimizer was used with a learning rate of 0.01, momentum of 0.9, and a weight decay of 0.0005.

All models were trained using the Ultralytics framework [44] in PyTorch [40]. The training was executed on a PC equipped with an Intel Xeon CPU, an NVIDIA Tesla K80 GPU with 16 GB of VRAM, and 12 GB of RAM.

The results, shown in Table 3, reveal that the YOLOv8 models achieved comparable performance to the models in the initial model evaluations, albeit not surpassing them. This underlines the effectiveness of the ensemble approach proposed in this study, which has shown superior performance in detecting *T. absoluta* infestations on tomato plants, as shown in Figure 4. The ensemble techniques enhanced the mAP_{50} scores, showcasing the advantage of this study's methodology in the realm of agricultural pest detection.

Table 3. Performance evaluation (mAP_{50}, recall, and precision) of YOLOv8 models.

Model	Depth	Width	mAP$_{50}$	Recall	Precision
YOLOv8n	0.33	0.25	0.55	0.53	0.60
YOLOv8s	0.33	0.50	0.57	0.55	0.61
YOLOv8m	0.67	0.75	0.58	0.54	0.61
YOLOv8l	1.00	1.00	0.58	0.57	0.62
YOLOv8x	1.00	1.25	0.58	0.54	0.65

Using ensemble techniques with Faster R-CNN and RetinaNet models allows this study to draw from the strengths of each model, overcoming the limitations of each when used alone. Unlike using a single model like YOLOv8, the ensemble method combines the detection capabilities of different models, achieving more reliable and accurate detection output. The shared information between Faster R-CNN and RetinaNet models, facilitated by ensemble techniques, provides a better understanding of the data, leading to more accurate detection of *T. absoluta* infestations even in challenging real-world agricultural scenarios.

Moreover, the ensemble approach balances the distinct attributes of Faster R-CNN and RetinaNet, which, when combined, contribute to a higher mAP_{50} score compared to using a single model like YOLOv8. This pairing, arranged by ensemble techniques, results in a detection pipeline that performs better than the YOLOv8 models. The comparative performance of YOLOv8 models in this benchmark analysis further supports the chosen methodology of this study, affirming the better performance and usefulness of ensemble techniques in real-world agricultural pest detection tasks.

5. Conclusions

In conclusion, the detection of *T. absoluta* infestation on tomato plants was addressed in this study using object detection models combined with ensemble techniques. This approach contrasts with previous works by focusing on the challenges of detecting *T. absoluta* infestations in real-life conditions, including open-field and greenhouse environments, and specifically identifying and localizing *T. absoluta* damage on tomato plants.

Initial evaluations of the Faster R-CNN and RetinaNet models, involving different backbones and heads, suggested a need for a refined strategy to enhance performance. The application of ensemble techniques, namely NMS, Soft NMS, NMW, and WBF, successfully improved detection predictions, leading to enhanced mAP scores. In particular, the employment of WBF led to a significant increment of 20% in the mAP score, from 0.58 to 0.70, highlighting the potential of ensemble methods to address the challenges associated with *T. absoluta* detection.

By enhancing the accuracy and reliability of *T. absoluta* detection using deep-learning-based object detection models combined with ensemble techniques, a contribution is made towards the development of more effective pest management strategies in tomato production. This study underscores the potential of these techniques in the field of agricultural pest detection and provides valuable insights for the development of advanced detection systems, aiding farmers in effectively managing *T. absoluta* infestations and minimizing economic losses in tomato production. Moreover, these technological strides resonate well with the objectives of integrated pest management, emphasizing a shift towards sustainable agriculture. This progression directly supports the Green Deal's ambitious aim of achieving a 50% reduction in pesticide use by 2030.

In future work, the refined models from this study could enable artificial-intelligence-enabled robotic traps to monitor and detect insects in real-time, integrating with decision support systems or mobile apps for immediate alerting to mitigate insect or moth attacks and spreads. Given that these traps capture images periodically, the enhanced accuracy of the models, even at the expense of minimal additional inference time, is a crucial attribute. Moreover, the models can be adapted for autonomous navigated mobile robots or unmanned aerial vehicles equipped for in-field detection and surveillance. Integrated with deep learning models, these robotic systems promise enhanced detection accuracy in precision agriculture, even in low fps modes, providing a sophisticated solution for real-time, on-site pest monitoring and management.

Author Contributions: Conceptualization, N.G. and E.-M.P.; methodology, N.G.; software, N.G.; validation, D.T.; formal analysis, E.-M.P.; investigation, E.-M.P. and N.G.; resources, E.-M.P.; data curation, N.F.; writing—original draft preparation, N.G. and E.-M.P.; writing—review and editing, N.F.; visualization, N.G.; supervision, D.T.; project administration, E.-M.P.; funding acquisition, E.-M.P. All authors have read and agreed to the published version of the manuscript.

Funding: This research was funded by "Horizon 2020 PestNu project, grant number 101037128". The APC was funded by the Centre for Research and Technology Hellas.

Institutional Review Board Statement: Not applicable.

Informed Consent Statement: Not applicable.

Data Availability Statement: The data presented in this study are available on request from the corresponding author. The data are not publicly available due to their purchase by the EDEN library.

Conflicts of Interest: The author Nikolaos Frangakis was employed by the company iKnowHow S.A. The remaining authors declare that the research was conducted in the absence of any commercial or financial relationships that could be construed as a potential conflict of interest.

Abbreviations

The following abbreviations are used in this manuscript:

C4	Convolutional Layer 4
COCO	Common Objects in Context
CLAHE	Contrast Limited Adaptive Histogram Equalization
CNN	Convolutional Neural Network
DC5	Dilated Convolutional Layer 5
FPN	Feature Pyramid Network
Faster R-CNN	Faster Region-Based Convolutional Neural Network
IoU	Intersection over Union

mAP	mean Average Precision
NMS	Non-Maximum Suppression
NMW	Non-Maximum Weighted
ResNet	Residual Network
SGD	Stochastic Gradient Descent
Soft NMS	Soft Non-Maximum Suppression
WBF	Weighted Boxes Fusion
YOLO	You Only Look Once

References

1. Zhang, S.; Griffiths, J.S.; Marchand, G.; Bernards, M.A.; Wang, A. Tomato brown rugose fruit virus: An emerging and rapidly spreading plant RNA virus that threatens tomato production worldwide. *Mol. Plant Pathol.* **2022**, *23*, 1262–1277. [CrossRef] [PubMed]
2. Tomato News. The Global Tomato Processing Industry. 2018. Available online: https://www.tomatonews.com/en/background_47.html (accessed on 1 May 2023).
3. Zekeya, N.; Chacha, M.; Ndakidemi, P.; Materu, C.; Chidege, M.; Mbega, E. Tomato Leafminer (*Tuta absoluta* Meyrick 1917): A Threat to Tomato Production in Africa. *J. Agric. Ecol. Res. Int.* **2017**, *10*, 1–10. [CrossRef]
4. Urbaneja, A.; Vercher, R.; Navarro, V.; Marí, F.G.; Porcuna, J.L. La polilla del tomate, *Tuta absoluta*. *Phytoma España* **2007**, *194*, 16–23.
5. Chidege, M.; Al-zaidi, S.; Hassan, N.; Julie, A.; Kaaya, E.; Mrogoro, S. First record of tomato leaf miner *Tuta absoluta* (Meyrick) (Lepidoptera: Gelechiidae) in Tanzania. *Agric. Food Secur.* **2016**, *5*, 17. [CrossRef]
6. Guimapi, R.Y.; Mohamed, S.A.; Okeyo, G.O.; Ndjomatchoua, F.T.; Ekesi, S.; Tonnang, H.E. Modeling the risk of invasion and spread of *Tuta absoluta* in Africa. *Ecol. Complex.* **2016**, *28*, 77–93. [CrossRef]
7. Guedes, R.N.C.; Picanço, M.C. The tomato boreri *Tuta absoluta*/iin South America: Pest status, management and insecticide resistance. *EPPO Bull.* **2012**, *42*, 211–216. [CrossRef]
8. Zahedi, S.R.; Zahedi, S.M. Role of Information and Communication Technologies in modern agriculture. *Int. J. Agric. Crop Sci.* **2012**, *4*, 1725–1728.
9. Singh, A.; Ganapathysubramanian, B.; Singh, A.K.; Sarkar, S. Machine Learning for High-Throughput Stress Phenotyping in Plants. *Trends Plant Sci.* **2016**, *21*, 110–124. [CrossRef]
10. Giakoumoglou, N.; Pechlivani, E.M.; Katsoulas, N.; Tzovaras, D. White Flies and Black Aphids Detection in Field Vegetable Crops using Deep Learning. In Proceedings of the 2022 IEEE 5th International Conference on Image Processing Applications and Systems (IPAS), Genova, Italy, 5–7 December 2022; Volume 5, pp. 1–6. [CrossRef]
11. Giakoumoglou, N.; Pechlivani, E.M.; Sakelliou, A.; Klaridopoulos, C.; Frangakis, N.; Tzovaras, D. Deep learning-based multi-spectral identification of grey mould. *Smart Agric. Technol.* **2023**, *4*, 100174. [CrossRef]
12. Giakoumoglou, N.; Pechlivani, E.M.; Tzovaras, D. Generate-Paste-Blend-Detect: Synthetic Dataset for Object Detection in the Agriculture Domain. *Smart Agric. Technol.* **2023**, *5*, 100258. [CrossRef]
13. Wang, Q.; Qi, F.; Sun, M.; Qu, J.; Xue, J. Identification of Tomato Disease Types and Detection of Infected Areas Based on Deep Convolutional Neural Networks and Object Detection Techniques. *Comput. Intell. Neurosci.* **2019**, *2019*, 9142753. [CrossRef] [PubMed]
14. Wang, X.; Liu, J.; Liu, G. Diseases Detection of Occlusion and Overlapping Tomato Leaves Based on Deep Learning. *Front. Plant Sci.* **2021**, *12*, 792244. [CrossRef]
15. Fuentes, A.; Yoon, S.; Kim, S.; Park, D. A Robust Deep-Learning-Based Detector for Real-Time Tomato Plant Diseases and Pests Recognition. *Sensors* **2017**, *17*, 2022. [CrossRef] [PubMed]
16. Zaborowicz, M.; Boniecki, P.; Koszela, K.; Przybylak, A.; Przybył, J. Application of neural image analysis in evaluating the quality of greenhouse tomatoes. *Sci. Hortic.* **2017**, *218*, 222–229. [CrossRef]
17. Zaborowicz, M.; Boniecki, P.; Koszela, K.; Przybył, J.; Mazur, R.; Kujawa, S.; Pilarski, K. Use of artificial neural networks in the identification and classification of tomatoes. In Proceedings of the SPIE Proceedings, Beijing, China, 21–22 April 2013. [CrossRef]
18. Brahimi, M.; Boukhalfa, K.; Moussaoui, A. Deep Learning for Tomato Diseases: Classification and Symptoms Visualization. *Appl. Artif. Intell.* **2017**, *31*, 299–315. [CrossRef]
19. Ferentinos, K.P. Deep learning models for plant disease detection and diagnosis. *Comput. Electron. Agric.* **2018**, *145*, 311–318. [CrossRef]
20. Liang, Q.; Xiang, S.; Hu, Y.; Coppola, G.; Zhang, D.; Sun, W. PD2SE-Net: Computer-assisted plant disease diagnosis and severity estimation network. *Comput. Electron. Agric.* **2019**, *157*, 518–529. [CrossRef]
21. Rangarajan, A.K.; Purushothaman, R.; Ramesh, A. Tomato crop disease classification using pre-trained deep learning algorithm. *Procedia Comput. Sci.* **2018**, *133*, 1040–1047. [CrossRef]
22. Trivedi, N.K.; Gautam, V.; Anand, A.; Aljahdali, H.M.; Villar, S.G.; Anand, D.; Goyal, N.; Kadry, S. Early Detection and Classification of Tomato Leaf Disease Using High-Performance Deep Neural Network. *Sensors* **2021**, *21*, 7987. [CrossRef]
23. Loyani, L.K.; Bradshaw, K.; Machuve, D. Segmentation of *Tuta absoluta*'s Damage on Tomato Plants: A Computer Vision Approach. *Appl. Artif. Intell.* **2021**, *35*, 1107–1127. [CrossRef]

24. Georgantopoulos, P.; Papadimitriou, D.; Constantinopoulos, C.; Manios, T.; Daliakopoulos, I.; Kosmopoulos, D. A Multispectral Dataset for the Detection of *Tuta absoluta* and Leveillula Taurica in Tomato Plants. *Smart Agric. Technol.* **2023**, *4*, 100146. [CrossRef]
25. Rubanga, D.P.; Loyani, L.K.; Richard, M.; Shimada, S. A Deep Learning Approach for Determining Effects of *Tuta absoluta* in Tomato Plants. *arXiv* **2020**, arXiv:2004.04023.
26. Mkonyi, L.; Rubanga, D.; Richard, M.; Zekeya, N.; Sawahiko, S.; Maiseli, B.; Machuve, D. Early identification of *Tuta absoluta* in tomato plants using deep learning. *Sci. Afr.* **2020**, *10*, e00590. [CrossRef]
27. Tsiakas, K.; Papadimitriou, A.; Pechlivani, E.M.; Giakoumis, D.; Frangakis, N.; Gasteratos, A.; Tzovaras, D. An Autonomous Navigation Framework for Holonomic Mobile Robots in Confined Agricultural Environments. *Robotics* **2023**, *12*, 146. [CrossRef]
28. Pechlivani, E.M.; Gkogkos, G.; Giakoumoglou, N.; Hadjigeorgiou, I.; Tzovaras, D. Towards Sustainable Farming: A Robust Decision Support System's Architecture for Agriculture 4.0. In Proceedings of the 2023 24th International Conference on Digital Signal Processing (DSP), Rhodes, Greece, 11–13 June 2023. [CrossRef]
29. Ren, S.; He, K.; Girshick, R.; Sun, J. Faster R-CNN: Towards Real-Time Object Detection with Region Proposal Networks. *arXiv* **2015**, arXiv:1506.01497.
30. Lin, T.; Goyal, P.; Girshick, R.B.; He, K.; Dollár, P. Focal Loss for Dense Object Detection. *arXiv* **2017**, arXiv:1708.02002.
31. He, K.; Zhang, X.; Ren, S.; Sun, J. Deep Residual Learning for Image Recognition. *arXiv* **2015**, arXiv:1512.03385.
32. Lin, T.; Dollár, P.; Girshick, R.B.; He, K.; Hariharan, B.; Belongie, S.J. Feature Pyramid Networks for Object Detection. In Proceedings of the 2016 IEEE Conference on Computer Vision and Pattern Recognition (CVPR), Las Vegas, NV, USA, 27–30 June 2016.
33. Bodla, N.; Singh, B.; Chellappa, R.; Davis, L.S. Soft-NMS—Improving Object Detection With One Line of Code. *arXiv* **2017**, arXiv:cs.CV/1704.04503.
34. Zhou, H.; Li, Z.; Ning, C.; Tang, J. CAD: Scale Invariant Framework for Real-Time Object Detection. In Proceedings of the 2017 IEEE International Conference on Computer Vision Workshops (ICCVW), Venice, Italy, 22–29 October 2017; pp. 760–768. [CrossRef]
35. Solovyev, R.; Wang, W.; Gabruseva, T. Weighted boxes fusion: Ensembling boxes from different object detection models. *Image Vis. Comput.* **2021**, *107*, 104117. [CrossRef]
36. Mohammed, A.; Kora, R. A comprehensive review on ensemble deep learning: Opportunities and challenges. *J. King Saud Univ. Comput. Inf. Sci.* **2023**, *35*, 757–774. [CrossRef]
37. Lin, T.Y.; Maire, M.; Belongie, S.; Bourdev, L.; Girshick, R.; Hays, J.; Perona, P.; Ramanan, D.; Zitnick, C.L.; Dollár, P. Microsoft COCO: Common Objects in Context. *arXiv* **2014**, arXiv:1405.0312.
38. Pizer, S.; Johnston, R.; Ericksen, J.; Yankaskas, B.; Muller, K. Contrast-limited adaptive histogram equalization: Speed and effectiveness. In Proceedings of the 1990 First Conference on Visualization in Biomedical Computing, Atlanta, GA, USA, 22–25 May 1990. [CrossRef]
39. Wu, Y.; Kirillov, A.; Massa, F.; Lo, W.Y.; Girshick, R. Detectron2. 2019. Available online: https://github.com/facebookresearch/detectron2 (accessed on 1 May 2023).
40. Paszke, A.; Gross, S.; Massa, F.; Lerer, A.; Bradbury, J.; Chanan, G.; Killeen, T.; Lin, Z.; Gimelshein, N.; Antiga, L.; et al. PyTorch: An Imperative Style, High-Performance Deep Learning Library. In *Advances in Neural Information Processing Systems 32*; Curran Associates, Inc.: Red Hook, NY, USA, 2019; pp. 8024–8035.
41. Gong, Z.; Zhong, P.; Hu, W. Diversity in Machine Learning. *IEEE Access* **2019**, *7*, 64323–64350. [CrossRef]
42. Patrício, D.I.; Rieder, R. Computer vision and artificial intelligence in precision agriculture for grain crops: A systematic review. *Comput. Electron. Agric.* **2018**, *153*, 69–81. [CrossRef]
43. Sagi, O.; Rokach, L. Ensemble learning: A survey. *WIREs* **2018**, *8*, e1249. [CrossRef]
44. Jocher, G.; Chaurasia, A.; Qiu, J. YOLO by Ultralytics. 2023. Available online: https://github.com/ultralytics/ultralytics (accessed on 1 May 2023).

AI

MDPI

Article

AgriNAS: Neural Architecture Search with Adaptive Convolution and Spatial–Time Augmentation Method for Soybean Diseases

Oluwatoyin Joy Omole [1,†], Renata Lopes Rosa [1,†], Muhammad Saadi [2,†] and Demóstenes Zegarra Rodriguez [1,*,†]

1 Department of Computer Science, Federal University of Lavras, Lavras 37200-000, Minas Gerais, Brazil; oluwatoyin.omole@estudante.ufla.br (O.J.O.); renata.rosa@ufla.br (R.L.R.)

2 School of Science and Technology, Department of Computer Science, Nottingham Trent University, Nottingham NG11 8NS , UK; muhammad.saadi@ntu.ac.uk

* Correspondence: demostenes.zegarra@ufla.br

† These authors contributed equally to this work.

Abstract: Soybean is a critical agricultural commodity, serving as a vital source of protein and vegetable oil, and contributing significantly to the economies of producing nations. However, soybean yields are frequently compromised by disease and pest infestations, which, if not identified early, can lead to substantial production losses. To address this challenge, we propose AgriNAS, a method that integrates a Neural Architecture Search (NAS) framework with an adaptive convolutional architecture specifically designed for plant pathology. AgriNAS employs a novel data augmentation strategy and a Spatial–Time Augmentation (STA) method, and it utilizes a multi-stage convolutional network that dynamically adapts to the complexity of the input data. The proposed AgriNAS leverages powerful GPU resources to handle the intensive computational tasks involved in NAS and model training. The framework incorporates a bi-level optimization strategy and entropy-based regularization to enhance model robustness and prevent overfitting. AgriNAS achieves classification accuracies superior to VGG-19 and a transfer learning method using convolutional neural networks.

Keywords: neural architecture search; adaptive convolutional architecture; data augmentation; soybean disease and pest detection

1. Introduction

Soybean is a globally significant legume crop, renowned for its high protein and oil content, making it a vital resource for both human and animal consumption [1]. Soybean is recognized as the plant that yields the highest protein per cultivated area, playing a crucial role in the global agricultural economy [2,3]. It accounts for 48% of the world's oil crop market, underlining its economic importance [4]. Soybean is extensively cultivated in countries such as the United States and Brazil, which together contribute over 80% of global exports [5]. The crop's significance is further highlighted by its utility in producing protein meal and vegetable oil, both of which are essential in the global food industry [6]. However, soybean yields are frequently threatened by pests and diseases, which can cause substantial losses [7]. Estimates suggest that pests alone are responsible for yield losses ranging between 26 and 29%, significantly impacting both the quantity and quality of soybean production [8,9]. Early and accurate detection of these threats is crucial for implementing timely interventions and minimizing yield losses [10].

Given the global importance of soybean as a staple crop, the ability to protect and enhance its yield through technological innovations is crucial. This has led to the increasing adoption of advanced artificial intelligence (AI) [11,12] techniques in agriculture, particularly in the areas of pest and disease detection. The convergence of agriculture and AI represents a significant step forward in addressing the challenges of food security, resource optimization, and sustainable farming practices. Among these AI advancements, deep learning (DL) [13] has shown remarkable potential due to its capacity to analyze large datasets and identify complex patterns that are not easily discernible by traditional methods [14].

In recent years, deep learning models [15] have emerged as powerful tools for the detection of plant pests and diseases [16,17], offering significant advantages over traditional methods. Traditional techniques for detecting pests and diseases, such as manual inspection or simple rule-based image processing methods, are often labor-intensive, time-consuming, and prone to human error. Furthermore, these methods struggle to accurately detect subtle variations in symptoms or adapt to new disease manifestations without extensive reprogramming or manual intervention.

By contrast, deep learning models, specifically Convolutional Neural Networks (CNNs), excel in recognizing complex patterns in large datasets [18], making them particularly well-suited for tasks involving high variability and subtle symptoms in agricultural images. Quantitative comparisons have demonstrated that CNN-based methods typically achieve superior performance in classification tasks, with accuracy rates often exceeding 90% [19], significantly higher than the 70–80% accuracy commonly reported for traditional image processing techniques [20]. This is primarily due to CNNs' ability to automatically learn hierarchical features from data, ranging from low-level textures to high-level patterns, which traditional methods struggle to capture.

Moreover, deep learning models can generalize well across different datasets without the need for manual feature engineering, significantly reducing the time and expertise required for implementation [21,22]. For example, CNNs have been shown to reduce error rates by up to 25% compared to traditional methods in studies focused on plant disease detection [20]. This makes deep learning a far more scalable and adaptable approach for real-time agricultural applications, where new diseases or pest infestations can emerge rapidly.

However, as the complexity and size of datasets in agriculture continue to grow, the demand for more efficient and scalable deep learning models becomes paramount. Traditional methods of designing and training neural networks are often time-consuming and computationally expensive [23]. This is especially challenging in the agricultural domain, where data variability and environmental factors require models that are both robust and adaptable. The manual design of optimal CNN architectures for such tasks can be a daunting and error-prone process, often leading to suboptimal solutions that do not fully exploit the available data [24].

To address the challenges of accurately and efficiently detecting pests and diseases in agriculture, this work leverages the integration of Neural Architecture Search (NAS) frameworks, a promising solution for automating the design of high-performing neural networks tailored to specific tasks [25]. Traditional methods for neural network design often involve manual tuning and heuristic approaches, which can be time-consuming, error-prone, and suboptimal in performance. NAS systematically explores a vast search space of possible neural network configurations, identifying architectures that are specifically optimized for accuracy, efficiency, and robustness.

By automating the process of architecture design, NAS addresses key shortcomings in existing methods—namely, the reliance on manually designed models that may not generalize well across different agricultural environments or datasets. This work focuses on applying NAS to agricultural image analysis to develop neural networks that are better suited to handling the variability and complexity inherent in detecting plant pests and diseases. The proposed approach not only accelerates model development but also ensures that the resulting architectures are optimized for both performance and resource efficiency, which is crucial for real-world deployment in resource-constrained environments such as farms.

In agricultural contexts, the application of NAS offers several key advantages. Firstly, NAS can optimize models to handle the variability inherent in agricultural data, such as differences in pest and disease manifestations across regions, variations in crop health due to environmental factors, and the presence of noise in data collected from the field. Secondly, NAS can be used to develop models that are computationally efficient, making them suitable for deployment in resource-constrained environments, such as rural areas where access to high-performance computing infrastructure may be limited. Finally, the use of NAS can lead to the discovery of novel architectures that outperform manually designed models, providing a competitive edge in terms of accuracy and robustness [26].

One of the most recent advancements in data augmentation, which plays a critical role in enhancing the robustness of deep learning models, is the introduction of Spatial–Time Augmentation (STA). This novel method, distinct from traditional Generative Adversarial Networks (GANs) used for data augmentation, incorporates both spatial and temporal variations to simulate real-world conditions in agricultural environments. While GANs have been extensively used to generate realistic synthetic data [27,28], STA offers a more dynamic approach by applying transformations inspired by relativity theory, enhancing the model's ability to generalize across diverse scenarios. This method is particularly useful in agriculture, where temporal changes in pest and disease manifestations can significantly affect model accuracy. The STA method ensures that the augmented data not only increase the dataset size but also capture the variability seen in real-world applications, leading to superior model performance.

This paper introduces AgriNAS, a novel approach that combines NAS with adaptive convolutional architectures specifically designed for soybean pest and disease detection. AgriNAS leverages the power of NAS to automatically discover optimal CNN architectures that are tailored to the unique challenges of agricultural data. By integrating an advanced data augmentation strategy [29], AgriNAS enhances the detection of subtle disease features, improving the model's ability to generalize across diverse agricultural conditions. Additionally, AgriNAS incorporates a multi-stage convolutional network [30] that dynamically adapts to the complexity of the input data, further optimizing both accuracy and computational efficiency.

The use of NAS in this context represents a major advancement in the field of precision agriculture, enabling the development of models that can more effectively address the challenges of pest and disease detection. Moreover, the flexibility of the AgriNAS framework allows it to be extended to other crops and agricultural challenges, highlighting its potential as a versatile tool for modern agriculture.

The contributions of this work include the following:

1. Development of AgriNAS, a tailored Neural Architecture Search framework with adaptive convolutional networks for enhanced detection of soybean pests and diseases.
2. Implementation of a novel data augmentation strategy, a Spatial–Time Augmentation (STA) method, improving the model's ability to generalize across diverse agricultural conditions.
3. Exploration of the potential for expanding AgriNAS to other agricultural applications, showcasing its versatility and impact on the broader field of precision agriculture.

The motivation behind the use of STA lies in its ability to capture both spatial and temporal variability in agricultural data. While spatial transformations enhance the model's ability to generalize across different visual perspectives, the inclusion of temporal augmentation simulates changes in pest and disease manifestations depending on factors such as time of day, season, and environmental changes. This temporal dimension is critical for detecting pests and diseases at different stages of development, making the model more robust in real-world agricultural environments where these factors evolve dynamically.

Our experimental results demonstrate that AgriNAS not only surpasses traditional CNNs but also outperforms other NAS-based solutions in terms of both accuracy and computational efficiency. The framework's ability to adapt to different agricultural conditions and its scalability make it a promising solution for large-scale agricultural applications.

Furthermore, the successful implementation of NAS in AgriNAS underscores the growing importance of automated model development in AI, particularly in domains where data variability and complexity are significant challenges.

The remainder of this article is structured as follows: Section 2 reviews the existing literature on ML and DL methods for pest detection, focusing on advancements in NAS and its applications in agriculture. Section 3 outlines the methodology of AgriNAS, while Section 4 presents the results and discusses their implications. Finally, Section 5 concludes the paper and suggests potential directions for future research.

2. Related Works

This section provides an overview of key research areas relevant to this study, focusing on both the agricultural aspects of soybean pests and the computational algorithms employed for pest detection.

2.1. Soybean Caterpillars

Soybean crops are particularly vulnerable to a range of pests [31], with caterpillars being among the most damaging [32]. These pests can attack the soybean plant at various stages, leading to significant yield losses if not properly managed [33]. Research by [34] highlights the critical impact of bedbugs and caterpillars on soybean health. While soybeans can tolerate a certain level of defoliation—up to 33% during the vegetative stage—the tolerance decreases significantly during the pod set stage, where even 16% defoliation can result in notable yield losses [35]. Species such as the velvet bean caterpillar and the soybean looper are particularly aggressive, with the former capable of stripping an entire field within a week [36].

The dataset used in this study includes images of soybean plants affected by various caterpillar species, particularly those from the genus Spodoptera [37], which are known to cause significant damage during both the vegetative and reproductive stages of the plant [38]. The focus of this study is on the early detection of these pests to prevent irreversible damage.

In [39], the use of a modified Faster R-CNN model for detecting defoliation from canopy-level images was explored. However, the model's precision and recall were limited, emphasizing the need for more targeted approaches, such as using individual leaf images for better accuracy in detecting pest actions.

While recent advancements in Camouflaged Object Detection (COD) demonstrate potential for detecting objects that blend into their surroundings, applying these techniques directly to agricultural pest detection may not be ideal. One significant challenge is that COD methods, such as the Dual-Constraint Coarse-to-Fine Network [40], are typically designed for specific environments and may require extensive retraining to adapt to the unique complexities of agricultural settings. In agriculture, the variety of crops, pest types, and changing environmental conditions introduce considerable variability that COD techniques might not effectively accommodate. Furthermore, the computational complexity of COD methods could hinder real-time applications in resource-constrained environments, where swift detection is crucial for timely interventions.

2.2. Soybean Pests (Diabrotica Speciosa)

Diabrotica speciosa [38] is a significant pest in Latin America, particularly in soybean crops. As adults, these beetles act as defoliators, causing extensive damage to pods and flowers, and are particularly harmful to seedlings and young plants [41,42]. The study by [43] further confirmed that Diabrotica speciosa preferentially feeds on the tender leaves of soybean plants, making precise and early detection crucial for effective pest management.

Recent advances in the use of NAS for agricultural applications have shown significant promise in enhancing the detection and management of pests [44]. NAS frameworks automate the design of neural networks, optimizing their architectures for specific tasks such as image classification and object detection in agricultural settings. For example,

Ref. [45] introduced DARTS, a method that efficiently explores the space of possible network architectures, leading to optimized models that can be adapted for image detection tasks. In the context of image detection, studies have demonstrated that NAS-generated models outperform traditional CNN architectures by providing more accurate and computationally efficient solutions [46]. However, few studies have explored the use of DARTS in pest detection scenarios.

In [47], the NAS is used to develop a model specifically designed for detecting soybean diseases, showing significant improvements in accuracy compared to manually designed networks. This approach, while primarily focused on disease detection, highlights the potential of NAS to be adapted for broader pest management applications, including the detection of pests like Diabrotica speciosa. In [48], a multi-objective NAS framework is employed to balance the trade-offs between model complexity and detection accuracy in agricultural applications. The work emphasizes the adaptability of NAS frameworks to different agricultural environments and crop types, making it a promising approach to the early detection of pests such as Diabrotica speciosa in soybean crops.

The integration of NAS with advanced data augmentation techniques, as discussed in [49], further enhances the model's ability to generalize across varying environmental conditions, which is critical for detecting Diabrotica speciosa in diverse agricultural settings. These advancements underscore the growing role of NAS in developing robust, scalable, and efficient models for pest detection in agriculture. Although NAS offers optimized models for pest detection, it faces significant challenges such as high computational costs [50–52]. Additionally, practical implementation in agricultural environments requires advanced infrastructure and can be complex [53,54]. These limitations highlight the need for improvements to make NAS more accessible and robust [55].

2.3. Data Augmentation Techniques

Data augmentation has become a critical technique for improving the robustness and generalization of deep learning models, particularly in fields like agriculture where the variability in data can be immense. Traditional data augmentation techniques, such as random rotations, flips, scaling, and color adjustments, have been widely employed to artificially increase the size and diversity of training datasets, thus helping models generalize better to unseen data [56]. These techniques have proven effective in a variety of image-based tasks, including the detection of plant diseases and pests [57,58].

In the context of agricultural applications, where conditions and visual characteristics of crops can vary significantly due to environmental factors, data augmentation plays a pivotal role in ensuring that models perform reliably across different scenarios. For instance, Wang et al. [59] investigated the impact of various augmentation methods on the performance of Convolutional Neural Networks (CNNs) in plant disease detection. Their findings revealed that the implementation of comprehensive augmentation techniques, including random rotations, flips, and color adjustments, led to an improvement in model accuracy by over 15%, highlighting the necessity of augmenting training data to capture the variability inherent in agricultural settings.

More recently, advanced data augmentation methods have gained traction, notably the use of Generative Adversarial Networks (GANs) to generate synthetic training data [60]. GANs can produce highly realistic images that augment existing datasets, providing a robust solution to the limitations of traditional augmentation methods. For example, GAN-generated images have been shown to enhance the training set's diversity, resulting in performance improvements of up to 20% in CNNs tasked with disease identification. However, despite their potential, GANs introduce challenges, such as the risk of generating images that may not accurately represent the variations found in real agricultural data [61]. This discrepancy can lead to models that perform well on synthetic data but fail to generalize effectively to real-world conditions. Therefore, while GANs offer a promising avenue for data augmentation, careful consideration of their limitations is essential for practical agricultural applications.

In this study, the STA method is presented as an approach to data augmentation. Unlike traditional or GAN-based techniques, which often rely on static transformations such as rotations or flips, STA introduces both spatial and temporal variability by simulating the relative movement of objects, such as pests, within different time and space dimensions. This is achieved through the application of Lorentzian transformations to the spatial coordinates of the images, which alters the observational perspective dynamically. Additionally, STA introduces a noise component proportional to the simulated relative velocity of the objects, mimicking the inherent uncertainties and irregularities found in real-world agricultural environments.

The use of Lorentzian transformations ensures that the augmented images reflect realistic motion and changes in the position of objects over time, allowing the model to learn from more diverse and complex datasets. By incorporating temporal aspects, STA helps capture the evolving nature of pest infestations and disease symptoms, which is crucial for accurate detection under varying conditions. Furthermore, the addition of noise proportional to velocity enhances the robustness of the model by preparing it to handle data with potential observational inaccuracies.

2.4. Convolutional Neural Networks (CNNs) and Neural Architecture Search (NAS)

CNNs have become the standard for image classification tasks, particularly in agriculture, where precise identification of diseases and pests is critical [62]. CNNs are designed with multiple layers that progressively learn to detect various features, such as edges, shapes, and textures. This hierarchical learning makes them exceptionally well-suited for analyzing plant health from images [63,64].

Quantitatively, CNNs have shown remarkable performance improvements over traditional image classification methods. For instance, in comparison to manual feature extraction techniques that often yield accuracy rates below 80%, CNNs consistently achieve accuracies above 90% in various agricultural tasks. In a study by Bevers et al. [47], a CNN method was implemented, incorporating transfer learning, data engineering, and data augmentation, resulting in an impressive overall testing accuracy of 96.8%. This highlights the capacity of CNNs to learn complex patterns from large datasets, which is particularly beneficial in environments where subtle differences in plant health need to be identified.

However, while traditional CNN architectures have proven effective, they often require extensive computational resources [62], especially when dealing with large and complex datasets [65]. This necessitates the exploration of more efficient architectures through methods like Neural Architecture Search (NAS), which automates the design of neural networks. By optimizing architectures for specific tasks, NAS enables the development of CNNs that not only maintain high accuracy but also reduce computational costs, making them more practical for real-world agricultural applications.

To address these challenges, NAS has emerged as a powerful tool for automating the design of neural networks. NAS frameworks, such as DARTS (Differentiable Architecture Search), allow for the efficient exploration of different network architectures, optimizing them for specific tasks [25]. By automating the search process, NAS can identify architectures that are not only more accurate but also more computationally efficient.

In this study, AgriNAS is introduced, a method that integrates NAS with adaptive convolutional architectures specifically tailored for soybean pest and disease detection. AgriNAS leverages advanced data augmentation strategies to enhance the model's ability to detect subtle disease features, while dynamically adjusting the network architecture to handle varying complexities in the input data. This approach addresses the limitations of traditional CNNs by providing a scalable solution that is both robust and efficient.

3. Methodology

This section details the comprehensive steps involved in developing and validating the AgriNAS model, with in-depth explanations of each component and the software used for simulations.

Figure 1 presents the process, beginning with the collection of soybean leaf images, which then undergo preprocessing using Python 3.12 libraries such as OpenCV for size standardization. Following this, the Spatial–Time Augmentation (STA) method is applied, introducing both spatial and temporal variability into the images. This novel augmentation strategy simulates the relative movement of objects (e.g., pests) across different times and spaces by applying a Lorentzian transformation to the spatial coordinates of features, resulting in augmented images that reflect varying observational perspectives. Additionally, STA adds a noise factor proportional to the simulated relative velocity, ensuring that the augmented images are both realistic and diverse, providing a robust training set for the deep learning models. Subsequently, the NAS is employed within the AgriNAS framework using the DARTS algorithm, implemented in PyTorch. This search is guided by a bi-level optimization strategy, ensuring that the architecture is both effective and efficient. The selected architecture dynamically adjusts its convolutional layers based on the complexity of the input image, using TensorFlow 2.13.0 or PyTorch 2.2 to modify layer depth, convolution types, and filter sizes to capture relevant features.

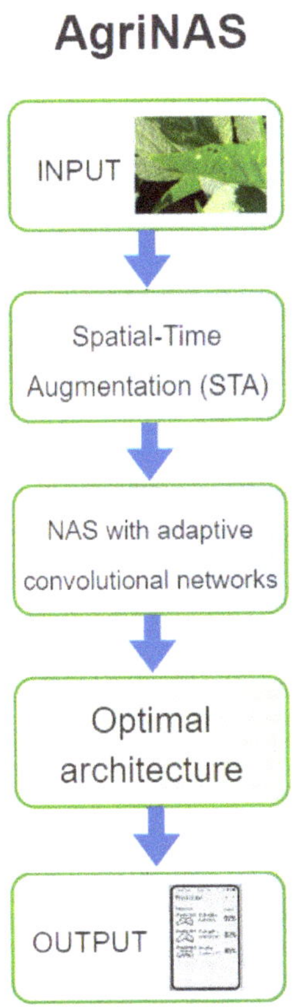

Figure 1. Steps involved in creating AgriNAS.

The AgriNAS framework is trained and optimized using Google Colab, which provides access to powerful NVIDIA Tesla K80 GPUs with 12GB of GPU memory. These advanced GPU resources significantly enhance the performance and efficiency of the system during the computationally intensive tasks associated with NAS and model training.

The parallel processing capabilities of the GPU accelerate the handling of large datasets and complex calculations, enabling real-time data augmentation and feature extraction.

Key parameters such as architecture depth, types of convolutions, and filter sizes are optimized by the NAS framework, and the use of powerful GPUs ensures faster model convergence, allowing the system to adapt and refine its architecture more quickly. Furthermore, this increased computational efficiency facilitates more extensive hyperparameter tuning and experimentation, contributing to improved model accuracy and robustness.

Initial testing and smaller-scale training were conducted on a local machine with an Intel Core i7 processor, 16 GB RAM, and an NVIDIA GeForce GTX 1060 GPU. The training procedure employed 5-fold cross-validation to evaluate the model's generalizability, with each model trained on 80% of the data and validated on the remaining 20%. Advanced regularization techniques, such as entropy-based regularization, were also employed to prevent overfitting and ensure the model's generalizability.

The trained model classifies the input images into categories such as Caterpillar, Diabrotica, and Healthy. The results are visualized using Matplotlib and Seaborn, providing clear outputs that can be used for monitoring and decision-making in agricultural practices.

3.1. Spatial–Time Augmentation (STA)

The Spatial–Time Augmentation (STA) method introduces a novel augmentation strategy that incorporates both spatial and temporal variability into agricultural images. Unlike traditional augmentation techniques that primarily focus on static transformations, STA simulates the relative movement of objects (e.g., pests) across different times and spaces. This is achieved by applying a Lorentzian transformation to the spatial coordinates of features, resulting in augmented images that reflect varying observational perspectives. Additionally, STA introduces a noise factor proportional to the simulated relative velocity, which accounts for observational uncertainties and enhances the realism of the augmented data. By generating a more diverse and realistic training set, STA contributes to improved model generalization and performance in deep learning tasks within agriculture.

Algorithm 1 describes the process of applying the Spatial–Time Augmentation (STA) technique to an original dataset D. The method begins by initializing key STA parameters: the velocity v, the Lorentz factor γ, and the noise factor ϵ.

Algorithm 1 AgriNAS: Neural Architecture Search Framework

Input: Initialize architecture parameters α, network weights w, learning rates η_α, η_w, and perturbation parameter ξ.
while not converged **do**
  Training Phase: Update network weights w using gradient descent:

$$w \leftarrow w - \eta_w \nabla_w \mathcal{L}_{\text{train}}(w, \alpha)$$

  Search Phase: Update architecture parameters α using second-order approximation:

$$\alpha \leftarrow \alpha - \eta_\alpha \nabla_\alpha \mathcal{L}_{\text{val}}(w - \xi \nabla_w \mathcal{L}_{\text{train}})$$

  Apply regularization to architecture parameters to avoid overfitting:

$$\mathcal{L}_{\text{val}} \leftarrow \mathcal{L}_{\text{val}} + \lambda \sum_i H(\alpha_i)$$

end while
Output: Optimal architecture α^* and corresponding weights w^*.

1. Spatial Transformation: For each image x_i in the dataset, a spatial transformation such as translation or rotation is applied. This transformation simulates different observational angles or positions.

2. Lorentz Transformation: The spatial coordinates of each image are then transformed using a Lorentzian model:
$$x'_i = \gamma(x_i - vt)$$
$$t'_i = \gamma\left(t - \frac{vx_i}{c^2}\right)$$
Here, x'_i represents the transformed spatial coordinates, and t'_i represents the transformed temporal coordinates. These transformations account for the relative movement of objects in space and time.
3. Adding Noise: To simulate realistic environmental conditions, a Gaussian noise factor with a variance of 0.01 is added to the transformed image x'_i. This step introduces variability and uncertainty, effectively mimicking real-world scenarios where observations may be affected by noise or imprecision. By incorporating this level of noise, we enhance the model's robustness, ensuring it can generalize better across diverse agricultural contexts.
4. Augmented Dataset: The augmented image x'_i is then appended to the new dataset D', which is used for training deep learning models.

Finally, the augmented dataset D' is returned, containing a diverse set of images that reflect different spatial and temporal scenarios, thus providing a robust training set for model development.

3.2. Dataset

The dataset used in this study, as described in [38], consists of images categorized into three folders: Caterpillar, Diabrotica, and Healthy. To enhance the dataset, additional images of healthy soybean leaves were sourced from Plant Village (https://plantvillage.org/, accessed on 12 March 2024), resulting in a total of 10,499 images. The final distribution of images across the categories is presented in Table 1.

Table 1. Number of images for Pest dataset.

Folder	Number of Images
Caterpillar	3309
Diabrotica Speciosa	2205
Healthy	4985
Total	10,499

The dataset used in this study consists of high-resolution images of soybean leaves, stems, and pods affected by various pests and diseases. These images were collected using a combination of high-definition field cameras. The images were captured under different environmental conditions, including varying light intensities and angles, to simulate real-world agricultural scenarios. This approach ensures diversity in the dataset, making the model more robust when detecting pests and diseases across different farming conditions. The dataset includes both healthy and infected samples, with careful attention to capturing early, middle, and advanced stages of pest and disease manifestation. Controlled lighting was used in some instances to improve image quality, while multi-angle shots were taken to ensure that the affected areas were adequately captured from different perspectives.

Additionally, the images were acquired using cameras with a minimum resolution of 12 megapixels in some instances, complementing the high-definition field cameras. This combination of imaging techniques not only enhances the dataset's resolution and detail but also allows for the capture of images in various field conditions. The use of cameras provides flexibility in accessing different areas of the fields while ensuring that high-quality images are obtained. Employing both high-definition cameras aims to create a comprehensive dataset that reflects the diversity and complexity of real agricultural environments, ultimately contributing to the model's ability to generalize effectively in practical applications. The dataset is balanced, ensuring that both healthy and infected samples are adequately represented, with no class being over- or under-represented.

The preprocessing phase includes an advanced data augmentation technique known as Spatial–Time Augmentation (STA) to enhance the dataset's variability and improve the model's robustness. STA introduces both spatial and temporal variability into the images by simulating the relative movement of objects across different spatial and temporal contexts. This is achieved by applying a Lorentzian transformation to the spatial coordinates of features, resulting in augmented images that reflect different observational perspectives. Additionally, STA introduces a noise factor proportional to the simulated relative velocity, accounting for observational uncertainties and further enriching the diversity of the training set.

This augmentation process generates new training examples that simulate potential variations in real-world conditions, such as different lighting, angles, and damage patterns on the leaves. This process helps prevent overfitting and enables the model to generalize better to unseen data. The augmented dataset is then used as input for the NAS process, ensuring that the selected architecture is optimized across a broad range of scenarios.

Figure 2 illustrates healthy soybean leaves as captured in the dataset. In contrast, Figure 3 presents soybean leaves damaged by caterpillar infestation, while Figure 4 demonstrates the specific impact of the Diabrotica Speciosa pest on the soybean plants.

Figure 2. Healthy soybean leaves from the dataset.

Figure 3. Soybean leaves damaged by caterpillars.

The feature analysis was performed by examining color, texture, and shape features in the soybean leaves. Color histograms were used to compare the color distribution between healthy and infested leaves, indicating stages of pest infestation. Texture features were assessed using local binary patterns and a gray level co-occurrence matrix, focusing on differences in contrast and homogeneity. Shape features, including size and irregularity, were analyzed using Hu Moments to detect morphological changes due to pests. We also considered how environmental factors, such as lighting variations, could affect feature extraction and model performance. Finally, class activation maps were employed to visualize the prioritized features during predictions, enhancing the interpretability of our results. This analysis provides valuable insights into the features contributing to the classification of soybean pest actions.

Figure 4. Soybean leaf damaged by Diabrotica Speciosa.

3.3. Neural Architecture Search (NAS) Framework

The core of the proposed methodology is the AgriNAS framework, an advanced NAS framework specifically tailored for optimizing CNNs in the context of pest and disease detection in soybean crops. AgriNAS leverages an enhanced extension of DARTS to enable an efficient and comprehensive exploration of a vast architectural search space, designed to meet the unique challenges of agricultural image analysis.

AgriNAS's search space $\mathcal{O}$ is meticulously crafted to include a diverse array of convolutional operations, such as standard convolutions, dilated convolutions, depthwise separable convolutions, pooling operations, and skip connections. Each operation $o_i \in \mathcal{O}$ is parameterized by learnable weights θ_{o_i}, allowing the framework to dynamically select and combine these operations to construct the optimal architecture for the specific task at hand.

The primary contribution of AgriNAS is its formulation of the architecture search as a bi-level optimization problem, which allows for a more structured and efficient approach to discovering optimal neural network configurations. In this framework, the goal is to minimize the validation loss $\mathcal{L}_{\text{val}}$ with respect to the architecture parameters α, while simultaneously optimizing the network weights w to minimize the training loss $\mathcal{L}_{\text{train}}$. This dual optimization problem can be expressed mathematically as follows:

$$\min_{\alpha} \mathcal{L}_{\text{val}}(w^*(\alpha), \alpha), \tag{1}$$

subject to the constraint:

$$w^*(\alpha) = \arg\min_{w} \mathcal{L}_{\text{train}}(w, \alpha), \tag{2}$$

where $w^*(\alpha)$ denotes the optimal network weights for a given architecture α.

This bi-level optimization strategy effectively decouples the architecture search from the training process, enabling AgriNAS to focus on finding the best architecture while ensuring that the associated weights are optimized for performance. By allowing the model to adaptively select architectures that minimize validation loss, AgriNAS enhances its ability to generalize across diverse datasets.

Additionally, AgriNAS employs entropy-based regularization techniques to improve model robustness and prevent overfitting. This regularization is incorporated directly into the optimization process by penalizing architectures that exhibit excessive reliance on specific operations. The regularization term is defined as follows:

$$\mathcal{R}(\alpha) = \lambda \sum_{i} H(\alpha_i), \tag{3}$$

where $H(\alpha_i)$ represents the entropy of the operation probabilities associated with the architecture parameters, and λ is a hyperparameter controlling the strength of the regularization.

By promoting diversity in the selected operations and discouraging overly complex architectures, the entropy-based regularization enhances the model's ability to generalize to unseen data. This dual approach of bi-level optimization combined with regularization techniques ensures that AgriNAS not only identifies effective neural architectures but also maintains robustness against overfitting, ultimately resulting in a more reliable model for pest and disease detection.

To efficiently address this complex bi-level optimization, AgriNAS employs a gradient-based search algorithm enhanced by a second-order approximation. This approach mitigates the computational burden typically associated with exact second-order methods, ensuring that the architecture search remains computationally feasible even in large-scale agricultural datasets. The update rule for the architecture parameters is given by the following:

$$\alpha \leftarrow \alpha - \eta_\alpha \nabla_\alpha \mathcal{L}_{\text{val}}(w - \xi \nabla_w \mathcal{L}_{\text{train}}), \tag{4}$$

where η_α represents the learning rate for α, and ξ is a small perturbation parameter employed to approximate the second-order gradients. This update mechanism is a key contribution of AgriNAS, as it refines the architecture parameters in a manner that effectively balances model performance on the validation set with computational efficiency, considering the complex interactions between architecture and network weights.

Algorithm 2 outlines the implementation of the AgriNAS framework. The process involves several key steps, each crucial for optimizing the architecture and weights of a neural network designed for detecting soybean pests and diseases.

Algorithm 2 Spatial–Time Augmentation (STA)

Input: Original dataset $D = \{x_i, y_i\}_{i=1}^{N}$
Output: Augmented dataset D'
Initialize: STA parameters v, γ, noise factor ϵ
for each image $x_i \in D$ **do**
  Apply spatial transformation to x_i (translation, rotation)
  Compute Lorentz transformation on spatial coordinates:

$$x_i' = \gamma(x_i - vt)$$

$$t_i' = \gamma\left(t - \frac{vx_i}{c^2}\right)$$

  Add relativistic noise to transformed image x_i'
  Append augmented image x_i' to D'
end for
Return: Augmented dataset D'

AgriNAS incorporates advanced regularization techniques to promote generalization and prevent overfitting. An entropy-based regularization term is applied directly to the architecture parameters α, expressed as follows:

$$\mathcal{R}(\alpha) = \lambda \sum_{i} H(\alpha_i), \quad \text{where} \quad H(\alpha_i) = -\sum_{o \in \mathcal{O}} \alpha_{io} \log \alpha_{io}. \tag{5}$$

This regularization term penalizes architectures that overly rely on a single operation, ensuring diversity in the selected operations and fostering robust performance across varying data distributions.

When employing NAS, the optimization process still requires careful tuning of key hyperparameters to ensure effective training and convergence of the selected architecture. The initial learning rate was set to 0.001, and a learning rate scheduler was employed to reduce the rate by a factor of 0.1 if no improvement in validation loss was observed over 5 epochs. This adjustment mechanism helped the NAS optimize architectures under varying learning conditions. A batch size of 32 was chosen, balancing memory constraints with computational efficiency, allowing the NAS to evaluate the performance of architectures on a sufficiently large subset of the data while maintaining manageable memory usage. The models were trained for up to 50 epochs, with early stopping based on validation loss to prevent overfitting, ensuring that the NAS-selected architectures were evaluated under conditions that allow for sufficient training without overfitting.

The Adam optimizer, known for its adaptive learning rate capabilities, was used, which was particularly beneficial for NAS as it allowed for the efficient training of diverse architectures with varying parameter sensitivities. Additionally, L2 regularization (weight decay) was applied with a factor of 0.01 to discourage overfitting, ensuring that the NAS framework prioritized architectures that generalize well. Dropout layers with a rate of 0.5 were incorporated to introduce robustness by randomly dropping units during training. These hyperparameters were integral to the NAS process, enabling the discovery of high-performing architectures tailored to the specific requirements of the dataset and task.

3.4. Adaptive Convolutional Architecture

The AgriNAS framework features an adaptive convolutional architecture that dynamically adjusts its complexity in response to the specific characteristics of the input data. This adaptability is vital for managing the variability inherent in agricultural image data, where manifestations of pests and diseases can differ significantly.

The core concept behind this adaptive architecture is to modulate the depth and configuration of the convolutional network based on the complexity of the input image x. Mathematically, the output of the convolutional network, parameterized by architecture parameters α and weights θ, is denoted as $f(x;\alpha,\theta)$. The network's adaptive behavior is governed by a function $\mathcal{F}(x)$, which determines the number of layers $L(x)$ and their respective configurations for each input image:

$$f(x;\alpha,\theta) = \sum_{i=1}^{L(x)} \mathcal{F}_i(x) \cdot f_i(x;\alpha_i,\theta_i), \tag{6}$$

where $L(x)$ represents the number of layers chosen by the adaptive function $\mathcal{F}(x)$, and $f_i(\cdot)$ corresponds to the operations performed in the i-th layer, parameterized by α_i and θ_i.

The adaptive function $\mathcal{F}(x)$ evaluates the complexity of the input image by assessing features such as texture, contrast, and the presence of multiple overlapping regions indicative of pest or disease severity. For more complex images, $\mathcal{F}(x)$ may increase the number of convolutional layers $L(x)$ and adjust filter sizes and pooling operations to capture more intricate patterns:

$$L(x) = \min(L_{\text{max}}, \max(L_{\text{min}}, \gamma \cdot \mathcal{C}(x))), \tag{7}$$

where L_{max} and L_{min} are predefined upper and lower bounds for the number of layers, γ is a scaling factor, and $\mathcal{C}(x)$ is a complexity measure derived from the input image. This complexity measure $\mathcal{C}(x)$ reflects a formal measure combining entropy, edge density, and pixel intensity variance of x, factors into a weighted function:

$$C(x) = \lambda_1 H(x) + \lambda_2 D(x) + \lambda_3 V(x), \tag{8}$$

where $H(x)$ is the entropy, $D(x)$ is the edge density, and $V(x)$ is the pixel intensity variance, with $\lambda_1, \lambda_2, \lambda_3$ representing the respective weights for these components.

In addition to dynamically adjusting the number of layers, AgriNAS adapts the configuration of each layer, including the type of convolution (standard, dilated, or depthwise separable), filter sizes, and activation functions. The adaptive function $\mathcal{F}_i(x)$ for each layer i is defined below.

The loss function $\mathcal{L}_{\text{train}}$ represents the training loss at layer i and is typically defined as a cross-entropy loss for classification tasks. The regularization term $\mathcal{R}(\alpha_i)$ penalizes overly complex architectures by introducing constraints on the architecture parameters α_i, preventing the model from becoming unnecessarily large or complex. This regularization is weighted by the hyperparameter λ, which controls the strength of the regularization, balancing between model complexity and performance.

$$\mathcal{F}_i(x) = \text{argmin}_{\alpha_i,\theta_i}\{\mathcal{L}_{\text{train}}(f_i(x;\alpha_i,\theta_i)) + \lambda \cdot \mathcal{R}(\alpha_i)\}, \tag{9}$$

where $\mathcal{L}_{\text{train}}$ measures the error between predicted and true labels, and $\mathcal{R}(\alpha_i)$ ensures the network remains computationally efficient and avoids overfitting.

Filter sizes are adaptively selected based on the complexity of the input image $C(x)$. Specifically, filter sizes increase with higher values of $C(x)$ to capture finer details in more complex images. This process involves dynamically scaling the filter size based on the computed complexity:

$$\text{Filter Size} = \alpha \cdot \text{Base Size} + \beta \cdot C(x) \tag{10}$$

where α and β are hyperparameters tuned during training.

It is important to note that the NAS framework explores various configurations of these parameters, selecting combinations that minimize training loss. Each network is trained with an optimized configuration tailored to the complexity of the input, ensuring a more efficient and adaptive model, rather than relying on predefined or static structures.

This adaptive architecture's optimization is integrated into the overall NAS framework, where both the architecture parameters α and the weights θ are updated iteratively. The gradient-based optimization of α and θ is performed using backpropagation through the dynamically adjusted network, ensuring the model converges to an architecture that is both effective for pest and disease detection and efficient in terms of computational resources.

The design of the multi-stage convolutional network incorporates various strategies for managing computational efficiency while ensuring effective feature extraction. Each layer is equipped with adaptive mechanisms that allow it to focus on relevant features based on the complexity of the input. For instance, in layers dealing with simpler images, the architecture might utilize fewer convolutional operations, while more complex images could trigger the use of multiple layers with varying filter types and sizes.

This multi-stage approach enables the model to dynamically allocate resources to layers that require more detailed processing, thereby optimizing the computational load. Additionally, the adaptive function $\mathcal{F}_i(x)$ not only determines the number of layers but also adapts the configuration of each layer, including the selection of activation functions and pooling strategies, ensuring that the network effectively captures intricate patterns indicative of pests and diseases. The flexibility inherent in this architecture allows AgriNAS to perform well across diverse agricultural conditions, improving both accuracy and efficiency in pest and disease detection tasks.

3.5. Performance Evaluation Metrics

The evaluation of the effectiveness of the AgriNAS algorithm is performed by measuring the results using several key performance metrics: accuracy, recall, precision, F-measure (F1 Score), and ROC AUC (Area Under the Curve).

1. Accuracy: Measures the ratio of correct predictions over the total number of instances evaluated. It is calculated as follows:

$$Accuracy = \frac{TP + TN}{TP + TN + FP + FN} \tag{11}$$

where True Positive (TP), True Negative (TN), False Positive (FP), and False Negative (FN) values are extracted from the confusion matrix.

2. Recall: Also known as sensitivity, it measures the fraction of actual positive instances that were correctly identified by the model. It is crucial for evaluating performance, particularly in cases of imbalanced datasets. Recall is calculated as follows:

$$Recall = \frac{TP}{TP + FN} \tag{12}$$

3. Precision: Precision measures the proportion of positive identifications that were actually correct. It is calculated as follows:

$$Precision = \frac{TP}{TP + FP} \tag{13}$$

4. F-Measure (F1 Score): The F1 Score represents the harmonic mean of precision and recall, providing a single metric that balances the two. It is calculated as follows:

$$F\text{-}measure = \frac{2 \times (precision \times recall)}{precision + recall} \tag{14}$$

5. ROC Curve and AUC: The ROC curve plots the true positive rate (Recall) against the false positive rate, providing a graphical representation of a model's diagnostic ability. The Area Under the Curve (AUC) indicates the overall ranking performance, with a higher AUC representing a better model.

These metrics collectively offer a robust evaluation of the model's effectiveness in classifying soybean pest and disease categories. They are essential for understanding the strengths and limitations of the AgriNAS model compared to other models, such as VGG-19 and another model using CNN, in recognizing leaf diseases [47].

4. Results and Discussion

This section presents a comprehensive analysis of the results obtained from the AgriNAS framework in classifying soybean pest actions on leaves. The models evaluated include the AgriNAS-optimized CNN architecture, VGG-19, and another CNN model [47], a state-of-the-art NAS-based model from related work. The performance metrics used in this analysis are accuracy, recall, precision, F-measure, and ROC AUC, providing a thorough evaluation of each model's effectiveness.

In this study, we compared the performance of the AgriNAS-optimized CNN architecture with the VGG-19 and another CNN model from related work [47]. The AgriNAS-optimized CNN utilizes advanced features from Neural Architecture Search (NAS) to enhance pest classification. In contrast, VGG-19 is a well-established model with 19 layers, known for its effectiveness in image classification, employing small receptive fields and max pooling layers. The state-of-the-art CNN model referenced in [47] follows a standard convolutional architecture designed for agricultural images.

For training, we employed a learning rate of 0.001 and a batch size of 32, using the Adam optimizer for all models.

4.1. Performance Comparison

The evaluation of the models was conducted using the aforementioned performance metrics, with the results summarized in Table 2. To ensure that the observed differences in model performance are statistically significant, we performed a t-test comparing the results of AgriNAS with those of VGG-19 and another CNN model [47]. The p-values obtained confirm that AgriNAS significantly outperforms the other models across all metrics ($p < 0.05$).

Table 2. Test accuracies and evaluation metrics for the models.

Model	Acc.	Recall	Precis.	F-Meas.	AUC ROC
VGG-19	0.94	0.94	0.94	0.94	0.93
CNN model [47]	0.96	0.96	0.96	0.96	0.95
AgriNAS	0.98	0.98	0.99	0.98	0.97

4.2. Ablation Study

To further understand the contribution of each component within the AgriNAS framework, we conducted an ablation study by systematically removing or modifying specific elements, such as data augmentation and adaptive convolutional layers. The results, as shown in Table 3, highlight the significant impact of data augmentation and adaptive layers on the overall performance of the model. The removal of data augmentation led to a decrease in accuracy by approximately 3%, while disabling adaptive layers resulted in a 2% drop in F-measure.

Table 3. Ablation study results for AgriNAS.

Config.	Acc.	Rec.	Precis.	F-Meas.	AUC ROC
AgriNAS (Full)	0.98	0.98	0.99	0.98	0.97
No Augment.	0.95	0.96	0.97	0.96	0.95
No Adapt. Layers	0.96	0.97	0.97	0.96	0.96

4.3. Feature Map Visualization

To gain deeper insights into how the AgriNAS model interprets input images, from Figure 5, it is possible to visualize the feature maps generated by different convolutional layers. Figure 5 shows the activation patterns in various layers, highlighting how the model progressively abstracts and focuses on critical features like texture and leaf venation, which are indicative of specific diseases and pests. The feature maps demonstrate the model's ability to capture intricate patterns that contribute to its high classification accuracy.

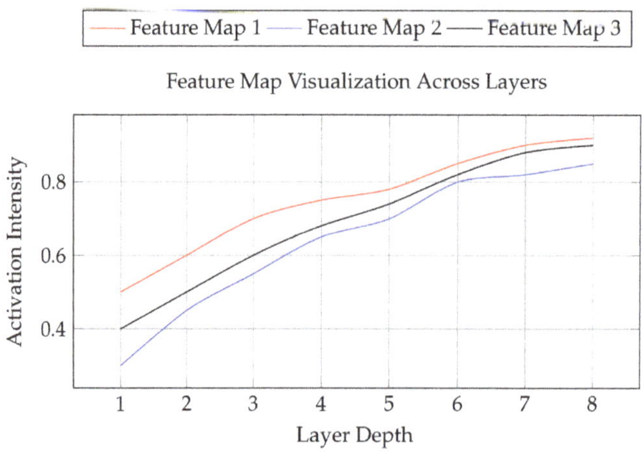

Figure 5. Visualization of feature maps from different layers of the AgriNAS model, showing how the model captures important features related to pest and disease detection.

4.4. Impact of Data Augmentation with STA

The data augmentation strategies employed in AgriNAS, with STA, play a crucial role in enhancing model robustness. Figure 6 illustrates the impact of data augmentation on model performance, comparing the accuracy, recall, and precision with and without STA. The results clearly show that data augmentation improved the model's ability to generalize across different conditions, leading to a 3% increase in overall accuracy.

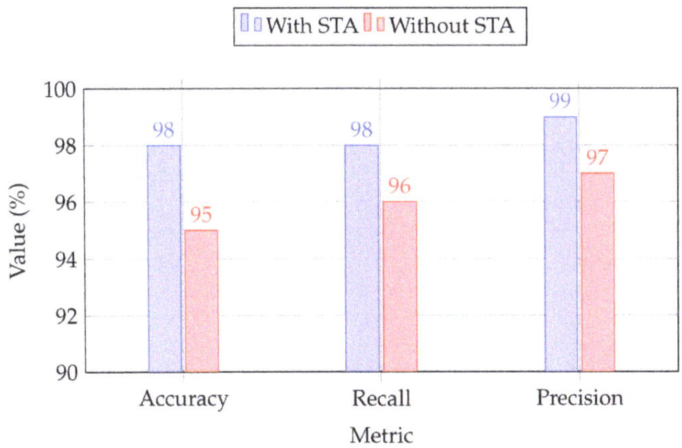

Figure 6. Impact of data augmentation on AgriNAS model performance. The results show significant improvements in accuracy, recall, and precision when data augmentation is applied.

4.5. Precision–Recall and ROC Curves

To further illustrate the performance differences among the models, Figure 7 shows the precision–recall curve for AgriNAS, another CNN model [47], and VGG-19. The adaptive convolutional layers in AgriNAS dynamically adjust based on input complexity, reducing unnecessary computational load. This, combined with effective regularization techniques and efficient search space exploration, leads to faster training times and lower GPU memory usage.

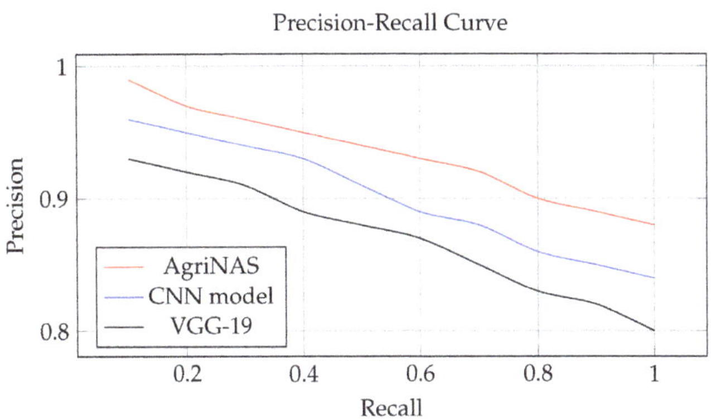

Figure 7. Precision–recall curve for AgriNAS, CNN model [47], and VGG-19.

4.6. Computation Time and Efficiency Analysis

Table 4 presents the time and computational resources required to train the AgriNAS model compared to other models. This is particularly important for real-time applications in agriculture.

Table 4. Computation time and resource utilization.

Model	Training Time (hours)	GPU Memory (GB)
VGG-19	12	8
CNN model [47]	15	10
AgriNAS	10	7

4.7. Error Analysis

In Figure 8, the confusion matrix illustrates the classification performance of the AgriNAS model across different pest and disease categories: Caterpillar, Diabrotica, and Healthy. Each cell in the matrix represents the proportion of predictions made by the model for each actual category. The diagonal elements of the matrix (i.e., $(1,1)$, $(2,2)$, and $(3,3)$) show the correctly classified instances for each category. For example, the model correctly classified 92% of the Caterpillar images, 94% of the Diabrotica images, and 95% of the Healthy images. These high values indicate that the model has learned to accurately distinguish between the different categories. However, the off-diagonal elements indicate misclassification. For instance, 3% of the Caterpillar images were incorrectly classified as Diabrotica, and 5% were misclassified as Healthy. Similarly, 4% of the Healthy images were misclassified as Diabrotica. These misclassifications are relatively low, suggesting that the model is highly reliable but still has some difficulty differentiating between similar visual features.

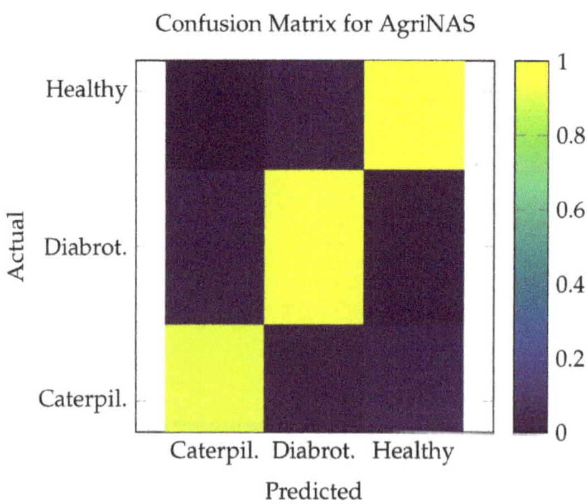

Figure 8. Confusion matrix for AgriNAS, illustrating the classification performance across different pest and disease categories.

The performance of the AgriNAS algorithm can be attributed to several factors:

- Adaptive Convolutional Architecture: The AgriNAS model dynamically adjusts its network complexity based on the input data, allowing it to capture intricate patterns in the images that are indicative of specific pests and diseases. This adaptability results in better feature extraction and, consequently, higher classification accuracy.
- Neural Architecture Search (NAS): The use of NAS enables the automatic discovery of optimal network architectures tailored to the task of pest and disease detection. By exploring a wide range of possible architectures, AgriNAS identifies the most effective configuration, which contributes to its superior performance.
- STA, Data Augmentation: The model employs the STA technique, which introduces spatial and temporal variability into the dataset by simulating the relative movement of objects, such as pests, across different conditions. This approach enhances the model's ability to generalize by creating a diverse and realistic set of training images

that reflect various observational perspectives. The effectiveness of STA is demonstrated in the confusion matrix, where the model consistently maintains high accuracy across all categories, showcasing its robustness.

- Regularization Techniques: The integration of advanced regularization methods, including entropy-based regularization, prevents the model from overfitting to specific features. This ensures that the model remains effective across a wide range of scenarios, as evidenced by the balanced classification performance shown in the confusion matrix.

Overall, the AgriNAS algorithm's ability to accurately classify a diverse set of images while minimizing misclassification errors highlights its effectiveness in the context of soybean pest and disease detection.

5. Conclusions

The results of this study demonstrate the effectiveness of the AgriNAS framework in optimizing CNN architectures for the detection of soybean diseases. The use of a NAS framework allowed for the automated discovery of highly specialized network architectures tailored specifically to the nuances of soybean leaf imagery. By dynamically adjusting the complexity of the network through adaptive layer configuration, AgriNAS was able to capture the intricate patterns associated with different pest and disease manifestations. Moreover, the integration of the STA method played an important role in improving the model's generalization capabilities. STA introduced spatial and temporal variability into the dataset, simulating realistic scenarios that pests might encounter across different times and locations. This approach to data augmentation ensured that the model was trained on a diverse and realistic set of images, which, in turn, enhanced its ability to maintain high accuracy across all categories, as reflected in the confusion matrix. Additionally, the advanced regularization techniques within the NAS framework were critical to maintaining model robustness. The entropy-based regularization applied to the architecture parameters helped in preventing the model from becoming overly reliant on specific operations, thereby encouraging a diverse set of features to be learned. Furthermore, the second-order gradient-based optimization employed by AgriNAS enabled more precise updates to the architecture parameters, leading to faster convergence and a more efficient exploration of the search space. This method reduced the computational overhead typically associated with NAS, allowing AgriNAS to maintain computational efficiency without sacrificing performance. The scalability challenges faced by AgriNAS in real agricultural settings primarily revolve around high computational costs and the necessity for advanced infrastructure, particularly in resource-constrained environments such as rural areas. To address these issues, several strategies can be implemented. First, enhancing the computational efficiency of AgriNAS is essential, which can be achieved through techniques like model pruning, quantization, and the use of lightweight architectures that require fewer resources without sacrificing performance. Second, leveraging cloud computing can help mitigate the need for local high-performance infrastructure by providing scalable resources on-demand, allowing for intensive processing tasks to be performed remotely. Additionally, implementing edge computing solutions can further enhance scalability by processing data closer to the source, thus reducing latency and bandwidth requirements. Developing more efficient data collection and processing methods, such as transfer learning, can also assist in minimizing the amount of data needed for effective model training, which is particularly beneficial in regions with limited data availability. Finally, providing training and support for farmers and agricultural workers on utilizing AgriNAS effectively can empower users with the knowledge necessary for broader adoption and more effective implementation in diverse agricultural contexts. By implementing these strategies, AgriNAS can overcome scalability challenges and become a more practical solution for pest and disease detection in various agricultural settings, ensuring accessibility and effectiveness even in resource-limited environments. Future research should explore the extension of this framework to other crops and pests, potentially leveraging additional data modalities such as multispectral or hyperspectral imagery. Additionally, the experimental evaluation of AgriNAS will be enhanced by incorporating more recent and advanced methods in the field of agricultural pest detection and deep learning. Specifically, we plan to include state-of-the-art techniques such as attention-based mechanisms, transformer networks, and more recent convolutional models. In future works, we will also expand our analysis to include a broader range of problematic cases affecting pest classification accuracy, such as overlapping visual features between different pest species, variations in pest life stages, and the impact of environmental factors like lighting conditions and occlusions on image quality. We also plan to explore the implications of these complexities on the confusion matrix results and consider incorporating more sophisticated evaluation metrics to capture model performance under challenging conditions. By addressing these shortcomings, we aim to enhance the robustness of our models and provide a more comprehensive framework for effectively classifying soybean pest actions in diverse agricultural environments.

Author Contributions: Conceptualization, O.J.O., R.L.R., M.S. and D.Z.R.; methodology, O.J.O., R.L.R., M.S. and D.Z.R.; software, O.J.O., R.L.R., M.S. and D.Z.R.; validation, O.J.O., R.L.R., M.S. and D.Z.R.; formal analysis, O.J.O., R.L.R., M.S. and D.Z.R.; investigation, O.J.O., R.L.R., M.S. and D.Z.R.; resources, O.J.O., R.L.R., M.S. and D.Z.R.; data curation, O.J.O., R.L.R., M.S. and D.Z.R.; writing—original draft preparation, O.J.O., R.L.R., M.S. and D.Z.R.; writing—review and editing, O.J.O., R.L.R., M.S. and D.Z.R.; visualization, O.J.O., R.L.R., M.S. and D.Z.R.; supervision, R.L.R. and D.Z.R.; project administration, D.Z.R.; funding acquisition, D.Z.R. All authors have read and agreed to the published version of the manuscript.

Funding: This research was funded by Conselho Nacional de Desenvolvimento Científico e Tecnológico (CNPq) under Proc. No.: 404764/2021-5; and Fundação de Desenvolvimento Científico e Cultural (FINEP) under Proc. No. 2817/22.

Data Availability Statement: The data presented in this study are available on request from the corresponding author.

Conflicts of Interest: The authors declare no conflict of interest.

References

1. Hamza, M.; Basit, A.W.; Shehzadi, I.; Tufail, U.; Hassan, A.; Hussain, T.; Siddique, M.U.; Hayat, H.M. Global impact of soybean production: A review. *Asian J. Biochem. Genet. Mol. Biol.* **2024**, *16*, 12–20. [CrossRef]
2. Medic, J.; Atkinson, C.; Hurburgh, C.R. Current knowledge in soybean composition. *J. Am. Oil Chem. Soc.* **2014**, *91*, 363–384. [CrossRef]
3. Liu, K. *Soybeans: Chemistry, Technology, and Utilization*; Springer: Berlin/Heidelberg, Germany, 2012.
4. Xiaoming, Z.; Qiong, L. A Brief Introduction of Main Diseases and Insect Pests in Soybean Production in the Global Top Five Soybean Producing Countries. *Plant Dis. Pests* **2018**, *9*, 17.
5. Gale, F.; Valdes, C.; Ash, M. Interdependence of China, United States, and Brazil in soybean trade. In *US Department of Agriculture's Economic Research Service (ERS) Report*; US Department of Agriculture: New York, NY, USA, 2019; pp. 1–48.
6. Hartman, G.L.; West, E.D.; Herman, T.K. Crops that feed the World 2. Soybean—Worldwide production, use, and constraints caused by pathogens and pests. *Food Secur.* **2011**, *3*, 5–17. [CrossRef]
7. Nendel, C.; Reckling, M.; Debaeke, P.; Schulz, S.; Berg-Mohnicke, M.; Constantin, J.; Fronzek, S.; Hoffmann, M.; Jakšić, S.; Kersebaum, K.C.; et al. Future area expansion outweighs increasing drought risk for soybean in Europe. *Glob. Chang. Biol.* **2023**, *29*, 1340–1358. [CrossRef]
8. Oerke, E.C. Crop losses to pests. *J. Agric. Sci.* **2006**, *144*, 31–43. [CrossRef]
9. Rupe, J.; Luttrell, R.G. Effect of pests and diseases on soybean quality. In *Soybeans*; Elsevier: Amsterdam, The Netherlands, 2008; pp. 93–116.
10. Thanh Noi, P.; Kappas, M. Comparison of random forest, k-nearest neighbor, and support vector machine classifiers for land cover classification using Sentinel-2 imagery. *Sensors* **2017**, *18*, 18. [CrossRef]
11. Lokner Lađević, A.; Kramberger, T.; Kramberger, R.; Vlahek, D. Detection of AI-Generated Synthetic Images with a Lightweight CNN. *AI* **2024**, *5*, 1575–1593. [CrossRef]
12. Farea, A.; Yli-Harja, O.; Emmert-Streib, F. Understanding Physics-Informed Neural Networks: Techniques, Applications, Trends, and Challenges. *AI* **2024**, *5*, 1534–1557. [CrossRef]
13. Payán-Serrano, O.; Bojórquez, E.; Carrillo, J.; Bojórquez, J.; Leyva, H.; Rodríguez-Castellanos, A.; Carvajal, J.; Torres, J. Seismic Performance Prediction of RC, BRB and SDOF Structures Using Deep Learning and the Intensity Measure INp. *AI* **2024**, *5*, 1496–1516. [CrossRef]

14. Ribeiro, D.A.; Silva, J.C.; Lopes Rosa, R.; Saadi, M.; Mumtaz, S.; Wuttisittikulkij, L.; Zegarra Rodriguez, D.; Al Otaibi, S. Light field image quality enhancement by a lightweight deformable deep learning framework for intelligent transportation systems. *Electronics* **2021**, *10*, 1136. [CrossRef]
15. Silva, J.C.; Saadi, M.; Wuttisittikulkij, L.; Militani, D.R.; Rosa, R.L.; Rodríguez, D.Z.; Al Otaibi, S. Light-field imaging reconstruction using deep learning enabling intelligent autonomous transportation system. *IEEE Trans. Intell. Transp. Syst.* **2021**, *23*, 1587–1595. [CrossRef]
16. Shoaib, M.; Shah, B.; Ei-Sappagh, S.; Ali, A.; Ullah, A.; Alenezi, F.; Gechev, T.; Hussain, T.; Ali, F. An advanced deep learning models-based plant disease detection: A review of recent research. *Front. Plant Sci.* **2023**, *14*, 1158933.
17. Omole, O.J.; Rosa, R.L.; Rodriguez, D.Z. Soybean Disease Detection by Deep Learning Algorithms. In Proceedings of the 2023 International Conference on Software, Telecommunications and Computer Networks (SoftCOM), Split, Croatia, 21–23 September 2023; pp. 1–5.
18. Ferentinos, K.P. Deep learning models for plant disease detection and diagnosis. *Comput. Electron. Agric.* **2018**, *145*, 311–318. [CrossRef]
19. Tammina, M.R.; Sumana, K.; Singh, P.P.; Lakshmi, T.V.; Pande, S.D. Prediction of Plant Disease Using Artificial Intelligence. In *Microbial Data Intelligence and Computational Techniques for Sustainable Computing*; Springer: Berlin/Heidelberg, Germany, 2024; pp. 25–48.
20. Ghazal, S.; Kommineni, N.; Munir, A. Comparative Analysis of Machine Learning Techniques Using RGB Imaging for Nitrogen Stress Detection in Maize. *AI* **2024**, *5*, 1286–1300. [CrossRef]
21. Shruthi, U.; Nagaveni, V.; Raghavendra, B. A review on machine learning classification techniques for plant disease detection. In Proceedings of the 2019 5th International Conference on Advanced Computing & Communication Systems (ICACCS), Coimbatore, India, 15–16 March 2019; pp. 281–284.
22. Janiesch, C.; Zschech, P.; Heinrich, K. Machine learning and deep learning. *Electron. Mark.* **2021**, *31*, 685–695. [CrossRef]
23. Mohaidat, T.; Khalil, K. A survey on neural network hardware accelerators. *IEEE Trans. Artif. Intell.* **2024**, *5*, 3801–3822. [CrossRef]
24. Malekloo, A.; Ozer, E.; AlHamaydeh, M.; Girolami, M. Machine learning and structural health monitoring overview with emerging technology and high-dimensional data source highlights. *Struct. Health Monit.* **2022**, *21*, 1906–1955. [CrossRef]
25. Elsken, T.; Metzen, J.H.; Hutter, F. Neural architecture search: A survey. *J. Mach. Learn. Res.* **2019**, *20*, 1–21.
26. Chitty-Venkata, K.T.; Emani, M.; Vishwanath, V.; Somani, A.K. Neural architecture search benchmarks: Insights and survey. *IEEE Access* **2023**, *11*, 25217–25236. [CrossRef]
27. Goodfellow, I.; Pouget-Abadie, J.; Mirza, M.; Xu, B.; Warde-Farley, D.; Ozair, S.; Courville, A.; Bengio, Y. Generative adversarial nets. *Adv. Neural Inf. Process. Syst.* **2014**, *27* .
28. Wang, X.; Yu, K.; Wu, S.; Gu, J.; Liu, Y.; Dong, C.; Qiao, Y.; Change Loy, C. Esrgan: Enhanced super-resolution generative adversarial networks. In Proceedings of the European Conference on Computer Vision (ECCV) Workshops, Munich, Germany, 8–14 September 2018.
29. Naveed, H.; Anwar, S.; Hayat, M.; Javed, K.; Mian, A. Survey: Image mixing and deleting for data augmentation. *Eng. Appl. Artif. Intell.* **2024**, *131*, 107791. [CrossRef]
30. Lu, M.; Jiao, S.; Deng, J.; Wang, C.; Zhang, Z. Efficient model updating of shaft-raft-hull system using multi-stage convolutional neural network combined with sensitivity analysis. *Ocean. Eng.* **2024**, *312*, 119041. [CrossRef]
31. Kheam, S.; Rubene, D.; Markovic, D.; Ith, S.; Uk, O.N.; Soung, S.; Ninkovic, V. The effects of cultivar mixtures on insect pest and natural enemy abundance, diseases, and yield in tropical soybean cropping system. *Biol. Control* **2024**, *196*, 105571. [CrossRef]
32. Romero, B.; Dillon, F.M.; Zavala, J.A. Different soybean cultivars respond differentially to damage in a herbivore-specific manner and decreas herbivore performance. *Arthropod-Plant Interact.* **2020**, *14*, 89–99. [CrossRef]
33. Yadav, V.; Khare, C.; Tiwari, P.; Srivastava, J. Important diseases of soybean crop and their management. In *Diseases of Field Crops Diagnosis and Management*; Apple Academic Press: Palm Bay, FL, USA, 2020; pp. 145–171.
34. Grande, M.L.M.; Rando, J.S.S. Integrated pest control adopted by soybean and corn farmers in Londrina, Paraná state, Brazil. *Arq. Inst. Biológico* **2018**, *85*, e0242015 . [CrossRef]
35. GRDC, Grains Research and Development Corporation. *Soybean Section 7 Insect Control*; GRDC, Grains Research and Development Corporation: Barton, Australia, 2016.
36. Hodgson, E.W.; Koch, R.L.; Davis, J.A.; Reisig, D.; Paula-Moraes, S.V. Identification and biology of common caterpillars in us soybean. *J. Integr. Pest Manag.* **2021**, *12*, 13. [CrossRef]
37. Justus, C.M.; Paula-Moraes, S.V.; Pasini, A.; Hoback, W.W.; Hayashida, R.; de Freitas Bueno, A. Simulated soybean pod and flower injuries and economic thresholds for Spodoptera eridania (Lepidoptera: Noctuidae) management decisions. *Crop Prot.* **2022**, *155*, 105936. [CrossRef]
38. Mignoni, M.E.; Honorato, A.; Kunst, R.; Righi, R.; Massuquetti, A. Soybean images dataset for caterpillar and Diabrotica speciosa pest detection and classification. *Data Brief* **2022**, *40*, 107756. [CrossRef]
39. Walker, H. Detection of Insect-Induced Defoliation in Soybeans with Deep Learning and Object Detection. Ph.D. Thesis, Kansas State University, Manhattan, KS, USA, 2021.
40. Yue, G.; Xiao, H.; Xie, H.; Zhou, T.; Zhou, W.; Yan, W.; Zhao, B.; Wang, T.; Jiang, Q. Dual-constraint coarse-to-fine network for camouflaged object detection. *IEEE Trans. Circuits Syst. Video Technol.* **2023**, *5*, 3286–3298. [CrossRef]

41. Cabrera Walsh, G.; Ávila, C.J.; Cabrera, N.; Nava, D.E.; de Sene Pinto, A.; Weber, D.C. Biology and management of pest Diabrotica species in South America. *Insects* **2020**, *11*, 421. [CrossRef]
42. Costa, E.N.; Nogueira, L.; De Souza, B.H.S.; Ribeiro, Z.A.; Louvandini, H.; Zukoff, S.N.; Júnior, A.L.B. Characterization of antibiosis to Diabrotica speciosa (Coleoptera: Chrysomelidae) in Brazilian maize landraces. *J. Econ. Entomol.* **2018**, *111*, 454–462. [CrossRef]
43. Ávila, C.J.; Bitencourt, D.R.; da Silva, I.F. Biology, Reproductive Capacity, and Foliar Consumption of Diabrotica Speciosa (Germar) (Coleoptera: Chrysomelidae) in Different Host Plants. *Embrapa Agropecuária-Oeste-Artig. Periódico Indexado (ALICE).* 2019. Available online: https://ccsenet.org/journal/index.php/jas/article/view/0/39111 (accessed on 8 September 2024).
44. Ye, Y.; Chen, Y.; Xiong, S. Field detection of pests based on adaptive feature fusion and evolutionary neural architecture search. *Comput. Electron. Agric.* **2024**, *221*, 108936. [CrossRef]
45. He, H.; Liu, L.; Zhang, H.; Zheng, N. IS-DARTS: Stabilizing DARTS through Precise Measurement on Candidate Importance. In Proceedings of the AAAI Conference on Artificial Intelligence, Washington, DC, USA, 20–27 February 2024; Volume 38, pp. 12367–12375.
46. Weng, Y.; Zhou, T.; Liu, L.; Xia, C. Automatic Convolutional Neural Architecture Search for Image Classification Under Different Scenes. *IEEE Access* **2019**, *7*, 38495–38506. [CrossRef]
47. Bevers, N.; Sikora, E.J.; Hardy, N.B. Soybean disease identification using original field images and transfer learning with convolutional neural networks. *Comput. Electron. Agric.* **2022**, *203*, 107449. [CrossRef]
48. Ma, A.; Wan, Y.; Zhong, Y.; Wang, J.; Zhang, L. SceneNet: Remote sensing scene classification deep learning network using multi-objective neural evolution architecture search. *ISPRS J. Photogramm. Remote Sens.* **2021**, *172*, 171–188. [CrossRef]
49. Tiwari, R.G.; Maheshwari, H.; Agarwal, A.K.; Jain, V. HECNNet: Hybrid Ensemble Convolutional Neural Network Model with Multi-Backbone Feature Extractors for Soybean Disease Classification. In Proceedings of the 2024 2nd International Conference on Intelligent Data Communication Technologies and Internet of Things (IDCIoT), Bengaluru, India, 4–6 January 2024; pp. 813–818.
50. Slimani, H.; El Mhamdi, J.; Jilbab, A. Optimizing Multi-Level Crop Disease Identification using Advanced Neural Architecture Search in Deep Transfer Learning. *Int. J. Comput. Digit. Syst.* **2024**, *16*, 1293–1306. [CrossRef]
51. Dewi, C.; Thiruvady, D.; Zaidi, N. Fruit Classification System with Deep Learning and Neural Architecture Search. *arXiv* **2024**, arXiv:2406.01869.
52. Chen, Y.; Liang, H.; Jiao, S. NAS-MFF: NAS-guided multiscale feature fusion network with pareto optimization for sonar images classification. *IEEE Sensors J.* **2024**, *24*, 14656–14667. [CrossRef]
53. Liu, Y.; Sun, Y.; Xue, B.; Zhang, M.; Yen, G.G.; Tan, K.C. A survey on evolutionary neural architecture search. *IEEE Trans. Neural Netw. Learn. Syst.* **2021**, *34*, 550–570. [CrossRef]
54. Gudzius, P.; Kurasova, O.; Darulis, V.; Filatovas, E. AutoML-based neural architecture search for object recognition in satellite imagery. *Remote Sens.* **2022**, *15*, 91. [CrossRef]
55. Ren, P.; Xiao, Y.; Chang, X.; Huang, P.Y.; Li, Z.; Chen, X.; Wang, X. A comprehensive survey of neural architecture search: Challenges and solutions. *ACM Comput. Surv.s (CSUR)* **2021**, *54*, 1–34. [CrossRef]
56. Maharana, K.; Mondal, S.; Nemade, B. A review: Data pre-processing and data augmentation techniques. *Glob. Transitions Proc.* **2022**, *3*, 91–99. [CrossRef]
57. Saleem, M.H.; Khanchi, S.; Potgieter, J.; Arif, K.M. Image-based plant disease identification by deep learning meta-architectures. *Plants* **2020**, *9*, 1451. [CrossRef]
58. Krishnaswamy Rangarajan, A.; Purushothaman, R. Disease classification in eggplant using pre-trained VGG16 and MSVM. *Sci. Rep.* **2020**, *10*, 2322. [CrossRef]
59. Wang, J.; Perez, L. The effectiveness of data augmentation in image classification using deep learning. *Convolutional Neural Netw. Vis. Recognit.* **2017**, *11*, 1–8.
60. Frid-Adar, M.; Diamant, I.; Klang, E.; Amitai, M.; Goldberger, J.; Greenspan, H. GAN-based synthetic medical image augmentation for increased CNN performance in liver lesion classification. *Neurocomputing* **2018**, *321*, 321–331. [CrossRef]
61. Makhlouf, A.; Maayah, M.; Abughanam, N.; Catal, C. The use of generative adversarial networks in medical image augmentation. *Neural Comput. Appl.* **2023**, *35*, 24055–24068. [CrossRef]
62. Alzubaidi, L.; Zhang, J.; Humaidi, A.J.; Al-Dujaili, A.; Duan, Y.; Al-Shamma, O.; Santamaría, J.; Fadhel, M.A.; Al-Amidie, M.; Farhan, L. Review of deep learning: Concepts, CNN architectures, challenges, applications, future directions. *J. Big Data* **2021**, *8*, 1–74. [CrossRef]
63. Krizhevsky, A.; Sutskever, I.; Hinton, G.E. Imagenet classification with deep convolutional neural networks. *Commun. ACM* **2017**, *60*, 84–90. [CrossRef]
64. Simonyan, K.; Zisserman, A. Very deep convolutional networks for large-scale image recognition. *arXiv* **2014**, arXiv:1409.1556.
65. Salehi, A.W.; Khan, S.; Gupta, G.; Alabduallah, B.I.; Almjally, A.; Alsolai, H.; Siddiqui, T.; Mellit, A. A study of CNN and transfer learning in medical imaging: Advantages, challenges, future scope. *Sustainability* **2023**, *15*, 5930. [CrossRef]

Article

Detection of Leaf Diseases in Banana Crops Using Deep Learning Techniques

Nixon Jiménez [1], Stefany Orellana [1], Bertha Mazon-Olivo [2,*], Wilmer Rivas-Asanza [2] and Iván Ramírez-Morales [3]

[1] Department of Information Technology, Facultad de Ingeniería Civil, Universidad Técnica de Machala, 5.5 km Pan-American Av., Machala 070150, Ecuador; njimenez4@utmachala.edu.ec (N.J.); sorellana2@utmachala.edu.ec (S.O.)

[2] AutoMathTIC Research Group, Facultad de Ingeniería Civil, Universidad Técnica de Machala, 5.5 km Pan-American Av., Machala 070150, Ecuador; wrivas@utmachala.edu.ec

[3] DINTA Research Group, Facultad de Ciencias Agropecuarias, Universidad Técnica de Machala, 5.5 km Pan-American Av., Machala 070150, Ecuador; iramirez@utmachala.edu.ec

* Correspondence: bmazon@utmachala.edu.ec

Abstract: Leaf diseases, such as Black Sigatoka and Cordana, represent a growing threat to banana crops in Ecuador. These diseases spread rapidly, impacting both leaf and fruit quality. Early detection is crucial for effective control measures. Recently, deep learning has proven to be a powerful tool in agriculture, enabling more accurate analysis and identification of crop diseases. This study applied the CRISP-DM methodology, consisting of six phases: business understanding, data understanding, data preparation, modeling, evaluation, and deployment. A dataset of 900 banana leaf images was collected—300 of Black Sigatoka, 300 of Cordana, and 300 of healthy leaves. Three pre-trained models (EfficientNetB0, ResNet50, and VGG19) were trained on this dataset. To improve performance, data augmentation techniques were applied using TensorFlow Keras's ImageDataGenerator class, expanding the dataset to 9000 images. Due to the high computational demands of ResNet50 and VGG19, training was performed with EfficientNetB0. The models—EfficientNetB0, ResNet50, and VGG19—demonstrated the ability to identify leaf diseases in bananas, with accuracies of 88.33%, 88.90%, and 87.22%, respectively. The data augmentation increased the performance of EfficientNetB0 to 87.83%, but did not significantly improve its accuracy. These findings highlight the value of deep learning techniques for early disease detection in banana crops, enhancing diagnostic accuracy and efficiency.

Keywords: banana disease detection; convolutional neural networks; Cordana; deep learning; EfficientNetB0; ResNet50; Sigatoka negra; VGG19

1. Introduction

The banana is one of Ecuador's main export products and plays a crucial role in its economy, representing 18% of its non-oil exports. According to Ciancio et al. [1], around 400 million people choose to consume bananas due to their health benefits and significant nutritional composition. Additionally, approximately 70 million people earn income from banana production, primarily in some countries in Africa, Asia, and Latin America. The trade of certified bananas under sustainability standards represents a significant part of the economy in several tropical countries, with an estimated market value of USD 2.4 billion [2].

However, the productivity of banana plantations is threatened by various foliar diseases, such as Black Sigatoka (*Pseudocercospora fijiensis*) and Cordana (*Cordana musae*), which have a direct impact on fruit quality and plant longevity [3]. Although there are other pathologies that also affect the leaves of this crop, such as Panama disease (*Fusarium oxysporum* f. sp. *Cubense*), moko (*Ralstonia solanacearum raza* 2), Yellow Sigatoka (*Mycosphaerella musicola*), Erwinia, and mosaic (*Tobacco mosaic* virus), our research mainly focuses on the two mentioned diseases. This is due to the low incidence of the other diseases in our area, which would have made it difficult to collect the images needed to create the dataset.

Black Sigatoka is one of the most aggressive diseases affecting banana crops, reducing photosynthesis by damaging leaves and causing the premature ripening of the fruit, thus affecting its quality [4]. This disease can decrease crop yields by up to 50%; to prevent its spread, fungicides must be applied up to 40 times a year, increasing the production costs [5]. In the case of *Cordana musae*, lesions develop as oval spots, characterized by a yellowish edge and a grayish center, exhibiting a pattern of concentric rings. This pathology, attributed to the fungus *Cordana musae*, presents visual characteristics similar to those of Sigatoka, complicating its identification [6].

In the agricultural context, the implementation of advanced technologies like deep learning in agriculture enables the early detection of plant diseases, helping farmers to make informed decisions and reduce costs. The timely detection of these diseases is crucial to prevent them from worsening, thus protecting both banana production and exports, while minimizing their impacts on producers in the country [7]. Research related to this topic is limited. The following are some important studies. Salehin et al. [8] focused on the detection of two diseases, Fusarium wilt (Panama disease) and Black Sigatoka, achieving accuracy of 99% using complex models like ResNet152 and Inception V3. The use of only two classes fails to represent the real complexity of banana crops, where multiple diseases can coexist, each requiring differentiated diagnosis.

Elinisa and Mduma [9] implemented a CNN model with four layers to classify two diseases, Fusarium wilt and Black Sigatoka, achieving accuracy of 91.17% with a dataset of 27,360 images. Thiagarajan et al. [10] focused on detecting diseases such as Black Sigatoka, Fusarium wilt (Panama disease), and mosaic disease in 937 images of banana leaves [11], using artificial neural networks (ANN) combined with feature extraction techniques such as the scale-invariant feature transform (SIFT), histogram of oriented gradients (HOG), and local binary patterns (LBP). The applied neural networks were convolutional (CNN) and pre-trained models like ResNet50 and Inception V3, achieving accuracy of 98.9% and 95.5%, respectively.

Rajalakshmi et al. [12] identified banana diseases such as Cordana and Pestalotiopsis in India with a dataset of 768 images. The proposed model, called 8C-DCNN, was a deep convolutional neural network that achieved average accuracy of 98.92%.

The models used offer high levels of precision but require significant computational capacities, which can be a practical limitation. Additionally, while the achieved accuracy seems very high, it may not reflect their true generalization, as the authors do not evaluate how the models perform in a multiclass environment. Furthermore, confusion analyses or errors between classes are often not exhaustive, hindering a comprehensive assessment of the model's performance in realistic scenarios.

After reviewing the existing studies, a significant lack of research focused on the detection of foliar diseases in banana crops within the Latin American context was identified.

The main contributions of this work are as follows:

- We propose an approach for early disease detection using pre-trained deep learning models;
- In response to the lack of banana leaf datasets contextualized to Latin America, a tailored dataset was collected and created specifically for this environment;

- We provide a comparison and evaluation of the performance of the models used (ResNet50, EfficientNetB0, VGG19);
- We describe the implementation of deep learning models in a mobile application to facilitate practical use.

Initially, the models were trained with a dataset of 900 images, achieving accuracy of 85% with ResNet50, 84% with EfficientNetB0, and 88% with VGG19. The EfficientNetB0 model stands out not only for its good accuracy but also for its efficiency in resource use, both computationally and in time, completing its training in approximately 30 min. In contrast, ResNet50 and VGG19 had longer training times, requiring around 1.5 h and 7 h, respectively, using computational resources such as an Intel Core i7 11th generation CPU with 32 GB of RAM and the Jupyter Notebook tool (6.4.12) with the Python language (3.9.13).

Given this difference in training times, it was decided not to train ResNet50 and VGG19 with the dataset of 9000 images due to the inability to perform multiple iterations of fine-tuning and comparative evaluation. Therefore, EfficientNetB0 was the only model trained with the complete set of images, achieving accuracy of 88%. The application of deep learning techniques not only facilitates the early detection of foliar diseases in bananas but also offers a tool for farmers, allowing them to apply corrective measures in time, improving the quality of the final product and reducing economic losses.

2. State of the Art

In addition to the studies reviewed in the Introduction section, related to the application of deep learning in banana disease detection, the work of Sanga, Mero, and Machuve [13] stands out, focusing on binary classification models; however, it does not include a multiclass analysis or a comparison between deep learning models. Nevertheless, related research on the application of deep learning in other crops has been identified, which is described below.

Blanc-Talon et al. [14] focused on applying deep learning techniques, presenting a deep meta-architecture and a refinement filter bank to address class imbalances and false positives. They evaluated their method on a dataset of tomato diseases, achieving improvements in disease detection under real field conditions.

Abbas et al. [15] proposed a method for the detection of diseases in tomato plants using transfer learning and synthetic image generation through conditional generative adversarial networks (C-GAN). To enhance the model's generalization, synthetic images were generated and combined with real data to train the DenseNet121 model. They focused on the multiclass classification of tomato diseases, achieving accuracy of 97.11% across 10 classes from the PlantVillage dataset [16]. This approach demonstrates the effectiveness of combining transfer learning and synthetic data to improve the performance in agricultural settings.

Atila et al. [17] utilized the EfficientNet model to classify diseases in maize, grape, orange, peach, potato, raspberry, soybean, strawberry, tomato, and pumpkin leaves, employing the PlantVillage dataset [16]. They trained different versions of EfficientNet (B0 to B7), comparing their performance with state-of-the-art models such as ResNet50, VGG16, and Inception V3. The EfficientNetB4 and B5 models achieved average accuracy of 99.97% and 99.91%, respectively, outperforming the other models. This study highlights the efficacy of EfficientNet for the multiclass classification of plant diseases and its potential for early detection.

Jadhav et al. [18] focused on identifying soybean plant diseases using pre-trained convolutional neural networks (CNN), specifically AlexNet and GoogleNet, with transfer learning to classify three common diseases in 649 images of soybean leaves, achieving accuracy of 98.75% with AlexNet and 96.25% with GoogleNet.

Anh et al. [19] concentrated on detecting foliar diseases in tomato, maize, potato, rice, wheat, soybean, cotton, grape, apple, and citrus using deep learning models on resource-limited devices like a Raspberry Pi 3, especially when utilizing large datasets. They found that MobileNetV3 was the most suitable for field implementation, achieving accuracy of 96.58%, with inference and initialization times of 127 ms and 11 ms, respectively, and requiring only 7.4 MB of memory.

Similarly, Andrew et al. [20] addressed the detection of foliar diseases in plants using deep learning models, specifically pre-trained convolutional neural networks (CNN) such as DenseNet-121, ResNet50, VGG-16, and Inception V4. The dataset used was PlantVillage [20], which consists of 38 classes and 54,305 images. They applied hyperparameter tuning and regularization techniques to improve the accuracy and reduce overfitting. DenseNet121 achieved the highest accuracy (99.81%) in the multiclass classification of diseases in agricultural crops, standing out as the most suitable model for this task.

3. Literature Review

In the search for information related to this topic, various authors have been identified who have provided relevant analyses regarding the methods and models applied in the detection of foliar diseases using deep learning.

Liu and Wang [21] present a review of disease and pest detection in plants using deep learning techniques, focusing on classifying, detecting, and segmenting images of diseases in crops. They compare these approaches with traditional methods and highlight the advantages, challenges, and potential solutions of deep neural networks in this field. Additionally, the study addresses the available datasets and evaluates the performance of different models in agricultural disease detection, providing a detailed analysis of the use of these technologies in various agricultural crops, although not specifically in bananas.

Hussein and Mousa [22] categorize and classify convolutional neural network (CNN) models, identifying tools and platforms while highlighting some implementation challenges in plant disease detection. The following presents a review of the state of the art in deep learning, which has served as a basis for the associated terminology and aided in carrying out our work.

3.1. Deep Learning

Deep learning, a subfield of machine learning and artificial intelligence, employs artificial neural networks with multiple layers. These architectures allow for the extraction of hierarchical representations from data, thereby optimizing the automation of complex tasks such as image recognition, natural language processing, and time series prediction. In the scientific literature, there has been a notable increase in research implementing deep learning techniques across various sectors, including the detection of plant diseases, such as convolutional neural networks (CNN), recurrent neural networks (RNN), and transfer learning [9,17,18].

3.2. Convolutional Neural Networks

Convolutional neural networks (CNNs) are models specifically designed to process data in image format. These networks have proven to be essential in the fields of computer vision and digital image processing due to their ability to model and adapt representations of relevant visual features [23]. The architecture is based on convolutional layers that perform convolution operations to extract hierarchies of spatial features, such as edges, textures, and shapes, at different levels of abstraction [24]; this design allows CNNs to excel in tasks such as image classification, object detection, and semantic segmentation, as well as in specific applications like facial recognition, autonomous vehicles, and medical diagnostics [25–28].

The structure of a CNN operates on data organized in matrices (images) through the implementation of convolutional layers, pooling layers, and fully connected layers. The convolutional layers are trained using a set of specifically optimized filters (kernels) for edge and texture detection, which are fundamental for visual interpretation [29]. The rectified linear unit (ReLU) activation function is incorporated in the hidden layers to introduce non-linearity into the model, thereby facilitating the learning of complex patterns in data representations. The pooling process is essential for dimensionality reduction while preserving the most important information from the detected features [30]. Finally, the fully connected layers perform the prediction stage based on the feature representations extracted in the previous layers, completing the inference process [31].

Throughout the development of CNNs [32,33], various architectures have emerged, designed to improve their performance in terms of accuracy, efficiency, and depth. For this research, three models have been considered, namely VGG, ResNet, and EfficientNet, due to their wide use in image classification tasks and their ability to extract relevant features in the analysis of crop diseases. The choice of these models was due to their performance in the detection of complex visual patterns, their adaptability to different image conditions, and their balance between accuracy and computational efficiency [34].

3.3. *EfficientNet B0*

EfficientNetB0 is a convolutional neural network that efficiently optimizes the depth, width, and resolution by using a composite coefficient. Unlike traditional approaches that adjust these dimensions without a defined criterion, this model uses preset scaling coefficients; this strategy allows the network to add more layers and channels as the input image resolution increases, which in turn expands the receptive field and facilitates the capture of more detailed patterns in the image [35,36].

3.4. *ResNet50*

ResNet is a convolutional neural network with 50 layers, and its main advantage lies in its lower memory consumption compared to other convolutional neural network models. This is due to the use of a layer called GlobalAveragePooling in the classification phase, instead of using dense layers. This layer transforms the two-dimensional feature maps generated in the last feature extraction stage into a vector of n-classes, which is used to calculate the probability of belonging to each category [37–39].

3.5. *VGG19*

VGG19 is a convolutional neural network with a deep architecture that improves feature extraction in images, uses alternating convolutional layers with non-linear activations and MaxPooling to reduce the dimensionality. In addition, it uses the ReLU to select the most representative values, which optimizes image processing by preserving key features and reducing the number of parameters [40,41].

3.6. *Transfer Learning*

Transfer learning reuses pre-trained models from a source domain to solve tasks in a related target domain, mitigating overfitting in small datasets. In convolutional neural networks (CNNs), it optimizes the feature selector of the target domain, adjusting only the fully connected layers, which reduces the complexity and computational costs without compromising the accuracy [30].

The implementation and development of advanced deep learning architectures primarily apply to large and diverse datasets and require high-performance computational platforms. This is particularly true for convolutional neural networks (CNNs), which require a substantial number of labeled image data to carry out the model training process for classification. These models possess a large number of parameters, whose optimal adjustment occurs through complex matrix operations during the training phase. Therefore, the adoption of specialized hardware, such as graphics processing units (GPUs), is common practice to significantly increase the efficiency and speed of these processes [30–32]. If such capabilities are not available for model construction, the training process may become unfeasible. A reduced volume of data may prevent models from generalizing to new cases; similarly, if the computational capacity is insufficient, training may require excessive time, hindering the effective use of the models [42].

In light of the inherent complexity of the issues at hand, the concept of "transfer learning" emerged, based on the premise that one task can leverage the results and knowledge acquired from another task, thereby facilitating the learning process and optimizing the transferred knowledge [43]. In the field of image processing, the most recognized transfer learning technique was introduced by Yosinski [44]. The proposed methodology facilitates the transfer of learned attributes from one model to another distinct model. However, to carry out this adaptation of the recipient model, a fine-tuning process must be executed on the interconnected layers of the latter [45].

3.7. Confusion Matrix and Associated Metrics for Classification Problems

In the context of convolutional neural networks (CNNs), the confusion matrix is a tool used to assess a classifier's performance by analyzing the relationships between classes within a dataset. It is generated from a test dataset with a known ground truth, where each class is compared to the others to identify the number of misclassified samples [46]. These improvements help to enhance both the accuracy and computational efficiency of the model. For instance, as mentioned in the work by Chipindu et al. [47], in an image detection scenario, the confusion matrix can be applied to binary classification tasks, such as distinguishing between two outputs: 0 (no abortion) and 1 (abortion).

The confusion matrix is used in neural networks to evaluate feature attribution methods, classifying data as "relevant" or "not relevant". Metrics such as precision, recall, and the F1 score are adopted, as visualized in Figure 1, where true positives represent instances correctly classified as belonging to the target class and false positives represent incorrect classifications.

CONFUSION MATRIX		PREDICTED BY THE MODEL		
		Positive	Negative	
REAL	Positive	TP	FN	**Sensitivity, Recall o TPR** $TPR = \frac{TP}{(TP+FN)}$
	Negative	FP	TN	**Specificity, Selectivity o TNR** $TNR = \frac{TN}{(TN+FP)}$
Accuracy o PPV $Accuracy = PPV = \frac{TP}{(TP+FP)}$		**NPV** $NPV = \frac{TN}{(TN+FN)}$		**Accuracy** $Accuracy = \frac{TP+TN}{(TP+TN+FP+FN)}$
		Harmonic mean between precision and recall, or F1-score $F1Score = \frac{2TP}{(2TP+FP+FN)}$		

Figure 1. Confusion matrix and model evaluation metrics.

This approach allows for a quantitative evaluation of the effectiveness of attribution methods, thereby facilitating the selection of more reliable and effective techniques to explain model behavior [48].

3.8. Performance Metrics for CNN Models

Zapeta Hernández et al. [49] define performance metrics used to quantify the accuracy and precision in image classification through convolutional neural networks. These metrics facilitate the identification of errors, such as incorrect predictions, underfitting, and overfitting during the machine learning process, thereby optimizing the model's ability to perform correct classification. The evaluation metrics include indicators such as accuracy, precision, recall (or completeness), specificity, and the F1 score, as well as the rates of true positives and false positives [50].

As a first point, it is essential to understand the meanings of the variables involved, which are described below.

- TP: True Positives
- TN: True Negatives
- FP: False Positives
- FN: False Negatives

Accuracy is defined as the proportion of predictions that a model correctly classifies and is represented by Equation (1). This metric is valuable due to its direct interpretation; however, it can be misleading in unbalanced datasets, as it may present high levels of accuracy even if the model does not adequately classify minority classes [51].

$$Accuracy = \frac{TP + TN}{(TP + TN + FP + FP)} \tag{1}$$

Precision measures the accuracy of the model in correctly classifying a positive class and is calculated using Equation (2).

$$Precision = \frac{VP}{VP + FP} \tag{2}$$

Specificity, also known as the true negative rate (TNR) and represented by Equation (5), is defined as the proportion of true negatives that are correctly identified compared to the total number of actual negative outcomes [51].

$$Specificity = \frac{TP}{(TP + FN)} \tag{3}$$

Recall, represented by Equation (4), indicates how many of the truly positive examples were classified. A model with high recall is effective in recognizing the majority of the positive cases in the dataset. On the other hand, a model with low recall has limitations in identifying all positive cases present in the data [52].

$$Recall = \frac{TP}{TP + FN} \tag{4}$$

The F1 score seeks a balance between precision and recall, as it penalizes both false positives and false negatives, and is calculated using Equation (5):

$$F1 - score = 2 \times \frac{Accuracy \times Recall}{Accuracy + Recall} \tag{5}$$

In addition to the previously mentioned metrics, there are advanced metrics that expand the evaluation of convolutional neural networks. Among these, the geometric precision, decisiveness, and robustness stand out, aimed at analyzing the model's confidence in different probability regions. Additionally, the index balanced accuracy (IBA) and G_mean allow for the assessment of its performance in unbalanced datasets. Finally, metrics such as the NetScore and information density integrate precision with architectural and computational complexity, being especially relevant for applications on resource-constrained devices [44–46].

3.9. Loss Function

The loss function used by a model is undoubtedly a key element in evaluating its performance. Algorithms adjust their parameters through this function, which measures the difference between the model's predictions and the actual data. When the loss function yields a high value, it indicates a significant discrepancy between the predictions and the observed results [53].

The categorical cross-entropy loss function is extensively used in classification tasks due to its effectiveness in quantifying the divergence between the predictions generated by the model and the true labels represented in one-hot format. Hui and Belkin [54] state that, although this function is prominent in widely used implementations, such as those in Hugging Face Transformers, its superiority is not clearly defined across all domains. They also highlight the limitation of this function in generating well-calibrated probabilities, indicating that the softmax activation at the end of the model does not always adequately represent the actual confidence probability. Each of these metrics is suitable for different purposes, such as preserving local structures in images or optimizing the perception of visual quality [55,56].

4. Materials and Methods

The methodology applied in this work is CRISP-DM, which consists of 6 phases [57]: bussiness understanding, data understanding, data preparation, Modeling, evaluation, deployment.

Each of the phases of the applied methodology is described below:

4.1. Business Understanding

The understanding of the problem in this work focuses on identifying the need to improve the early detection of leaf diseases in banana crops using deep learning models, as described in Section 1.

4.2. Data Understanding

As we can see in Figure 2, 900 images of banana leaves were collected, classified into three categories: 300 images of leaves affected by Black Sigatoka, 300 images of those affected by Cordana, and 300 images of healthy leaves.

Figure 2. Authors' own dataset, collected from Orense farms.

In order to collect images, we visited several banana farms in the province of El Oro. With the collaboration of an agronomist, we were able to identify and classify the previously mentioned diseases, ensuring that the collected images were representative and relevant for the study.

4.3. Data Preparation

Next, data preparation was carried out, which involved data cleaning to remove images that contained noise or where the leaf was not clearly visible due to blurriness or poor quality. Additionally, the images were separated into three distinct classes: healthy, affected by Black Sigatoka, and affected by Cordana.

Once classified, the images were resized to a standard size of 224 × 224 pixels to ensure consistency in the model's input. Regarding data augmentation, although initially 900 images were used (300 per class), it was decided to increase the number of images using data augmentation techniques with the TensorFlow Keras ImageDataGenerator class. Transformations included rotations of up to 45 degrees, horizontal and vertical shifts of up to 20%, cropping transformations with a range of 0.2, random zoom of up to 20%, horizontal flipping with a probability of 50%, and brightness adjustments ranging from 50% to 150%. These techniques increased the variability of the images, generating an expanded dataset of 9000 images to improve the model's representativeness.

To ensure that the model could generalize well and avoid underfitting, the dataset was divided into two parts: the training set, which represented 80% of the images, and the validation set, which included the remaining 20%. This division allowed the model to train on a wide variety of examples while maintaining a separate dataset to evaluate its performance on images not seen during training. This strategy helped to obtain a more accurate measure of the model's generalization ability, preventing the model from overfitting the training data and promoting a more objective assessment of its performance.

4.4. Modeling

A deep learning analysis was conducted, focusing on convolutional neural networks (CNNs) applied to the classification of leaf diseases. These architectures have been proven to be robust in handling large volumes of data, allowing us to develop generalizable systems [58–60].

In this study, not only were three main models evaluated, but additional architectures were also tested, such as NasNetLarge, Inception ResNetV2, AlexNet, and MobileNet. However, due to overfitting issues, achieving accuracy of around 33% without any improvement, these latter models had to be discarded. As a result, the three models that demonstrated better performance and generalization capabilities were selected.

In the following sections, the architectures of the deep learning models used in this study are described in detail. Their structural components are analyzed, from the input layers to the output layers, emphasizing the specific configurations of each model and the modifications implemented to optimize the classification of leaf diseases in bananas.

As shown in Figure 3, the architecture begins with the input of images sized 224 × 224 pixels and progresses through a series of convolutional layers, highlighting the use of MBConv mobile layers. These layers progressively reduce the spatial resolution while increasing the depth of the extracted features, facilitating the learning of patterns present in the leaves.

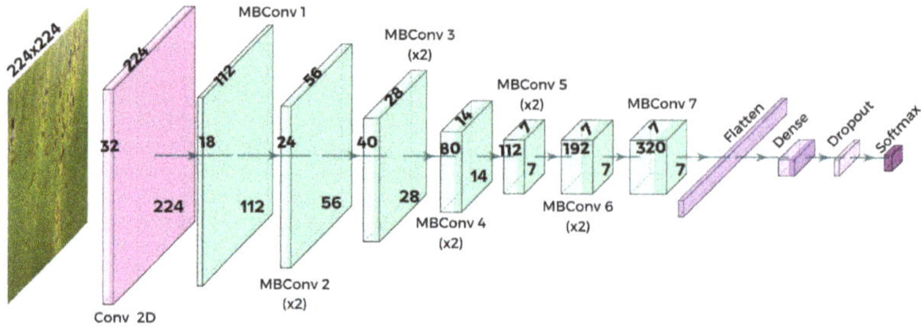

Figure 3. Integrated EfficientNetB0 architecture with custom extension layers.

The network consists of seven MBConv blocks applied repetitively, designed to enhance the detection of visually similar features. Subsequently, a flattening layer transforms the features into a one-dimensional vector, followed by a dense layer with 32 neurons, responsible for classification. To reduce the risk of overfitting, a dropout layer is included, and the architecture concludes with a softmax layer that classifies the inputs into three categories.

As shown in Figure 3, which describes the data input and the development of the architecture, the model presented in Figure 4 maintains the general structure previously described.

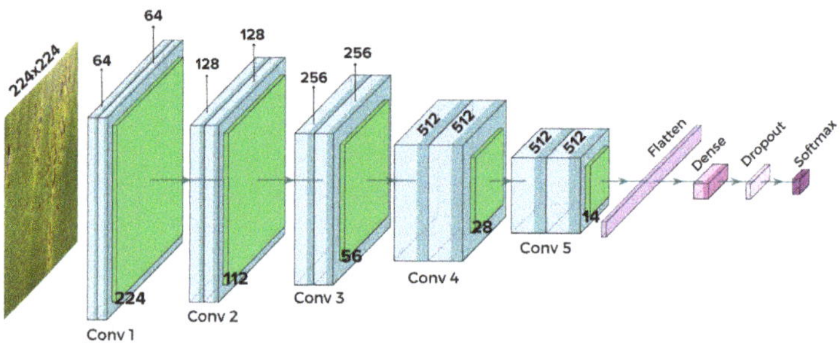

Figure 4. Integrated architecture of ResNet50 with custom extension layers.

However, unlike its predecessor, this version consists of five convolutional blocks. The model begins with the conv1 layer, which has 64 filters, and progresses through the conv2, conv3, conv4, and conv5 layers, progressively increasing the number of filters to 128, 256, and 512, respectively, while the spatial resolution gradually decreases.

Similarly, as observed in Figures 3 and 4, the model in Figure 5 presents an analogous structure; however, it differs at the start by incorporating the conv1 layer with 64 filters.

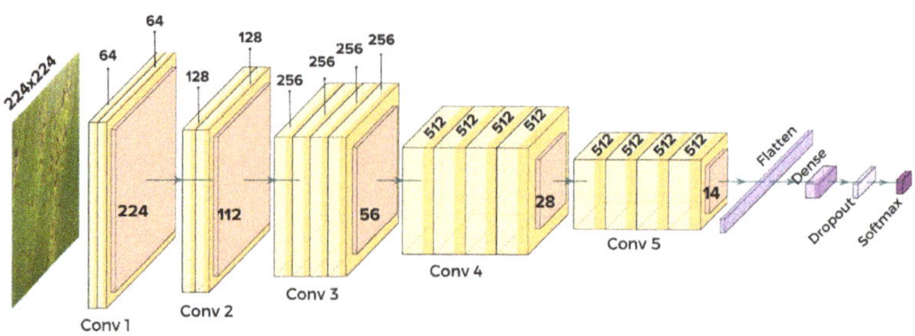

Figure 5. Integrated architecture of VGG19 with custom extension layers.

From this layer, the model progresses through conv2, conv3, conv4, and conv5, gradually increasing the number of filters to 128, 256, and 512. This modification allows for the capture of more complex features in the leaf images, while the optimized design of the blocks, with the inclusion of multiple filters, enhances the model's learning capacity.

Next, the training and comparison of the models were conducted using three pre-trained deep learning architectures, namely ResNet50, EfficientNetB0, and VGG19, all with initial weights pre-trained on ImageNet. To ensure a fair comparison, the hyperparameter configuration (Table 1) was kept constant across all three cases, varying only the pre-trained neural network.

Table 1. Parameter configuration.

Parameter	Value
Input size	224×224
Batch size	64
Number of epochs	100
Optimizer	Adam
Learning rate	1×10^{-3}
Include_top	False
Weights	ImageNet
Pooling	Flatten
Classes	3
Classifier activation	Softmax

A sequential architecture was used, excluding the original top layer, which was replaced with task-specific layers. The layers of the pre-trained network were kept frozen; in other words, their weights were not updated during training, preserving the knowledge previously acquired with the ImageNet dataset. Following the pre-trained network, custom layers were integrated: a flatten layer to flatten the data, followed by a dense layer with 32 neurons and a ReLU activation function, to which L2 regularization was applied to prevent overfitting. Subsequently, a dropout layer with a rate of 50% was added to further reduce the risk of overfitting. Finally, a dense layer with softmax activation was responsible for classifying the inputs into the three corresponding categories: healthy, Black Sigatoka, and Cordana.

The model was compiled using the Adam optimizer with a learning rate of 0.001 and the loss function (categorical_crossentropy), suitable for multi-class classification problems. Training was conducted for 100 epochs, with a batch size of 64. The training results were monitored using TensorBoard, allowing for the visualization of loss and accuracy curves for both training and validation data, thus facilitating a detailed analysis of the model's performance.

In addition, an input size of 224×224 pixels was chosen, in line with standard pre-trained models such as EfficientNetB0, ResNet50, and VGG19. This choice allowed us to leverage pre-trained ImageNet weights, improving the model's performance and accelerating convergence by providing better parameter initialization.

A batch size of 64 was selected for training, balancing stability and computational efficiency without demanding excessive memory usage. One hundred epochs were set to ensure adequate convergence and minimize the risk of premature overfitting.

To optimize the training process, the Adam optimizer was used, known for dynamically adjusting the learning rates of each parameter, facilitating more efficient convergence. The learning rate of 0.001 was chosen based on empirical evidence showing its effectiveness for these models, ensuring a balance between rapid convergence and stability.

The top layer of the pre-trained model was deactivated (Include_top = False) to allow for a custom final classification layer tailored to the specific task. A flatten layer was used for dimensionality reduction before classification, preserving key spatial information without losing important features.

Finally, a softmax activation function was employed in the classification layer, which is ideal for multi-class classification problems, enabling the efficient handling of multiple categories.

5. Results

5.1. Evaluation

The results obtained from the three convolutional neural network models (EfficientNetB0, ResNet50, and VGG19) used in this project are presented below. To evaluate their performance in classifying leaf diseases in banana crops, key evaluation metrics such as accuracy, recall, and the F1-score were used; these metrics reflect the effectiveness of each model in correctly classifying the images, providing an objective basis for the comparison their disease detection capabilities, including diseases such as Black Sigatoka and Cordana.

Presenting these metrics is essential to identify which of the evaluated architectures offers better performance in this specific task, considering both their accuracy and generalization abilities.

In Figure 6, the confusion matrix of the ResNet50 model is presented, which illustrates its performance in classifying healthy banana leaves and those affected by Cordana and Black Sigatoka. The model successfully classified 57 out of 60 leaves with Cordana, presenting an error rate of three samples, which were incorrectly classified as Black Sigatoka. In the case of healthy leaves, the performance was very good, with 59 out of 60 correct classifications and a single error regarding Cordana. However, the performance in classifying leaves affected by Black Sigatoka was lower, achieving 44 out of 60 correct predictions, while 16 leaves were incorrectly classified as Cordana, indicating significant confusion between these two diseases.

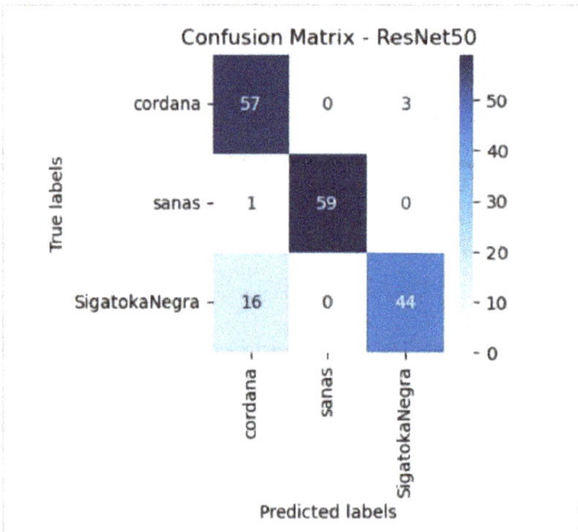

Figure 6. Confusion matrix—ResNet50.

In general terms, the confusion matrix observed in Figure 6 reveals that most of the model's errors are concentrated in the classification of Black Sigatoka. This suggests that the features extracted by the ResNet50 model may not be sufficiently discriminative to differentiate between Cordana and Black Sigatoka. This difficulty could be related to the visual similarity of the lesions caused by both diseases, especially in the advanced stages, which significantly complicates their differentiation.

Figure 7 shows the evolution of the accuracy per epoch for the ResNet50 model during training and validation on a dataset of leaf images. The curves reflect how the accuracy changes over 100 epochs. The accuracy gradually increases, reaching nearly 90%, but decreases after epoch 60.

Figure 7. Learning evolution curve for ResNet50 model: fuchsia line (training) vs. yellow line (validation).

The accuracy stabilizes at around 90% from epoch 30, showing more consistent behavior than during training. Visually, we can say that the discrepancy between the two curves suggests that the model memorizes the training data rather than generalizing correctly.

Figure 8 illustrates the evolution of the loss for the ResNet50 model during training and validation over 100 epochs. A rapid decrease in loss is observed in the early epochs, indicating that the model learns effectively at the beginning of the process. From epoch 10, the loss stabilizes at low values, remaining close to that of validation, suggesting a model without signs of overfitting.

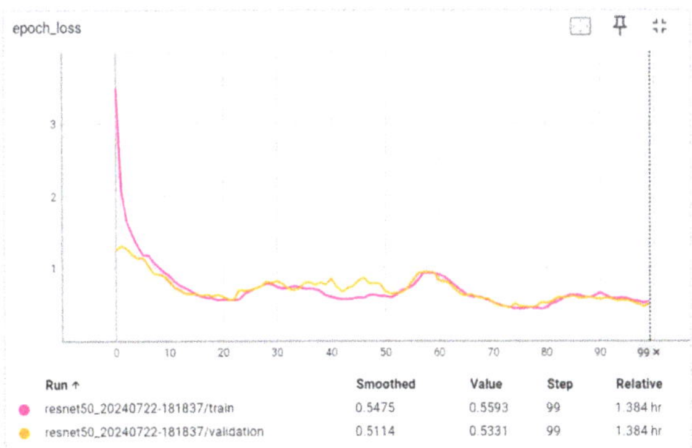

Figure 8. Loss evolution curve for ResNet50 model: fuchsia line (training) vs. yellow line (validation).

The validation loss follows a similar pattern to the training loss, with a rapid decrease in the early epochs. Subsequently, both curves show slight fluctuations but remain close to each other, indicating the consistent evolution of the model. From epoch 50, both curves stabilize and present low and similar loss values, suggesting that the model has adequately converged on both the training and validation data.

The training process lasted approximately one and a half hours, indicating that the model achieved good optimization in a reasonable time. In conclusion, the overall behavior of the loss suggests that the model was well fitted, as the loss curves for training and validation were similar throughout the process. Additionally, there were no signs of overfitting, as the validation loss did not show significant increases in the final epochs.

The confusion matrix in Figure 9 reveals a significant pattern of errors in classifying leaves with diseases, particularly in the classes of Cordana and Black Sigatoka. The EfficientNetB0 model shows a tendency to confuse these two classes, indicating the inherent difficulty of the model in identifying distinctive features between both diseases. This confusion is likely due to the visual similarity in the symptoms that they present, such as leaf spots, which complicates their differentiation. Nevertheless, the model achieves the accurate classification of healthy leaves, indicating that it can clearly recognize the absence of symptoms.

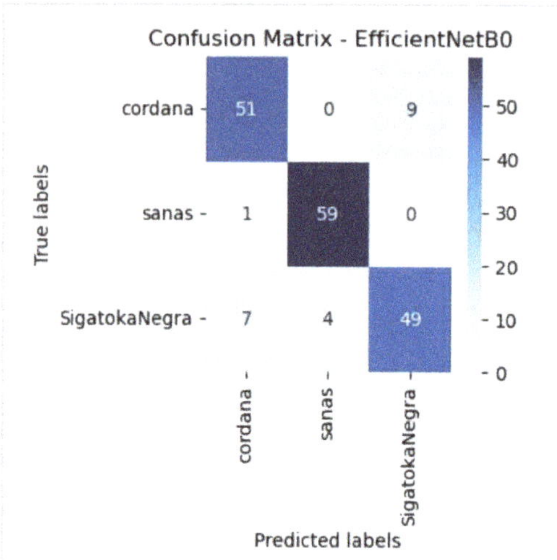

Figure 9. Confusion matrix—EfficientNetB0.

Figure 10 shows the evolution of the accuracy of the EfficientNetB0 model during training and validation over 100 epochs. The model's accuracy improves rapidly in the first 10 epochs, reaching a value close to 95% in the training set, where it stabilizes with slight fluctuations, indicating that the model learns effectively from the training dataset.

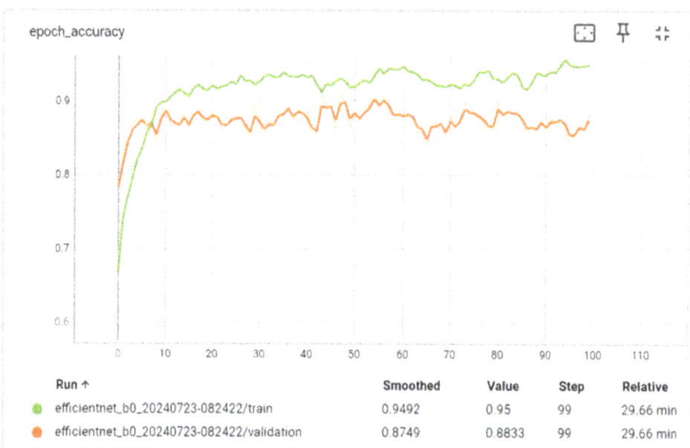

Figure 10. Learning evolution curve of the EfficientNetB0 model: green line (training) vs. orange line (validation).

The validation accuracy also experiences a rapid increase during the early epochs but stabilizes at around 88%, showing consistent performance, although slightly lower than that obtained in training. This suggests that the model generalizes well, although there is a slight gap between the two accuracy values.

Similarly, Figure 11 shows the evolution of the loss of the EfficientNetB0 model during training and validation over 100 epochs. The loss drops quickly in the first 10 epochs, reaching low values of around 0.5, which suggests that the model learns efficiently from the training data. After this point, the loss stabilizes, with slight fluctuations around this value, indicating that the model has achieved a good fit on the training data. The validation loss also decreases rapidly at first but stabilizes around higher values, close to 0.8, reflecting somewhat lower performance compared to that on the training data.

Figure 11. Loss evolution curve of the EfficientNetB0 model: green line (training) vs. orange line (validation).

The EfficientNetB0 model shows overall good performance and a proper fit to the training data, completing the process in about half an hour. Similarly, in Figures 7 and 8, we observe that the accuracy and loss functions over the 100 epochs demonstrate the favorable evolution of learning. The accuracy in both the training and validation sets converges at around 90%, while the loss function decreases rapidly in the early epochs before stabilizing, indicating an effective fit to the data.

The confusion matrix for the VGG19 model, shown in Figure 12, reflects a trend similar to that observed in the previous models, where the highest number of errors is concentrated in classifying the Black Sigatoka class. Specifically, the model struggles to differentiate between Cordana and Black Sigatoka, resulting in a considerable amount of confusion between these two categories. This may be related to the visual similarity of the lesions on the leaves affected by both diseases. However, it is important to highlight that the VGG19 model demonstrates high accuracy in classifying healthy leaves, as it shows no errors in this category, indicating that the model is effective in detecting the absence of symptoms.

Figure 13 shows the evolution of the accuracy of the VGG19 model during training and validation over 100 epochs. The accuracy increases rapidly during the first 10 epochs, reaching values close to 88%. From this point on, the accuracy stabilizes with slight fluctuations, indicating a good fit to the training data.

The validation accuracy follows a similar pattern to that of the training, with a rapid improvement in the first 10 epochs. Subsequently, it remains stable at around 87%, with some minor variations, suggesting that the model generalizes well to unseen data.

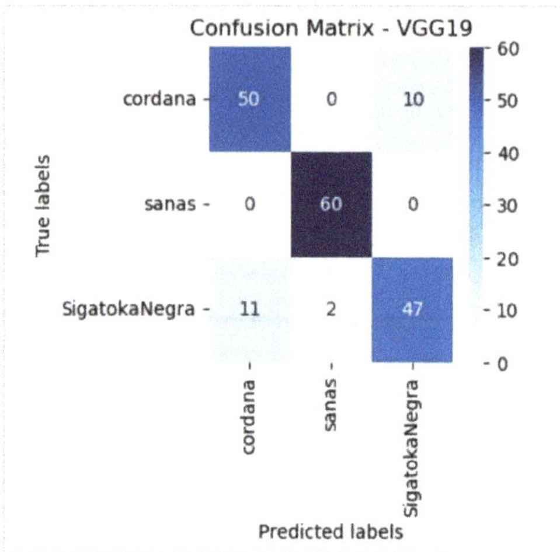

Figure 12. Confusion matrix for VGG19.

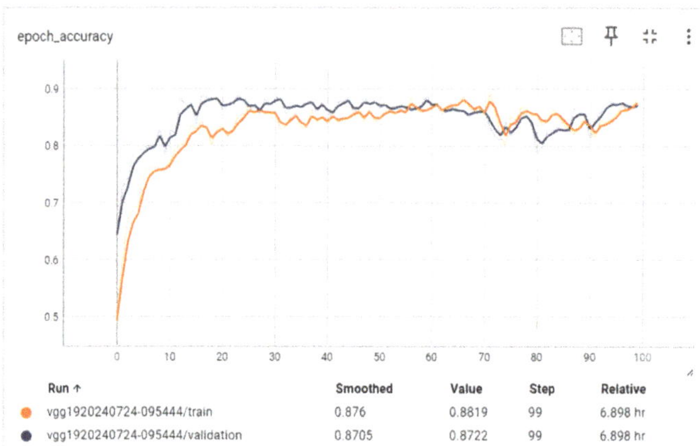

Figure 13. Learning evolution curve of the VGG19 model: orange line (training) vs. dark blue line (validation).

Figure 14 shows the evolution of the loss of the VGG19 model during training and validation over 100 epochs. The loss decreases rapidly during the early epochs, dropping from over 4 to values close to 0.5, suggesting that the model learns effectively from the training data. After epoch 10, the loss stabilizes, remaining at low values, indicating the proper fit of the model to the training data.

The validation loss also decreases quickly at first but stabilizes at around 1 starting from epoch 10, with minor fluctuations throughout the training. This suggests that the model generalizes well to the validation data, although it shows a greater loss compared to the training data.

In Figures 13 and 14, we can observe that the accuracy and loss functions over the 100 epochs show the favorable evolution of learning. It is important to highlight the training time, which was around 7 h, reflecting the time required for this model due to its greater complexity and the size of the data. The accuracy in both the training and validation sets converges at around 90%, while the loss function decreases rapidly in the early epochs before stabilizing, indicating an effective fit to the data.

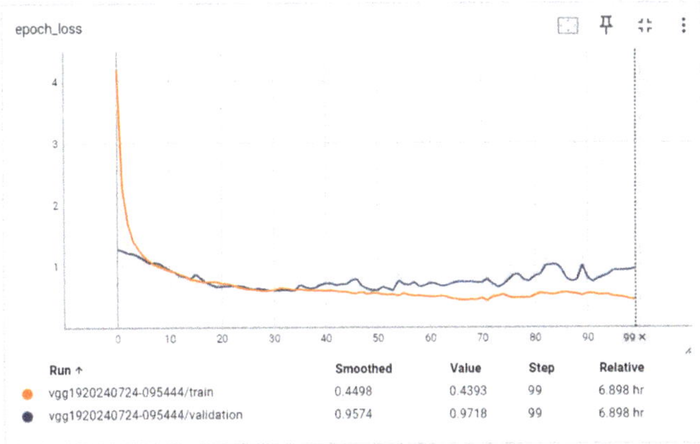

Figure 14. Loss evolution curve of the VGG19 model: orange line (training) vs. dark blue line (validation).

5.2. *Evaluation of the Best Model*

In this study, although an expanded dataset of 9000 images was available, obtained by applying data augmentation techniques to an original set of 900 images, the complete training of all models using the entirety of these images was not carried out due to processing time limitations. Initial tests showed that the training time varied significantly among the selected models. EfficientNetB0, being a lighter and more efficient model, completed its training in a relatively short time, approximately 30 min, allowing the model to be used with the complete set of 9000 images.

In contrast, ResNet50 and VGG19 exhibited significantly longer training times, with ResNet50 taking about 1.5 h and VGG19 nearly 7 h when using the initial set of 900 images. Given this situation, it was concluded that training both models with the 9000 images would not be feasible within the project's time frame, as this would have considerably extended the total processing time, impacted the study's efficiency, and limited the opportunity for multiple fine-tuning iterations and comparative evaluations.

Therefore, the EfficientNetB0 model was the only one trained with the complete set of images due to its shorter training time, making it suitable for working with larger datasets without affecting the total experiment time. This strategy facilitated quicker results and allowed for a more thorough evaluation of its performance in a larger data context. In contrast, the other models were trained using a reduced set of 900 images, enabling a general comparison of their performance without compromising the available processing time.

As we can see in Figure 15, we visualized the confusion matrix showing the model's performance in classifying leaves affected by Cordana and Black Sigatoka and healthy leaves. The model correctly identified 71.67% of the leaves with Cordana, although it confused 26.67% of cases with Black Sigatoka. In classifying healthy leaves, the model performed excellently, with accuracy of 99.5%. Regarding Black Sigatoka, the model correctly classified 92.33% of the leaves but confused 6% of cases with Cordana. Overall, the model showed good performance, with high accuracy in identifying healthy leaves, although it struggled to differentiate between Cordana and Black Sigatoka, suggesting possible visual similarities between the diseases.

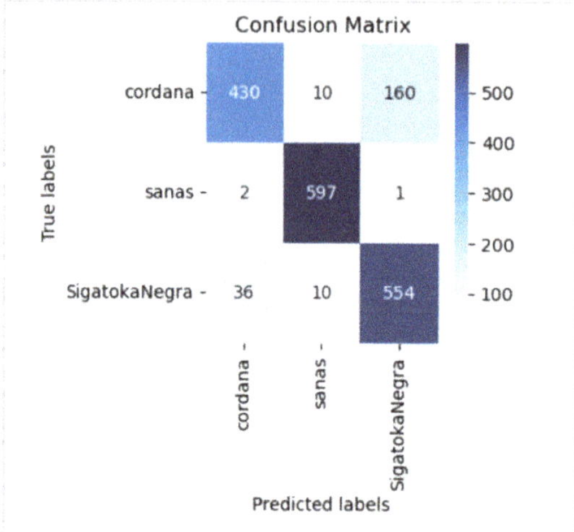

Figure 15. Confusion matrix of EfficientNetB0 model with 9000 images.

Figure 16 illustrates the accuracy achieved during the training and validation process of the EfficientNetB0 model, using a dataset that was expanded through augmentation techniques. This analysis covered a total of 100 epochs, where the vertical axis represents the accuracy and the horizontal axis indicates the corresponding epochs.

Figure 16. Learning curve of the EfficientNetB0 model with 9000 dataset points: sky blue line (training) vs. fuchsia line (validation).

Two curves can be observed in the graph. The blue line denotes the accuracy obtained on the training set. Initially, there is a notable increase in accuracy, which stabilizes at around 0.85, showing slight fluctuations as the training progresses. The pink line represents the accuracy on the validation set. This curve consistently remains above the training curve, reaching values close to 0.89. This behavior suggests the robust performance of the model on data that were not used during training.

At the end of the training, the accuracy achieved was 87.83% on the training set and 86.32% on the validation set. These results indicate that the model learned effectively, with no significant evidence of overfitting, as the difference between these accuracy values was moderate. The training process was completed in approximately 3.13 h.

Figure 17 illustrates the evolution of the loss of the EfficientNetB0 model during the training and validation process on the augmented dataset, over a total of 100 epochs. In this context, the vertical axis represents the loss value, while the horizontal axis indicates the epochs.

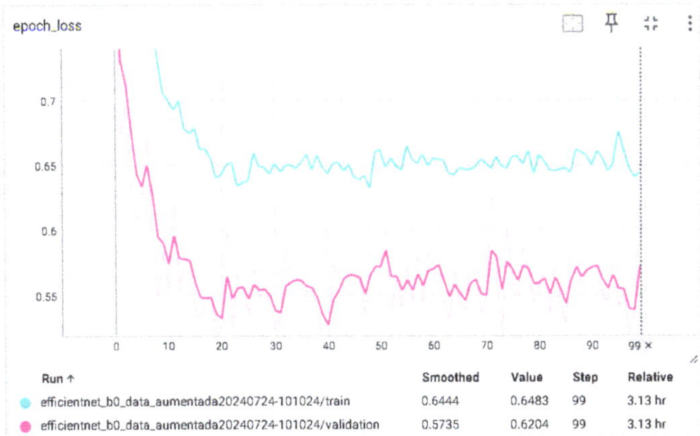

Figure 17. Loss curve of the EfficientNetB0 model with 9000 dataset points: sky blue line (training) vs. fuchsia line (validation).

The blue line reflects the loss on the training set, showing a rapid decrease during the initial stages of the process and stabilizing around a value of 0.65, with moderate fluctuations throughout the epochs. This behavior suggests that the model effectively reduces its prediction errors in the early phases, although the rate of improvement slows over time.

On the other hand, the pink line represents the loss on the validation set, which consistently remains lower than the training loss. This indicates that the model performs better on the validation data, with the loss values stabilizing near 0.57.

The final results reveal a training set loss of 0.6444 and a validation set loss of 0.5735. Both values demonstrate a significant reduction from the start of the training process, reflecting the model's effective learning. It is worth noting that the training process was completed in approximately 3.13 h.

5.3. Deployment

Following the principles of the CRISP-DM methodology [61], the final stage of the project focused on developing a mobile application designed for the deployment of the previously trained neural network. This step allowed the model's results to be transferred to a practical environment, evaluating its functionality under conditions similar to a production environment and demonstrating its adaptability in more dynamic scenarios. The main interface (Figure 18) of the application includes a scanner that enables the capture of leaf images for analysis, classifying them into three categories: Black Sigatoka, Cordana, and healthy leaves.

The application was developed using Angular as the framework to build a modular and efficient user interface, while Firebase was utilized for data management, authentication, and hosting. Additionally, a section of the application (Figure 19) provides specific recommendations based on the results of the disease prediction.

The pre-trained neural network was integrated into the application, enabling predictions to be executed directly from a mobile device. This not only validates the model's performance under conditions closer to a production environment but also demonstrates the possibility of implementing artificial intelligence solutions outside controlled settings, such as laboratories.

Figure 18. Scanning system for the identification of two diseases (Black Sigatoka and Cordana) and healthy leaves.

Figure 19. Recommendations for management the detected disease based on its classification.

This deployment marks a significant step within the research framework by transitioning the model's capabilities to a production environment, while focusing on evaluating and documenting its behavior, accuracy, and potential limitations in applied scenarios.

6. Discussion

6.1. Discussion of the Results Obtained

The models ResNet50, EfficientNetB0, and VGG19 were evaluated in terms of their accuracy, loss, generalization capabilities, and training time, yielding diverse results that can help in selecting the most suitable model according to the specific requirements of the leaf disease detection system.

In Table 2, the results in terms of all metrics used during the training of each neural network model are summarized.

Table 2. Evaluation metrics for the models.

Class	Metric	ResNet50	EfficientNetB0	VGG19
Black Sigatoka	Accuracy	94%	84%	82%
	Recall	73%	82%	78%
	F1 score	82%	83%	80%
Cordana	Accuracy	73%	82%	78%
	Recall	82%	83%	80%
	F1 score	77%	86%	82%
Healthy	Accuracy	100%	94%	97%
	Recall	98%	98%	100%
	F1 score	99%	96%	98%
Global	Accuracy	88.90%	88.33%	87.22%

Table 2 shows the comparative performance of three deep learning models (ResNet50, EfficientNetB0, and VGG19) in classifying three categories of banana leaves: Black Sigatoka, Cordana, and healthy leaves. They were evaluated using key metrics such as precision, recall, the F1 score, and the overall accuracy.

ResNet50 stands out for its high precision in identifying healthy leaves and cases of Black Sigatoka, meaning that it correctly classifies the detected cases. However, its recall is lower for the Black Sigatoka class, indicating that it may fail to detect all existing cases of this disease. Overall, this model achieves the highest global accuracy (88.90%), positioning it as the most accurate model among the three.

EfficientNetB0 exhibits balanced performance, with more consistent metrics across all classes. Although its global accuracy is slightly lower than that of ResNet50 (88.33%), it maintains a better balance between precision and recall, indicating that it can detect more cases without significantly compromising correct classifications. This suggests it is a versatile and reliable model for this type of classification.

VGG19 demonstrates acceptable performance across all metrics, excelling in detecting healthy leaves, where it achieves perfect recall (1.00). However, its global accuracy is the lowest among the three models (87.22%), indicating that it is less effective in overall classification.

6.2. *Training and Validation Precision and Loss*

The VGG19 model achieved the best training set accuracy, reaching values close to 88%, with a loss that rapidly decreased in the initial epochs and stabilized at around 0.44.

EfficientNetB0, on the other hand, showed more balanced performance, with the validation accuracy converging at around 88% and the loss stabilizing at approximately 0.80 for validation and 0.45 for training. This behavior indicates a lower tendency to overfit and a higher capacity for generalization.

In comparison, ResNet50 demonstrated a good fit on the training set but showed a significant discrepancy in the validation accuracy and loss, revealing challenges in generalizing, particularly in the classification of Black Sigatoka.

6.3. Generalization Capabilities

In terms of generalization, EfficientNetB0 proved to be the most robust model, with minimal differences between the training and validation curves, indicating a good balance between both. VGG19 exhibited the largest gap between the loss curves, signaling lower generalization capabilities. Similarly, ResNet50 displayed a tendency toward overfitting compared to the other models, which primarily impacted the classification of leaves affected by Black Sigatoka.

6.4. Training Time

The training time was a critical factor in the model comparison. EfficientNetB0 emerged as the most efficient, completing training in approximately 30 min, making it a viable option for systems requiring rapid implementation without sacrificing accuracy. In contrast, VGG19 took nearly 7 h to complete training, which, despite achieving high precision, may not be ideal in terms of the computational cost. ResNet50 required 1.5 h of training, representing a middle ground in terms of time but with inferior generalization performance compared to EfficientNetB0.

6.5. Disease Classification Performance

Regarding the disease classification performance, VGG19 excelled in identifying healthy leaves and those affected by Cordana. However, like ResNet50, it faced challenges in correctly distinguishing leaves affected by Black Sigatoka. EfficientNetB0, although also prone to errors in Black Sigatoka classification, demonstrated a more consistent balance across the classes, highlighting its ability to handle complex datasets and reduce the confusion among similar categories.

6.6. Comparison with the Results of Other Studies

The use of neural networks for the identification of diseases such as Cordana and Black Sigatoka in banana crops has generated growing interest in the scientific community. However, studies in this area still face significant challenges, which underlines the need to continue refining these methodologies for their application in real agricultural scenarios. A fundamental aspect to consider is the similarity in the appearance of the spots caused by both diseases, which can complicate accurate differentiation through deep learning models. Below are some of the main challenges identified in the literature:

- The scarcity of specific studies on the detection of Cordana and Black Sigatoka using neural networks;
- Few works that compare the performance of different neural network architectures in the detection of these pathologies;
- Complexity in the collection and annotation of high-quality images, considering the similarity in the foliar lesions of these diseases.

In the following section, a comparative analysis is presented between the results obtained in this study and those reported in previous research. Table 3 summarizes these findings, allowing an assessment of the progress and the main limitations identified in this field.

In our study, the pre-trained architectures ResNet50, EfficientNetB0, and VGG19 achieved accuracy scores of 88.90%, 88.33%, and 87.22%, respectively, with ResNet50 showing the best performance. To improve the generalizability of the model, data augmentation with the TensorFlow Keras ImageDataGenerator was applied. However, due to the long processing times, EfficientNetB0 was chosen, which, with 88.33% accuracy, offered a balance between performance and computational efficiency. Compared to the DCNN model of Rajalakshmi et al. [12], which reports 99% accuracy, it is important to note that this study does not show evidence of class balancing or techniques to avoid underfitting, which could affect the model's generalizability. Our results confirm that pre-trained models optimized with data augmentation strategies and efficient architecture selection can offer competitive and viable performance for field applications.

Table 3. Comparative table of various studies using convolutional neural network models applied to banana diseases.

Authors	Classes Considered	# Images	Model Used	Accuracy
Linero-Ramos et al. [61]	**Total images** Black Sigatoka Healthy	3180 1890 1290	EfficientNetV2B3 VGG19 MobileNetV2	87.33% 83.94% 77.20%
Sanga et al. [13]	Fusarium wilt race 1 Black Sigatoka	3000 *	ResNet152 Inceptionv3	99.20% 95.41%
Yan K., Chowdhury K., and Jin S. [62]	**Total images** Fusarium wilt race Healthy	156 72 84	ResNet50	98.00%
Elinisa, Mduma [9]	Fusarium wilt race Black Sigatok Healthy	27.360 *	CNN model	91.17%
Rajalakshmi N. et al. [12]	**Total images** Cordana Healthy Pestalotiopsis Sigatoka	803 86 164 131 422	DCNN	98.92%
Models applied in our proposed work	**Total images** Black Sigatoka Cordana Healthy	900 300 300 300	EfficientNetB0 ResNet50 VGG19	88.33% 88.90% 87.22%

* Does not specify the size of the data subset for each class.

7. Conclusions

This study evidences the effectiveness of deep learning techniques in the detection of foliar diseases in banana crops, focusing on Black Sigatoka and Cordana. Among the models analyzed, EfficientNetB0 proved to be the most efficient, reaching accuracy of 88.33% and completing its training in only 30 min. Although ResNet50 showed slightly higher accuracy (88.90%), its training time was significantly longer (1.5 h), as was that of VGG19, which achieved 87.22% accuracy in 7 h. These results position EfficientNetB0 as the most suitable choice for early disease diagnosis systems, thanks to its ability to balance accuracy and computational costs. This finding underlines the importance of efficient models in the agricultural context, enabling the implementation of scalable and sustainable solutions for disease management in banana crops.

This research focused on the diseases Black Sigatoka and Cordana; however, its approach can be extended to other diseases, such as moko, Panama disease, or mosaic. Additionally, the implementation of advanced techniques can be considered, such as the coarse-to-fine double constraint network for camouflaged object detection, as well as the investigation of algorithms applied to multispectral images.

Looking ahead, innovative approaches such as multispectral and near-infrared (NIR) imaging could be considered, which would open up new possibilities for the detection of diseases that are not visible in the traditional spectrum, improving the diagnostic accuracy. In addition, the integration of IoT sensors to monitor key environmental variables, such as the temperature and humidity, could contribute to the more accurate prediction of disease outbreaks, correlating the environmental conditions with the appearance of pathologies in real time. Finally, the development of autonomous monitoring platforms, such as drones with advanced sensors, would allow the automated, real-time monitoring of large areas of plantations, optimizing diagnosis and information management.

Author Contributions: Conceptualization, N.J. and S.O.; methodology, S.O.; software, N.J.; validation, N.J. and S.O.; investigation, N.J. and S.O.; resources, N.J. and S.O.; data curation, N.J.; writing—original draft preparation, S.O.; writing—review and editing, B.M.-O. and W.R.-A.; supervision, B.M.-O., W.R.-A. and I.R.-M. All authors have read and agreed to the published version of the manuscript.

Funding: This research was funded by Universidad Técnica de Machala, AutoMathTIC Research Group, Research Project with Resolution No. 0244-2024-CU-SO-13: "Adopción de las Tecnologías de Inteligencia Artificial e Internet de las Cosas en el Sector Agropecuario de la Provincia de El Oro".

Institutional Review Board Statement: Not applicable.

Informed Consent Statement: Not applicable.

Data Availability Statement: The original data presented in the study are openly available at https://github.com/NixonJimenez02/deep-learning-banana-diseases (accessed on 10 March 2025).

Acknowledgments: We thank the Universidad Técnica de Machala for funding this research work.

Conflicts of Interest: The authors declare no conflicts of interest.

Abbreviations

The following abbreviations are used in this manuscript:

CNN	Convolutional Neural Network
ResNet	Residual Network
VGG	Visual Geometry Group
PPV	Positive Predictive Value
NPV	Negative Predictive Value
TPR	True Positive Rate
TNR	True Negative Rate
TP	True Positives
TF	True Negatives
FP	False Positives
FN	False Negatives

References

1. Ciancio, A.; Rosso, L.C.; Lopez-Cepero, J.; Colagiero, M. Rhizosphere 16S-ITS Metabarcoding Profiles in Banana Crops Are Affected by Nematodes, Cultivation, and Local Climatic Variations. *Front. Microbiol.* **2022**, *13*, 855110. [CrossRef] [PubMed]
2. Voora, V.; Larrea, C.; Bermudez, S. *Global Market Report: Bananas*; International Institute for Sustainable Development: Winnipeg, MB, Canada, 2020.
3. Esguera, J.G.; Balendres, M.A.; Paguntalan, D.P. Overview of the *Sigatoka* Leaf Spot Complex in Banana and Its Current Management. *Trop. Plants* **2024**, *3*, e002. [CrossRef]
4. Strobl, E.; Mohan, P. Climate and the Global Spread and Impact of Bananas' Black Leaf Sigatoka Disease. *Atmosphere* **2020**, *11*, 947. [CrossRef]
5. Yonow, T.; Ramirez-Villegas, J.; Abadie, C.; Darnell, R.E.; Ota, N.; Kriticos, D.J. Black Sigatoka in Bananas: Ecoclimatic Suitability and Disease Pressure Assessments. *PLoS ONE* **2019**, *14*, e0220601. [CrossRef]
6. Mathew, D.; Kumar, C.S.; Anita Cherian, K. Classification of Leaf Spot Diseases in Banana Using Pre-Trained Convolutional Neural Networks. In Proceedings of the 2023 International Conference on Control, Communication and Computing (ICCC), Thiruvananthapuram, India, 19–21 May 2023; pp. 1–5.
7. Lin, H.; Zhou, G.; Chen, A.; Li, J.; Li, M.; Zhang, W.; Hu, Y.; Yu, W.T. EM-ERNet for Image-Based Banana Disease Recognition. *J. Food Meas. Charact.* **2021**, *15*, 4696–4710. [CrossRef]

8. Salehin, Y.; Siddique, A.; Nafisa, A.T.; Jahan, I.; Priyanka, M.M.; Ul Islam, R.; Hasan, M.; Rahman, A.; Rashid, M.R.A. A Comparative Analysis on Transfer Learning Models to Classify Banana Diseases- Fusarium Wilt and Black Sigatoka. In Proceedings of the 2024 2nd International Conference on Advancement in Computation & Computer Technologies (InCACCT), Gharuan, India, 2–3 May 2024; pp. 208–213.
9. Elinisa, C.A.; Mduma, N. Mobile-Based Convolutional Neural Network Model for the Early Identification of Banana Diseases. *Smart Agric. Technol.* **2024**, *7*, 100423. [CrossRef]
10. Thiagarajan, J.D.; Kulkarni, S.V.; Jadhav, S.A.; Waghe, A.A.; Raja, S.P.; Rajagopal, S.; Poddar, H.; Subramaniam, S. Analysis of Banana Plant Health Using Machine Learning Techniques. *Sci. Rep.* **2024**, *14*, 15041. [CrossRef]
11. Banana Leaf Spot Diseases (BananaLSD) Dataset. Available online: https://www.kaggle.com/datasets/shifatearman/bananalsd (accessed on 19 December 2024).
12. Rajalakshmi, N.R.; Saravanan, S.; Arunpandian, J.; Mathivanan, S.K.; Jayagopal, P.; Mallik, S.; Qin, G. Early Detection of Banana Leaf Disease Using Novel Deep Convolutional Neural Network. *J. Data Sci. Intell. Syst.* **2024**. [CrossRef]
13. Sanga, S.L.; Machuve, D.; Jomanga, K. Mobile-Based Deep Learning Models for Banana Disease Detection. *Eng. Technol. Appl. Sci. Res.* **2020**, *10*, 5674–5677. [CrossRef]
14. Blanc-Talon, J.; Delmas, P.; Philips, W.; Popescu, D.; Scheunders, P. (Eds.) *Advanced Concepts for Intelligent Vision Systems: 20th International Conference, ACIVS 2020, Auckland, New Zealand, 10–14 February 2020, Proceedings*; Lecture Notes in Computer Science; Springer International Publishing: Cham, Switzerland, 2020; Volume 12002, ISBN 978-3-030-40604-2.
15. Abbas, A.; Jain, S.; Gour, M.; Vankudothu, S. Tomato Plant Disease Detection Using Transfer Learning with C-GAN Synthetic Images. *Comput. Electron. Agric.* **2021**, *187*, 106279. [CrossRef]
16. PlantVillage Dataset. Available online: https://www.kaggle.com/datasets/abdallahalidev/plantvillage-dataset (accessed on 20 November 2024).
17. Atila, Ü.; Uçar, M.; Akyol, K.; Uçar, E. Plant Leaf Disease Classification Using EfficientNet Deep Learning Model. *Ecol. Inform.* **2021**, *61*, 101182. [CrossRef]
18. Jadhav, S.B.; Udupi, V.R.; Patil, S.B. Identification of Plant Diseases Using Convolutional Neural Networks. *Int. J. Inf. Technol.* **2021**, *13*, 2461–2470. [CrossRef]
19. Anh, P.T.; Duc, H.T.M. A Benchmark of Deep Learning Models for Multi-Leaf Diseases for Edge Devices. In Proceedings of the 2021 International Conference on Advanced Technologies for Communications (ATC), Ho Chi Minh City, Vietnam, 14–16 October 2021; pp. 318–323.
20. PlantVillage. Available online: https://www.kaggle.com/datasets/mohitsingh1804/plantvillage (accessed on 20 November 2024).
21. Liu, J.; Wang, X. Plant Diseases and Pests Detection Based on Deep Learning: A Review. *Plant Methods* **2021**, *17*, 22. [CrossRef] [PubMed]
22. Hamed, B.S.; Hussein, M.M.; Mousa, A.M. Plant Disease Detection Using Deep Learning. *Int. J. Intell. Syst. Appl.* **2023**, *15*, 38–50. [CrossRef]
23. Amjoud, A.B.; Amrouch, M. Object Detection Using Deep Learning, CNNs and Vision Transformers: A Review. *IEEE Access* **2023**, *11*, 35479–35516. [CrossRef]
24. Bangalore Vijayakumar, S.; Chitty-Venkata, K.T.; Arya, K.; Somani, A.K. ConVision Benchmark: A Contemporary Framework to Benchmark CNN and ViT Models. *AI* **2024**, *5*, 1132–1171. [CrossRef]
25. Rivas-Asanza, W.; Mazon-Olivo, B.; Tusa-Jumbo, E. Reconocimiento de Patrones En Imágenes. In *Redes Neuronales Artificiales Aplicadas al Reconocimiento de Patrones*; Rivas-Asanza, W., Mazon-Olivo, B., Eds.; Universidad Técnica de Machala: Machala, Ecuador, 2018; pp. 61–126, ISBN 978-9942-24-100.
26. Moya, E.; Campoverde, E.; Tusa, E.; Ramirez-Morales, I.; Rivas, W.; Mazon, B. Multi-Category Classification of Mammograms by Using Convolutional Neural Networks. In Proceedings of the 2017 International Conference on Information Systems and Computer Science (INCISCOS), Quito, Ecuador, 23–25 November 2017; pp. 133–140.
27. Rivas-Asanza, W.; Mazon-Olivo, B.; Mejía-Peñafiel, E. Generalidades de Las Redes Neuronales Artificiales. In *Redes Neuronales Artificiales Aplicadas al Reconocimiento de Patrones*; Rivas-Asanza, W., Mazon-Olivo, B., Eds.; Universidad Técnica de Machala: Machala, Ecuador, 2018; pp. 11–35, ISBN 978-9942-24-100.
28. González, Y.; Rivas-Asanza, W.; Mazon-Olivo, B.; Tusa, E. Revision of Classification Schemes in Tissue Carcinogenic Mammography Images. *Alternativas* **2018**, *19*, 72–83. [CrossRef]
29. Turay, T.; Vladimirova, T. Toward Performing Image Classification and Object Detection with Convolutional Neural Networks in Autonomous Driving Systems: A Survey. *IEEE Access* **2022**, *10*, 14076–14119. [CrossRef]
30. Chen, L.; Li, S.; Bai, Q.; Yang, J.; Jiang, S.; Miao, Y. Review of Image Classification Algorithms Based on Convolutional Neural Networks. *Remote Sens.* **2021**, *13*, 4712. [CrossRef]
31. Diwan, T.; Anirudh, G.; Tembhurne, J.V. Object Detection Using YOLO: Challenges, Architectural Successors, Datasets and Applications. *Multimed. Tools Appl.* **2023**, *82*, 9243–9275. [CrossRef]

32. Elaraby, A.; Hamdy, W.; Alruwaili, M. Optimization of Deep Learning Model for Plant Disease Detection Using Particle Swarm Optimizer. *Comput. Mater. Contin.* **2022**, *71*, 4019–4031. [CrossRef]
33. Ali, A.H.; Yaseen, M.G.; Aljanabi, M.; Abed, S.A.; Gpt, C. Transfer Learning: A New Promising Techniques. *Mesopotamian J. Big Data* **2023**, *2023*, 29–30. [CrossRef]
34. Zhang, X.; Gao, J. Measuring Feature Importance of Convolutional Neural Networks. *IEEE Access* **2020**, *8*, 196062–196074. [CrossRef]
35. Amin, H.; Darwish, A.; Hassanien, A.E.; Soliman, M. End-to-End Deep Learning Model for Corn Leaf Disease Classification. *IEEE Access* **2022**, *10*, 31103–31115. [CrossRef]
36. Khomidov, M.; Lee, J.-H. The Novel EfficientNet Architecture-Based System and Algorithm to Predict Complex Human Emotions. *Algorithms* **2024**, *17*, 285. [CrossRef]
37. Yapici, M.M.; Tekerek, A.; Topaloglu, N. Performance Comparison of Convolutional Neural Network Models on GPU. In Proceedings of the 2019 IEEE 13th International Conference on Application of Information and Communication Technologies (AICT), 23–25 October 2019; IEEE: Baku, Azerbaijan; pp. 1–4.
38. Singh, V.; Baral, A.; Kumar, R.; Tummala, S.; Noori, M.; Yadav, S.V.; Kang, S.; Zhao, W. A Hybrid Deep Learning Model for Enhanced Structural Damage Detection: Integrating ResNet50, GoogLeNet, and Attention Mechanisms. *Sensors* **2024**, *24*, 7249. [CrossRef]
39. J., A.; Eunice, J.; Popescu, D.E.; Chowdary, M.K.; Hemanth, J. Deep Learning-Based Leaf Disease Detection in Crops Using Images for Agricultural Applications. *Agronomy* **2022**, *12*, 2395. [CrossRef]
40. Nguyen, T.-H.; Nguyen, T.-N.; Ngo, B.-V. A VGG-19 Model with Transfer Learning and Image Segmentation for Classification of Tomato Leaf Disease. *AgriEngineering* **2022**, *4*, 871–887. [CrossRef]
41. Awan, M.J.; Masood, O.A.; Mohammed, M.A.; Yasin, A.; Zain, A.M.; Damaševičius, R.; Abdulkareem, K.H. Image-Based Malware Classification Using VGG19 Network and Spatial Convolutional Attention. *Electronics* **2021**, *10*, 2444. [CrossRef]
42. Rodrigues, I.; Santos, G.L.; Sadok, D.F.H.; Endo, P.T. Classifying COVID-19 positive X-ray using deep learning models. *IEEE Lat. Am. Trans.* **2021**, *19*, 884–892. [CrossRef]
43. Iman, M.; Arabnia, H.R.; Rasheed, K. A Review of Deep Transfer Learning and Recent Advancements. *Technologies* **2023**, *11*, 40. [CrossRef]
44. Yosinski, J.; Clune, J.; Bengio, Y.; Lipson, H. How Transferable Are Features in Deep Neural Networks? In Proceedings of the Advances in Neural Information Processing Systems 27 (NIPS'14), Montreal, QC, Canada, 8–13 December 2014.
45. Ardalan, Z.; Subbian, V. Transfer Learning Approaches for Neuroimaging Analysis: A Scoping Review. *Front. Artif. Intell.* **2022**, *5*, 780405. [CrossRef] [PubMed]
46. Ahmad, S.; Ansari, S.U.; Haider, U.; Javed, K.; Rahman, J.U.; Anwar, S. Confusion Matrix-Based Modularity Induction into Pretrained CNN. *Multimed. Tools Appl.* **2022**, *81*, 23311–23337. [CrossRef]
47. Chipindu, L.; Mupangwa, W.; Mtsilizah, J.; Nyagumbo, I.; Zaman-Allah, M. Maize Kernel Abortion Recognition and Classification Using Binary Classification Machine Learning Algorithms and Deep Convolutional Neural Networks. *AI* **2020**, *1*, 361–375. [CrossRef]
48. Arias-Duart, A.; Mariotti, E.; Garcia-Gasulla, D.; Alonso-Moral, J.M. A Confusion Matrix for Evaluating Feature Attribution Methods. In Proceedings of the 2023 IEEE/CVF Conference on Computer Vision and Pattern Recognition Workshops (CVPR), Vancouver, BC, Canada, 17–24 June 2023; pp. 3709–3714.
49. Zapeta Hernández, A.; Galindo Rosales, G.A.; Juan Santiago, H.J.; Martínez Lee, M. Métricas de rendimiento para evaluar el aprendizaje automático en la clasificación de imágenes petroleras utilizando redes neuronales convolucionales. *Cienc. Lat. Rev. Cient. Multidiscip.* **2022**, *6*, 4624–4637. [CrossRef]
50. *Machine Learning for Brain Disorders*; Colliot, O., Ed.; Neuromethods; Springer: New York, NY, USA, 2023; Volume 197, ISBN 978-1-07-163194-2.
51. Strelcenia, E.; Prakoonwit, S. Improving Classification Performance in Credit Card Fraud Detection by Using New Data Augmentation. *AI* **2023**, *4*, 172–198. [CrossRef]
52. George, C.A.; Barrera, E.A.; Nelson, K.P. Applying the Decisiveness and Robustness Metrics to Convolutional Neural Networks. *arXiv* **2020**, arXiv:2006.00058.
53. Wong, A. NetScore: Towards Universal Metrics for Large-Scale Performance Analysis of Deep Neural Networks for Practical On-Device Edge Usage. In Proceedings of the Image Analysis and Recognition: 16th International Conference, ICIAR 2019, Waterloo, ON, Canada, 27–29 August 2019.
54. Hui, L.; Belkin, M. Evaluation of Neural Architectures Trained with Square Loss vs Cross-Entropy in Classification Tasks. *arXiv* **2020**, arXiv:2006.07322.
55. Zhao, H.; Gallo, O.; Frosio, I.; Kautz, J. Loss Functions for Neural Networks for Image Processing. *arXiv* **2018**, arXiv:1511.08861v3.
56. Lopez-Betancur, D.; Bosco Duran, R.; Guerrero-Mendez, C.; Zambrano Rodríguez, R.; Saucedo Anaya, T. Comparación de arquitecturas de redes neuronales convolucionales para el diagnóstico de COVID-19. *Comput. Sist.* **2021**, *25*, 601–615. [CrossRef]

57. Schröer, C.; Kruse, F.; Gómez, J.M. A Systematic Literature Review on Applying CRISP-DM Process Model. *Procedia Comput. Sci.* **2021**, *181*, 526–534. [CrossRef]
58. Zhang, Y.; Wa, S.; Liu, Y.; Zhou, X.; Sun, P.; Ma, Q. High-Accuracy Detection of Maize Leaf Diseases CNN Based on Multi-Pathway Activation Function Module. *Remote Sens.* **2021**, *13*, 4218. [CrossRef]
59. Balafas, V.; Karantoumanis, E.; Louta, M.; Ploskas, N. Machine Learning and Deep Learning for Plant Disease Classification and Detection. *IEEE Access* **2023**, *11*, 114352–114377. [CrossRef]
60. Mahouachi, D.; Akhloufi, M.A. Recent Advances in Infrared Face Analysis and Recognition with Deep Learning. *AI* **2023**, *4*, 199–233. [CrossRef]
61. Linero-Ramos, R.; Parra-Rodríguez, C.; Espinosa-Valdez, A.; Gómez-Rojas, J.; Gongora, M. Assessment of Dataset Scalability for Classification of Black Sigatoka in Banana Crops Using UAV-Based Multispectral Images and Deep Learning Techniques. *Drones* **2024**, *8*, 503. [CrossRef]
62. Yan, K.; Shisher, M.K.C.; Sun, Y. A Transfer Learning-Based Deep Convolutional Neural Network for Detection of Fusarium Wilt in Banana Crops. *AgriEngineering* **2003**, *5*, 2381–2394. [CrossRef]

AI

MDPI

Article

Development and Comparison of Artificial Neural Networks and Gradient Boosting Regressors for Predicting Topsoil Moisture Using Forecast Data

Miriam Zambudio Martínez [1,*], Larissa Haringer Martins da Silveira [1], Rafael Marin-Perez [1] and Antonio Fernando Skarmeta Gomez [2]

1 Odin Solutions SL, 30009 Murcia, Spain; lharinger@odins.es (L.H.M.d.S.); rmarin@odins.es (R.M.-P.)
2 Department of Information and Communication Engineering, University of Murcia, 30100 Murcia, Spain; skarmeta@um.es
* Correspondence: mzambudio@odins.es

Abstract: Introduction: The Earth's growing population is increasing resource consumption, heavily pressuring agriculture, which, currently, uses 70% of the world's freshwater from rivers and lakes, which, themselves, comprise only 1% of the Earth's water reserves. Combined with climate change, the situation is alarming. These challenges drive Agriculture 4.0, which is focused on sustainable agricultural processes to optimise water use. **Objective:** Given this context, this study proposes a model, based on Artificial Intelligence (AI) techniques to predict topsoil moisture in a study area located in the south of the Iberian Peninsula, primarily an agricultural region facing recurrent droughts and water scarcity. **Methods:** To develop the model, a comparison between Artificial Neural Networks (ANNs) and Gradient Booster Regressors (GBRs) was conducted, and topsoil moisture data from seven probes distributed over the study area were used, in addition to several variables (temperature, relative humidity, solar radiation, wind speed, precipitation and evapotranspiration) from a selection of weather stations and ensemble forecasts from meteorological models. **Results:** The final GBR model, with a 0.01 learning rate, 5 max depth, and 350 estimators, predicted topsoil moisture with an average mean squared error (MSE) of 0.027 and a maximum difference between observed and predicted data of 20.09% in a two-year series (May 2022–June 2024).

Keywords: topsoil moisture; Artificial Neural Network; Gradient Boosting Regressor; intelligent water management; precision agriculture; environmental monitoring; machine learning

1. Introduction

The significant and continuous growth of the global population has led to an ever-increasing demand for natural resources, particularly raw materials, intensifying pressure on production systems. Furthermore, the rising need for food places substantial challenges on ecosystems and agriculture [1,2].

Intense anthropogenic activities affect the quality of natural resources and also contribute to the acceleration of climate change, favouring complex, non-linear processes that put terrestrial ecosystems at risk of major changes in their composition and structure.

A remarkable example was observed during the COVID-19 pandemic, when the significant reduction in anthropogenic activities on a global scale led to a notable decrease in environmental pressure. During this period, a significant direct relationship was observed

 https://doi.org/10.3390/10.3390/ai6020041

between the decrease in human activity and the increase in air quality, associated with a decrease in observed concentrations of gases from the burning of fossil fuels [3] as well as an improvement in water quality, specifically in the reduction of suspended particulate matter observed in Vembanad, India's largest freshwater lake [4].

Among the natural resources most impacted by anthropogenic activities and climate change, water stands out as one of the most essential and vulnerable. Vital for life, this resource is under increasing pressure due to intensive consumption, particularly in agriculture, which accounts for approximately 70% of the planet's available freshwater—this is primarily drawn from rivers and lakes, which make up only 1% of the Earth's total water resources [5,6]. The intensification of climate change exacerbate this already precarious situation, with extreme events such as more frequent and severe droughts posing a serious threat to water availability [7,8], underscoring the urgency of adopting sustainable management practices.

The increasing frequency and severity of droughts further intensify the challenges faced by agricultural producers who rely heavily on these limited resources [9,10], particularly in semi-arid regions where agricultural demands are already high [11]. One such region is the semi-arid Mediterranean, which frequently experiences droughts that severely impact water resources by depleting groundwater levels and reducing water stored in dams and reservoirs [12]. These conditions compromise agricultural production and threaten the survival of agroecosystems during dry summers [13].

Moreover, rising global temperatures are projected to further reduce water availability, complicating the water management strategies required for sustainable agriculture [14].

A report released by the United Nations indicates that by the year 2050, global food production must increase by 50% to meet population demands [5]. However, limited water availability, unsustainable agricultural practices, and inefficient irrigation threaten the ability to meet these needs, especially in already water-stressed regions such as the Middle East [15], parts of Africa [16,17], southeastern Spain [18], and South Asia [19].

The Mediterranean stands out as one of the regions with the highest socio-economic vulnerability to droughts, the severity of which is expected to increase due to climate change [20]. Moreover, the uneven distribution of its water resources exacerbates this vulnerability, leading to international and supranational conflicts, particularly in transboundary basins and regions dependent on inter-basin water transfers. These conflicts are exacerbated by recurrent droughts and changes in the provision of ecosystem services [21,22].

In this context, understanding and managing surface soil moisture (SSM) is becoming a critical factor in addressing agricultural water challenges. Recognised as an Essential Climate Variable (ECV) by the European Union in 2010 [23], SSM refers to the water content of the top layer of soil. This factor plays a key role in hydrological processes, influencing the exchange of water and energy between the land and the atmosphere.

In addition, SSM directly affects plant growth and agricultural yields by determining the availability of water for crops [24]. It is also a key variable in hydrological models, influencing streamflow generation and ecosystem health [25]. Understanding and monitoring SSM is therefore essential for intelligent water resource management, improving agricultural productivity and formulating climate adaptation strategies.

To address these challenges, advanced tools such as machine learning algorithms, including Artificial Neural Networks (ANNs) and Gradient Boosting Regressors (GBRs) have proven effective in analysing moisture data and improving predictive accuracy for resource management.

Due to these growing issues, effective agricultural water management practices—such as improving irrigation efficiency and enhancing drought preparedness—are essential to mitigating the impacts of climate change and ensuring global food security [14,26].

This study focuses on comparing GBR and ANN methodologies for predicting surface soil moisture using weather station data and meteorological forecasts, providing a robust approach to support smarter water resource management and foster more sustainable agricultural practices.

2. Related Works

A recent investigation in the prediction of SSM utilised sophisticated methodologies, including machine learning, which effectively capture spatial and temporal patterns in soil moisture data, achieving high predictive accuracy [27]. Furthermore, other studies have also highlighted the importance of certain machine learning techniques for soil moisture prediction. For example, recent studies [28,29] have experimented with Random Forest in conjunction with data from the Global Navigation Satellite System (GNSS) to enhance soil moisture retrieval accuracy. Its effectiveness has also been tested in sensitive ecosystems such as peatlands [30], and, in general, in estimating vertical soil moisture profiles [31]. Other studied techniques include Long Short-Term Memory (LSTM) networks, which have been combined with the attention mechanism for predicting soil moisture and temperature [32], and with residual learning [33]; these networks have also been used in hybrid models with wavelet decomposition or particle swarm optimisation [34] and Convolutional Neural Networks (CNN) [35]. In this section, we present a literature review of various studies that have explored soil moisture estimation using neural networks, especially ANNs and GBRs, highlighting their methodologies, advantages, and limitations, as these models have proved to be effective in this field of study, as detailed in the following two subsections. Moreover, these models do not consume an excessive amount of computational resources, unlike more complex models such as LSTMs. Additionally, focusing on these two specific techniques allowed for a more in-depth and exhaustive exploration of their capabilities.

2.1. Neural Networks

Neural networks have emerged as powerful tools for predicting topsoil moisture, leveraging various data sources and methodologies. Recent studies highlight their effectiveness in integrating remote sensing data and machine learning techniques to enhance prediction accuracy, as demonstrated in [36,37]. Other approaches combined ANNs with physical models and equations, such as Richards' equation, demonstrating the potential of integrating physical principles with machine learning [38,39]. Moreover, a fully connected feed-forward ANN model was developed in [40], which outperformed ten other machine learning algorithms.

In addition to these advancements, high-resolution aerial imagery has been explored as a valuable input for soil moisture estimation. For instance, a study in [41] developed an ANN model that leverages spectral images from the AggieAir™ UAV platform to estimate surface soil moisture.

Therefore, as highlighted by the provided studies, ANNs exhibits several strengths in predicting soil moisture, with one of these primary strengths being their ability to handle complex, multi-scale data inputs, such as satellite and in situ data, which allows for high-resolution soil moisture predictions. For instance, in [42] a multi-scale deep learning model achieved a median correlation of 0.901 and a root mean square error of 0.034 m^3/m^3, outperforming traditional land surface models and satellite products. Moreover, as stated before, ANNs can integrate physics-based principles, such as the water balance principle, to enhance prediction accuracy by capturing the underlying physical processes, while reducing the need for extensive datasets and training time [43]. However, ANNs also face limitations, including challenges in processing structurally diverse characteristics and meteorological influences, which complicate the development of optimal computational formulas for soil moisture predictions [44].

On the other hand, and with the aim to address some of the issues of the ANNs, such as a lack of spatial autocorrelation and vague feature representation, which can hinder model performance, further studies demonstrate that Long Short-Term Memory (LSTM) networks effectively capture soil moisture dynamics across different scales, utilising extensive in situ data to enhance predictive accuracy [45–47]. Moreover, incorporating attention in models like CONV1D+Attention has improved the ability to handle complex time series data, achieving significant predictive performance [48]. Lastly, optimised models, such as a GRU recurrent neural network, have shown high accuracy in short-term predictions [49].

2.2. Gradient Boosting Models and Ensemble Techniques

Gradient boosting regressors effectively predict surface soil moisture by leveraging atmospheric and soil factors, enhancing prediction accuracy through advanced modelling techniques [50,51]. Particularly, Extreme Gradient Boosting (XGBoost) has been highlighted for its superior performance in predicting soil moisture, achieving high correlation coefficients and low RMSE values in provincial-level studies [52,53]. In a multi-sensor data approach, the XGBR-GA model outperformed other models like Random Forest and Support Vector Machines, as stated in [54].

Furthermore, ensemble methods, including stacking and Boruta-GBDT, have been shown to enhance prediction accuracy by combining multiple algorithms and integrating various machine learning techniques, leading to improved RMSE and correlation metrics, as well as proving effective in managing the spatial variability of soil moisture [55–57].

Similarly to ANNs, GBR models also present notable strengths and some limitations in soil moisture prediction. One significant advantage is their high accuracy and robustness, as demonstrated by the GBR-RF algorithm, which outperformed traditional methods with R^2 values reaching 0.8838 [58]. In addition, these models effectively integrate various data sources, such as satellite imagery, enhancing predictive capabilities through multi-band data combinations [58]. Furthermore, its ability to handle missing data through imputation techniques enhances overall model performance [59]. Nonetheless, regarding limitations, they include the requirement for extensive data preprocessing and feature engineering to ensure model reliability, as well as potential overfitting if the hyperparameters are not optimally tuned [60,61]. Furthermore, while gradient boosting models outperform traditional methods, they may still struggle with extreme outlier conditions or when faced with highly variable soil types [52,61].

Therefore, and as a general conclusion of this section, it is worth mentioning that recent advancements in soil moisture prediction have demonstrated the effectiveness of both neural networks (NNs) and GBRs in integrating diverse data sources, such as remote sensing, in situ measurements, and multi-scale atmospheric variables. NNs, particularly ANNs and LSTMs, have shown strong capabilities in handling complex, multi-temporal inputs, achieving high predictive accuracy by incorporating physics-based principles and deep learning architectures. Meanwhile, GBR models, including XGBoost and ensemble techniques, have proven robust in capturing spatial variability, efficiently integrating heterogeneous data, and delivering competitive results with minimal computational costs.

The study presented in this paper builds upon the work presented in [36], expanding the scope and refining the prediction model for surface soil moisture, incorporating a comparison between ANNs and GBRs. Additionally, the new model utilises a more comprehensive dataset, spanning (daily) from May 2022 to June 2024 (unlike the previous study, which considered data solely from the year 2023), since it is the maximum period of time available for the moisture data collected by the seven installed probes, and integrates additional variables from selected weather stations and the ensemble of numerical weather forecasting models, enhancing the robustness of the predictions.

In this section, we have reviewed recent studies utilising neural networks and gradient boosting techniques to predict surface soil moisture, demonstrating their effectiveness in capturing spatial and temporal patterns and improving predictive accuracy. The rest of the paper is structured as follows: Section 3 is divided in two subsections, with Section 3.1 describing the data used, including probe measurements and meteorological variables, as well as statistical analysis methods and model development; Section 3.2 details the study area, dataset construction, and the development and cross-validation of neural network and gradient boosting models. Furthermore, Section 4 presents and discusses the obtained results, comparing model accuracy and analysing performance in different scenarios; finally, Section 5 summarises the main findings of the study, discusses implications for water management in agriculture, and suggests possible future research directions.

3. Materials and Methods

3.1. Background

3.1.1. Probes' Data

Currently, the installation of probes in crop fields is becoming increasingly popular and widely used. The reason lies in different key aspects. First of all, probes provide real-time data on several soil variables, allowing farmers to make informed decisions immediately, optimising the use of resources such as water and fertilisers, and therefore reducing the costs and the environmental impact. Moreover, they can detect anomalous changes in the soil, helping in identifying potential problems before they become severe, thereby allowing for early intervention.

Furthermore, the data measured by probes are of paramount importance for training AI models, as they provide large volumes of historical real, precise, and accurate data on specific variables for specific study areas, allowing AI models to make more reliable predictions, and reducing the human error that can be committed making manual measurements. Moreover, probes can be installed in multiple locations within a field, providing a detailed spatial coverage through a granular view of the crop or area of interest.

Seven Drill & Drop SDI-12 probes, which are Sentek Technologies products [62], have been used for this study, the same ones employed in [36]. Although this type of probe collects different parameters, thanks to the integration of multiple sensors, only measurements taken for surface soil moisture have been used for this research. These measurements have a precision of ±0.03% vol and a resolution of 1:10,000, demonstrating that the probes are capable of detecting small changes in soil moisture. The distribution of the probes employed in the study area is shown in Figure 1, where the white rectangle defines the ROI and each yellow pin represents a probe.

3.1.2. Weather Stations Data

Weather stations offer similar advantages to probes, but they are applied to environmental variables. For this study, four meteorological stations were selected, distributed throughout the study area and located at points closest to each of the seven probes used. Since the defined basin of interest spans several provinces, stations from different public and open networks were used, depending on the location of the probes. For the three probes situated within the Region of Murcia, data were sourced from the stations of the Agrometeorological Network of IMIDA (Murcian Institute for Agricultural and Environmental Research and Development), which were accessed through the SIAM (Agricultural Information System of Murcia) online platform [63]. Specifically, two stations from this network, positioned in Cieza and Cehegín, were utilised. In the case of the two probes located in the province of Almería, a meteorological station established in the municipality of Chirivel, part of the METEO Network, was employed. Access to the data from this station was facilitated by the web portal [64] of the FrostSE project, which was developed by the Department of Geography at the University of Murcia [65]. For the probes installed in Granada, a meteorological station situated in Puebla de Don Fadrique, designated with the code GR02, and operated by the Andalusian Institute of Agricultural and Fisheries Research and Training (IFAPA), was utilised. To retrieve the environmental information documented by this station, the web portal of the Agroclimatic Information Network of Andalusia (RIA) [66] was referenced.

Figure 1. Distribution on the study area of the seven probes employed for the research.

Figure 2 shows the distribution of the meteorological stations used, marked with red pins, in relation to the probes in the study area.

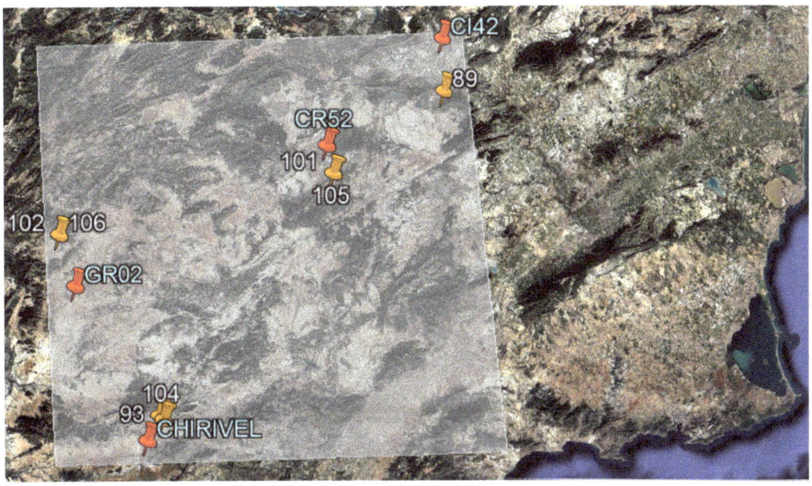

Figure 2. Distribution of the meteorological stations used in the study area.

Daily data on maximum, minimum, and average temperatures; maximum, minimum, and average relative humidity; average wind speed; precipitation; and average solar radiation were collected from the stations.

3.1.3. Meteorological Models' Forecasts

Weather forecast models allow future weather conditions to be estimated based on initial and contour data, providing forecasts up to 15 days ahead.

Although they are not as accurate as observed data, such as that from meteorological stations or in situ probes, which collect data in near real time, they provide one of the inputs needed for artificial intelligence models, such as the one developed in this study, to be able to make predictions of the variable of interest. In this research, the Visual Crossing (VC) API [67] has been used, as it offers a free limited version from which data can be accessed. From its API, two types of data can be retrieved:

- **Historical data**: these data are retrieved from multiple sources, such as weather stations, satellites, and radars. Normally, to collect historical records from a concrete latitude and longitude, three weather stations in a range of 50 km around the latitude and longitude specified are used [68,69]. Due to this low spatial resolution, the data retrieved are less accurate and reliable than those obtained from the weather stations mentioned in Section 3.1.2.
- **Weather forecast**: current data are combined with mathematical and physics algorithms to obtain weather forecasts using meteorological models; in this case, these are the models of ECMWF (European Centre for Medium-Range Weather Forecasts), GFS (Global Forecast System), ICON (Icosahedral Nonhydrostatic Model from the German Meteorological Office), NAM (North American Mesoscale Model), and HRRR (High-Resolution Rapid Refresh). The models are continuously adjusted with new data to improve accuracy, through the calibration and validation of the generated data using real data to ensure their precision and coincidence with the real conditions [69,70].

Daily data on maximum, minimum, and average temperatures; maximum, minimum, and average relative humidity; average wind speed; precipitation; and average solar radiation were collected from Visual Crossing.

3.1.4. Evapotranspiration

Evapotranspiration (ET) is the process by which water is transferred from land to the atmosphere through evaporation and plant transpiration. The Penman–Monteith Equation (1), which has been employed in this study, is a widely used method for calculating reference evapotranspiration (ETo), integrating several meteorological parameters to provide accurate estimates. This equation combines energy balance and mass transfer principles, requiring inputs such as temperature, solar radiation, humidity, and wind speed [71]. It is particularly effective for estimating ETo under standard conditions, as demonstrated in [72], which utilised local weather data to validate its accuracy. Moreover, the equation accounts for stomatal conductance and the vapor pressure deficit, which are crucial for understanding plant water use [71].

$$\mathrm{ETo} = \frac{0.408\Delta(R_n - G) + \gamma\frac{900}{T+273}u_2(e_s - e_a)}{\Delta + \gamma(1 + 0.34u_2)} \tag{1}$$

3.1.5. Artificial Neural Networks

Artificial Neural Networks (ANNs) are computational models inspired by the human brain, designed to recognise patterns and make predictions from large datasets. These networks consist of layers of interconnected nodes (neurons) that process input data and adjust their connections based on errors, thereby improving predictions over time [73].

ANNs are particularly valuable in predicting soil properties due to their ability to handle the complex, non-linear relationships inherent in soil data. They have been effectively applied in various soil science domains, including predicting soil–water characteristics, assessing soil quality, and estimating compressive strength in geotechnical engineering [74–77]. Their capacity to learn from extensive datasets enables accurate predictions of soil characteristics such as pH, organic carbon concentration, and moisture content, which are crucial for land use planning and sustainable agriculture [78]. Additionally, ANNs can quantify uncertainty in predictions, enhancing the reliability of soil assessments, especially in areas with limited data [74].

3.1.6. Gradient Boosting Regressors

Gradient Boosting Regressors are a class of machine learning algorithms that build predictive models by combining multiple weak learners, typically decision trees, in a sequential manner [79]. The process begins with an initial model, and subsequent models are trained to correct the errors made by the previous ones, effectively minimising a loss function through gradient descent techniques. This approach allows for the creation of a robust model that can handle complex data patterns and improve accuracy over time [80]. The adaptability of gradient boosting to stochastic data streams allows it to efficiently process varying soil moisture conditions, making it suitable for real-time applications [81].

3.2. *Proposed Solution*

The methodology is structured into subsections that constitute the main steps taken to obtain the solution.

3.2.1. Area of Study

The geographical area of interest for this study, depicted in Figure 1, was primarily identified based on the spatial distribution of the seven probes available for in situ soil moisture assessment. This area is located in the southeastern part of the Iberian Peninsula, encompassing regions within the provinces of Murcia, Almería, Albacete, and Granada. The terrain covers a total area of approximately 7539 km^2 and is defined by two specific vertices with the following coordinates:

- Vertex 1: Latitude: 38°19′58.65″ N—Longitude: 2°26′12.26″ W;
- Vertex 2: Latitude: 37°33′52.96″ N—Longitude: 1°23′4.54″ W.

The study area is characterised by a semi-arid subtropical Mediterranean climate, which features elevated summer temperatures and relatively mild winter conditions. In terms of thermal measurements, it is noteworthy that the mean annual temperature hovers around 19 °C, with pronounced climatic variances observed between coastal and inland regions, where altitude significantly influences these differences, complemented by an average annual sunshine duration that surpasses 3014.3 h [82]. Conversely, precipitation is limited and erratic, exhibiting a propensity for torrential occurrences that can result in flooding, predominantly during the autumn and spring seasons [83]. Furthermore, in light of the region's topographical characteristics, nearly 27% of the area is composed of mountainous terrains, 35% consists of plains and plateaus, while the remaining 38% encompasses intramountain depressions and valleys. The median elevation is approximately 645 m [84]. Additionally, in the province of Murcia alone, 47.44% of the total land area is allocated for agricultural purposes [82]. Given that nearly half of the province's area is committed to agriculture, it is imperative to monitor soil moisture, as it represents one of the most significant determinants of plant growth and development.

3.2.2. Dataset Construction

Although the detailed creation of the datasets used in this study will be explained throughout this section, a flowchart is presented first in Figure 3 to visually provide an initial overview of the main steps followed.

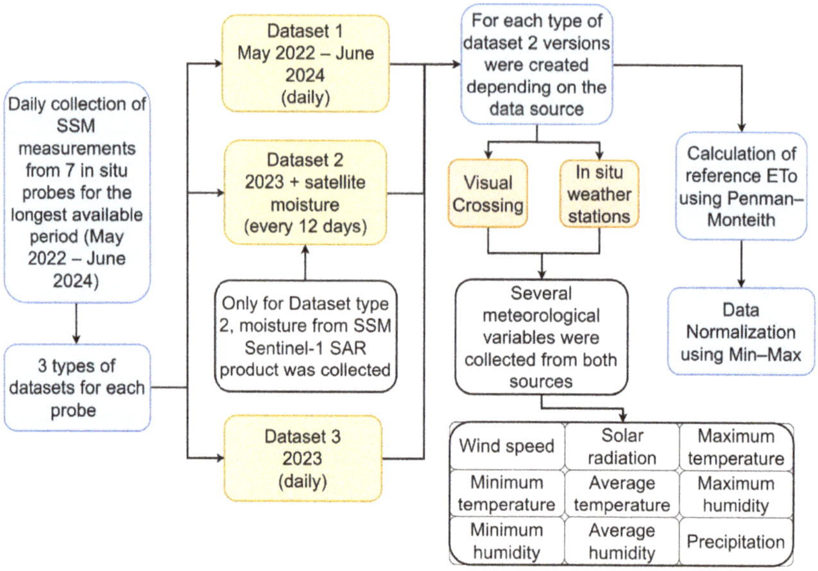

Figure 3. Flowchart illustrating the main steps followed in the creation of the three datasets used in this study. This visual representation provides an initial overview before delving into the detailed explanations presented in this section.

First of all, the longest possible time series of SMM measurements, recorded by each of the seven probes, was obtained, covering, daily, the period from May 2022 to June 2024. The rest of the data were retrieved according to these time limits. After this, two types of data were collected:

- **Real data from the closest weather stations to each of the probes**: from each weather station, daily data were collected for nine variables: minimum temperature, average temperature, maximum temperature, minimum relative humidity, average relative humidity, maximum relative humidity, average wind speed, accumulated precipitation, and average solar radiation.
- **Historical data from Visual Crossing**: using the limited free access API, daily data from seven variables were obtained: minimum temperature, maximum temperature, minimum relative humidity, maximum relative humidity, average wind speed, accumulated precipitation, and average solar radiation.

Both of these variables are essential for the Penman–Monteith formula, which is used to calculate evapotranspiration. Although precipitation is not required in the Penman–Monteith formula, it is important to use it to better estimate the water balance, where, in a simplified way, precipitation represents the contribution of water, some of which infiltrates into the soil and contributes to an increase in soil surface moisture, and evapotranspiration represents the loss of water to the atmosphere, where it contributes to a decrease in soil moisture.

Moreover, data from the SSM Sentinel-1 SAR product, derived from Sentinel-1 SAR, were collected for the year 2023. These data have a temporal limitation, as they only fully cover the selected study area every 12 days.

Once all the data were obtained, three datasets were created for each type of data and probe, where the different environmental variables were the independent variables that would be used for predicting the SSM measured by the probe (the dependent variable):

- The first dataset covers data from May 2022 to June 2024 and takes into account all the environmental variables (minimum temperature, average temperature, maximum temperature, minimum relative humidity, average relative humidity, maximum relative humidity, average wind speed, accumulated precipitation, and average solar radiation), either from Visual Crossing or from weather stations.
- The second dataset covers the entire year 2023 in 12 day intervals and, in addition to the environmental variables, also includes satellite moisture.
- The third dataset also covers the year 2023, but with records for each day of the year. In this case, only the environmental variables are included, without the satellite moisture. This third dataset is constructed as an intermediate degree between the previous two.

The aim of this division was twofold. On one hand, the aim was to compare the results obtained by the different models on each of the datasets, in order to determine, especially, the impact of satellite data on the dependent variable. The goal was to elucidate what was better: to use a longer time series of data or to renounce having a 'large' amount of data in favour of having a variable that offers a broader satellite perspective on soil moisture. On the other hand, the objective was to compare which models offer better and more accurate results: those trained with real data measured by the weather station closest to each probe, or those trained with the historical data from Visual Crossing, which offers measurements with a lower spatial resolution.

Once all the datasets were constructed, the different environmental variables were used to calculate the evapotranspiration (ETo) associated with each daily record. To carry this out, the Penman–Monteith formula was programmed in Python (version 3.12.6, with libraries Pandas (2.2.2) and NumPy (1.26.4)), and the calculation of this variable was automated for each record in each dataset.

After incorporating this final variable, all datasets were normalised using Min–Max normalisation. This method is advantageous when variables do not follow a normal distribution (as is the case here, which will be elaborated on in the statistical analysis section), as it preserves the original data distribution while scaling all values within a fixed range [0, 1] [85]. Additionally, Min–Max normalisation prevents data structure distortion that other techniques, such as Z-score normalisation, might introduce when dealing with skewed or heavy-tailed distributions. Furthermore, it enhances feature weighting by emphasising features with strong correlations to the target variable, thereby improving classification performance across various machine learning classifiers [86]. It is also worth noting that the global maximums and minimums of each variable (i.e., the maximum and minimum values of each variable across all datasets) were used, ensuring a more robust and general normalisation for the entire study area.

Furthermore, applying normalisation is crucial for the performance of the models (the ANN and the GBR) that will be trained on these data. In the case of ANNs, normalisation helps to prevent issues related to different scales among input variables, improving convergence speed and stability during training by avoiding saturation in activation functions [85,87–89]. For GBR, while tree-based models are generally less sensitive to feature scaling than gradient-based methods, normalisation can still be beneficial when dealing with heterogeneous data distributions, ensuring a more balanced contribution of features to the model and improving accuracy [90].

Finally, each dataset was divided into two parts: training (constituting 80%), and the validation (the remaining 20%) of the different models that were developed.

3.2.3. Statistical Analysis and Comparison Between Visual Crossing Data and Weather Stations Data

With the aim of deciding which environmental variables were the most significant for predicting topsoil moisture, a statistical analysis and comparison between Visual Crossing and weather stations data was conducted. Several steps were followed.

- First of all, an exploratory data analysis was conducted. For this, the mean, median, standard deviation, and interquartile range of the data were calculated, and outliers were identified using box plots. For both types of data, and generally for all probes, the results were very similar.
  Next, the distributions followed by the different variables for each probe were analysed, which again, were very similar both for all probes and between the VC data and the station data. It is noteworthy in this case that **none of the variables, in any data or probe, followed a normal distribution**, as demonstrated by the Shapiro–Wilk test. This statistical test determines, for a small or medium number of records, as is the case here, whether the distribution followed by a variable is normal or not, based on whether a certain threshold (having a p-value greater than 0.05) is exceeded.
- The next phase involved conducting a correlation analysis. Given that the previous step demonstrated that the variables did not follow normal distributions, Spearman's correlation was employed. This method measures the statistical dependence between the ranks of two variables and is particularly useful when the relationships between variables are not linear but follow a monotonic trend. In most cases, a significant correlation was observed between the different variables and the moisture recorded by the probes, notably highlighting the inverse relationship between humidity and either ETo or solar radiation. It is also noteworthy that precipitation data obtained from Visual Crossing did not show a clear correlation with the dependent variable, unlike the data recorded by most of the stations. This discrepancy is due to the low resolution of the VC data, which averages data from three stations within a 50-km range, whereas the manually selected weather station is, at worst, less than 10 km away from its respective probe. These results are partially illustrated in Figures 4 and 5. These figures represent average correlation matrices, showing the mean of all correlations between the different independent variables and the dependent variable 'probe_moisture'. Thus, the values presented provide a general representation of each particular correlation matrix, which may be influenced by outlier correlation values from some of the probes.
- The final phase consisted of detecting systematic errors in the data. This was carried out in three steps. First, a bias analysis was conducted by calculating the difference between the means of the station data and those calculated by VC. In this initial step, it was discovered that VC tended to slightly overestimate ETo, precipitation, and minimum temperature, and notably overestimate average wind speed and minimum relative humidity. Conversely, it perceptibly underestimated maximum relative humidity. The next step involved performing a residual analysis, which is the difference between the VC values and the station values, where the results obtained in the previous bias analysis were reaffirmed. Finally, Bland–Altman plots were used, which serve to visually compare two measurement techniques and to assess the agreement between two data sets.

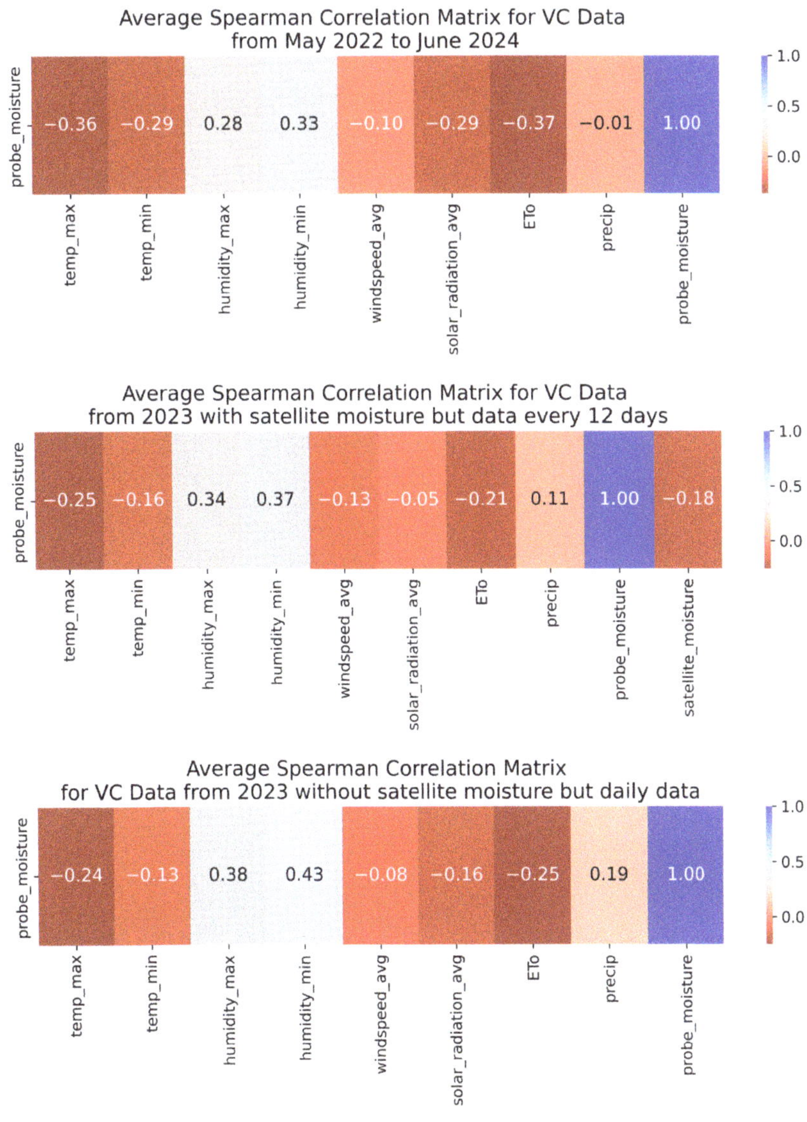

Figure 4. Spearman correlation of the different environmental variables (minimum temperature, average temperature, maximum temperature, minimum relative humidity, average relative humidity, maximum relative humidity, average wind speed, accumulated precipitation, and average solar radiation) collected in each of the three datasets (dataset from May 2022 to June 2024 with daily but without satellite soil moisture measurements, dataset from the year 2023 in 12-day intervals with satellite soil moisture measurements, and dataset from the year 2023 with daily data and without the satellite soil moisture measurements) for VC data, with respect to the dependent variable *probe_moisture*.

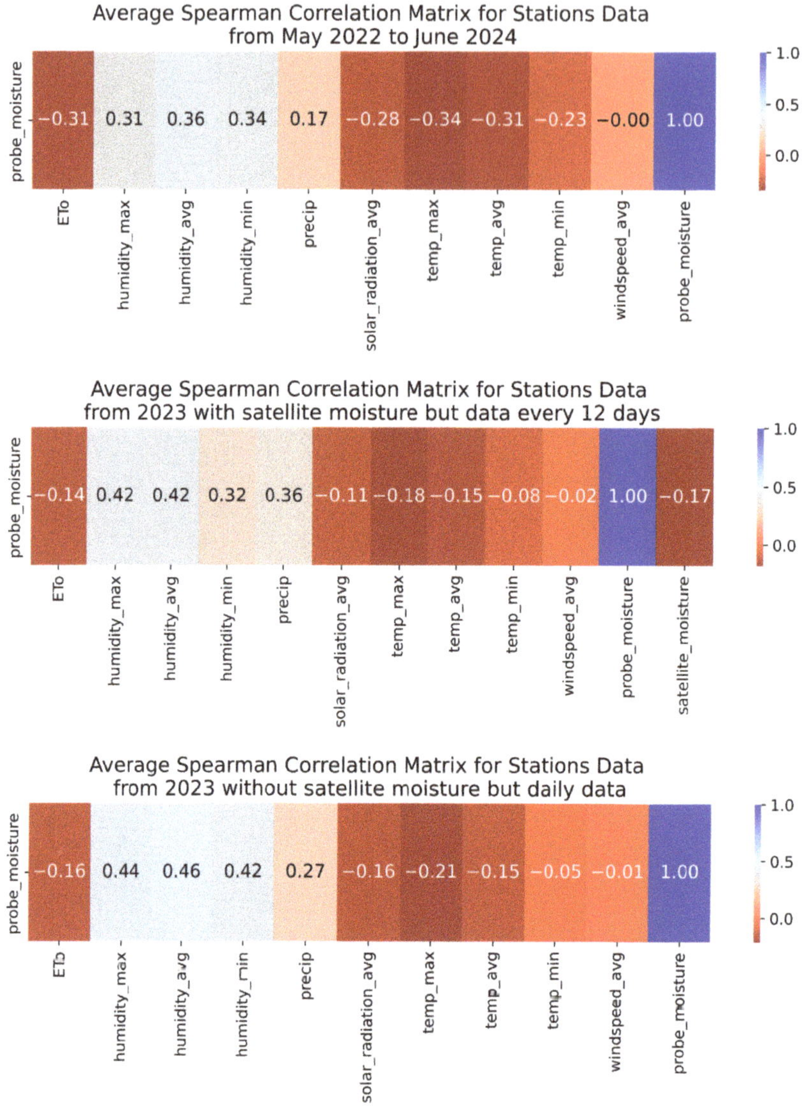

Figure 5. Spearman correlation of the different environmental variables collected (minimum temperature, average temperature, maximum temperature, minimum relative humidity, average relative humidity, maximum relative humidity, average wind speed, accumulated precipitation, and average solar radiation) in each of the three datasets (dataset from May 2022 to June 2024 with daily but without satellite soil moisture measurements, dataset from the year 2023 in 12-day intervals with satellite soil moisture measurements, and dataset from the year 2023 with daily data and without the satellite soil moisture measurements) for weather stations data, with respect to the dependent variable *probe_moisture*.

3.2.4. Statistical Analysis of the Influence of the Satellite Moisture on the Dependent Variable

This analysis is an extension of the previous one, although it is specifically focused on determining the impact of satellite moisture on the moisture recorded by the probes. The most notable phases of this analysis are presented below:

- Firstly, a Spearman correlation test was conducted (it was previously verified that satellite moisture also did not follow a normal distribution). Here, it was observed that the data covering the period from 2022 to 2024 showed a slightly higher correlation with the dependent variable than the data covering the year 2023 daily, and that the latter had a higher correlation than the datasets covering 2023 in 12-day intervals. Additionally, it was found that the correlation of satellite moisture with the moisture measured by the probes was null or even significantly inverse. Nevertheless, it was retained in the second type of dataset explained in Section 3.2.2 to ensure that models trained using this additional variable performed worse than those using the other two types of datasets also explained in Section 3.2.2. These results are shown in the Figures 4 and 5.
- Next, a regression analysis was performed, as a good way to understand how the independent variables affect the dependent variable is to fit a multiple regression model and interpret the obtained coefficients, evaluating their statistical significance through p-values. The most notable observations from this analysis are, firstly, that as the time series of the data is reduced, the p-value associated with each variable increases significantly, making them less significant in the predictions. It is noteworthy that the average p-value associated with satellite moisture in the different regression models trained, both for VC and station data, is above 0.3. However, as the data time series is reduced, the R^2 coefficient of the models increases. This could be explained by considering that we are using linear models to obtain an initial idea of the influences of the independent variables on the probe moisture, when the relationships between these variables do not have to be linear, especially as the volume of data increases, since these models are too simple and the relationships between the variables are complex.
- Lastly, and especially since ETo was calculated from the other variables using the Penman–Monteith formula, a collinearity analysis was conducted using the Variance Inflation Factor (VIF), where it was observed that all datasets had variables with very high collinearity indices. To address this, Principal Component Analysis (PCA) was employed to extract the principal components that explained 95% of the variance of the dependent variable. However, after applying PCA, the linear models were repeated, resulting in a significant decrease in the R^2 coefficient in all linear models, especially those trained on the 2023 dataset with satellite moisture. Therefore, all variables (except for satellite moisture) were left in the final datasets.

More details on the following two phases are given in Section 4, where the results are presented.

3.2.5. Development and Cross-Validation of the ANNs

The following flowchart in Figure 6 summarises the key aspects of the neural network development and cross-validation process. Although detailed explanations are provided throughout this section, this visual representation offers an initial overview of the main steps involved.

The first type of model trained was the ANNs. For each of the three datasets, within each type of data, several ANNs were trained (i.e., with different structures, number of layers, and neurons), using the Grid Search technique with a wide grid of hyperparameters with 90 combinations. The hyperparameters varied primarily included the batch size, number of epochs, and optimiser. Specifically, the batch sizes tested were 1, 4, and 8; the number of epochs tested were 50, 100, 200, 300, 400, and 500; and the optimisers tested included Adam, Nadam, SGD, RMSprop, and AdamW. The batch size was varied to balance between computational efficiency and model performance, the number of epochs was adjusted to ensure sufficient training without overfitting, and different optimisers were tested to find the most effective one for each dataset. A fivefold cross-validation was used to find the best combination of these hyperparameters, a method extensively recognised and employed in the literature [91–93]. In this cross-validation, the corresponding training data were divided into five folds, with each fold containing temporally contiguous data. Each model was then trained five times, using four folds for training and validating on the remaining fold each time.

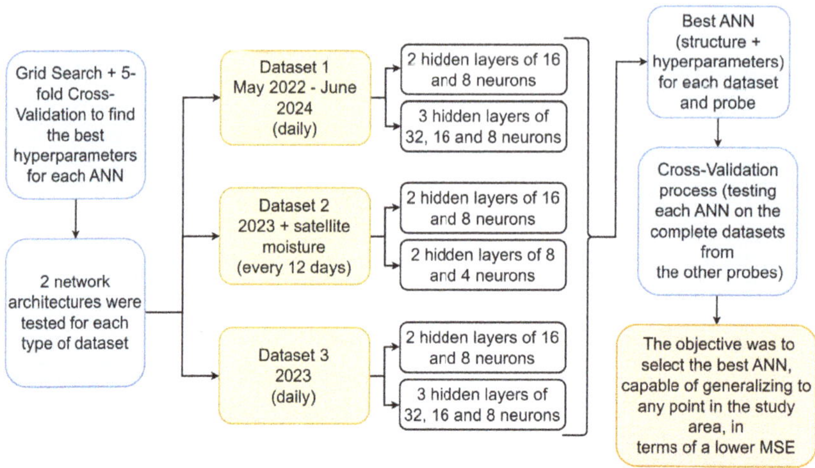

Figure 6. Flowchart summarising the key aspects of the neural network development and cross-validation process.

Moreover, for all datasets, a base structure was tested, consisting of two hidden layers, the first with 16 neurons and the second with 8. This base structure was chosen as a starting point due to its simplicity and effectiveness in initial tests. Additionally, for the datasets comprising data from 2022 to 2024, and 2023 daily, a more complex neural network structure was also tested, consisting of three hidden layers with 32, 16, and 8 neurons, respectively. This more complex structure was expected to capture more intricate patterns in the data. For the datasets covering only 2023 in 12-day intervals, a slightly simpler structure was tested, consisting of two hidden layers with eight and four neurons, as the preliminary results indicated that a simpler model was sufficient for this dataset.

Over all the trained neural networks, a cross-validation process was carried out. In this case, cross-validation means testing each neural network on the complete datasets from the other probes. As an example, the two neural network models created for probe 89 with data from 2022 to 2024 were used to predict the datasets of the same type (from 2022 to 2024) from the other six probes. It is worth noting that this process was carried out only on the neural networks trained with data extracted from weather stations since they obtained the best results, as will be discussed in Section 4. The objective of this cross-validation is to select the best neural network, capable of generalising to any point in the study area, in terms of a lower MSE.

3.2.6. Development and Cross-Validation of the GBRs

Once the development, training, and validation of the neural networks were completed, GBRs were developed solely for the station's data, as they demonstrated better results in almost all the previous analyses and models. Similar to the ANNs, an extensive grid of hyperparameters was used to find the best configuration for each dataset and each probe. In this case, the grid consisted of 420 combinations, varying primarily between the number of estimators, learning rate, and maximum depth. Specifically, the number of estimators tested were 5, 10, 15, 20, 25, 50, 75, 100, 150, 200, 250, 300, 350, and 400; the learning rates tested were 0.1, 0.01, 0.001, and 0.0001; and the maximum depths tested were 3, 5, 10, 20, 50, and 75. The number of estimators was varied to control the number of trees in the ensemble, the learning rate was adjusted to balance the speed of learning and the risk of overfitting, and the maximum depth was tested to determine the complexity of the individual trees.

The same process of cross-validation employed to find the ANN that best generalised for the entire study area was also applied to the GBRs.

4. Results and Discussion

4.1. Main Results from the Statistical Analysis of the Data

First, analysing the **results obtained from the statistical analysis of the data**, the following key points can be highlighted:

- The data collected by the weather stations showed a higher correlation with the SSM recorded by the probes, especially in the case of precipitation and average wind speed. This is due to the higher spatial resolution of the data taken by the weather stations and their closer proximity to the probes. The VC data consist of an average of the data collected by three weather stations in an area of 50 km around the point where the probe is located, and therefore, the data are more biased and inaccurate, as changes in the value of the environmental variables can be lost in the averaging process; the data collected by the manually-selected weather stations are never more than 10 kilometres away from the point where the probe is located, enhancing the accuracy of the data.
- Regarding the influence of the satellite data on the dependent variable, it is worth noting that it was null and even inverse in terms of correlation; this seems to be due to the differences in depth and spatial resolution. While the satellite data cover wide regions of a square kilometre, probe data are collected locally, which are highly accurate but less general. Moreover, satellite data comprise the soil moisture in a vertical column that is 5 cm deep, while probes collect the soil moisture at a concrete point at 10 cm deep. Moreover, the temporal restrictions imposed by including this variable in the data significantly reduce the time series, and, therefore, the models' ability to extract a better behaviour pattern in topsoil moisture, since it requires records every 12 days (due to spatial coverage restrictions) instead of making daily measurements. Furthermore, it was demonstrated that the data showed a higher correlation with the dependent variable the longer that the chosen time series was.

4.2. Model Performance Comparison

According to the **results obtained in the development and cross-validation of the ANNs**, it is worth noting that the best neural network, in terms of the lowest MSE reached in the cross-validation process, was the one trained with the data from the weather station closest to probe 93 that covered the period from May 2022 to June 2024. The MSE reached by this model in the respective cross-validation process was 0.041591 (denormalised it is 6.67%). This ANN consists of two hidden layers with 16 and 8 neurons, respectively, and the HeNormal kernel initialiser. It was trained with a batch size of 1 and the RMSprop optimiser, during 50 epochs.

Furthermore, regarding the **results in the development and cross-validation of the GBRs**, the model trained with the same data as the best ANN once again stood out as the best model developed with this machine learning technique. In fact, this model outperformed the best neural network, showing an average MSE of 0.027231 (denormalised it is 6.05%) in the cross-validation process. This model was trained with a learning rate of 0.01, a max depth of 5, and with 350 estimators (decision trees).

The reason why GBRs outperform ANNs in predicting surface soil moisture can be attributed to their ability to effectively manage bias and variance through the aggregation of multiple predictors. Moreover, as the GBR technique specifically enhances model performance by sequentially correcting errors from previous simpler models, it allows for a more robust prediction in complex environments like soil moisture estimation. In contrast, ANNs can be sensitive to overfitting, especially when the training data are limited or noisy, leading to less reliable predictions.

4.2.1. Implications of Feature Importance

Furthermore, to better understand the reasons behind the models' performance and the variables that most influence their predictions, the feature importance graphs of the two best models, the ANN and the GBR, are presented in Figures 7 and 8, respectively. In these Figures, the differences in how each model calculates feature importance become evident:

- For the ANN, feature importance is derived using permutation importance, where each feature is randomly shuffled to measure the degradation (or improvement) in model performance. This method can result in both positive and negative importance values. Positive values, as in evapotranspiration, minimum and maximum temperature, precipitation, and maximum, minimum, and average humidity, indicate that the features contribute positively to the model's accuracy (i.e., by using these features, the model performance improves), while negative values (in average temperature, windspeed and solar radiation) suggest that permuting the feature paradoxically improves performance. These negative values are likely because the feature is noisy or irrelevant (especially in the case of average windspeed, as shown in Figures 4 and 5, where the correlation of this variable with the soil moisture was null) or because the model might be overfitting on that feature.
- In contrast, the GBR calculates feature importance based on the reduction in the loss function achieved when a feature is used to split data in its decision trees. This method inherently produces only positive values, as it aggregates the contributions of features to reducing the overall error, and, therefore, it primarily reflects direct contributions to prediction. In this case, Figure 8 shows that features in the GBR, such as maximum temperature or average humidity that were not as relevant for the ANN, are of great importance. This is due to the way GBR prioritises features that create effective splits in the data, capturing linear and monotonic relationships more effectively, being, therefore, contrary to the ANN, which is better in detecting complex, non-linear interactions.

4.2.2. Data Source Evaluation: Weather Station Data vs. Visual Crossing Data

The two best final models were used to estimate the **difference in predicting soil surface moisture using data from weather stations and Visual Crossing**. The maximum difference obtained by the best ANN was 32.35%, while the best GBR showed a maximum difference of 20.09% between using Visual Crossing data and weather station data.

Figure 9 presents a bar chart comparing the MSE values obtained by the two best models (GBR and ANN), both in the Visual Crossing and in the meteorological station datasets from May 2022 to June 2024 of each of the seven probes. It can be seen, that, in general, the GBR obtained lower MSEs for all the probes' data, demonstrating, once again, to outperform the ANN.

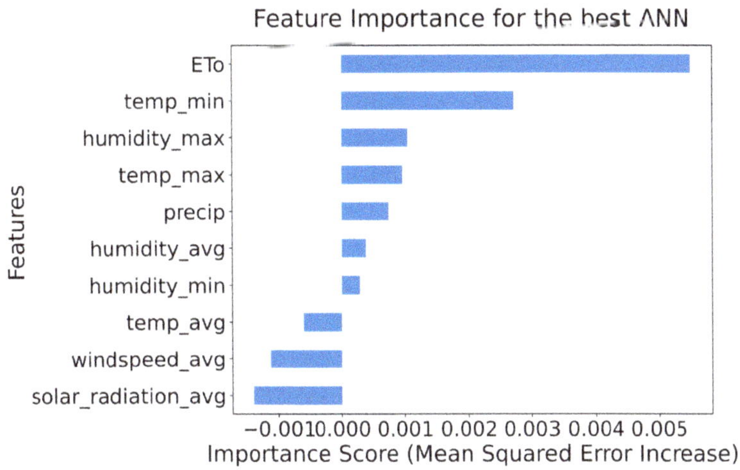

Figure 7. Feature importance of the ANN, derived using permutation importance. Positive values indicate features that improve model performance, while negative values suggest noisy or irrelevant features.

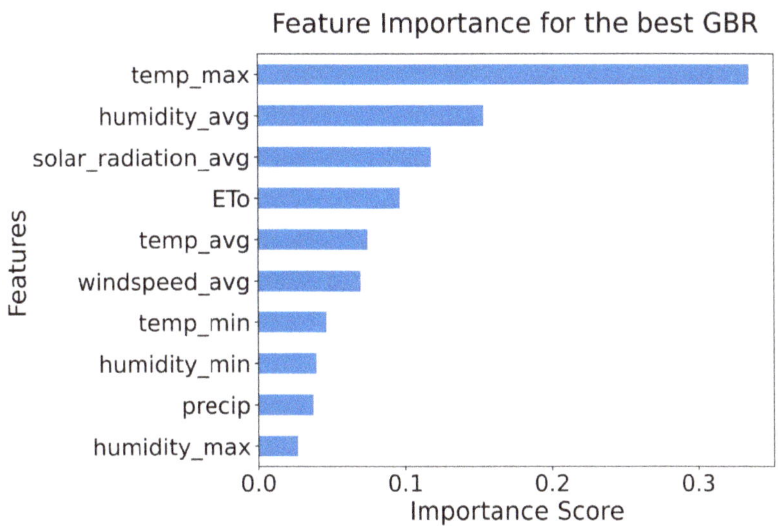

Figure 8. Feature importance of the GBR, based on reduction in the loss function. Positive values indicate features that directly contribute to prediction accuracy, with maximum temperature being particularly important.

As demonstrated, the GBR model trained with weather station data and probe 93 data achieves superior results, evidenced by the lowest average MSE in the cross-validation process and the lowest maximum absolute difference in predicting surface moisture using data from weather stations and Visual Crossing. Figure 10 illustrates the performance of the best model for predicting topsoil moisture over a reserved test period from 23 August 2023 to 22 January 2024. The predictions are based on data from Visual Crossing and the nearest weather station to probe 93. Additionally, the graph includes a line representing the 10-day moving average for the actual topsoil moisture during this period. It can be concluded from this graph that both predictions made by the model (using Visual Crossing weather data and the nearest weather station data) generally follow the trend of the actual moisture. However, both predictions exhibit more fluctuations but tend to remain within a relatively stable range, unlike the actual moisture, which shows abrupt peaks on specific days (e.g., during a very punctual torrential rain). The predictions based on Visual Crossing weather data tend to be higher than those based on the weather station data, although they are very close to each other. Additionally, except for periods of abrupt peaks in actual moisture, both types of predictions tend to slightly overestimate topsoil moisture.

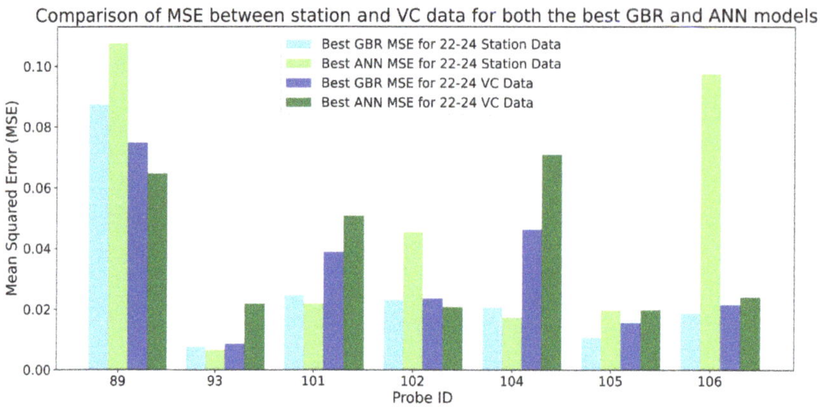

Figure 9. Comparison of the mean squared error obtained between weather station and Visual Crossing data for both the best Gradient Boosting Regressor and Artificial Neural Network models in a cross-validation using all the probes' data.

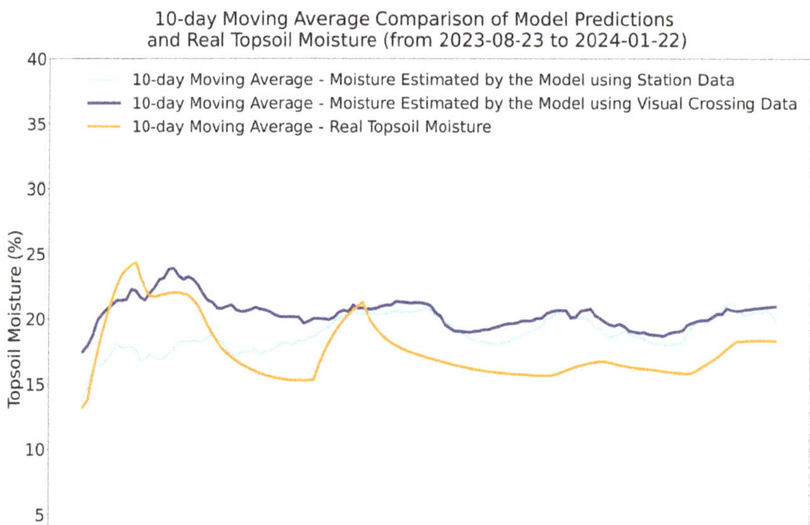

Figure 10. Comparison of the 10-day moving average for model predictions and real topsoil moisture from 23 August 2023 to 22 January 2024. The plot shows the 10-day moving average for the moisture estimated by the model using station data and Visual Crossing data, along with the real topsoil moisture.

4.3. Surface Soil Moisture Prediction: Comparison of Forecasted and Actual Data

Furthermore, to demonstrate the model's capability to accurately predict topsoil moisture with a certain lead time, two maps were created. The first map, shown in Figure 11, displays the surface soil moisture predicted by the model using forecast data retrieved from Visual Crossing for the period from 30 November 2024 to 9 December 2024, inclusively. Due to the limitations of the free version of the Visual Crossing API, only 10 days of weather forecast data could be collected. These data had to be gathered all at once to avoid any alterations, resulting in a large volume of requests for all points in the study area. Conversely, Figure 12 shows a map with the moisture predictions made by the model using actual data collected from Visual Crossing's historical records after the same 10-day period. Both maps represent average values, calculated by taking the daily moisture predictions made by the model for each latitude and longitude and averaging them over the entire period. In both maps, bluer shades correspond to areas with higher moisture, while redder shades indicate areas with lower moisture.

Predicted topsoil moisture by the best GBR with Visual Crossing forecast weather data from 30 November 2024 to 9 December 2024 (10 days)

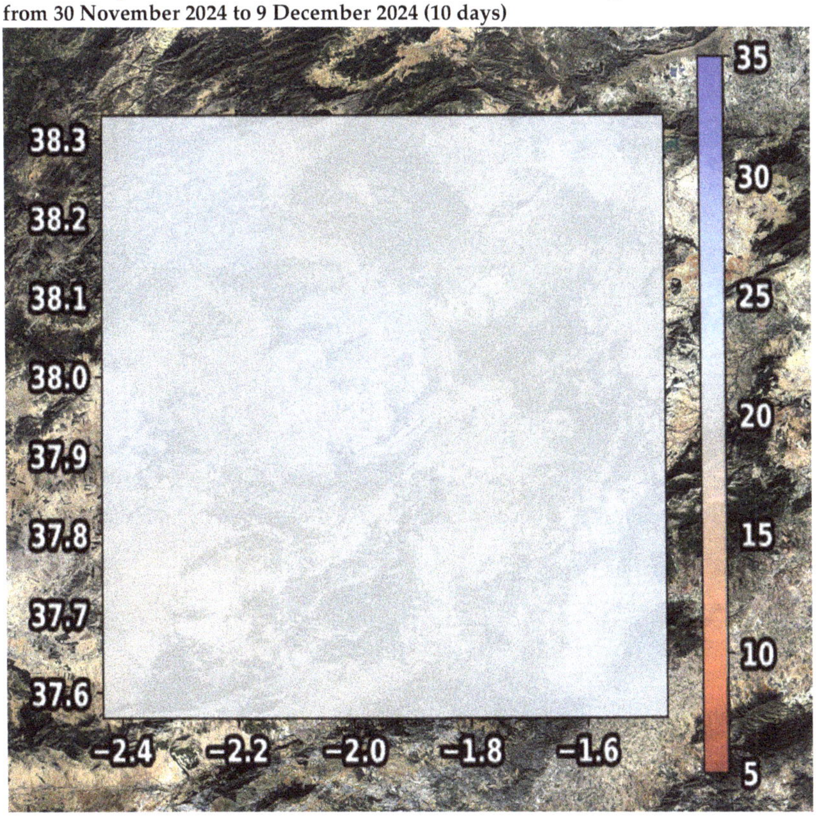

Figure 11. The map shows the surface soil moisture predicted by the best GBR using forecast data from Visual Crossing for the period from 30 November 2024 to 9 December 2024. Bluer shades indicate higher moisture levels, while redder shades indicate lower moisture levels.

Predicted topsoil moisture by the best GBR with Visual Crossing real weather data from 30 November 2024 to 9 December 2024 (10 days)

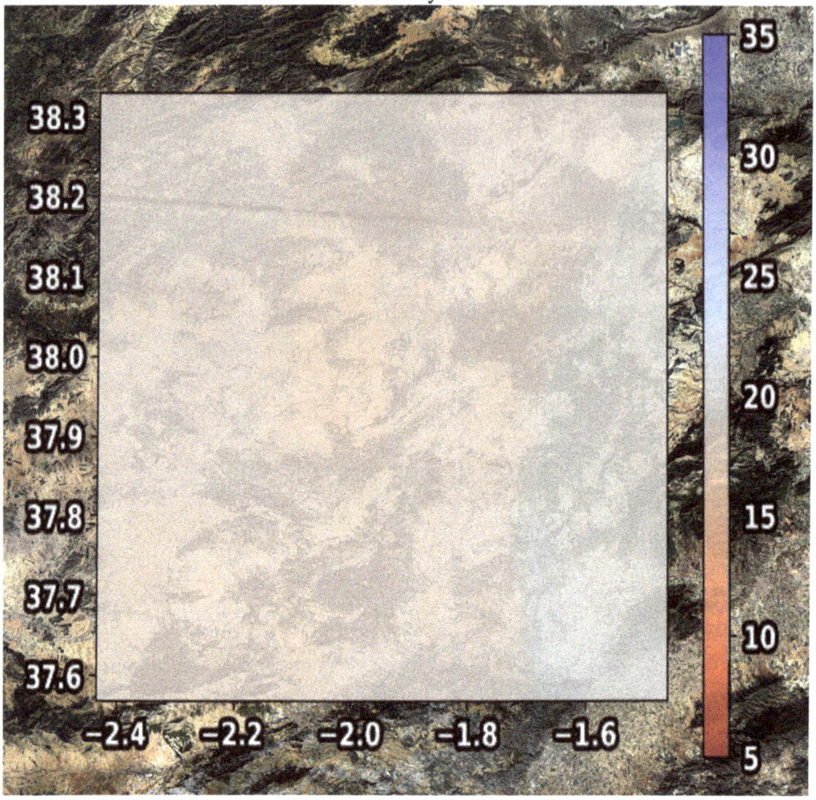

Figure 12. The map displays the surface soil moisture predictions made by the best GBR using actual data collected from Visual Crossing's historical records for the same period from 30 November 2024 to 9 December 2024. Bluer shades represent areas with higher moisture, and redder shades represent areas with lower moisture.

Based on the comparison of both maps, it can be observed that there is little visual difference between the soil moisture predictions made by the model using forecast data from Visual Crossing and the predictions made using actual data from Visual Crossing for the same period. Specifically, it is worth noting that using forecast data, the model tends to overestimate the soil moisture. In fact, the maximum difference between the soil moisture map predicted by the model using forecast data and the map with moisture estimates using actual data is 2.96%, demonstrating that the model is capable of predicting surface soil moisture with a certain lead time. This small discrepancy highlights the model's accuracy and reliability in predicting soil moisture levels, even when using forecast data instead of real-time measurements. It is also noteworthy that, given that the study area has a very similar climate throughout and the period comprises only 10 days, the differences between the maximum and minimum moisture estimated by the model are not very high.

4.4. Discussion

Among the studies most similar to this one, we analysed the results of [94], which presented a multiple linear regression model with an MSE of 0.14 for soil moisture forecasts one day ahead. In the same context, our study obtained a lower average MSE of 0.027 for the Gradient Boosting Regressor (GBR) during cross-validation, where the model was evaluated on complete datasets, spanning two years of data, from all probes, also making one-day-ahead predictions. Moreover, in [95], a root mean squared error (RMSE) of 4.6% was reported for their Random Forest model, which translates to an RMSE of approximately 21.16%. However, the GBR model developed in this study performed with an average RMSE of 6.05% (which is equivalent to 0.027 but denormalised) during the same cross-validation process.

Furthermore, after an extensive literature review, no previous study was found that specifically compared the results of soil moisture prediction using observed meteorological data and predicted meteorological data. The difference over a 10-day period found in this study was 2.96%, demonstrating the validity and feasibility of using a machine learning model on forecasted meteorological data, even when trained on observed meteorological data, and also highlighting the accuracy of meteorological predictions and their proximity to actual values. Additionally, the study not only analysed and evaluated the impact of using forecasted versus observed meteorological data on soil moisture predictions, but also assessed and compared the effects of using meteorological data obtained directly from weather stations located near the soil moisture probes versus data obtained from third-party services such as Visual Crossing. While third-party services offer more convenient and universal data access without requiring researchers to set up their own weather stations, their spatial resolution is often lower, as the weather stations they rely on may not be near the regions of interest.

Beyond these methodological contributions, the practical implications of this study are particularly relevant for real-time irrigation management systems. The high accuracy and low latency of the GBR model make it a strong candidate for integration into smart agriculture platforms.

Notably, integrating GBR involves real-time data assimilation from meteorological forecasts and soil moisture sensors, enabling dynamic irrigation recommendations tailored to specific field conditions. This approach supports sustainable water management by preventing over-irrigation and ensuring that crops receive the necessary moisture based on future conditions rather than reactive responses to soil dryness.

5. Conclusions

The study concluded that data from weather stations had a higher correlation with soil moisture measured by the probes, primarily due to their higher resolution and proximity compared to Visual Crossing data, which averaged data from three stations located 50 km away. In contrast, satellite data did not show a significant correlation with the dependent variable, mainly because of differences in depth and spatial resolution, as well as temporal constraints that limit the data series. Among the models tested, the best neural network (ANN) was trained with data from the weather station closest to probe 93, achieving a mean squared error (MSE) of 0.041591 in cross-validation. However, the Gradient Boosting Regressor (GBR) model trained with the same data outperformed the ANN, with an average MSE of 0.027231, demonstrating superior predictive capability in estimating soil surface moisture. Additionally, the maximum absolute difference in predicting soil surface moisture using Visual Crossing data versus weather station data was 32.35% for the best ANN and 20.09% for the best GBR, indicating that GBR models exhibit less variation in this aspect. This comprehensive analysis underscores the importance of selecting appropriate data sources and models for accurate soil moisture prediction. Lastly, the comparison between the predicted soil moisture map using forecast data and the map using actual data showed a maximum difference of only 2.96%. This small difference demonstrates the model's ability to accurately predict soil moisture with a certain lead time, confirming its reliability and effectiveness.

These findings hold significant implications for agriculture in semi-arid regions, where efficient water resource management is crucial for ensuring sustainable crop production. The ability of the GBR model to provide more accurate and stable predictions of soil moisture can help to optimise irrigation practices, reduce water waste, and improve crop yields. Moreover, ongoing efforts are focused on integrating the final GBR model into a smart agriculture platform specifically designed for the southern region of the Iberian Peninsula, characterised by a semi-arid climate and increasing water resource stress. Therefore, our model will help to mitigate the risks associated with drought and water scarcity by utilising the on-field sensor data provided by the platform, allowing farmers to monitor soil moisture levels and receive irrigation recommendations tailored to their specific location and crop type.

It is worth noting that this study is affected by some limitations. In particular, it should be highlighted that the models have been trained with data recorded for the selected basin, thus being limited to the dry climate of this southern Mediterranean region of the Iberian Peninsula. This means that they will not perform well in areas with humid climates, where storms are much more frequent. Although this limitation stands as the main future direction of this research, it is worth noting that the developed model can be recalibrated with data from different regions since the main variables related to the processes of water accumulation and loss in the soil are included in the model. This would allow the methodology and input variables to be correctly generalised to other climatic regions. Nonetheless, another approach to address this limitation is to develop specific models for several regions with different climate types, installing probes in each of these regions to collect in situ soil moisture measurements to be used as references for training and testing the models. Then, each specific model can be integrated into a larger model capable of selecting the appropriate prediction based on the latitude and longitude of the desired geographic area by the user.

Author Contributions: M.Z.M.: Conceptualisation, methodology, software, formal analysis, investigation, resources, data curation, writing—original draft, writing—review and editing, visualisation. L.H.M.d.S.: Conceptualisation, methodology, formal analysis, investigation, resources, data curation, writing—original draft, writing—review and editing, visualisation, supervision. R.M.-P.: Writing—review and editing, supervision, project administration, funding acquisition. A.F.S.G.: Writing—review and editing, supervision, project administration, funding acquisition. All authors have read and agreed to the published version of the manuscript.

Funding: This paper has been supported by the EU Commission under the following projects: NEPHELE (GA 101070487) and HYPER-AI (GA 101135982). This paper was also supported by the Innovation Spanish Agency under the following projects: EDEN (Reference CPP2021-009146), OPEN2CLOSE (Reference CPP2021-008538), PROMETEO (Reference TSI-064200-2022-003), GEMINI (Reference TED2021-129767B-I00), and Agro6GSense (Reference TSI-064200-2023-8).

Institutional Review Board Statement: Not applicable.

Informed Consent Statement: Not applicable.

Data Availability Statement: The original data presented in the study are openly available in GitHub at https://github.com/miriozzy/Development_and_Comparison_of_ANNs_and_GBRs_for_Predicting_Topsoil_Moisture_Using_Forecast_Data (accessed on 16 February 2025), commit: a6bce44db885cb533767636147dffee3e8dfd1ff.

Conflicts of Interest: The authors declare no conflicts of interest.

References

1. Lianos, T.P. Population and Environment. In *Capitalism, Degrowth and the Steady State Economy: Debating Future Economic Models*; Springer: Berlin/Heidelberg, Germany, 2024; pp. 45–56. [CrossRef]
2. Wang, J.; Azam, W. Natural resource scarcity, fossil fuel energy consumption, and total greenhouse gas emissions in top emitting countries. *Geosci. Front.* **2024**, *15*, 101757. [CrossRef]
3. da Silveira, L.H.M.; Cataldi, M.; de Farias, W.C.M. Development of multi-scale indices of human mobility restriction during the COVID-19 based on air quality from local and global NO2 concentration. *iScience* **2023**, *26*, 107599. [CrossRef]
4. Yunus, A.P.; Masago, Y.; Hijioka, Y. COVID-19 and surface water quality: Improved lake water quality during the lockdown. *Sci. Total Environ.* **2020**, *731*, 139012. [CrossRef] [PubMed]
5. Programme, U.W.W.A.; Koncagül, E.; Connor, R.; Abete, V. *The United Nations World Water Development Report 2024: Water for Prosperity and Peace; Facts, Figures and Action Examples*; UNESCO Publishing: Paris, France, 2024.
6. Tolba, M.K. Recursos de agua dulce y calidad del agua. In *Salvemos El Planeta: Problemas y Esperanzas*; Springer: Dordrecht, The Netherlands, 1992; pp. 45–56. [CrossRef]
7. Eze, J.N.; Salihu, B.Z.; Isong, A.; Aliyu, U.; Ibrahim, P.A.; Gbanguba, A.U.; Ayanniyi, N.N.; Alfa, N.; Alfa, M.; Aremu, P.A.; et al. Climate Change Impact on Agriculture and Water Resources—A Review. *Badeggi J. Agric. Res. Environ.* **2022**, *4*, 43–53. [CrossRef]
8. Augusto Getirana, R.L.; Cataldi, M. Brazil is in water crisis—It needs a drought plan. *Nature* **2021**, *600*, 218–220. [CrossRef]
9. Srivastav, A.L.; Dhyani, R.; Ranjan, M.; Madhav, S.; Sillanpää, M. Climate-resilient strategies for sustainable management of water resources and agriculture. *Environ. Sci. Pollut. Res.* **2021**, *28*, 41576–41595. [CrossRef] [PubMed]
10. Singh, V.P.; Mishra, A.K.; Chowdhary, H.; Khedun, C.P. Climate change and its impact on water resources. In *Modern Water Resources Engineering*; Humana Press: Totowa, NJ, USA , 2013; pp. 525–569. [CrossRef]
11. Kanae, S. Global warming and the water crisis. *J. Health Sci.* **2009**, *55*, 860–864. [CrossRef]
12. Lorenzo-Lacruz, J.; Garcia, C.; Morán-Tejeda, E. Groundwater level responses to precipitation variability in Mediterranean insular aquifers. *J. Hydrol.* **2017**, *552*, 516–531. [CrossRef]
13. Guerrero-Baena, M.; Gómez-Limón, J. Insuring Water Supply in Irrigated Agriculture: A Proposal for Hydrological Drought Index-Based Insurance in Spain. *Water* **2019**, *11*, 686. [CrossRef]
14. O'Neill, M.P.; Michael, D.P. Water and agriculture in a changing climate. *HortScience* **2011**, *46*, 155–157. [CrossRef]
15. Sraïri, M. IWater uses in sustainable agriculture practices: Reconsidering the priorities in water scarce areas. *Adv. Plants Agric. Res.* **2018**, *8*, 333–334. [CrossRef]
16. Du Plessis, A.; du Plessis, A. Current and future water scarcity and stress. In *Water as an Inescapable Risk: Current Global Water Availability, Quality and Risks with a Specific Focus on South Africa*; Springer: Cham, Switzerland, 2018; pp. 13–25. [CrossRef]
17. Kotze, H.C.; Qotoyi, M.S.; Bahta, Y.T.; Jordaan, H.; Monteiro, M.A. A Systematic Review and Meta-Analysis of Factors Influencing Water Use Behaviour and the Efficiency of Agricultural Production in South Africa. *Resources* **2024**, *13*, 94. [CrossRef]
18. Martínez-Alvarez, V.; González-Ortega, M.; Martin-Gorriz, B.; Soto-García, M.; Maestre-Valero, J. The use of desalinated seawater for crop irrigation in the Segura River Basin (south-eastern Spain). *Desalination* **2017**, *422*, 153–164. [CrossRef]
19. Mehta, P.; Siebert, S.; Kummu, M.; Deng, Q.; Ali, T.; Marston, L.; Xie, W.; Davis, K.F. Half of twenty-first century global irrigation expansion has been in water-stressed regions. *Nat. Water* **2024**, *2*, 254–261. [CrossRef]
20. Gu, L.; Chen, J.; Yin, J.; Sullivan, S.C.; Wang, H.; Guo, S.; Zhang, L.; Kim, J. Projected increases in magnitude and socioeconomic exposure of global droughts in 1.5 and 2 °C warmer climates. *Hydrol. Earth Syst. Sci.* **2020**, *24*, 451–472. [CrossRef]
21. Schleussner, C.; Lissner, T.K.; Fischer, E.M.; Wohland, J.; Perrette, M.; Golly, A.; Rogelj, J.; Childers, K.; Schewe, J.; Frieler, K.; et al. Differential climate impacts for policy-relevant limits to global warming: The case of 1.5 °C and 2 °C. *Earth Syst. Dyn.* **2016**, *7*, 327–351. [CrossRef]
22. Cramer, W.; Guiot, J.; Fader, M.; Garrabou, J.; Gattuso, J.; Iglesias, A.; Lange, M.A.; Lionello, P.; Llasat, M.C.; Paz, S.; et al. Climate change and interconnected risks to sustainable development in the Mediterranean. *Nat. Clim. Change* **2018**, *8*, 972–980. [CrossRef]
23. Wagner, W.; Dorigo, W.; de Jeu, R.; Fernandez, D.; Benveniste, J.; Haas, E.; Ertl, M. Fusion of Active and Passive Microwave Observations to Create an Essential Climate Variable Data Record on Soil Moisture. *ISPRS Ann. Photogramm. Remote Sens. Spat. Inf. Sci.* **2012**, *I-7*, 315–321. [CrossRef]
24. Stańczyk, T.; Kasperska-Wołowicz, W.; Szatyłowicz, J.; Gnatowski, T.; Papierowska, E. Surface Soil Moisture Determination of Irrigated and Drained Agricultural Lands with the OPTRAM Method and Sentinel-2 Observations. *Remote Sens.* **2023**, *15*, 5576. [CrossRef]
25. World Meteorological Organization. *State of Global Water Resources 2022*; United Nations: New York, NY, USA, 2023; pp. 14–15. [CrossRef]
26. de Fraiture, C.; Smakhtin, V.; Bossio, D.; McCornick, P.; Hoanh, C.; Noble, A.; Molden, D.; Gichuki, F.; Giordano, M.; Finlayson, M.; et al. Facing climate change by securing water for food, livelihoods and ecosystems. *J. Semi-Arid Trop. Agric. Res.* **2007**, *4*, 21.

27. ElSaadani, M.; Habib, E.; Abdelhameed, A.M.; Bayoumi, M. Assessment of a spatiotemporal deep learning approach for soil moisture prediction and filling the gaps in between soil moisture observations. *Front. Artif. Intell.* **2021**, *4*, 636234. [CrossRef] [PubMed]
28. Jiang, Y.; Zhang, R.; Sun, B.; Wang, T.; Zhang, B.; Tu, J.; Nie, S.; Jiang, H.; Chen, K. GNSS-IR Soil Moisture Retrieval Using Multi-Satellite Data Fusion Based on Random Forest. *Remote Sens.* **2024**, *16*, 3428. [CrossRef]
29. Wilson, M.; Datta, R.; Savarimuthu, S.; Moller, D.; Ruf, C. Prediction of Soil Moisture From Near-Global Cygnss Gnss-Reflectometry Using a Random Forest Machine Learning Model. In Proceedings of the IGARSS 2024—2024 IEEE International Geoscience and Remote Sensing Symposium, Athens, Greece, 7–12 July 2024; IEEE: Piscataway, NJ, USA, 2024; pp. 4465–4471. [CrossRef]
30. Taihuttu, H.Y.; Sitanggang, I.S.; Syaufina, L. Soil Moisture Prediction Model in Peatland Using Random Forest Regressor. *BAREKENG J. Ilmu Mat. Dan Terap.* **2024**, *18*, 2505–2516. [CrossRef]
31. Brakhasi, F.; Walker, J.P.; Judge, J.; Liu, P.W.; Shen, X.; Ye, N.; Wu, X.; Yeo, I.Y.; Prajapati, R.; Kim, E.; et al. Multi-Layer Soil Moisture Estimation Using Combined L-and P-Band Radiometry: An Application of Machine Learning Algorithms. In Proceedings of the IGARSS 2024—2024 IEEE International Geoscience and Remote Sensing Symposium, Athens, Greece, 7–12 July 2024; IEEE: Piscataway, NJ, USA, 2024; pp. 4411–4415. [CrossRef]
32. Li, Q.; Zhu, Y.; Shangguan, W.; Wang, X.; Li, L.; Yu, F. An attention-aware LSTM model for soil moisture and soil temperature prediction. *Geoderma* **2022**, *409*, 115651. [CrossRef]
33. Li, X.; Zhang, Z.; Li, Q.; Zhu, J. Enhancing Soil Moisture Forecasting Accuracy with REDF-LSTM: Integrating Residual En-Decoding and Feature Attention Mechanisms. *Water* **2024**, *16*, 1376. [CrossRef]
34. Jayasinghe, W.; Deo, R.C.; Raj, N.; Ghimire, S.; Yaseen, Z.M.; Nguyen-Huy, T.; Ghahramani, A. Forecasting Multi-Step Soil Moisture with Three-Phase Hybrid Wavelet-Least Absolute Shrinkage Selection Operator-Long Short-Term Memory Network (moDWT-Lasso-LSTM) Model. *Water* **2024**, *16*, 3133. [CrossRef]
35. Zhou, G.; Li, G. Forecasting Soil Moisture Using PSO-CNN-LSTM Model. In Proceedings of the 2024 IEEE Congress on Evolutionary Computation (CEC), Yokohama, Japan, 30 June–5 July 2024; IEEE: Piscataway, NJ, USA, 2024; pp. 1–7. [CrossRef]
36. Martínez, M.Z.; Da Silveira, L.H.M.; Marin-Perez, R.; Gomez, A.F.S. Development of a Neural Network System for Predicting Topsoil Moisture Using Remote Sensing and Rainfall Forecast Data. In Proceedings of the 2024 4th International Conference on Embedded & Distributed Systems (EDiS), Bechar, Algeria, 3–5 November 2024; pp. 249–254. [CrossRef]
37. Hrushikesh, R.; Pathak, A.A.; Punithraj, G. Quantifying Surface Soil Moisture Variability Through Synergistic Applications of SAR and Machine Learning Techniques. In Proceedings of the 2023 IEEE India Geoscience and Remote Sensing Symposium (InGARSS), Bangalore, India, 10–13 December 2023; IEEE: Piscataway, NJ, USA, 2023; pp. 1–4. [CrossRef]
38. Singh, A.; Gaurav, K. A physics-informed machine learning approach to estimate surface soil moisture. In Proceedings of the EGU General Assembly Conference Abstracts, Vienna, Austria, 24–28 April 2023. [CrossRef]
39. Guan, J.; Bragdon, S.; Clausen, J. *Predicting Soil Moisture Content Using Physics-Informed Neural Networks (PINNs)*; Engineer Research and Development Center: Vicksburg, MS, USA, 2024. [CrossRef]
40. Singh, A.; Gaurav, K. Deep learning and data fusion to estimate surface soil moisture from multi-sensor satellite images. *Sci. Rep.* **2023**, *13*, 2251. [CrossRef] [PubMed]
41. Hassan-Esfahani, L.; Torres-Rua, A.; Jensen, A.; McKee, M. Assessment of surface soil moisture using high-resolution multi-spectral imagery and artificial neural networks. *Remote Sens.* **2015**, *7*, 2627–2646. [CrossRef]
42. Liu, J.; Shen, C.; Rahmani, F.; Lawson, K. A multiscale deep learning model integrating satellite-based and in-situ data for high-resolution soil moisture predictions. In Proceedings of the EGU General Assembly Conference Abstracts, Vienna, Austria, 23–28 April 2023. [CrossRef]
43. Ma, Z.; Wu, B.; Chang, S.; Yan, N.; Zhu, W. Developing a physics-guided neural network to predict soil moisture with remote sensing evapotranspiration and weather forecasting. In Proceedings of the EGU General Assembly Conference Abstracts, Vienna, Austria, 24–28 April 2023. [CrossRef]
44. Meenakshi, M.; Naresh, R. Prediction of soil moisture root zone health in Artificial Neural Network. In Proceedings of the 2021 4th International Conference on Recent Trends in Computer Science and Technology (ICRTCST), Jamshedpur, India, 11–12 February 2022; IEEE: Piscataway, NJ, USA, 2022; pp. 1–5. [CrossRef]
45. Bischof, B.; Zehe, E.; Loritz, R. Using neural networks for predicting soil water storage based in situ soil moisture observations. techreport. In Proceedings of the Copernicus Meetings, Vienna, Austria, 14–19 April 2024. [CrossRef]
46. Wang, Y.; Shi, L.; Hu, Y.; Hu, X.; Song, W.; Wang, L. A comprehensive study of deep learning for soil moisture prediction. *Hydrol. Earth Syst. Sci.* **2024**, *28*, 917–943. [CrossRef]
47. Li, L.; Dai, Y.; Shangguan, W.; Wei, N.; Wei, Z.; Gupta, S. Multistep forecasting of soil moisture using spatiotemporal deep encoder–decoder networks. *J. Hydrometeorol.* **2022**, *23*, 337–350. [CrossRef]
48. Grubišić, V.; Vasić, D.; Ljubić, H.; Rozić, R.; Volarić, T. Soil Moisture Prediction with Attention-Enhanced Models: A Deep Learning Approach. *Authorea Prepr.* **2024**. [CrossRef]

49. Wang, G.; Wei, C.; Yan, L.; Li, J. Soil Moisture Prediction Model Based on Improved GRU Recurrent Neural Network. *Strateg. Plan. Energy Environ.* **2024**, *43*, 381–400. [CrossRef]
50. Islam, M.N.; Logofatu, D.; Haque, M.Z. A Comparative Study on Machine Learning Methods Through Evaluating the Impact of Contributing Factors on The Accuracy of Soil Moisture Prediction. In Proceedings of the 2023 International Conference on Innovations in Intelligent Systems and Applications (INISTA), Hammamet, Tunisia, 20–23 September 2023; IEEE: Piscataway, NJ, USA, 2023; pp. 1–6. [CrossRef]
51. Divya, A.; Josphineleela, R.; Sheela, L.J. A machine learning based approach for prediction and interpretation of soil properties from soil spectral data. *J. Environ. Biol.* **2024**, *45*, 96–105. [CrossRef]
52. Ren, Y.; Ling, F.; Wang, Y. Research on Provincial-Level Soil Moisture Prediction Based on Extreme Gradient Boosting Model. *Agriculture* **2023**, *13*, 927. [CrossRef]
53. Zhu, Y.; Jing, X.; Ding, A. Prediction of soil moisture in Inner Mongolia's League based on machine learning. In Proceedings of the Fourth International Conference on Signal Processing and Computer Science (SPCS 2023), Guilin, China, 25–27 August 2023; SPIE: Bellingham, WA, USA, 2023; Volume 12970, pp. 17–20. [CrossRef]
54. Nguyen, T.T.; Ngo, H.H.; Guo, W.; Chang, S.W.; Nguyen, D.D.; Nguyen, C.T.; Zhang, J.; Liang, S.; Bui, X.T.; Hoang, N.B. A low-cost approach for soil moisture prediction using multi-sensor data and machine learning algorithm. *Sci. Total Environ.* **2022**, *833*, 155066. [CrossRef]
55. Jamei, M.; Ali, M.; Karbasi, M.; Sharma, E.; Jamei, M.; Chu, X.; Yaseen, Z.M. A high dimensional features-based cascaded forward neural network coupled with MVMD and Boruta-GBDT for multi-step ahead forecasting of surface soil moisture. *Eng. Appl. Artif. Intell.* **2023**, *120*, 105895. [CrossRef]
56. Cheng, Y.; Li, Y.; Wu, H.; Li, F.; Li, Y.; He, L. Soil Moisture Retrieval Using Stacked Generalization: An Ensemble Machine Learning Method. In Proceedings of the 2021 IEEE International Geoscience and Remote Sensing Symposium IGARSS, Brussels, Belgium, 11–16 July 2021; IEEE: Piscataway, NJ, USA, 2021; pp. 6984–6987. [CrossRef]
57. Han, Q.; Zeng, Y.; Zhang, L.; Cira, C.I.; Prikaziuk, E.; Duan, T.; Wang, C.; Szabó, B.; Manfreda, S.; Zhuang, R.; et al. Ensemble of optimised machine learning algorithms for predicting surface soil moisture content at a global scale. *Geosci. Model Dev.* **2023**, *16*, 5825–5845. [CrossRef]
58. Li, X.; Wu, J.; Yu, J.; Zhou, Z.; Wang, Q.; Zhao, W.; Hu, L. Inversion of Soil Moisture Content in Cotton Fields Using GBR-RF Algorithm Combined with Sentinel-2 Satellite Spectral Data. *Agronomy* **2024**, *14*, 784. [CrossRef]
59. Sharma, S.; Singh, G. Cultivating Precision: Integrating XGBoost Imputation with Random Forest Regression for Accurate Soil Moisture Prediction. In Proceedings of the 2024 10th International Conference on Advanced Computing and Communication Systems (ICACCS), Coimbatore, India, 14–15 March 2024; IEEE: Piscataway, NJ, USA, 2024; Volume 1, pp. 156–161. [CrossRef]
60. Kumar, A.; Kaushik, K.; Singh, G. Predicting Soil Moisture Levels Using Ensemble Machine Learning Methods. In Proceedings of the 2024 10th International Conference on Advanced Computing and Communication Systems (ICACCS), Coimbatore, India, 14–15 March 2024; IEEE: Piscataway, NJ, USA, 2024; Volume 1, pp. 127–132. [CrossRef]
61. Acharya, U.; Daigh, A.L.; Oduor, P.G. Machine learning for predicting field soil moisture using soil, crop, and nearby weather station data in the Red River Valley of the North. *Soil Syst.* **2021**, *5*, 57. [CrossRef]
62. Technologies, S. Sentek SDI-12 Series II Manual (Ver 1.1). 2022. Available online: https://www.sentektechnologies.com/manuals/sdi-12-series-ii (accessed on 27 March 2024).
63. IMIDA. Informe Agrometeorológico Personalizado. 2024. Available online: http://siam.imida.es/apex/f?p=101:46:344545242704833:::: (accessed on 25 March 2024).
64. Department of Geography of the University of Murcia. SUREMET. 2024. Available online: https://suremet.es/index.php (accessed on 25 March 2024).
65. Department of Geography of the University of Murcia. Proyecto FrostSE. 2021. Available online: https://frostsureste.wordpress.com/proyecto (accessed on 25 March 2024).
66. IFAPA. Datos de la estación|Instituto de Investigación y Formación Agraria y Pesquera (IFAPA). 2024. Available online: https://www.juntadeandalucia.es/agriculturaypesca/ifapa/riaweb/web/estacion/18/2 (accessed on 25 March 2024).
67. Visual Crossing. Visual Crossing Weather API. 2020. Available online: https://www.visualcrossing.com/weather-api (accessed on 16 February 2025).
68. Visual Crossing. FAQs for Visual Crossing Weather Data. 2020. Available online: https://www.visualcrossing.com/resources/documentation/weather-data/frequently-asked-questions-faq-for-visual-crossing-weather-data/ (accessed on 16 February 2025).
69. Visual Crossing. Available Data for Visual Crossing Weather. 2020. Available online: https://www.visualcrossing.com/resources/documentation/weather-data/available-data-for-visual-crossing-weather-updated-january-2020/ (accessed on 16 February 2025).
70. Visual Crossing. Weather Data Documentation for Visual Crossing. 2023. Available online: https://www.visualcrossing.com/resources/documentation/weather-data/weather-data-documentation/ (accessed on 16 February 2025).

71. Han, Y.; Calabrese, S.; Du, H.; Yin, J. Evaluating biases in Penman and Penman–Monteith evapotranspiration rates at different timescales. *J. Hydrol.* **2024**, *638*, 131534. [CrossRef]
72. Cárdenas, O.L.; Gastélum, R.D.E.; Campos, M.N.; Galaviz, R.E.P.; Serrano, L.A.G.; Montoya, J.M. Penman–Monteith Reference Evapotranspiration Estimation Models, Using Latitude–Temperature Data, in the State of Sinaloa, Mexico. *Preprints* **2024**. [CrossRef]
73. Wang, S.C. Artificial neural network. *Interdiscip. Comput. Java Program.* **2003**, *743*, 81–100. [CrossRef]
74. Rau, K.; Eggensperger, K.; Schneider, F.; Hennig, P.; Scholten, T. How can we quantify, explain, and apply the uncertainty of complex soil maps predicted with neural networks? *Sci. Total Environ.* **2024**, *944*, 173720. [CrossRef] [PubMed]
75. Acharjee, P.; Souliman, M.; Isied, M. Artificial neural network-based prediction model for soil-water characteristics curve coefficients from soil index properties. In *Bituminous Mixtures and Pavements VIII*; CRC Press: Boca Raton, FL, USA, 2024; p. 129. [CrossRef]
76. Pacci, S.; Dengiz, O.; Alaboz, P.; Saygın, F. Artificial neural networks in soil quality prediction: Significance for sustainable tea cultivation. *Sci. Total Environ.* **2024**, *947*, 174447. [CrossRef] [PubMed]
77. Uzer, A.U. Efficient prediction of compressive strength in geotechnical engineering using artificial neural networks. *Turk. J. Eng.* **2024**, *8*, 457–468. [CrossRef]
78. Elakiya, N.; Keerthana, G. Application of Artificial Neural Networks in Soil Science Research. *Arch. Curr. Res. Int.* **2024**, *24*, 1–15. [CrossRef]
79. Natekin, A.; Knoll, A. Gradient boosting machines, a tutorial. *Front. Neurorobot.* **2013**, *7*, 21. [CrossRef]
80. Friedman, J.H. Greedy function approximation: A gradient boosting machine. *Ann. Stat.* **2001**, *29*, 1189–1232. [CrossRef]
81. Hu, H.; Sun, W.; Venkatraman, A.; Hebert, M.; Bagnell, A. Gradient boosting on stochastic data streams. In Proceedings of the Artificial Intelligence and Statistics, Ft. Lauderdale, FL, USA, 20–22 April 2017; Singh, A., Zhu, J., Eds.; PMLR, Proceedings of Machine Learning Research; pp. 595–603.
82. CARM. Data of the Region of Murcia. 2024. Available online: https://www.carm.es/web/pagina?IDCONTENIDO=1619&IDTIPO=100&RASTRO=c$m25987,127,1604 (accessed on 21 February 2024).
83. CARM. Geographic Data of the Region of Murcia. 2024. Available online: https://www.carm.es/web/pagina?IDCONTENIDO=1613&IDTIPO=100&RASTRO=c\protect\T1\textdollarm25987,127,1604 (accessed on 17 February 2024).
84. topographic map.com. Topographic Map of the Region of Murcia, Altitude, Relief . 2024. Available online: https://es-es.topographic-map.com/map-7lkf3/Regi%C3%B3n-de-Murcia (accessed on 9 October 2024).
85. Chepino, B.G.; Yacoub, R.R.; Aula, A.; Saleh, M.; Sanjaya, B.W. Effect of MinMax Normalization on ORB Data for Improved ANN Accuracy. *J. Electr. Eng. Energy Inf. Technol. (J3EIT)* **2023**, *11*, 29–35. [CrossRef]
86. Shantal, M.; Othman, Z.; Bakar, A.A. A novel approach for data feature weighting using correlation coefficients and min–max normalization. *Symmetry* **2023**, *15*, 2185. [CrossRef]
87. Huang, L. Motivation and Overview of Normalization in DNNs. In *Normalization Techniques in Deep Learning*; Springer: Cham, Switzerland, 2022; pp. 11–18. [CrossRef]
88. Nawi, N.M.; Atomi, W.H.; Rehman, M.Z. The effect of data pre-processing on optimized training of artificial neural networks. *Procedia Technol.* **2013**, *11*, 32–39. [CrossRef]
89. Liew, S.S.; Khalil-Hani, M.; Bakhteri, R. Bounded activation functions for enhanced training stability of deep neural networks on visual pattern recognition problems. *Neurocomputing* **2016**, *216*, 718–734. [CrossRef]
90. Upadhyay, D.; Manero, J.; Zaman, M.; Sampalli, S. Gradient boosting feature selection with machine learning classifiers for intrusion detection on power grids. *IEEE Trans. Netw. Serv. Manag.* **2020**, *18*, 1104–1116. [CrossRef]
91. Shekar, B.; Dagnew, G. Grid search-based hyperparameter tuning and classification of microarray cancer data. In Proceedings of the 2019 Second International Conference on Advanced Computational and Communication Paradigms (ICACCP), Gangtok, India, 25–28 February 2019; IEEE: Piscataway, NJ, USA, 2019; pp. 1–8. [CrossRef]
92. Abas, M.A.H.; Ismail, N.; Ali, N.; Tajuddin, S.; Tahir, N.M. Agarwood oil quality classification using support vector classifier and grid search cross validation hyperparameter tuning. *Int. J. Emerg. Trends Eng. Res.* **2020**, *8*, 2551–2556. [CrossRef]
93. Sah, S.; Surendiran, B.; Dhanalakshmi, R.; Yamin, M. COVID-19 cases prediction using SARIMAX Model by tuning hyperparameter through grid search cross-validation approach. *Expert Syst.* **2023**, *40*, e13086. [CrossRef] [PubMed]
94. Prakash, S.; Sharma, A.; Sahu, S.S. Soil moisture prediction using machine learning. In Proceedings of the 2018 Second International Conference on Inventive Communication and Computational Technologies (ICICCT), Coimbatore, India, 20–21 April 2018; IEEE: Piscataway, NJ, USA, 2018; pp. 1–6. [CrossRef]
95. Adab, H.; Morbidelli, R.; Saltalippi, C.; Moradian, M.; Ghalhari, G.A.F. Machine learning to estimate surface soil moisture from remote sensing data. *Water* **2020**, *12*, 3223. [CrossRef]

AI

Article

Data Synthesis for Alfalfa Biomass Yield Estimation

Jonathan Vance [1], Khaled Rasheed [1,2,*], Ali Missaoui [3] and Frederick W. Maier [2]

[1] School of Computing, University of Georgia, 415 Boyd Graduate Studies, 200 D. W. Brooks Drive, Athens, GA 30602, USA
[2] Institute for Artificial Intelligence, University of Georgia, 515 Boyd Graduate Studies, 200 D. W. Brooks Drive, Athens, GA 30602, USA
[3] Department of Crop and Soil Sciences, Institute of Plant Breeding Genetics and Genomics, University of Georgia, 4317 Miller Plant Science, Athens, GA 30602, USA
* Correspondence: khaled@uga.edu

Abstract: Alfalfa is critical to global food security, and its data is abundant in the U.S. nationally, but often scarce locally, limiting the potential performance of machine learning (ML) models in predicting alfalfa biomass yields. Training ML models on local-only data results in very low estimation accuracy when the datasets are very small. Therefore, we explore synthesizing non-local data to estimate biomass yields labeled as high, medium, or low. One option to remedy scarce local data is to train models using non-local data; however, this only works about as well as using local data. Therefore, we propose a novel pipeline that trains models using data synthesized from non-local data to estimate local crop yields. Our pipeline, synthesized non-local training (SNLT pronounced like sunlight), achieves a gain of 42.9% accuracy over the best results from regular non-local and local training on our very small target dataset. This pipeline produced the highest accuracy of 85.7% with a decision tree classifier. From these results, we conclude that SNLT can be a useful tool in helping to estimate crop yields with ML. Furthermore, we propose a software application called Predict Your CropS (PYCS pronounced like Pisces) designed to help farmers and researchers estimate and predict crop yields based on pretrained models.

Keywords: machine learning; data synthesis; generative models; alfalfa; biomass; precision agriculture; classification; climate change; yield prediction; deep learning

1. Introduction

The alfalfa crop is an important livestock feed and is crucial to global food security. In previous work, we used climate data to estimate alfalfa biomass yields. We compared the accuracies of feature selection techniques and machine learning (ML) models for this task. We obtained promising results using local training data with R^2 values over 0.90, as we had access to rich curated datasets from state university variety trials [1]. However, since our team is developing a software application to aid real-world farmers, whose datasets may be much smaller, the current work addresses the problem of estimating yields for very small target datasets. We find that local training on very small target datasets results in very low accuracy, while, non-local training on much larger datasets performs only about as well as local training. Our solution combines ideas inspired by [2], which shows success using pretrained models and sparse datasets, with ideas inspired by [3,4], which show the promise of deep learning generative models like the adversarial autoencoder (AAE) [3] and generative adversarial networks (GANs) [4]. We propose a novel pipeline where models are trained with data generated or synthesized by other deep learning (DL) models. In this pipeline, the synthesized training data are synthesized from non-local sources, and the resulting classifiers estimate local targets. We call this synthesized non-local training (SNLT pronounced like sunlight), and we show it consistently achieves better accuracy than both local and non-local training. We extend the work of Xu et al. [5] by using their conditional tabular GAN (CTGAN) and tabular variational autoencoder (TVAE) synthesizers in our pipeline. The highest accuracy we obtained from SNLT was 85.7% using a decision tree classifier (DT) with CTGAN. We also obtained good results using data synthesized by TVAE, especially when training extreme gradient boosting models (XGBoost), scoring as high as 75.0% and 70.0% classification accuracy. One long-term goal of this work is to develop a software application called Predict Your CropS (PYCS, pronounced like Pisces) [6]. PYCS is a research software application and what-if tool that farmers and others can use to build, train, and run ML models to potentially predict future crop yields based on climate or weather data. Though the current iteration of PYCS is not based on time series, and therefore does not forecast future yields or trends, that is the ultimate goal for the application. However, the current iteration is a potentially useful what-if tool. For example, a farmer could design worst-case and best-case scenarios and input these hypothetical features into PYCS to create what-if predictions. This could help the farmer understand how large or small their yields might be, so they could prepare in advance to handle good, fair, and bad years appropriately and manage their resources wisely.

Alfalfa is so crucial to food security that many U.S. universities support initiatives to research this crop and publish data on the experimental growth, cultivation and harvesting of available varieties of alfalfa. These variety trial reports often include crop yield data for multiple cuts each year, contributing to the richness of the nationally available yield data. Very recent work by [7] highlights the relevance of alfalfa research, as their data come from the same variety trial reports as ours. They focus on the effects of rain on yield gaps and, like us, primarily look at rainfed rather than irrigated sites [7]. The importance of alfalfa and this abundance of alfalfa data, along with recent strides forward in ML, motivate our team to apply ML to the problem of estimating and predicting crop yields, with a focus on the alfalfa crop. As in our team's previous work [1], we use aggregated climate and yield data detailed in Tables 1 and 2.

Table 1. Shows the five features used to train all KY and GA models plus year harvested and class, assigned according to standard deviation and yield; these are the 7 records in the very small GA target dataset.

Year	Yield (tons/acre)	Class	Solar Radiation (MJ/m^2)	Total Rainfall (mm)	Avg Min Temp (°C)	Avg Max Temp (°C)	Avg Soil Moisture (%)
2009	2.40	0	3400.10	740.67	16.10	27.88	0.19
2008	3.33	1	3810.06	664.88	18.44	29.26	0.12
2008	3.35	1	3545.32	413.28	12.54	24.64	0.13
2008	5.20	1	4463.92	599.97	15.77	28.37	0.19
2009	5.92	1	5320.75	1323.90	15.69	26.19	0.14
2009	6.26	2	3915.63	925.56	16.31	27.22	0.13
2010	6.50	2	4092.75	847.82	16.28	28.33	0.13

Mitigating the effects of climate change is another motivator behind this work. Evidence suggests that traditional approaches to forecasting crop yields are becoming less reliable as the Earth's climate becomes less predictable [8–15]. As the global effort to combat climate change continues, led in part by the United Nations [16], we hope this work is a positive force toward these efforts.

This work is part of an ongoing project that explores the benefits of using ML to predict biomass yields for crops. We focus on the alfalfa crop, but this research could be expanded to other crops. Our first paper concentrates on feature selection, as it presents results from a battery of tests of the following feature selection techniques: correlation-based feature selection (CFS), ReliefF method, and a wrapper method. The data for that paper is a curated mixture of alfalfa biomass yields from variety trials in Georgia (GA) and Kentucky

(KY) combined with weather and soil moisture sourced from the National Oceanic and Atmospheric Association (NOAA) [1].

Table 2. Shows the four features used to train all SD and OH models plus year harvested and class, assigned according to standard deviation and yield; these are the 10 records in one very small OH target dataset.

Year	Annual Avg Yield (tons/acre)	Annual Avg Min Temp (°C)	Annual Avg Max Temp (°C)	Total Accumulated Rain (mm)	Total Accumulated Radiation (W/m^2)	Class
2019	1.36	7.56	17.82	7462.16	552,381.90	1
2019	1.54	6.32	16.84	13,854.04	1,009,929.62	1
2010	1.80	6.76	17.44	3764.83	561,106.56	2
2010	1.24	5.88	16.98	9814.53	1,054,700.52	0
2011	1.28	6.16	17.26	6155.62	574,282.91	0
2011	1.89	6.19	16.88	7295.27	1,033,772.80	2
2011	0.88	5.74	16.81	15,084.74	1,518,744.53	0
2012	1.30	6.31	17.48	10,412.91	1,014,862.70	0
2012	1.57	6.37	17.09	11,656.76	1,459,165.91	1
2012	1.59	5.10	16.46	4568.04	400,766.84	1

Our second paper focuses on comparing the ML models themselves, presenting R, R^2, and MAE scores for most of the same models in the current and previous works. That paper expands the dataset to include the U.S. states of Pennsylvania (PA), Wisconsin (WI), and Mississippi (MS). Those results show that larger datasets lead to higher accuracies when the states are all combined into one large dataset. That work also revealed that while data from multiple sites can be combined to train one model with good R^2 scores above 0.90, training a model with strictly non-local data to estimate yields in a separate locality did not produce usable results with regression models [17]. This motivated us to reframe this as a classification problem in the current work, and the results are much better.

We reframed our regression problem from [17] as a classification problem with three tiers for yields: high, medium, and low. We swapped the regression models in our original application programmer interface (API) from [1,17] for analogous classification models from ScikitLearn [18]. For this work, as we are interested in estimating very small target datasets, we reduce the original target GA dataset in [1] to one alfalfa variety per year, and we annualize the cuts as detailed in Section 3, resulting in only seven records detailed in Table 1. We repeat this procedure for Ohio (OH) and South Dakota (SD), producing very small OH target datasets of 8 and 10 records and producing very small datasets for one SD town, also of 8 and 10 records. Table 2 shows our annualized, one-variety per year target data from OH with its four features. OH's features are slightly different from GA's and KY's in that OH lacks soil moisture data. OH and SD follow the same schema. In Tables 1 and 2, class labels 0, 1, and 2 correspond to low, medium, and high yields, respectively.

Though CTGAN consistently outperforms TVAE in [5], TVAE was competitive in the current work, especially when combined with XGBoost in the SNLT pipeline. The synthetic datasets generated by CTGAN train our most accurate predictor when training with KY and classifying GA, but those generated by TVAE beat CTGAN in our SD and OH experiments. SNLT trained with TVAE data produced highest accuracies of 75.0% and 70.0% using random forest (RF) and XGBoost models. SNLT trained with CTGAN data, however, produces the highest accuracy of 87.5%. Other accuracies using synthetic datasets achieved modest accuracies of 62.5% and 60.0%, but this is a significant improvement over local-only and non-local training. We determined through experimentation that accuracies increased as sample sizes increased up to 1000 or 2000, but we noticed diminishing returns after that on all models except XGBoost, so we chose 1000 or 2000 samples for all other models. For XGBoost, which trains far faster than all the other models, we experimented with synthetic sample sizes up to 200,000, though 5000 usually delivered similar accuracies faster; therefore, all but one of our XGBoost models generate 5000 samples, and the other generates 200,000, as summarized in Section 4′s results tables. Overall, non-local and local training result in low classification accuracies, with no clear winner between the two; however, our SNLT accuracies are significantly higher than non-local training or local training.

2. Related Work

While results in this team's previous papers are very promising and show very high accuracies using rich datasets collected by scientists and researchers, the current work considers the problem of very small datasets potentially collected by farmers at real-world farms. In previous experiments, our high accuracies are aided by the coexistence of many varieties of alfalfa growing in the same location, providing plenty of data to train and predict locally or on datasets of combined locations. On the other hand, the current work's end goal is to create a practical application that is useful to actual farmers, and we contend that farms in the real world will typically have much smaller datasets. This presents a problem where strong, accurate models are difficult or impossible to train. As a solution, PYCS offers the option to use already-trained models to estimate users' yields, even if their datasets are very small.

Our previous work has detailed some of the related work in the domain of ML and precision agriculture. Overall, the literature shows many complex techniques involving remote sensors, unmanned aerial vehicles (UAVs) or drones, computer vision, and image processing to develop vegetation indices to measure crop health [1,17]. Our work diverges from this trend in that our data come from public sources and simple weather data. More recently, Baral et al. at Kansas State University published a study of the effects of rain gaps on alfalfa yields across the United States. They aggregated alfalfa yield and weather data for 10 states, and like the current work, they obtained their yield data from university variety trials [7]. This helped inform the current work's use of data from South Dakota (SD) State University and Ohio (OH) State University, and as [7] is only interested in rainfed alfalfa, it provides a helpful guide of rainfed locations for our team to use in future experiments. The exhaustiveness of [7] inspires us to explore as many states as possible. Furthermore, they measure alfalfa yield over growing degree days (GDD), which means they do not consider days where the temperature did not meet some threshold alfalfa requires to grow, so that concept may help inform our future work [7]. Otherwise, the current work bares little similarity to [7]. That work essentially analyzed the cost in alfalfa of dry periods (rain gaps) using frontier function analysis and boundary function analysis. They also used conditional inference trees (CITs) to determine which weather features most affect crop yields. They did not compare ML models' ability to predict or estimate alfalfa yields, and they do not mention a goal of forecasting future yields using time series and ML, as the current work does.

The current work takes advantage of one DL model that has shown great promise in recent years—the generative adversarial network (GAN). Training a GAN requires finding the Nash equilibrium in a minimax game, an idea which precipitated a breakthrough in economics in the 20th century [19]. The GAN applies game theory to DL training, as a competition between two entities drives up the performance of each. One entity, called the generator, creates simulated, fictitious data that resembles the real training data. The other entity, called the discriminator, looks at real and simulated data and tries to determine which are which. As the discriminator improves at guessing, this causes the generator to generate more convincing fakes, which causes the discriminator to continue to become a better guesser, and so on. This equates to a game where each player maximizes its own chances of winning by minimizing its opponent's chances, also known as a minimax game [4]. Though mainly leading to breakthroughs in the computer vision domain, GANs have also shown promise in other applications such as differential privacy with DPGAN [20], generating private tabular health record data with medical GAN (MedGAN) [21], and synthesizing table data with TableGAN in [22]. Refs. [20–22] demonstrate a GAN's ability to generate data that is statistically similar to the training data but introduces enough noise that connections between real people and the data are broken, protecting individuals' privacy. MedGAN combines a GAN with an autoencoder to apply this concept to health records, where privacy is a clear concern that may inhibit individuals' willingness to share data, but also where abundant data can make ML more useful in helping cure illness [21].

CTGAN is called a conditional GAN because it uses a conditional generator and training-by-sampling to address class imbalance in discrete columns. CTGAN introduces a technique called "mode-specific normalization" to normalize columns individually, and its underlying networks are fully connected [5].

We also investigate the usefulness of variational autoencoders (VAEs) in synthesizing table data, as VAEs are another class of generative neural network that has received a lot of attention and shown promising results in recent years [5]. Specifically, this work uses the tabular variational autoencoder (TVAE) proposed in [5], which adapts a plain VAE to address distribution concerns particular to generating table data.

Another somewhat recent development that helped our pipeline produce some of its best results is extreme gradient boosting or the XGBoost model. Generally, boosting combines an ensemble of weak learners to create one strong learner. One popular boosting algorithm called Adaboost often works well with random forest (RF) or decision tree (DT) models trained iteratively, where each iteration improves on the errors of the previous by weighting incorrect predictions higher in subsequent iterations, until some predefined depth or maximum accuracy is reached [23]. In XGBoost, DTs are trained to minimize prediction errors using gradient descent, creating an optimized strong learner, and this results in extremely fast training times, facilitating training on very large datasets [24].

3. Materials and Methods

Our alfalfa yield data come from publicly available alfalfa variety trials conducted by the University of Georgia [25], University of Kentucky [26], South Dakota State University [27], and Ohio State University [28]. Our weather data in GA and KY come from various NOAA resources, while our weather data for SD and OH comes from the Daymet tool, provided by Oak Ridge National Laboratory in Oak Ridge, TN and supported by NASA [29]. We provide our curated, aggregated datasets along with source code at the Github repository for this project: www.github.com/thejonathanvancetrance/SNLT (accessed on 29 November 2022). Every record in the original KY and GA datasets include the following 15 features: State, City, Date Sown, Variety, Date of Cut, Julian Day, Yield (tons/acre), Time Since Sown (Days), Time Since Last Harvest (Days), Total Radiation (MJ/m^2), Total Rainfall (mm), Avg Air Temp (°C), Avg Min Temp (°C), Avg Max Temp (°C), Avg Soil Moisture (%). Table 1 depicts the five most salient features we settled on for training for KY and GA experiments, minus class, year, and yield: average minimum temperature, average maximum temperature, total rainfall, solar radiation, and soil moisture. In SD and OH experiments, we omitted soil moisture, because we cannot consistently obtain soil moisture data for most locations as explained in Section 5. We set the thresholds of our yield classes according to standard deviation, where we label yields less than -1 standard deviations below the mean as low (class 1), yields between -1 and +1 standard deviations as medium (class 2), and yields more than +1 standard deviations above the mean as high yield in GA and KY. Our yield thresholds for all locations are summarized in Table 3. We settled on the hyperparameters shown in Appendix A via experimentation.

Table 3. We calculated the standard deviations (stdvs) of the yields to assign class labels for yields in each location; we used three different thresholds (1, 0.8, 0.5) to avoid high class imbalance.

Location	Class 1 (Low Yield)	Class 2 (Medium Yield)	Class 3 (High Yield)
KY & GA	<−1 stdvs below mean	−1 < +1 stdvs below/above mean	>+1 stdvs above mean
SD & OH	<−0.5 stdvs below mean	−0.5 < +0.5 stdvs below/above mean	>+0.5 stdvs above mean
SD & Highmore, SD	<−0.8 stdvs below mean	−0.8 < +0.8 stdvs below/above mean	>+0.8 stdvs above mean

We began the current work with the same GA and KY data, plus the code base and API that we used in [1,17]. We reduced the target GA dataset to seven records by including only one variety per year and annualizing the data. A real-world farm's dataset may not include multiple varieties, so we attempt to simulate this in our target datasets. Additionally, annualizing our data helps reduce the influence of short-term anomalous weather conditions on our models. Furthermore, some variety trial data, such as PA data, are only reported annually, so some of our data were originally annualized to be compatible with those. We expect this annualized data to simulate accuracies we might expect from PYCS when using already-trained models to estimate very small datasets. In GA and KY, the alfalfa yields are total annual yields. As this project is ongoing and evolving, we used total yields originally to match data from other states that report annually. As we expanded our data to include OH and SD, we decided that the average yield from all cuts in a year may be more informative when multiple cut information is available, mainly because some years have more cuts than others, potentially inflating the total annual yield. The current work is intended as a step toward forecasting future yields, but we are not truly forecasting yet, as we train with data from the entire year up to the date of the final cut in GA and KY. Therefore, our total rainfall, total solar radiation, average minimum and maximum temperature, and average soil moisture data typically represent about the first eight or nine months of that year in GA and KY. In SD and OH, those features represent the totals and averages over the lifetime of one planting, which may be several years, up to the last cut of that year. Table 1 lists "Solar Radiation" and "Total Rainfall" while Table 2 lists "Total Accumulated Radiation" and "Total Accumulated Rain" because of these differing approaches, where Table 1 reflects one year's weather until the final cut, and Table 2 reflects weather totals over the lifetime of that planting up to the final cut of that year. The average minimum and maximum daily temperatures are sums of each day's minimum and sum of each day's maximum temperatures, divided by the number of days considered. Table 1 weather data is sourced from the NOAA, which reports solar radiation as solar irradiation in MJ/m^2, while Table 2 weather data is sourced from Daymet, which reports solar radiation as solar irradiance in W/m^2. Though Tables 1 and 2 reflect slightly different annualization processes, we think both are reasonable approaches.

The annualization process reflected in Table 2 resulted in a target dataset of 10 records of OH data, then we removed two arbitrary records to create an eight-record dataset for better comparison with GA results. Once again, we followed this process with SD data, creating a local target ten-record dataset of data from only Highmore, SD, leaving the other SD locations for non-local training and synthesis; then, we again arbitrarily removed two records to create an eight-record Highmore dataset for comparison. We also annualized the SD training dataset.

We modified the original code base to build classification models instead of their regression counterparts, using the ScikitLearn [18] ML library for the Python programming language along with NumPy [30]. We swapped [18]'s k-nearest neighbors (KNN) regressor for a KNN classifier, random forest (RF) regressor for a RF classifier, linear regression for logistic regression (LR), support vector regressor for a support vector classifier (SVC), decision tree regressor for decision tree classifier (DT), and the multilayer perceptron (MLP) regressor for an MLP classifier (which we refer as the artificial neural network or ANN interchangeably). We added ScikitLearn's XGBoost classifier, not tried in our previous work, to our cast of models. Since the task is now one of classification, we modified the API used in [1] to provide accuracy percentage scores indicating percent of estimates correct, rather than correlation coefficients or mean absolute errors (MAEs). We used Matplotlib to generate confusion matrices [31].

For local training, we trained all models on our very small GA, OH, and SD datasets reflected in Tables 1 and 2, then we used those models to classify those same datasets, resulting in low accuracies as shown in Tables 4–8. For regular non-local training, we trained all models on: the KY dataset with 183 records from Lexington, KY; the complete SD dataset with 767 records; and the SD dataset minus Highmore with 604 records. Next, we used those models to classify the local GA, OH, and Highmore, SD datasets, resulting in the different but mostly unimproved accuracies shown in Tables 4–8.

Our third and featured set of experiments explored our SNLT pipeline, depicted in Figure 1. SNLT begins by choosing a dataset that is not local to the target area that is being classified as high, medium, or low yield. That non-local dataset is fed to a data synthesizer, which in the case of the current work is either CTGAN or TVAE. Figure 1 depicts SNLT using a fully connected deep CTGAN, but other synthesis techniques could be used in place of this in the SNLT pipeline. The synthesizer outputs a larger training dataset than the original non-local dataset, and the size of the new dataset is an adjustable parameter of SNLT. As our experiments confirmed, these larger synthesized datasets usually train more accurate classifiers than those trained by the non-local data. Next, a model is selected for training with the synthesized data. We experimented with ANN, KNN, RF, DT, SVC, LR, and XGBoost, and since DT and XGBoost produced the best results, Figure 1 depicts SNLT as using a tree-style model, though the others are also valid. Finally, the trained model is used to classify the target dataset's alfalfa yield. This dataset would typically be too small to adequately train most models or effectively split into test and training. Based on sun, rain, temperature, and soil moisture when available, the model classifies target alfalfa yields as low, medium, or high, and SNLT produces accuracy scores as compared to the true class labels.

Our experiments begin with the same labeled KY and SD non-local training data as before, but we feed them through a generative model that synthesizes the data, resulting in larger datasets of 1000 to 200,000 records. The generative model in the pipeline is interchangeable with any generative model or synthesis technique, but the current work explores CTGAN and TVAE [5]. Next, we use these synthetic source states' data to train all our models, and we again use those trained models to classify labels for our small GA, OH, and SD targets, resulting in the promising accuracies depicted in Tables 4–8.

Table 4. Train KY, test GA: best accuracies from GA local training, non-local training (KY), and SNLT with CTGAN and TVAE (KY), each for three class labels (high, medium, and low yield), with synthetic dataset size of 1000 samples, except XGBoost, which is 200,000 samples. Highest accuracy is bold.

Model	Local	Non-Local	SNLT (TVAE)	SNLT (CTGAN)
ANN	28.6%	42.8%	57.1%	57.1%
KNN	28.6%	28.6%	42.8%	57.1%
LR	NA *	14.3%	14.3%	57.1%
DT	42.8%	28.6%	57.1%	**85.7%**
SVC	14.3%	14.3%	57.1%	57.1%
RF	28.6%	42.8%	42.8%	57.1%
XGB	28.6%	28.6%	28.6%	57.1%

* failed to train, omitted.

Table 5. Train SD, test OH with 8 samples: best accuracies (bold) from OH local training, non-local training (SD), and SNLT with CTGAN and TVAE (SD), each for three class labels (high, medium, and low yield), plus size of synthesized dataset.

Model	Local	Non-Local	SNLT (TVAE)	SNLT (CTGAN)	Sample Size
ANN	25.0%	12.5%	62.5%	62.5%	5000
KNN	25.0%	12.5%	62.5%	50.0%	2000
LR	25.0%	NA *	25.0%	50.0%	2000
DT	37.5%	12.5%	62.5%	62.5%	3000
SVC	25.0%	12.5%	50.0%	50.0%	2000
RF	12.5%	12.5%	**75.0%**	62.5%	2000
XGB	NA *	50.0%	**75.0%**	62.5%	5000

* failed to train, omitted.

Table 6. Train SD, test OH with 10 samples: best accuracies (bold) from OH local training, non-local training (SD), and SNLT with CTGAN and TVAE (SD), each for three class labels (high, medium, and low yield), plus size of synthesized dataset.

Model	Local	Non-Local	SNLT (TVAE)	SNLT (CTGAN)	Sample Size
ANN	20.0%	10.0%	50.0%	40.0%	2000
KNN	10.0%	40.0%	50.0%	40.0%	2000
LR	10.0%	10.0%	30.0%	40.0%	2000
DT	10.0%	20.0%	50.0%	50.0%	2000
SVC	20.0%	20.0%	40.0%	60.0%	2000
RF	30.0%	30.0%	50.0%	50.0%	2000
XGB	10.0%	50.0%	**70.0%**	60.0%	5000

Table 7. Train SD, test Highmore, SD only one variety per year resulting in 10 target samples: best accuracies (bold) from Highmore local training, non-local training (rest of SD), and SNLT with CTGAN and TVAE (rest of SD), each for three class labels (high, medium, and low yield), plus size of synthesized dataset.

Model	Local	Non-Local	SNLT (TVAE)	SNLT (CTGAN)	Sample Size
ANN	20.0%	60.0%	50.0%	50.0%	2000
KNN	40.0%	40.0%	50.0%	50.0%	2000
LR	30.0%	50.0%	50.0%	60.0%	2000
DT	30.0%	50.0%	60.0%	50.0%	2000
SVC	40.0%	50.0%	60.0%	50.0%	2000
RF	20.0%	60.0%	60.0%	50.0%	2000
XGB	30.0%	60.0%	**70.0%**	50.0%	5000

Table 8. Train SD, test Highmore, SD only one variety per year and 8 target samples: best accuracies (bold) from Highmore local training, non-local training (rest of SD), and SNLT with CTGAN and TVAE (rest of SD), each for three class labels (high, medium, and low yield), plus size of synthesized dataset.

Model	Local	Non-Local	SNLT (TVAE)	SNLT (CTGAN)	Sample Size
ANN	25.0%	37.5%	62.5%	50.0%	2000
KNN	12.5%	50.0%	50.0%	50.0%	2000
LR	25.0%	50.0%	50.0%	50.0%	2000
DT	12.5%	25.0%	62.5%	50.0%	2000
SVC	37.5%	37.5%	62.5%	50.0%	2000
RF	12.5%	37.5%	62.5%	50.0%	2000
XGB	25.0%	50.0%	**75.0%**	**75.0%**	5000

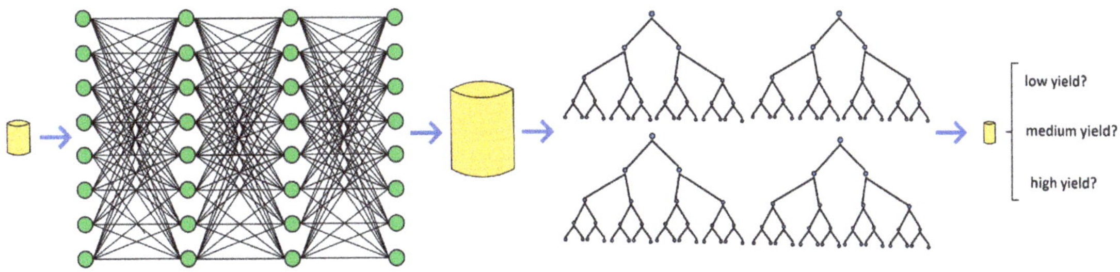

Figure 1. The SNLT Pipeline. Original data fed to a data synthesizer like CTGAN, which outputs lager dataset. Synthesized dataset used to train models like XGBoost. Trained model classifies very small target dataset.

4. Results

Whereas we previously found non-local training unviable for regression [17], our classification version of non-local training resulted in meaningful, though low accuracies which do not always beat local training, as indicated in Tables 4–8. Our SNLT pipeline produced the highest accuracy score of 85.7% (Table 4) out of these three general approaches, training a DT model with 1000 samples of data synthesized from KY using CTGAN, and classifying a very small target GA dataset of seven records. Figure 2 depicts these results as a confusion matrix. On other states' datasets of sizes 8 and 10 samples, SNLT achieved top accuracies of 75.0% and 70.0%, respectively. As Table 5 shows, SNLT achieved 75.0% accuracy training on data synthesized from SD and classifying OH, training RF and XGBoost models with 2000 and 5000 records, respectively, using TVAE, and classifying a very small target OH dataset of eight records. Figure 3 depicts these results as a confusion matrix. As Table 6 shows, SNLT achieved 70.0% accuracy training on data synthesized from SD and classifying OH, training an XGBoost model with 5000 records, using TVAE, and classifying a very small target OH dataset of ten records. As Table 7 shows, SNLT achieved 70.0% accuracy training on data synthesized from most of SD and classifying Highmore, SD, training an XGBoost model with 5000 records, using TVAE, and classifying a very small target Highmore, SD dataset of ten records. Figure 4 depicts these results as a confusion matrix. Finally, as Table 8 shows, SNLT achieved 75.0% accuracy training on data synthesized from most of SD and classifying Highmore, SD, training an XGBoost model with 5000 records, using TVAE and CTGAN, and classifying a very small target Highmore, SD dataset of eight records. Even when SNLT resulted in underwhelming accuracies of 60.0% and 62.5%, it beat regular non-local training and local training soundly and consistently as indicated in Tables 4–8. We were pleased to see XGBoost surface as a consistent top performer, as this was the only model in our cast that would train in an acceptable timeframe on tens to hundreds of thousands of records, and the only model that did not drop in accuracy with larger sample sizes. Almost everywhere our results indicate good scores using SNLT with XGBoost using 5000 samples, we received the same results using 100,000 and 200,000 samples, though we emphasize the smaller sample sizes of 5000 because they are effectively cheaper. Each run of SNLT produces a new synthetic dataset and a new trained model, and while not all runs are equally successful, once we produce a relatively successful model and dataset, we can store them for repeated classifications.

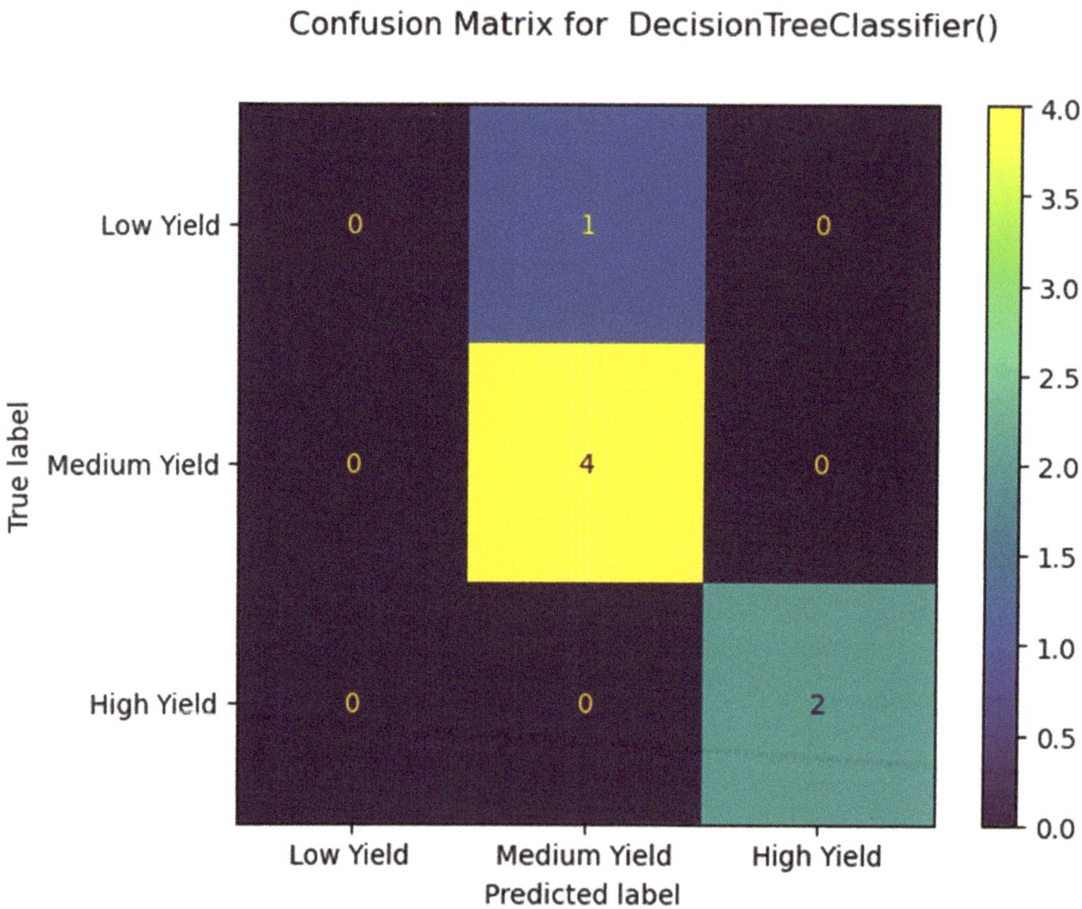

Figure 2. Confusion matrix for Table 4 best accuracy of 85.7% using DT and CTG with 1000 samples.

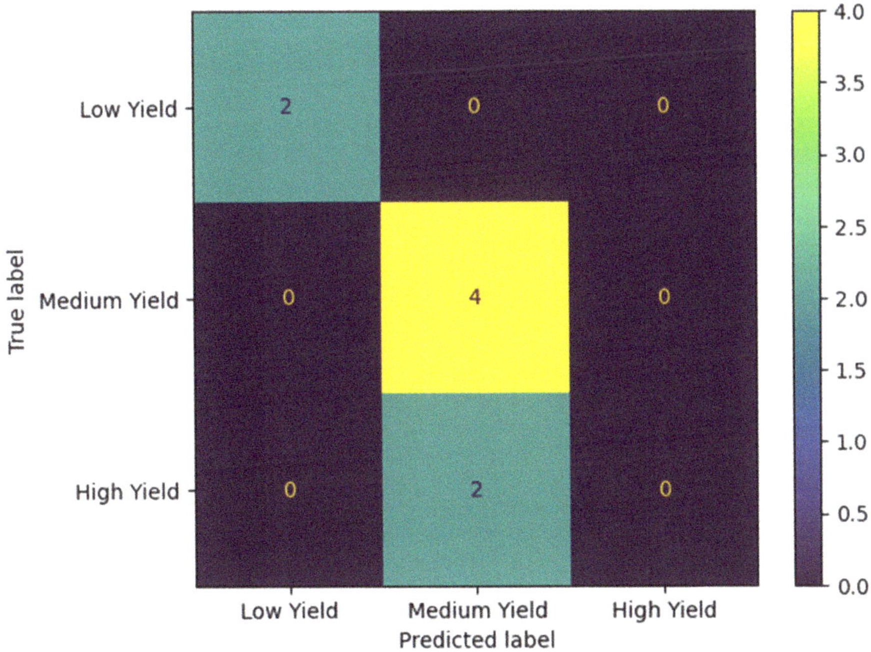

Figure 3. Confusion matrix for Table 5 best accuracy of 75.0% using RF and TVAE with 2000 samples.

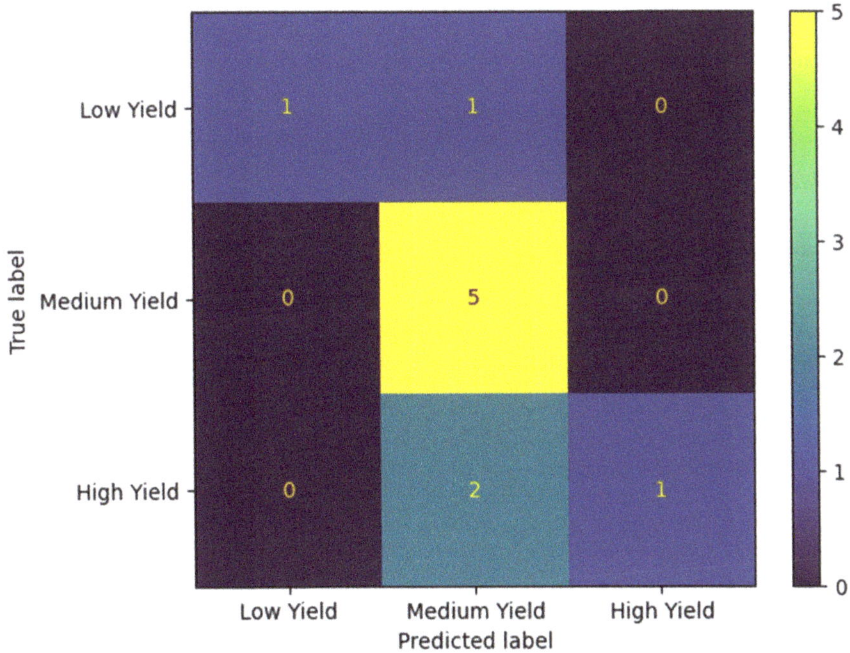

Figure 4. Confusion matrix for Table 7 best accuracy of 70.0% using XGB and TVAE with 5000 samples.

5. Discussion

Our results show that we can significantly improve estimation accuracy using our SNLT pipeline over training with non-local data or with training and estimating on the same local target dataset if it is very small. It was key for us to convert our regression problem to a classification problem. Our results are encouraging, and we think further efforts may increase our accuracies. For example, our best accuracy is higher in KY and GA experiments, where the soil moisture feature is included, and this leaves us wondering whether soil moisture is key to higher accuracies. Unfortunately, we found it difficult to obtain soil moisture data consistently throughout the U.S. Compared to our previous work, the current work aggregates yield data with much more precise weather data for SD and OH based on latitude and longitude, using the Daymet tool [30]. Before, our weather and soil moisture data came from those NOAA stations nearest to where each variety trial was conducted, which is often dozens of miles away or more. These NOAA data often include soil moisture data, but it is not always present or complete. Ideally, we would harvest data from a source that allows us to look up weather data by geolocation and would also include soil moisture data for that precise location, so we note this gap in publicly available data as well as this opportunity for someone to create such a tool.

Our results are promising enough to warrant further development of PYCS, which is still in its infancy, and the results facilitate a better understanding of how PYCS might perform and be useful in the real world. For future work, we are continuing to develop PYCS, which will provide predictions for future yields based on the past, thus providing actual crop forecasting. We have reported our best results, which come from combinations of our most accurate models with our best synthesized datasets. Our pipeline features storable models and datasets for reusability, so the live PYCS tool can estimate unseen locations using these top performers once they are created.

We are also working on gradually expanding our dataset to cover the entire contiguous United States (CONUS), or at least those where alfalfa variety trials are performed, especially where the crops are rainfed and not exclusively "Roundup Ready", as explained below. Using the Daymet [30] tool has vastly increased our efficiency with curating data and allows us to obtain weather data for any precise location where we have access to variety trials, so we expect to accelerate the growth of our datasets going forward. Since we are trying to contribute to mitigating the effects of climate change, we are more interested in rainfed crops, as rain is a climate factor, not to mention that artificial irrigation might cloud the significance of rain as a feature.

Glyphosate herbicide, known under the trade name Roundup, is a very common tool for battling pests in many crops, including alfalfa. Glyphosate kills virtually any plant it contacts, except those whose seeds are genetically engineered to be immune to glyphosate, and such modified seeds are called "Roundup-Ready". Crops grown from Roundup-Ready seeds are tolerant to glyphosate, meaning that fields where these crops are grown can be treated with the chemical, which will kill surrounding plants and weeds but not the crop itself [32]. However, in recent years, the World Health Organization determined that glyphosate is a "probable carcinogen" that may negatively affect the health of humans, animals, and our ecosystem, and that more research is necessary to determine the health consequences of such a ubiquitous product [33]. While research connecting glyphosate and related pesticides to cancer is not proven, our team generally subscribes to a non-invasive approach to precision farming, adopting the perspective that we can more effectively mitigate climate change by adapting to our environment than manipulating it. Noting this recent controversy surrounding Monsanto's Roundup herbicide, its active chemical glyphosate, and its potential carcinogenic qualities and negative environmental impacts, our work considers this product to be potentially at odds with efforts to combat climate change in a non-invasive way; therefore, our work focuses on crops that do not use glyphosate.

The next phase of our team's research will focus on true forecasting of future alfalfa yields based on time series data and models. We are exploring known time series models like autoregressive integrated moving average (ARIMA) and vectorized autoregression (VAR), but we are also investigating a pretraining technique where we retune models already trained according to a sliding time window. We are happy to provide our data and code publicly at https://www.github.com/thejonathanvancetrance/SNLT (accessed on 29 November 2022) so that others may reproduce and extend this work.

6. Conclusions

The main contributions of this work are: we show that training classifiers with synthesized data often produces more accurate classifiers than training with real data, when the synthesized dataset is larger than the real one, and the target dataset is very small; we propose the novel SNLT pipeline, which trains more accurate models than local and non-local training; we propose the PYCS application, which offers a graphical user interface to perform crop yield estimation; we provide a publicly available, growing dataset of aggregated alfalfa yield and weather data. Furthermore, this work sets the stage for true forecasting of future yields, which we plan to include in future iterations of PYCS. SNLT produced the highest accuracy of 85.7% with a decision tree classifier, scored above 70% accuracy with an XGBoost classifier, and beat non-local and local training by over 40%. These results indicate that the SNLT pipeline is useful, and that it is a good feature to include in the PYCS application. When a farmer using PYCS wants to estimate alfalfa yields for a location where data is too scarce to train ML models, and when non-local training data from a nearby location is more abundant, yet too scarce to train strong models, we can synthesize larger datasets, which greatly improve the classification accuracy of our ML models. Furthermore, as researchers, we can expand the data and train and store more models as we use SNLT to conduct further experiments on new locations. Admittedly, we would like to see higher accuracies on larger target datasets with greater consistency, but we may have merely gotten our foot in the door with demonstrating this technique's usefulness. XGBoost has consistently produced encouraging results and trained very fast on very large datasets, so we plan to include it and similar models in future work, moving beyond the traditional models in our early work. A deeper study of XGBoost may reveal tuning or modifications we should perform to achieve better results. Most importantly, our results show 70% or greater accuracy in multiple locations on multiple runs, fairly well convincing us that our pipeline is useful and merits further exploration.

Author Contributions: Conceptualization, J.V., K.R., A.M. and F.W.M.; methodology, J.V., K.R., A.M. and F.W.M.; software, J.V.; validation, J.V., K.R., A.M. and F.W.M.; formal analysis, J.V., K.R., A.M. and F.W.M.; investigation, J.V., K.R., A.M. and F.W.M.; resources, J.V., K.R., A.M. and F.W.M.; data curation, J.V.; writing—original draft preparation, J.V.; writing—review and editing, J.V., K.R., A.M. and F.W.M.; visualization, J.V.; supervision, K.R., A.M. and F.W.M.; project administration, J.V., K.R., A.M. and F.W.M.; funding acquisition, J.V., K.R., A.M. and F.W.M. All authors have read and agreed to the published version of the manuscript.

Funding: This research received no external funding.

Informed Consent Statement: Not applicable.

Data Availability Statement: Our aggregated climate and alfalfa yield data from KY, GA, and WI, which we used in these experiments, and the Jupyter Notebooks [34] with already-run code and results are available at www.github.com/thejonathanvancetrance/SNLT (accessed on 29 November 2022).

Acknowledgments: Jonathan would like to thank his Ph.D. Committee members Hamid Arabnia and John Miller for their support and advice during this project.

Conflicts of Interest: The authors declare no conflict of interest.

Appendix A

The grid for the hyperparameters of each model is as follows:
Decision Tree

- 'criterion': ['gini'];
- 'max_depth': [5, 10, 25, 50, 100].

Random forest-

- 'n_estimators': [5, 10, 25, 50, 100];
- 'max_depth': [5, 10, 15, 20];
- 'criterion': ['gini'].

K-nearest neighbors

- 'n_neighbors': [2, 5, 10];
- 'weights': ['uniform', 'distance'];
- 'leaf_size': [5, 10, 30, 50].

Support vector classifier

- 'kernel': ['linear', 'poly', 'rbf', 'sigmoid'];
- 'C': [0.1, 1.0, 5.0, 10.0];
- 'gamma': ['scale', 'auto'];
- 'degree': [2, 3, 4, 5].

Neural Network

- 'hidden_layer_sizes': [(3), (5), (10), (3,3), (5,5), (7,7)];
- ''solver': ['sgd', 'adam'];
- 'learning_rate': ['constant', 'invscaling', 'adaptive'];
- 'learning_rate_init': [0.1, 0.01, 0.001].

Logistic Regression—default parameters; XGBoost—default parameters.

References

1. Whitmire, C.D.; Vance, J.M.; Rasheed, H.K.; Missaoui, A.; Rasheed, K.M.; Maier, F.W. Using Machine Learning and Feature Selection for Alfalfa Yield Prediction. *AI* **2021**, *2*, 71–88. [CrossRef]
2. Hendrycks, D.; Lee, K.; Mazeika, M. Using Pre-training Can Improve Model Robustness and Uncertainty. In Proceedings of the 36th International Conference on Machine Learning, Long Beach, CA, USA, 9–15 June 2019; Chaudhuri, K., Ruslan, S., Eds.; PMLR: Long Beach, CA, USA, 2019; pp. 2712–2721.
3. Makhzani, A.; Shlens, J.; Jaitly, N.; Goodfellow, I.; Frey, B. Adversarial autoencoders. *arXiv* **2015**, arXiv:1511.05644.
4. Goodfellow, I. Nips 2016 tutorial: Generative adversarial networks. *arXiv* **2016**, arXiv:1701.00160.
5. Xu, L.; Skoularidou, M.; Cuesta-Infante, A.; Veeramachaneni, K. Modeling tabular data using conditional gan. *Adv. Neural Inf. Process. Syst.* **2019**, *32*, 7335–7345.
6. PYCS. Available online: www.jonathanvance.online/pycs (accessed on 10 October 2022).
7. Baral, R.; Lollato, R.P.; Bhandari, K.; Min, D. Yield gap analysis of rainfed alfalfa in the United States. *Front. Plant Sci.* **2022**, *13*, 2492. [CrossRef] [PubMed]
8. Matouq, M.; El-Hasan, T.; Al-Bilbisi, H.; Abdelhadi, M.; Hindiyeh, M.; Eslamian, S.; Duheisat, S. The climate change implication on Jordan: A case study using GIS and Artificial Neural Networks for weather forecasting. *J. Taibah Univ. Sci.* **2013**, *7*, 44–55. [CrossRef]
9. Feleke, H.G. Assessing Weather Forecasting Needs of Smallholder Farmers for Climate Change Adaptation in the Central Rift Valley of Ethiopia. *J. Earth Sci. Clim. Chang.* **2015**, *6*, 1–8. [CrossRef]
10. Scher, S.; Messori, G. How Global Warming Changes the Difficulty of Synoptic Weather Forecasting. *Geophys. Res. Lett.* **2019**, *46*, 2931–2939. [CrossRef]
11. Yahya, B.M.; Seker, D.Z. Designing Weather Forecasting Model Using Computational Intelligence Tools. *Appl. Artif. Intell.* **2018**, *33*, 137–151. [CrossRef]
12. Schlenker, W.; Lobell, D.B. Robust negative impacts of climate change on African agriculture. *Environ. Res. Lett.* **2010**, *5*, 014010. [CrossRef]
13. Jeong, J.W.; Kim, S.; Lee, I.K.; So, N.; Ko, H.S. Negative effect of abnormal climate on the fruits productivity-focusing on the special weather report. *Korean J. Agric. For. Meteorol.* **2018**, *20*, 305–312.
14. Dhore, A.; Byakude, A.; Sonar, B.; Waste, M. Weather prediction using the data mining Techniques. *Int. Res. J. Eng. Technol.* **2017**, *4*, 2562–2565.

15. Steve, M.; Johns, T.C.; Easterbrook. Engineering the software for understanding climate change. *Comput. Sci. Eng.* **2009**, *11*, 65–74. [CrossRef]
16. United Nations. Transforming our world: The 2030 agenda for sustainable development. In *Resolution Adopted by the General Assembly*; United Nations: New York, NY, USA, 2015.
17. Vance, J.; Rasheed, K.; Missaoui, A.; Maier, F.; Adkins, C.; Whitmire, C. Comparing Machine Learning Techniques for Alfalfa Biomass Yield Prediction. *arXiv* **2022**, arXiv:2210.11226.
18. Pedregosa, F.; Varoquaux, G.; Gramfort, A.; Michel, V.; Thirion, B.; Grisel, O.; Blondel, M.; Prettenhofer, P.; Weiss, R.; Vanderplas, J.; et al. Scikit-learn: Machine learning in Python. *J. Mach. Learn. Res.* **2011**, *12*, 2825–2830.
19. Myerson, R.B. Nash Equilibrium and the History of Economic Theory. *J. Econ. Lit.* **1999**, *37*, 1067–1082. [CrossRef]
20. Xie, L.; Lin, K.; Wang, S.; Wang, F.; Zhou, J. Differentially private generative adversarial network. *arXiv* **2018**, arXiv:1802.06739.
21. Choi, E.; Biswal, S.; Malin, B.; Duke, J.; Stewart, W.F.; Sun, J. Generating multi-label discrete patient records using generative adversarial networks. In Proceedings of the 2nd Machine Learning for Healthcare Conference, Boston, MA, USA, 18–19 August 2017; Doshi-Velez, F., Fackler, J., Kale, D., Ranganath, R., Wallace, B., Wiens, J., Eds.; PMLR: Boston, MA, USA, 2017; Volume 68, pp. 286–305.
22. Park, N.; Mohammadi, M.; Gorde, K.; Jajodia, S.; Park, H.; Kim, Y. Data synthesis based on generative adversarial networks. *arXiv* **2018**, arXiv:1806.03384. [CrossRef]
23. Freund, Y.; Schapire, R.E. Experiments with a new boosting algorithm. In Proceedings of the 13th International Conference on International Conference on Machine Learning, Bari, Italy, 3 July 1996; ICML: Bari, Italy, 1996; Volume 96, pp. 148–156.
24. Chen, T.; Guestrin, C. Xgboost: A scalable tree boosting system. In Proceedings of the 22nd ACM SIGKDD International Conference on Knowledge Discovery and Data Mining, San Francisco, CA, USA, 13–17 August 2016; pp. 785–794. [CrossRef]
25. UGA Variety Trials. Available online: https://georgiaforages.caes.uga.edu/content/dam/caes-subsite/forages/docs/species/alfalfa-variety-trials-2008-2010.pdf (accessed on 29 November 2022).
26. UK Forage Variety Trials. Available online: https://forages.ca.uky.edu/variety_trials (accessed on 29 November 2022).
27. SDSU Extension Publications Archine. Available online: https://openprairie.sdstate.edu/extension_pubs/4/ (accessed on 11 November 2022).
28. OSU Crop Performance Trials. Available online: https://u.osu.edu/perf/archive/ (accessed on 11 November 2022).
29. Daymet. Available online: https://daymet.ornl.gov/ (accessed on 12 October 2022).
30. Oliphant, T.E. *A Guide to NumPy*; Trelgol Publishing: Wilmington, DE, USA, 2006; Volume 1, p. 85.
31. Hunter, J.D. Matplotlib: A 2D graphics environment. *Comput. Sci. Eng.* **2007**, *9*, 90–95. [CrossRef]
32. Duke, S.O.; Powles, S.B. Glyphosate: A once-in-a-century herbicide. *Pest Manag. Sci. Former. Pestic. Sci.* **2008**, *64*, 319–325. [CrossRef]
33. Cressey, D. Widely used herbicide linked to cancer. *Nature* **2015**, *24*, 1–3. [CrossRef]
34. Kluyver, T.; Ragan-Kelley, B.; Pérez, F.; Granger, B.E.; Bussonnier, M.; Frederic, J.; Kelley, K.; Hamrick, J.B.; Grout, J.; Corlay, S.; et al. Jupyter Notebooks-a publishing format for reproducible computational workflows. In *Positioning and Power in Academic Publishing: Players, Agents and Agendas*; Loizides, F., Schmidt, B., Eds.; IOS Press: Amsterdam, The Netherlands, 2016; pp. 87–90.

AI

MDPI

Article

Leveraging Spectral Neighborhood Information for Corn Yield Prediction with Spatial-Lagged Machine Learning Modeling: Can Neighborhood Information Outperform Vegetation Indices?

Efrain Noa-Yarasca [1,*], Javier M. Osorio Leyton [1], Chad B. Hajda [2], Kabindra Adhikari [2] and Douglas R. Smith [2]

[1] Texas A&M AgriLife Research, Blackland Research and Extension Center, Temple, TX 76502, USA; javier.osorio@ag.tamu.edu

[2] Grassland Soil and Water Research Laboratory, United States Department of Agriculture–Agriculture Research Service, Temple, TX 76502, USA; chad.hajda@usda.gov (C.B.H.); kabindra.adhikari@usda.gov (K.A.); douglas.r.smith@usda.gov (D.R.S.)

* Correspondence: efrain.noa-yarasca@ag.tamu.edu

Abstract: Accurate and reliable crop yield prediction is essential for optimizing agricultural management, resource allocation, and decision-making, while also supporting farmers and stakeholders in adapting to climate change and increasing global demand. This study introduces an innovative approach to crop yield prediction by incorporating spatially lagged spectral data (SLSD) through the spatial-lagged machine learning (SLML) model, an enhanced version of the spatial lag X (SLX) model. The research aims to show that SLSD improves prediction compared to traditional vegetation index (VI)-based methods. Conducted on a 19-hectare cornfield at the ARS Grassland, Soil, and Water Research Laboratory during the 2023 growing season, this study used five-band multispectral image data and 8581 yield measurements ranging from 1.69 to 15.86 Mg/Ha. Four predictor sets were evaluated: Set 1 (spectral bands), Set 2 (spectral bands + neighborhood data), Set 3 (spectral bands + VIs), and Set 4 (spectral bands + top VIs + neighborhood data). These were evaluated using the SLX model and four decision-tree-based SLML models (RF, XGB, ET, GBR), with performance assessed using R^2 and RMSE. Results showed that incorporating spatial neighborhood data (Set 2) outperformed VI-based approaches (Set 3), emphasizing the importance of spatial context. SLML models, particularly XGB, RF, and ET, performed best with 4–8 neighbors, while excessive neighbors slightly reduced accuracy. In Set 3, VIs improved predictions, but a smaller subset (10–15 indices) was sufficient for optimal yield prediction. Set 4 showed slight gains over Sets 2 and 3, with XGB and RF achieving the highest R^2 values. Key predictors included spatially lagged spectral bands (e.g., Green_lag, NIR_lag, RedEdge_lag) and VIs (e.g., CREI, GCI, NCPI, ARI, CCCI), highlighting the value of integrating neighborhood data for improved corn yield prediction. This study underscores the importance of spatial context in corn yield prediction and lays the foundation for future research across diverse agricultural settings, focusing on optimizing neighborhood size, integrating spatial and spectral data, and refining spatial dependencies through localized search algorithms.

Keywords: corn yield prediction; spatial-lagged machine learning model; spectral neighborhood data; vegetation indices (VIs); spatial autocorrelation

1. Introduction

As the primary grain crop in the United States, corn production significantly impacts both local economies and global food supply chains [1]. Accurate yield estimation is therefore essential for efficient agricultural management, decision-making, resource allocation, crop insurance, and policy planning [1]. Moreover, reliable yield forecasts are critical for guiding decisions on harvesting, marketing, and resource management [2]. With growing pressures of climate change and rising demand, reliable forecasting models are becoming increasingly essential for guiding farmers and stakeholders in crop management and resource allocation [3]. Enhanced prediction capabilities can further optimize planting practices, improve pest and nutrient management, and increase agricultural productivity [3]. Thus, advancing yield prediction methods is not only a scientific challenge but a critical need to support the production agriculture sector in adapting to environmental changes.

The integration of machine learning (ML) into agricultural practices has transformed traditional yield prediction methodologies. Unlike classical statistical models, which often rely on linear assumptions, ML algorithms can capture complex, non-linear relationships within high-dimensional data through automated transformations and large datasets [4,5]. Studies using these algorithms have used extensive datasets, including agronomic variables, remote sensing spectral data (satellite and UAV imagery), and environmental factors, to model crop yield responses more effectively [6]. In this context, crop yield prediction with remote sensing data typically correlates spectral information from specific spatial coordinates with corresponding yield values.

Previous studies have explored various spectral band transformations into vegetation indices (VIs), ranging from simple to complex mathematical formulations [7,8]. However, this approach, which often relies on VIs derived from reflectance data, has limitations, as it primarily considers localized information and overlooks neighborhood spatial patterns that may influence yield outcomes [9,10]. In response, spatial regression analysis has increasingly highlighted the importance of spatial autocorrelation, where the dependent variable at (x, y) is influenced not only by direct predictors at that point in (x, y) but also by neighboring values ($x \pm i$, $y \pm i$). Here, spatial metrics like Moran's I index provide a quantitative measure of autocorrelation, enhancing our understanding of spatial dependencies among target variable values. This insight highlights the value of incorporating neighborhood information into predictive models, illustrating the critical role of spatial regression in capturing the interplay among spatial data points [11].

Spatial regression models have been widely used across various fields, including mixed forest production [12], urbanization's impact on air quality [13], and milk production [14]. They have also been applied to predict flood frequencies [15], assess agricultural impacts on property values [16], and analyze soil carbon stocks [17]. These diverse applications highlight the versatility of spatial regression in addressing complex, location-dependent phenomena. In agriculture, spatial regression analysis has notably enhanced crop yield predictions by integrating spatial data with agronomic variables. For instance, spatially structured data have improved model performance in assessing weather and climate impacts on yields [18], while accounting for spatial dependencies has yielded more accurate crop yield estimates [19]. Furthermore, spatially varying coefficients have enabled better predictions of temperature effects on yields [20]. Recent innovations include using UAV spectral reflectance to analyze the spatial structure of wheat grain yield and protein content [21], as well as applying Bayesian spatial regression models with satellite imagery to predict soybean yields and evaluate computational efficiency [22]. These studies highlight the growing importance of spatial regression in agricultural modeling.

The typical spatial regression approaches include the spatial autoregressive (SAR), spatial error (SEM), and spatial lag of X (SLX) models, each treating spatial interactions differently [23,24]. The SAR model captures spatial dependence in the dependent variable, assuming that outcomes in one location are influenced by outcomes in neighboring locations. The SEM, on the other hand, accounts for spatial dependence in the error terms, attributing spatial autocorrelation to unobserved factors that vary spatially. The SLX model includes spatially lagged predictors, allowing evaluation of how neighboring explanatory variables influence the dependent variable. These models are often paired with spatial weight matrices to define spatial relationships. While SAR and SEM models are effective, they are less practical for predictive tasks where the dependent variable is unknown, such as in predictions involving new or future data. In contrast, the SLX model, which does not require spatially lagged dependent variables, is better suited for prediction, making it ideal for forecasting at new locations [24]. By explicitly incorporating spatial dependencies, these models improve prediction accuracy and offer valuable insights across fields like agriculture, environmental science, and urban planning.

Despite the growing interest in incorporating spatial information into agricultural models, there remains a significant gap in research focused on the integration of neighboring spectral data within machine learning frameworks for crop yield prediction. Existing studies predominantly concentrate on optimizing VIs [10,25], with little attention given to spatially integrated methods that leverage the relationships between neighboring spectral values. This study aims to address this gap by introducing a novel approach that integrates spatial dependencies directly into machine learning models. Specifically, we propose the spatially lagged machine learning (SLML) approach, which combines the power of spatially lagged predictors with decision-tree-based algorithms to enhance the predictive accuracy of corn yield models.

The SLML approach involves incorporating spatially lagged data—specifically, the spectral values at neighboring coordinates—into machine learning models such as random forest (RF), extreme gradient boosting (XGB), extra trees regressor (ET), and gradient boosting regression (GBR). These decision-tree-based ensemble models were chosen for their ability to capture complex, non-linear relationships within high-dimensional data and their strong performance with moderate-sized datasets, which are common in agricultural studies [4,26]. In addition, decision trees are highly interpretable, offering direct insights into feature importance, which is crucial for evaluating how spatially lagged spectral data improves yield predictions. This interpretability is a key advantage over more complex models, such as neural networks or support vector machines, which require additional techniques for interpretation and often demand large datasets—typically in the tens of thousands to millions of data points—to perform effectively due to their complex architectures [27–29].

The spatial lag X (SLX) model and the proposed SLML model both incorporate spatial dependencies, but they differ significantly in their ability to capture complex spatial patterns. The SLX model, a linear approach, explicitly includes spatially lagged independent variables but remains limited by its linear assumptions, which may overlook intricate spatial interactions common in agricultural systems. In contrast, SLML models use machine learning algorithms to capture non-linear relationships and model complex spatial dependencies that the linear SLX model cannot detect. This non-linearity allows SLML models to identify and process spatial patterns more effectively, which is especially crucial in agricultural settings, where spatial interactions can be highly complex and variable. The flexibility of SLML models in handling these complex spatial relationships makes them particularly valuable for applications like crop yield prediction, where the spatial structure of the data plays a critical role in model accuracy.

By including neighborhood spectral data, this approach aims to capture broader spatial interactions that are often overlooked in traditional models relying solely on point-specific data. The research hypothesizes that these extended spatial relationships will improve predictive performance by better accounting for the spatial structure and autocorrelation inherent in agricultural landscapes. To assess the effectiveness of the SLML approach, this study compares its performance against conventional machine learning models that do not include spatially lagged predictors, as well as models including raw spectral data + VIs, and SLX models. By evaluating these approaches, this study seeks to determine whether incorporating spatial neighborhood information significantly outperforms standard modeling techniques in corn yield prediction. Ultimately, this research aims to contribute to the field of agricultural prediction by demonstrating the potential benefits of integrating spatially lagged data within machine learning frameworks, offering new avenues for more accurate and reliable yield forecasts.

2. Methodology

2.1. Study Area

The research was conducted at the Texas A&M AgriLife Blackland Research and Extension Center in Temple, Bell County, Texas (Figure 1). The study site encompassed a 19.03 ha rainfed field, collaboratively managed by Texas A&M AgriLife's Blackland Research and Extension Center and the USDA-ARS Grassland, Soil, and Water Research Laboratory in Temple, Texas (31.059444° N, 97.345833° W; 192 m elevation). Corn was sown on 28 February 2023 and fertilized with 448 kg/ha of a 32.5N–16.2P–0K blend, which was broadcasted prior to planting. No-till practices were employed, with a planting density of 4.8 plants/m^2. The corn was harvested on 14 August 2023. The topography features a relatively flat landscape, with an 8 m difference in elevation between its highest and lowest points. The predominant soil types are Houston Black clay (fine, thermic, smectitic, Udic Haplusterts) and Austin clay (fine-silty, thermic, carbonatic, Udorthentic Haplustolls), both characteristic of the Blackland Prairies region [30]. These soils are known for their high fertility and strong moisture retention capacity, though they are also prone to compaction. During the growing season, the field received a total of 389.4 mm of precipitation.

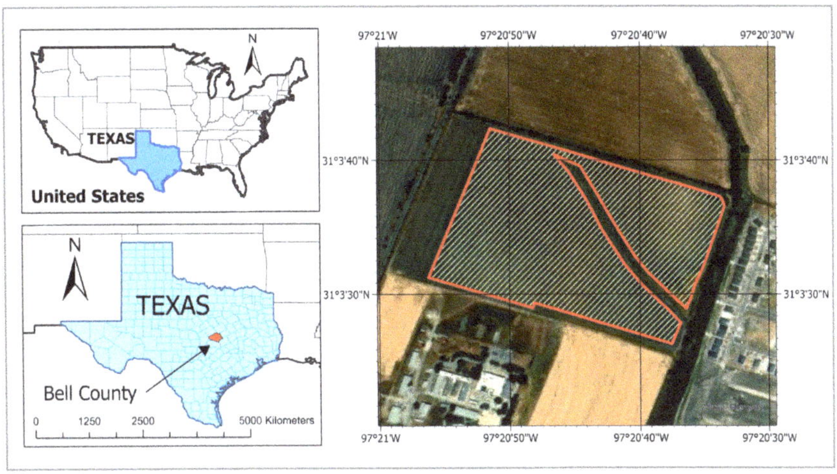

Figure 1. Study area location.

Harvesting was conducted using a John Deere 9510 combine (Deere and Company, Moline, IL, USA) fitted with an AgLeader corn yield monitor (AgLeader Technology Inc., Ames, IA, USA) to record yield data. The yield-monitor recorded data for six-row harvests (16.9 m^2 per section) and logged GPS coordinates for the centroid of each harvested area using an integrated GPS recorder. To ensure measurement accuracy, pre-harvest and in-field calibrations were conducted for load size, weighing, and GPS data. The raw yield data were then processed using Yield Editor software Version 2.0.7 to correct for factors such as speed and pass delays, overlaps, moisture variability, and outliers [31]. The refined yield data, paired with their corresponding GPS coordinates, were used as ground truth for model training and validation.

2.2. Data Collection and Processing

The corn imagery for this study was collected on 22 May, corresponding to Day 83 and the tasseling/silking (VT/R1) stage. A WingtraOne GenII fixed-wing UAV (Wingtra, Zurich, Switzerland) equipped with a MicaSense RedEdge-P multispectral sensor (AgEagle Aerial Systems, Seattle, WA, USA) was used for image acquisition. The sensor captured five spectral bands: red (R), green (G), blue (B), red-edge (RE), and near-infrared (NIR) (Figure 2a), each with a resolution of 1456 × 1088 pixels (1.6 MP), enabling detailed monitoring of crop health. The UAV flew at an altitude of 60 m with 75% overlap for both forward and sideways directions, ensuring high-quality data for analysis. The flight was conducted autonomously using pre-defined mission routes, with the flight operations being controlled through the WingtraHub 1.0 software (Wingtra, Zurich, Switzerland).

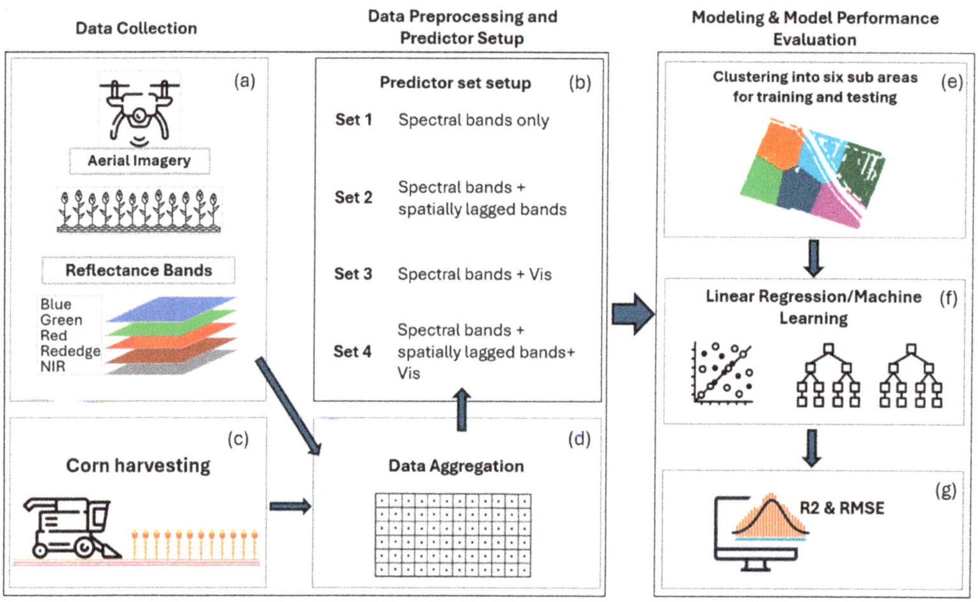

Figure 2. Graphical abstract of the methodology: (**a**) Crop imagery collection, (**b**) Predictor set setup, (**c**) Corn yield data collection, (**d**) Spectral and corn yield data aggregation, (**e**) Spatial clustering of data into six subareas for training and testing, (**f**) Modeling, and (**g**) Model performance evaluation.

For accurate georeferencing, the UAV utilized its built-in GPS system for navigation, nadir image acquisition, and recording coordinates of individual images. The images were processed and orthomosaiced using Pix4Dmapper v4.8.4 (Prilly, Switzerland), resulting in multispectral rasters with 6 cm resolution for each spectral band.

2.2.1. Corn Yield Data Collection Processing

Corn yield data were collected using a John Deere 9510 harvester equipped with an AgLeader sensor (Figure 2c), aggregating yield values across six rows and a length of approximately 3.6 m, covering an area of about 21.6 square meters per measurement. These yield values were mapped in ArcGIS, where influence polygons were generated for each measurement. Irregular polygons and those at the field edges were excluded to improve data accuracy, resulting in 8581 yield values, each precisely linked to its corresponding influence polygon as a shapefile (Figure 2d). Corn yield varied between 1.69 and 15.86 Mg/ha (Figure 3), with a mean yield of 10.19 Mg/ha, which exceeded the Texas state average of 6.62 Mg/ha for the 2022/2023 growing season [32,33]. The interquartile range (Q1: 9.46, Q3: 10.95) was 1.49 Mg/Ha, covering 50% of the values. A coefficient of variation of 12.11% indicates consistent yields with moderate variability, enabling reliable estimation.

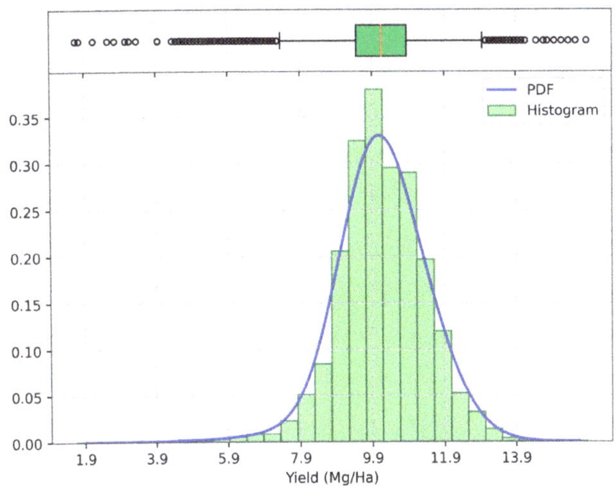

Figure 3. Frequency distribution of corn yield.

The box plot displays Q1 (25th percentile), the median (Q2), and Q3 (75th percentile), illustrating data distribution. Whiskers extend beyond the interquartile range (IQR = Q3 − Q1), with lower and upper limits set at Q1 − 1.5 × IQR and Q3 + 1.5 × IQR, respectively. Data points beyond these limits were not excluded or classified as outliers, as variability in agricultural yield often reflects real environmental factors such as soil heterogeneity, water availability (particularly in rain-fed systems), and topography. Removing such values could lead to information loss and bias. Potential biases from events like pest outbreaks were not explicitly identified or processed, as the yield data showed no significant clustering or unusual patterns. Additionally, the machine learning models used—RF, XGB, ET, and GBR—are robust to extreme values due to their non-parametric and ensemble structures, mitigating the influence of outliers more effectively than linear regression. Data normalization was not required, as decision-tree-based ML models split data based on feature thresholds rather than computing distances or gradients, making them inherently robust to varying data magnitudes.

2.2.2. Imagery Processing

To correlate UAV-derived image data with corn yield data, multispectral images were segmented and aggregated using the corn yield polygon shapefile, producing 8581 polygons with spectral band information. Pixels representing bare ground were excluded, retaining only those corresponding to corn vegetation. Using the zonal statistics function in ArcGIS, mean reflectance values for each band (B, G, R, RE, and NIR) were calculated within each polygon. This process generated a dataset of average reflectance values aligned with corresponding corn yield values, enabling a consistent analysis of the relationship between spectral reflectance and crop yield. To ensure representativeness, the Shapiro–Wilk test showed that over 98% of the spectral values in the 8581 polygons had p-values > 0.05, indicating normality. This aligns with expectations, as only vegetation pixels were included.

2.3. Spatial Autocorrelation Evaluation

Spatial autocorrelation of corn yield data was assessed using Moran's I, a widely used spatial statistic for evaluating autocorrelation [34]. This method involved cross-product statistics between the target variable and its spatial lag, with the variable expressed as deviations from its mean ($z_i = x_i - \overline{x}$). Moran's I statistic is calculated as:

$$I = \frac{n}{S_o} \frac{\sum_{i=1}^{n} \sum_{j=1}^{n} w_{ij} z_i z_j}{\sum_{i=1}^{n} z_i^2}$$

where w_{ij} represents spatial weights in a matrix computed with $w_{ij} = 1/d_{ij}^2$, n is the total features, and $S_0 = \sum_{i=1}^{n} \sum_{j=1}^{n} w_{ij}$ is the sum of all weights. Global Moran's index values range between -1 and 1, where a value of 1 signifies a high degree of positive spatial autocorrelation, -1 indicates strong negative autocorrelation, and 0 reflects a random spatial distribution. Perfect autocorrelation is theoretically unattainable, but an index above 0.3 or below -0.3 is considered strong autocorrelation [35].

The spatial autocorrelation of corn yield was computed for four neighborhood levels to evaluate the trend with increasing neighbor inclusion. These levels involved the immediate surrounding 8 cells and progressively larger neighborhoods encompassing 24, 48, and 80 cells around each target cell. Figure 4 illustrates this structure, with the target cell at the center and neighborhood levels expanding outward. The first level comprises the eight adjacent cells, while subsequent levels incorporate additional surrounding layers, reaching up to 80 neighboring cells in the fourth level.

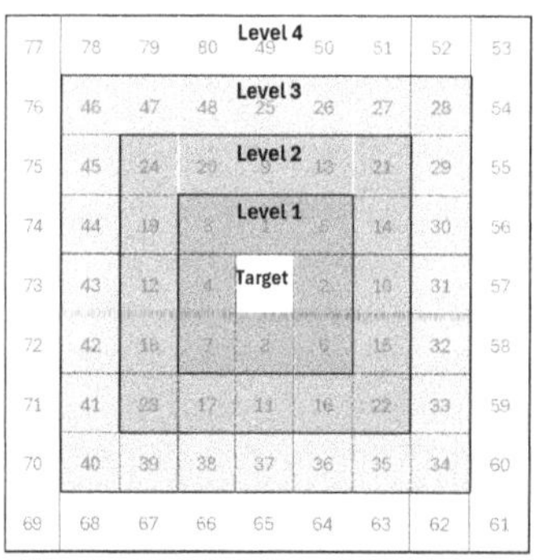

Figure 4. Neighborhood levels for spatial autocorrelation analysis, showing the target cell and expansions to 8, 24, 48, and 80 neighboring cells, each shaded differently for clarity.

2.4. Setting up Predictor Sets for Modeling

The predictors were grouped into four sets, some with subsets (Table 1). Set 1 consisted of the five spectral bands (R, G, B, RE, NIR) as the baseline. Set 2 expanded on Set 1 by adding neighboring information in four variations: Set 2A (4 neighbors), Set 2B (8 neighbors), Set 2C (20 neighbors), and Set 2D (24 neighbors). Figure 5 shows how neighborhood data were considered around the dependent or target corn yield value. Set 3 included the five spectral bands plus eight variations of the most commonly used vegetation indices (VIs) (Table 2). VIs were grouped into subsets through a pre-computed process (pre-process) using a backward stepwise approach. The backward stepwise approach is a method used in regression analysis to improve model performance by systematically removing the least significant predictors. The process begins with a full model and then iteratively eliminates the least important variables based on feature importance criteria. After each removal, the model is re-fitted, and the process is repeated until only the most significant predictors remain.

Table 1. Overview of predictor sets and subsets used in corn yield regression modeling.

Set	Description	Sub-Set	Details
Set 1 (S-1)	Spectral bands only (Baseline)	S-1	Blue (B), green (G), red (R), red-edge (RE), near infrared (NIR)
Set 2 (S-2)	Spectral bands + spatially lagged bands	S-2A	S-1 + 4 neighbors
		S-2B	S-1 + 8 neighbors
		S-2C	S-1 + 20 neighbors
		S-2D	S-1 + 24 neighbors
Set 3 (S-3)	Spectral bands + vegetation indices	S-3A	S-1 + CREI
		S-3B	S-3A + GCI
		S-3C	S-3B + NPCI
		S-3D	S-3C + ARI, CCCI
		S-3E	S-3D + EVI, MCARI, MCCI, NDRE, NG
		S-3F	S-3E + BGI, NGRDI, PPR, PSRI, TVI
		S-3G	S-3F + GNDVI, MTVI2, NDVI, B-NDVI, TCI
		S-3H	S-3G + MSAVI, RDVI, SAVI, TrVI, TSAVI
Set 4 (S-4)	Spectral bands + spatially lagged bands + VIs	S-4	S-2B + 20 VIs

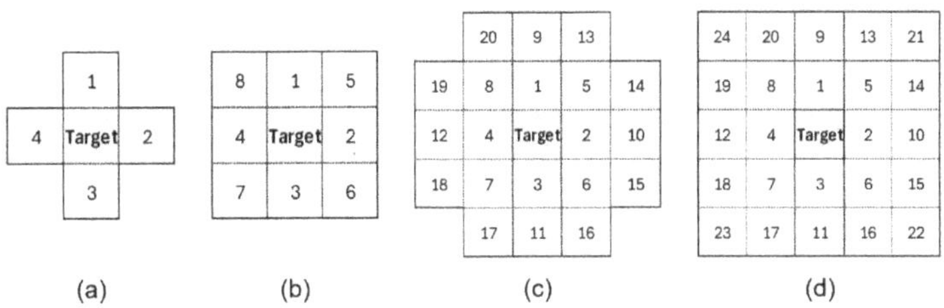

Figure 5. Spatial neighborhood configurations in Set 2 predictors for assessing lagged spectral band values in corn yield regression modeling: (**a**) four neighbors, (**b**) eight neighbors, (**c**) twenty neighbors, and (**d**) twenty-four neighbors.

In this study, the process began with a full set of 25 VIs, and models were initially run with all variables. The five least important VIs (MSAVI, RDVI, SAVI, TrVI, TSAVI) were removed based on feature importance, leaving 20 VIs. This process was repeated, with the five least important VIs removed at each iteration. In the last three iterations, fewer than five predictors were removed to better control model performance, as previous studies have shown that prediction accuracy increases rapidly with the addition of initial variables but diminishes asymptotically as more predictors are added [36]. Throughout the process, the most important VIs for the SLML regression models were retained. For example, in Set 3, the most important VI was CREI (S-3A), followed by GCI (S-3B). Set 4 combined the five spectral bands, the 20 optimal VIs, and data from eight neighboring cells (Figure 2b).

Table 2. Vegetation indices included in the subsets of Set 3 predictors for corn yield regression modeling.

Vegetation Index	Name	Equation	Ref.
CREI	Chlorophyll red-edge index	$\text{NIR}/\text{RE} - 1$	[37]
GCI	Green chlorophyll index	$\text{NIR}/\text{G} - 1$	[37]
NPCI	Normalized pigment chlorophyll index	$(\text{R} - \text{B})/(\text{R} + \text{B})$	[38]
ARI	Anthocyanin reflectance index	$1/\text{G} - 1/\text{R}$	[39]
CCCI	Canopy chlorophyll content index	$\left(\frac{\text{NIR}-\text{RE}}{\text{NIR}+\text{RE}}\right)\Big/\left(\frac{\text{NIR}-\text{R}}{\text{NIR}+\text{R}}\right)$	[40]
EVI	Enhanced vegetation index	$\frac{2.5(\text{NIR}-\text{R})}{\text{NIR}+6\text{R}-7.5\text{B}+1}$	[41]
MCARI	Modified chlorophyll absorption in reflectance index (red)	$((\text{RE} - \text{R}) - 0.2(\text{RE} - \text{G}))\left(\frac{\text{RE}}{\text{R}}\right)$	[42]
MCCI	Modified chlorophyll content index	$\left(\frac{\text{RE}-\text{R}}{\text{RE}+\text{R}}\right) - \left(\frac{\text{RE}-\text{G}}{\text{RE}+\text{G}}\right)$	
NDRE	Normalized difference red-edge index	$(\text{NIR} - \text{RE})/(\text{NIR} + \text{RE})$	[40]
NG	Normalized green index	$\text{G}/(\text{NIR} + \text{R} + \text{G})$	[43]
BGI	Blue green pigment index	B/G	[44]
NGRDI	Normalized green red difference index	$\frac{\text{G}-\text{R}}{\text{G}+\text{R}}$	[45]
PPR	Plant pigment ratio	$(\text{G} - \text{B})/(\text{G} + \text{B})$	[46]
PSRI	Plant senescence reflectance index	$(\text{R} - \text{B})/\text{RE}$	[47]
TVI	Triangular vegetation index	$0.5(120(\text{NIR} - \text{G}) - 200(\text{R} - \text{G}))$	[48]
GNDVI	Green normalized difference vegetation index	$\frac{\text{NIR}-\text{G}}{\text{NIR}+\text{G}}$	[49]
MTVI2	Modified triangular vegetation index (TVI) 2	$\frac{1.5[1.2(\text{NIR}-\text{G})-2.5(\text{R}-\text{G})]}{\sqrt{(2\text{NIR}+1)^2-(6\text{NIR}-5\sqrt{\text{R}})-0.5}}$	[50]
NDVI	Normalized difference vegetation index	$(\text{NIR} - \text{R})/(\text{NIR} + \text{R})$	[51]
B-NDVI	Blue normalized difference vegetation index	$\frac{\text{NIR}-\text{B}}{\text{NIR}+\text{B}}$	[52]
TCI	Triangular chlorophyl index	$1.2(\text{RE} - \text{G}) - 1.5(\text{R} - \text{G})\sqrt{\text{RE}/\text{R}}$	[50]
MSAVI	Modified soil-adjusted vegetation index	$\frac{2\text{NIR}+1-\sqrt{(2\text{NIR}+1)^2-8(\text{NIR}-\text{R})}}{2}$	[53]
RDVI	Renormalized difference vegetation index	$(\text{NIR} - \text{R})/\sqrt{\text{NIR} + \text{R}}$	[54]
SAVI	Soil-adjusted vegetation index	$(1 + 0.5)\left(\frac{\text{NIR}-\text{R}}{\text{NIR}+\text{R}+0.5}\right)$	[41]
TrVI	Transformed vegetation index	$\sqrt{\frac{\text{NIR}-\text{R}}{\text{NIR}+\text{R}} + 0.5}$	[51]
TSAVI	Transformed soil-adjusted vegetation index	$\frac{0.5(\text{NIR}-0.5\text{R}-0.5)}{0.5\text{NIR}+\text{R}+0.75}$	[55]

2.5. Spatial Regression Modeling

Spatially lagged models are advanced extensions of linear regression designed to account for spatial dependencies and interactions among observations in geographical data [23,24]. These models incorporate spatial lag terms to capture the influence of neighboring locations on the dependent variable, improving both the accuracy and interpretability of the analysis. Spatial regression models, such as spatial autoregressive (SAR), spatial error (SEM), and spatial lag of X (SLX), handle spatial interactions differently: SAR captures spatial dependence in the dependent variable, SEM addresses spatial dependence in the error terms, and SLX includes spatially lagged predictors. While SAR and SEM are useful, they are less practical for predicting the dependent variable when it is unknown. In contrast, the SLX model, which does not require spatially lagged dependent variables, is better suited for forecasting.

2.5.1. Spatial Lag X Model (SLX)

The SLX model incorporates spatially lagged independent variables to examine the impact of neighboring observations on each observation's outcome [24]. The model is defined as:

$$y_i = X_i\beta + \sum_{j=1}^{n} W_{ij}X_j\theta + \epsilon_i$$

where y_i is the dependent variable at location i, X_i represents the value of the independent variables at location i, W_{ij} represents the spatial weight between locations i and j, X_j is the value of the independent variable at neighboring location j, β is the vector of coefficients ($n \times 1$), while θ is the coefficient vector of size $k \times 1$ associated with the spatially lagged independent variables. The error term ϵ is a vector of size $n \times 1$, assumed to follow independent and identically distributed (i.i.d.) errors with a mean of zero and constant variance, n is the number of observations (or spatial units) in the dataset, and k is the number of independent variables in the model. W_{ij} determines which observations actually contribute to the sum. If $W_{ij} = 0$ for a give j, the observation j does not influence i (i.e., it is not a "neighbor"). In practice, for immediate or fixed distance, W_{ij} would be nonzero only for those locations within the specified neighborhood or distance.

2.5.2. Machine Learning Models

Random Forest

Random forest is an ensemble learning technique that generates a series of decision trees to improve the accuracy of predictions. Each tree uses a bootstrap sample D_t from the original dataset D (where $D_t \subseteq D$ and $|D_t| = |D|$). At each split, a random subset of features Θ is considered, and the best split θ^* is selected to maximize the split criterion G [56]:

$$\theta^* = \underset{\theta \in \Theta}{\operatorname{argmax}} G(D_t, \theta)$$

For regression analysis, G measures the reduction in variance:

$$G(D_t, \theta) = \mathrm{Var}(D_t) - \left(\frac{|D_{t,L}|}{|D_t|}\mathrm{Var}(D_{t,L}) + \frac{|D_{t,R}|}{|D_t|}\mathrm{Var}(D_{t,R})\right)$$

where $D_{t,L}$ and $D_{t,R}$ are the subsets after the split. The final prediction is the average of all trees: $\hat{y}_{RF} = \frac{1}{T}\sum_{t=1}^{T} f_t^{RF}(x)$, where T is the total number of trees, and $f_t^{RF}(x)$ is the prediction from the tth tree. This ensemble approach reduces overfitting and enhances generalization by averaging the outputs of multiple independently trained trees.

Extreme Gradient Boosting (XGB)

XGB constructs an ensemble of decision trees in a sequential manner, with each subsequent tree aiming to correct the errors made by the previous ones by optimizing a regularized objective function [57]:

$$L^{(t)} = \sum_{i=1}^{n} l\left(y_i, \hat{y}_i^{(t-1)} + f_t(x_i)\right) + \Omega(f_t)$$

where l is the loss function (e.g., mean squared error), y_i is the true label, $\hat{y}_i^{(t-1)}$ is the prediction at iteration $t-1$, and $f_t(x_i)$ is the new tree's prediction. The regularization term $\Omega(f_t)$ penalizes the model complexity to avoid overfitting and is defined as:

$$\Omega(f_t) = \gamma T + \frac{1}{2}\lambda \sum_{j=1}^{T} w_j^2$$

where T is the number of leaves in the tree, w_j are the leaf weights, γ is the penalty for adding a new leaf, and λ is the regularization parameter. The prediction is updated as: $\hat{y}_i^{(t)} = \hat{y}_i^{(t-1)} + f_t(x_i)$. XGB's efficient handling of missing data, parallelized tree construction, and robust regularization make it highly effective for preventing overfitting.

Extremely Randomized Tree Regression (ET)

ET constructs multiple decision trees by introducing randomness in both data sampling and feature splits to enhance model generalization and reduce overfitting [58]. Unlike traditional decision trees, ET uses the entire dataset D for each tree and split nodes using random thresholds. Specifically, at each split, a random subset of features Θ_{random} is considered, and a random split point θ_{random} is chosen from each feature's range $(\theta_{random}\Theta_{random})$. The split criterion G evaluates the chosen split $G(D, \theta_{random})$. For regression analysis, G measures the reduction in variance:

$$G(D, \theta_{random}) = Var(D) - \left(\frac{|D_L|}{|D|} Var(D_L) + \frac{|D_R|}{|D|} Var(D_R) \right)$$

where D_L and D_R are the subsets after the split. The final prediction is the average of all tree predictions: $\hat{y}_{ET} = \frac{1}{T}\sum_{t=1}^{T} f_t^{ET}(x)$, where T is the total number of trees, and $f_t^{ET}(x)$ is the prediction from the tth tree. By averaging the predictions from multiple randomized trees, this approach achieves lower variance and improved generalization compared to individual decision trees.

Gradient Boosting Regressor (GBR)

GBR combines multiple weak decision-tree learners to produce a strong predictive model. It operates iteratively, where each new tree corrects the residual errors of the previous ones by fitting to the negative gradient of the loss function [59,60]. The procedure starts with an initial estimation, typically represented by the mean value of the target variable, which is expressed as

$$F_0(x) = \underset{\gamma}{\operatorname{argmin}} \sum_{i=1}^{N} L(y_i, \gamma)$$

where $L(y_i, \gamma)$ is the loss function (e.g., squared error), and y_i represents the observed values (true value). At each iteration m, a decision tree $h_m(x)$ is trained on the residuals $(r_i^{(m-1)} = y_i - F_{m-1}(x_i))$, and the model is updated as

$$F_m(x) = F_{m-1}(x) + \eta h_m(x)$$

where η is the learning rate controlling the step size. After M iterations, the final prediction is $\hat{y} = F_M(x)$. By progressively reducing errors, GBR achieves high predictive accuracy, with regularization techniques such as limiting tree depth and applying shrinkage (η) to prevent overfitting [61].

2.6. Model Training and Testing

For model training and testing, the corn field area was partitioned into six subareas. Five subareas were used for training while the remaining subarea served as the test data. Like the leave-one-out cross-validation approach, each subarea was designated as the testing set at least once. Therefore, six combinations of training/testing sets were conducted.

2.6.1. Computation Tools

All computations, including the calculation of Moran's I and the development of SLX and SLML models, were performed using Python (version 3.8). The Python environment facilitated comprehensive data analysis and modeling through libraries such as NumPy, pandas, and scikit-learn [62–65]. The code for modeling and calculations is available in the repository at https://github.com/noayarae/Spatial_Lagged_ML, accessed on 11 December 2024.

The models were trained and validated on a Lenovo workstation running Windows 11 Enterprise (23H2, build 22631.4751), equipped with an AMD Ryzen Threadripper PRO 5975WX processor (32 cores, 64 threads, 3.6 GHz), 128GB RAM, and an NVIDIA GeForce RTX 4090 GPU (24GB VRAM, CUDA 12.0, Driver Version 528.24).

2.6.2. Hyperparameter Tunning

To ensure the robustness and reliability of the machine learning models, hyperparameters were optimized using a 6-fold cross-validation approach combined with grid search. In this process, the dataset was partitioned into six subsets, with each subset serving as a validation set while the remaining five were used for training. This iterative procedure was repeated six times, allowing every data point to be used for both training and validation, thereby reducing the risk of overfitting and ensuring that the models' performance was not overly sensitive to specific data divisions. By systematically evaluating all combinations of specified hyperparameter values, this approach identified the optimal configurations that maximized predictive performance while enhancing model generalization. The final hyperparameter settings derived from this process are detailed in Table 3.

Table 3. Tuned hyperparameters, ranges, and values for machine learning models in corn yield regression.

Model	Hyperparameter	Range or List of Values	Pace Step	Tunned Value
RF	n_estimators	100–900	100	500
	max_depth	1–40	5	11
	max_features	1–20	5	11
XGB	n_estimators	[100, 200, 300, 400]		200
	max_depth	1–12	3	7
	learning_rate	0–2	0.05	0.05
	subsample	0.4–0.8	0.1	0.7
	gamma	0.1–0.4	0.1	0.3
	colsample_bytree	[0.7, 0.8, 0.9]		0.9
ET	n_estimators	100–900	100	300
	max_depth	1–40	5	21
	max_features	1–21	4	21
DT	n_estimators	100–900	100	300
	max_depth	1–15	3	12
	max_features	1–21	4	21
GBR	n_estimators	100–900	100	200
	max_depth	1–13	4	5
	learning_rate	0–2	0.05	0.1
	subsample	0.4–0.8	0.1	0.7

2.7. Model Performance Evaluation

Model performance is assessed using two key metrics: root mean square error (RMSE) and coefficient of determination (R^2) (Figure 2g). RMSE indicates the average magnitude of prediction errors, reflecting model accuracy.

$$\mathrm{RMSE} = \sqrt{\frac{1}{n}\sum_{i=1}^{n}(y_i - \hat{y}_i)^2},$$

where y_i are the observed values, $\hat{y}_i$ are the predicted values, and n is the number of observations. A lower RMSE indicates better model performance.

The R^2 value represents the proportion of variation in the dependent variable that can be explained by the independent variables:

$$R^2 = 1 - \frac{\sum_{i=1}^{n}(y_i - \hat{y}_i)^2}{\sum_{i=1}^{n}(y_i - \bar{y}_i)^2},$$

where $\bar{y}$ is the mean of the observed values. An R^2 value closer to 1 indicates a better fit.

3. Results

3.1. Spatial Autocorrelation

The corn yield autocorrelation results show significant positive spatial autocorrelation, with Moran's I values decreasing as the neighborhood size increases (Figure 6). For the immediate neighborhood (8 neighbors), Moran's I was 0.48, dropping to 0.34 at 24 neighbors, 0.31 at 48 neighbors, and 0.28 at 80 neighbors. These results indicate that the predicted corn yield values are strongly correlated within local areas. As the neighborhood radius expands, the spatial dependency diminishes, suggesting that the influence of spatial autocorrelation is most pronounced within the closest cells and gradually reduces over larger distances. The observed spatial autocorrelation supports the consideration of spatial lag effects in predictive modeling, as high Moran's I values in close neighborhoods highlight the potential benefit of integrating nearby yield data to enhance model accuracy.

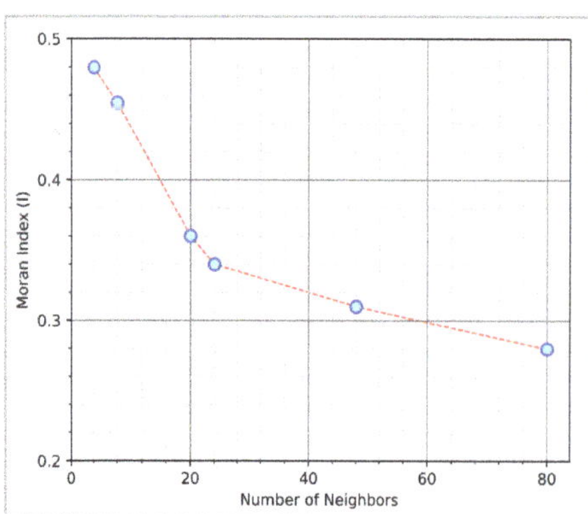

Figure 6. Variation in corn yield autocorrelation (Moran's I) with the number of neighbors, showing a significant drop after 8 neighbors. Blue dots indicate Moran's I, with a dashed trend line.

3.2. Model Performance With and Without Neighbor Information

Figure 7 summarizes the performance of SLX and SLML models (R^2 and RMSE) for predictor sets S-1 (baseline) and S-2 (with neighbor information) (The markers and lines are color-modelled as follows: blue for LR, green for RF, red for XGB, cyan for ET, and magenta for GBR). Incorporating neighborhood data consistently improved model performance compared to the baseline. For example, SLX showed a notable R^2 increase from 0.19 (S-1) to 0.48 with 20 neighbors (S-2C), demonstrating the utility of spatial data in enhancing even simpler models. However, its performance plateaued with the inclusion of additional neighbors.

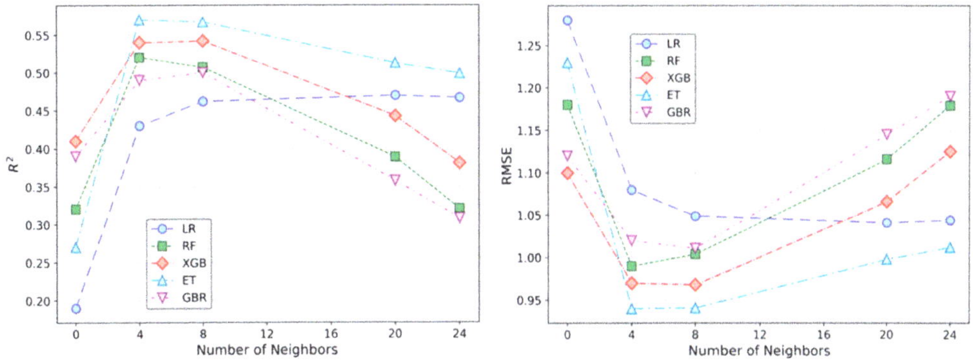

(a) (b)

Figure 7. Performance metrics of regression models for corn yield prediction: (**a**) R^2 and (**b**) RMSE for Set S-1 (baseline) and Set S-2 (incorporating neighborhood information).

Among SLML models, ET achieved the highest R^2 (0.57) with four and eight neighbors, while XGB reached 0.54 with the same neighbor configurations. Random forest (RF) and gradient boosting regression (GBR) performed best with fewer neighbors (four or eight) but showed declining accuracy with larger neighbor sets (S-2C, S-2D), likely due to overfitting or noise. These results highlight the importance of balancing spatial information to optimize model performance, as excessive inclusion of neighbors can diminish accuracy in some models.

The RMSE values, in Figure 7b, further confirm the benefits of incorporating spatially lagged neighbors into predictive modeling. RMSE decreased significantly across all models when moving from the baseline set (S-1) to spatially enhanced sets (S-2). ET had the lowest RMSE (0.94) with four and eight neighbors, while XGB performed best at 0.97 RMSE for the same neighbor counts. SLX showed a steady RMSE decline, reaching 1.03 with 20 neighbors. RF and GBR showed more variable responses, with RMSE improving at fewer neighbors but increasing with excessive ones. Overall, the RMSE results reinforce the R^2 findings, showcasing the value of spatially lagged data in improving corn yield predictions.

Among the spatial-lagged machine learning (SLML) models incorporating neighborhood information, ET achieved the highest predictive performance, followed by XGB, RF, and GBR. This ranking remained consistent across different neighborhood sizes, suggesting that spatially lagged features introduce complexity that models handle differently. ET's randomized split mechanism likely enhances its ability to capture spatial heterogeneity, while XGB benefits more from structured, engineered features such as vegetation indices.

3.3. Model Performance With and Without Vegetation Indices

Figure 8 summarizes the R^2 and RMSE values for SLX and SLML models using predictor sets S-1 (baseline) and S-3 (including VIs) (The markers and lines are color-modelled as follows: blue for LR, green for RF, red for XGB, cyan for ET, and magenta for GBR). These values markedly show the role of VIs in improving model performance. The incremental inclusion of VIs consistently enhanced the coefficient of determination (R^2)

across all SLML models and SLX, with varying levels of improvement depending on the modeling approach. The five top VIs that showed the best performance in the regression models were CREI, GCI, NCPI, ARI, and CCCI. As mentioned above, VIs were selected using a backward stepwise approach.

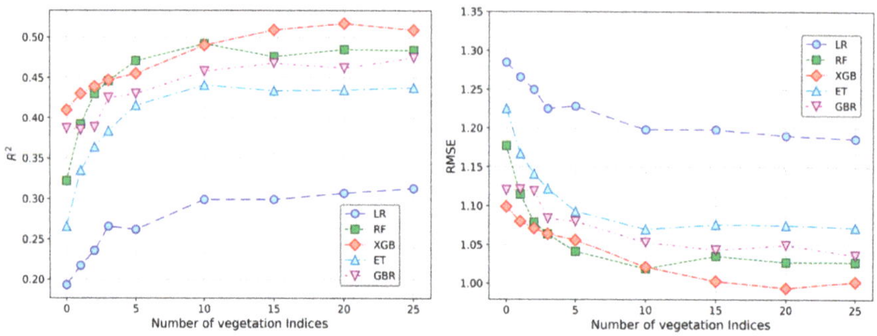

(a) (b)

Figure 8. Performance metrics of regression models for corn yield prediction: (**a**) R^2 and (**b**) RMSE for Set S-1 (baseline) and Set S-3 (incorporating vegetation indices).

Among the SLML models, XGB exhibited the highest R^2 of 0.52 when using 20 VIs, demonstrating a substantial improvement over its baseline (R^2 = 0.41). Similarly, RF achieved its peak performance (R^2 = 0.49) with 10 VIs, a marked enhancement compared to its baseline of 0.32. GBR reached a similar R^2 of 0.48 when all 25 VIs were included but showed no meaningful improvements as fewer indices were removed, indicating a lower sensitivity to feature reduction. SLX demonstrated limited predictive power overall, but its R^2 improved steadily from 0.19 at baseline to a maximum of 0.31 with 25 VIs. In contrast, ET showed moderate performance gains, with R^2 plateauing at 0.44 when 10–25 vegetation indices were included.

After a certain number of VIs were included, improvements in R^2 became marginal across all models. For example, XGB and RF showed negligible changes in R^2 beyond 10–20 VIs, with ET and GBR exhibiting similar trends. This plateauing effect highlights the reduced contribution of additional VIs to predictive performance. These results suggest that while VIs provide valuable spectral information, an optimal subset (e.g., 10–15 indices) is sufficient to capture most of the relevant variability in corn yield. Including more indices beyond this threshold adds unnecessary complexity and computational cost without meaningful performance gains. Thus, optimizing the number of indices is essential to balance model complexity and performance.

The RMSE results in Figure 8b corroborate the trends observed in the R^2 values, highlighting the impact of vegetation indices (VIs) on model performance. Across all regression models, including VIs (S-3) consistently reduces RMSE compared to baseline models (Set S-1), which use only spectral bands (B, G, R, RE, NIR). For example, the XGB model shows a notable RMSE reduction from 1.10 in S-1 to 1.00 in S-3 with 25 VIs. However, beyond 15–20 VIs, further RMSE reductions are negligible, indicating diminishing returns in predictive accuracy as less critical indices are added.

When evaluating various vegetation indices in ML models (without neighborhood information), XGB outperformed RF, GBR, and ET, consistently with more than 10 indices. With 10 or fewer indices, performance varied among XGB, RF, and GBR, while ET consistently lagged. This performance could be attributed to how each model processes input features. XGB, RF, and GBR capture complex relationships within structured data, benefiting from multiple indices. XGB excels with boosting and engineered features, but its performance becomes more variable with fewer indices. ET, relying on randomized splits instead of boosting or feature interactions, struggles to extract meaningful patterns, resulting in lower performance. These results highlight that XGB performs best with vegetation indices, while ET is more effective with spatially lagged spectral data.

3.4. Model Performance Combining Neighborhood Information and Vegetation Indices

Regression models with Set 4 predictors (five bands, 20 VIs, and eight neighbors) slightly outperformed those with sets S-1, S-2, and S-3. Among the SLML models, XGB achieved the highest R^2 value (0.57), followed closely by random forest (RF, R^2 = 0.56) and extra trees (ET, R^2 = 0.55). Gradient boosting regressor (GBR) showed a moderate R^2 of 0.50, while SLX had the lowest performance in this group with an R^2 of 0.46. Furthermore, it can be highlighted that the SLML models showed superior performance over the LR.

The RMSE values support this ranking, reflecting the model's predictive accuracy. XGB showed the lowest RMSE (0.94), indicating superior precision, with RF (0.95) and ET (0.96) performing similarly. GBR had a slightly higher RMSE (1.01), and SLX again showed the least accurate predictions with an RMSE of 1.05. The consistent superiority of XGB, RF, and ET across both metrics underscores the ability of SLML approaches to effectively capture non-linear relationships and complex interactions within the data, including spectral bands, neighborhood information, and vegetation indices.

When neighborhood information was integrated with vegetation indices, XGB outperformed the other models, followed by RF, ET, and GBR. This ranking reflects how models handle Vis and spatially correlated data. XGB and RF effectively utilize vegetation indices while also capturing spatial dependencies, with XGB refining weak predictions through boosting and RF benefiting from ensemble averaging. Although ET performs best when using neighborhood data, it remains less efficient when incorporating both vegetation indices and spatial information due to its reliance on randomized splits. GBR, relying on sequential learning, struggles with spatial autocorrelation, leading to the lowest performance. These findings highlight the varying abilities of ML models to integrate Vis and spatial information for crop yield prediction.

3.5. Feature Importance When Combining All the Predictors

The feature importance analysis for Set 4 predictors showed that spatially lagged spectral bands consistently ranked among the top predictors across various SLML models (Figure 9). In the RF, ET, and GBR models, all spatially lagged features (Green_lag, NIR_lag, RedEdge_lag, Blue_lag, and Red_lag) were within the top 11 features out of 30. In the RF and GBR models, for example, Green_lag and NIR_lag ranked as the top two predictors, surpassing all VIs. In the ET model, these features ranked third and fourth, respectively, highlighting their significance among the large number of VIs evaluated. Meanwhile, in the XGB model, although ranked lower, Green_lag and NIR_lag still stood out for their high importance. These results emphasize the value of spatially lagged features in capturing spatial variability and improving the corn yield prediction.

Regarding vegetation indices, NDRE, CCCI, and CREI consistently emerged as significant contributors across all models, reinforcing their relevance for yield estimation. In contrast, raw spectral bands such as green, red, and rededge ranked lower, which can be attributed to the fact that VIs synthesize and enhance spectral information, making them more relevant for predicting crop yield. Thus, the information from spectral bands is more effectively utilized through VIs, leading to more accurate yield predictions.

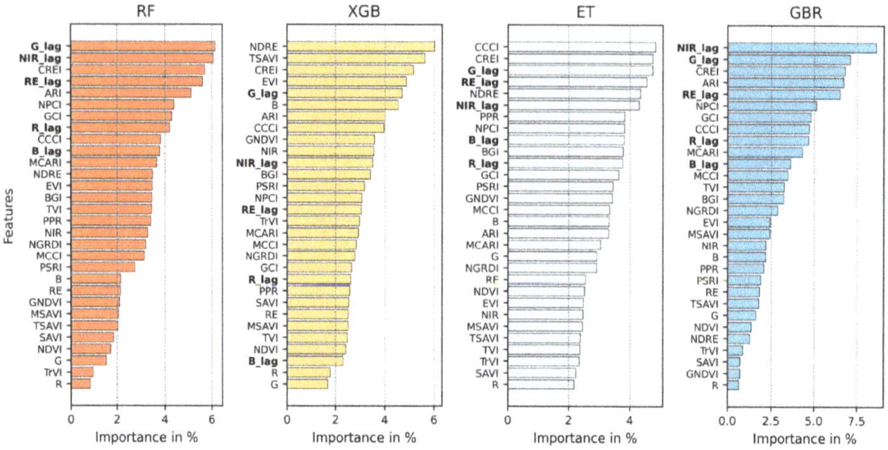

Figure 9. Feature importance of ML modeling in corn yield prediction. The five spatially lagged spectral band features (Green_lag (G-lag), NIR_lag, RedEdge_lag (RE-lag), Blue_lag (B-lag), and Red_lag (R-lag)) are in bold to highlight their significant role among all evaluated predictors.

3.6. Model Performance Comparison Across Predictor Sets

Table 4 summarizes the performance of various corn yield prediction models across the four predictor sets. For Sets 2 (S-2) and Set 3 (S-3), the optimal number of neighbors and vegetation indices, respectively, were considered. The results show that the model performance using the baseline predictors was outperformed by the other sets (S-2, S-3, and S-4), with lower R^2 values compared to the other sets.

Table 4. Best R^2 and RMSE values of regression models for the four predictors sets evaluated in corn yield prediction, with top-performing values in each model highlighted in bold.

Model	Coefficient of Determination (R^2)				Root Mean Square Error (RMSE)			
	S-1	Best S-2	Best S-3	S-4	S-1	Best S-2	Best S-3	S-4
LR	0.19	**0.48**	0.31	0.46	1.28	**1.03**	1.19	1.05
RF	0.32	0.52	0.49	**0.56**	1.18	0.99	1.02	**0.95**
XGB	0.41	0.54	0.52	**0.57**	1.10	0.97	0.99	**0.94**
ET	0.27	**0.57**	0.44	0.55	1.23	**0.94**	1.07	0.96
GBR	0.39	**0.50**	0.48	0.50	1.12	**1.01**	1.04	1.01

Results also show that incorporating spatial neighborhood data (S-2) enhances model performance more significantly than including VIs (S-3). The R^2 values using optimal neighborhood information were always higher than using VIs (S-3). For example, the XGB model achieved an R^2 of 0.54 with eight neighbors (S-2B), outperforming its performance when using both spectral bands and VIs (Set S-3G, R^2 = 0.52 with 20 VIs). Similarly, the ET model showed an R^2 of 0.57 with neighborhood data, compared to 0.44 with VIs. A similar trend was observed in the other models (RF, GBR, and LR) when comparing these approaches. The RMSE values further support this trend. For instance, the XGB model achieved an RMSE of 0.97 with eight neighbors (S-2B), compared to 0.99 when 20 VIs were used (S-3). A similar pattern was observed for RF, with an RMSE of 0.99 when using neighborhood data, compared to 1.02 when incorporating VIs. These results highlight that spatial relationships captured through neighborhood information are more effective for accurate yield prediction than the inclusion of additional spectral indices. These results highlight the greater predictive power of spatial data in corn yield prediction.

The performance of regression models using the S-4 predictor set generally demonstrated comparable or slightly superior results relative to other predictor-sets, particularly with RF and XGB. For example, RF and XGB achieved their highest R^2 values with the S-4 set (0.56 and 0.57, respectively), outperforming their results with S-2 and S-3. In contrast, the ET, GBR, and SLX models showed no significant advantage with S-4 over S-2 but maintained competitive performance. These findings highlight the robustness of the S-4 predictors, especially for ensemble methods like RF and XGB, which effectively utilized the combined spectral and spatial data

4. Discussion

4.1. Spatial Autocorrelation of Corn Yield

The findings confirm a strong positive spatial autocorrelation of corn yield, particularly within the immediate neighborhood, as indicated by Moran's I values. This reinforces the idea that yield variability in agricultural fields is significantly influenced by localized spatial factors. The gradual decrease in autocorrelation with increasing neighborhood size further supports the notion that proximity plays a central role in yield patterns, aligning with the first law of geography [66,67].

The results highlight the importance of incorporating spatial lag effects in yield prediction models. Capturing local spatial dependencies significantly improved model performance, as reflected in the enhanced R^2 and RMSE values. Notably, the spatially lagged predictors consistently ranked among the top features in feature importance analyses, highlighting their predictive value. This finding aligns with previous studies emphasizing the relevance of spatial patterns in agricultural predictions [68]. Additionally, the integration of lagged variables, as suggested by [11], proved essential in improving model accuracy.

However, a limitation of this approach is that spatial autocorrelation weakens as neighborhood size increases, suggesting that incorporating too many distant neighbors may not offer additional predictive benefits [11]. This could be due to the diminishing relevance of spatial dependencies over larger distances.

This finding is particularly relevant for precision agriculture, where optimizing the spatial scale of prediction models is crucial for accurate decision-making and resource allocation. By demonstrating the potential of spatially lagged spectral information, our study establishes a strong foundation for advancing predictive modeling across diverse agricultural settings. Recognizing its significance, we identify this as a key area for future research and plan to investigate it further in upcoming studies.

4.2. Neighborhood Information and Model Performance

Incorporating neighborhood information significantly improved corn yield prediction across all regression techniques. The benefits were most pronounced in spatially lagged machine learning (SLML) models, such as ET and XGB, emphasizing the importance of spatial context in understanding crop yield variability. This finding is consistent with research that underscores the role of spatially structured variables in agricultural modeling [69] and the recognition that agricultural production is inherently spatial, necessitating spatial dependence models for accurate predictions [70].

Despite the positive impact of neighborhood data, R^2 did not scale linearly with the number of neighbors. This suggests that spatial dependencies are primarily localized, with diminishing returns beyond a certain spatial extent. Similar patterns have been observed in other spatial prediction studies [71,72], but are less studied in the context of crop yield prediction. This could be attributed to the variability in agricultural practices and environmental factors, which may limit the benefits of extending spatial dependencies beyond a certain threshold [73,74].

Our results contribute to this gap by demonstrating that an optimal balance exists between model complexity and predictive performance. Specifically, models incorporating four to eight neighboring points provided the highest predictive accuracy, while increasing the number of neighbors led to diminishing returns due to potential overfitting or the inclusion of less relevant spatial information. We recognize that the optimal neighborhood size can vary based on factors such as soil type, climate, management practices, and irrigation, which all affect the relationship between crop yield, spectral data, and spatial dependencies. This highlights the importance of optimizing neighborhood parameters, a challenge that remains underexplored in agricultural research [75,76].

Moreover, while this study used a fixed number of neighbors for neighborhood selection, we acknowledge that a more scientific approach to determining the optimal neighborhood size could improve accuracy. However, our findings provide a solid foundation for further research into systematically identifying both the number and type of neighbors. Future work could explore localized optimal neighbor search algorithms, integrated with spatial-lagged machine learning (SLML), to refine spatial dependency estimates, considering additional spatial features such as soil characteristics.

4.3. Vegetation Indices and Model Performance

The inclusion of vegetation indices (VIs) consistently enhanced regression model performance compared to using only spectral bands, reinforcing the importance of VIs in improving corn yield predictions. SLML models, particularly XGB and RF, showed the most substantial improvements, emphasizing the ability of these indices to capture crop variability effectively. These results align with previous research highlighting the predictive strength of spectral indices in agricultural modeling [77,78]. Particularly in this study, indices like CREI, GCI, NCPI, ARI, and CCCI emerged as the most influential.

A key finding was the plateauing effect when more than 10–15 VIs were included, indicating that only a subset of indices is needed to capture most of the relevant variability. The diminishing returns beyond this threshold suggest that additional indices contribute redundant or weakly relevant information, aligning with feature selection principles in machine learning [79]. These findings highlight the need for balancing model complexity and accuracy [76,80].

While adding VIs reduced prediction errors, the marginal improvements with additional indices also highlighted the challenges of computational inefficiency and potential overfitting. Furthermore, SLX, while showing slight gains, remained limited in its ability to model the complex, non-linear relationships inherent in crop yield predictions. These findings reinforce the importance of optimizing feature selection to achieve robust and efficient models.

4.4. Model Performance When Combining All Predictors

The combination of raw spectral bands, spatially lagged bands, and VIs yielded competitive predictive accuracy. However, improvements over predictor sets that included only spatial neighborhood data (S-2) were relatively modest. This finding suggests that while VIs are valuable, spatially lagged data remain the dominant factor in driving model performance. This aligns with studies emphasizing the role of spatial factors in agricultural modeling, challenging the traditional prioritization of VIs as primary predictors [69,70].

Feature importance analysis further reinforced this perspective, with spatially lagged predictors consistently ranking among the top predictors. This finding contributes to the evolving understanding of crop yield prediction by highlighting the interplay between spatial and spectral data. Unlike previous research, which has not studied the combination of neighborhood spectral values and VIs, our study demonstrates their combined effects, providing a more comprehensive framework for yield estimation. However, the modest accuracy gains from integrating these predictors suggest potential redundancies or collinearities, which warrant further investigation.

Additionally, raw spectral bands (B, G, R, RE, NIR) ranked lower in feature importance than VIs, reinforcing the value of synthesized indices in capturing meaningful spectral information [78,81]. These findings align with prior studies demonstrating that VIs outperform raw reflectance in biomass estimation and other crop productivity measures [10,25].

The overall performance of decision-tree-based models showed that no single ML algorithm demonstrated clear superiority; performance varied depending on whether neighborhood information, vegetation indices, or both were included. This suggests that model effectiveness in predicting corn yield is context-dependent, influenced by input type (spatially lagged data vs. vegetation indices) and the number of variables used. ET efficiently handles high-dimensional data and captures complex relationships [58], showing strong performance with spatially lagged data by using neighboring spectral features through randomized feature splits. In contrast, XGB excels with structured data, using boosting to refine weak learners [57], though its sensitivity to spatial correlations may limit its effectiveness when spatial autocorrelation is strong. RF, with its ensemble-based approach, balances predictive performance and robustness, making it effective across different input types but less specialized for spatial or structured data. GBR, relying on sequential learning [59], struggles with spatial autocorrelation, reducing its performance when neighborhood information is included. ET performs better with spatially lagged spectral data, while XGB thrives with engineered features like vegetation indices. These findings underscore the importance of context-dependent model selection. Future studies should assess multiple ML approaches based on dataset characteristics, as no single model is universally superior.

4.5. Implications for Future Research

These findings have several methodological and practical implications for agricultural modeling. First, the results highlight the need for context-specific optimization of spatial parameters, as the optimal neighborhood size and the relative importance of spatial and spectral predictors may vary across crops, regions, and data sources. Understanding these variations can help refine adaptive modeling strategies for different agricultural settings.

Second, the study highlights the value of high-resolution UAV data in capturing fine-scale spatial dependencies, suggesting that future research should explore how these findings scale to coarser resolutions or larger areas. It remains unclear whether the same spatial relationships hold at broader scales, necessitating multi-scale validation efforts.

Third, the observed trade-offs between model complexity and predictive performance emphasize the importance of balancing data richness with computational efficiency. While incorporating additional features generally improves accuracy, diminishing returns and the risk of overfitting must be carefully managed. These insights are particularly relevant for operational applications, where computational efficiency is critical for real-time decision support in precision agriculture.

Finally, this study emphasizes the value of neighborhood information in enhancing corn yield prediction, laying the groundwork for future research across diverse crops and agricultural settings with spatial dependencies. Future studies should explore how specific spectral bands (e.g., blue, green, red, NIR, rededge) and their interactions influence prediction accuracy, considering context-dependent factors such as agronomic conditions, crop types, and environmental stressors. Additionally, our fixed neighborhood size may not be optimal for all contexts. Research should focus on developing methods for determining the optimal neighborhood size through localized neighbor search algorithms integrated with spatial-lagged machine learning (SLML). Incorporating additional spatial features, such as soil characteristics, could further refine these models and improve prediction performance.

5. Conclusions

This study presents an innovative approach to crop yield prediction by integrating spectral neighborhood data using the spatial-lagged machine learning model (SLML), an enhanced version of the spatial lag X model (SLX). The research aims to demonstrate how integrating spatially lagged spectral information enhances predictive accuracy compared to traditional approaches that primarily rely on vegetation indices (VIs). The study hypothesizes that integrating neighborhood spectral data significantly enhances predictive accuracy, providing novel insights into the role of spatial dependencies in agricultural yield prediction.

This study was conducted on a 19-hectare cornfield at the ARS Grassland, Soil, and Water Research Laboratory during the 2023 growing season. Multispectral imagery was collected using a WintraOne UAV with a MicaSense RedEdge-P sensor, capturing five spectral bands (R, G, B, RE, NIR) at 6 cm resolution during the VT/R1 stage. Yield data covered 21.6 m^2 per measurement, resulting in 8581 yield values ranging from 1.69 to 15.86 Mg/Ha (M = 10.19 Mg/Ha). SLML models, using decision-tree-based techniques (RF, XGB, ET, and GBR) in addition to SLX, were applied to predict corn yield, incorporating spatially lagged predictors (neighborhood data). Four predictor sets were evaluated: Set 1 (spectral bands—baseline), Set 2 (spectral bands + neighborhood data), Set 3 (spectral bands + VIs), and Set 4 (spectral bands + top VIs + neighborhood data). Model performance was assessed using R^2 and RMSE.

The main finding revealed that the introduced approach (incorporating neighborhood data—Set 2), consistently outperformed the traditional approach using VIs (Set 3) in predicting corn yield. This highlights the superior value of spatial context, where the proximity of neighboring cells provides more relevant information about yield variability than the spectral reflectance captured by VIs. Additionally, this study demonstrated the significant role of spatial autocorrelation in corn yield data, with the strongest correlations observed in immediate neighborhoods (eight neighbors). Incorporating spatial neighborhood data consistently improved model performance, with SLML models such as XGB, RF, and ET performing best with four to eight neighbors. However, excessive inclusion of neighbors led to diminishing returns, underscoring the importance of optimizing neighbor selection.

Evaluation in Set 3 showed that while VIs are effective for assessing crop health and vigor; however, their predictive power for yield plateaus once an optimal set of indices is reached. In this regard, VIs enhanced model performance, particularly with XGB, although a smaller subset (10–15 indices) proved sufficient for optimal yield prediction.

Evaluation in Set 4, combination of spatial and spectral data showed slight performance improvements, with XGB and RF achieving the highest R^2 values, emphasizing the value of integrating both data types for enhanced predictive accuracy. In this case, combining spatial and spectral data, spatially lagged spectral bands (e.g., Green_lag, NIR_lag, RedEdge_lag) were ranked predictors, highlighting the importance of including neighborhood information into the regression model for corn yield prediction. Vegetation indices such as CREI, GCI, NCPI, ARI, and CCCI were also key predictors, competing with spatially lagged data and consistently outperforming raw spectral data alone.

In addition, the performance of decision-tree-based ML models in predicting corn yield varied depending on the inclusion of neighborhood information, vegetation indices, or both. No single model emerged as superior, with effectiveness being context dependent. ET excelled with spatially lagged data, XGB performed best with structured data, RF balanced performance across data types, and GBR struggled with spatial autocorrelation. These findings highlight the need for context-driven model selection and suggest future studies should evaluate multiple ML approaches based on dataset characteristics.

This study emphasizes the significance of spatial context and neighborhood information in improving corn yield prediction. It highlights the need for optimizing spatial parameters, feature selection, and neighborhood size to enhance model accuracy. Future research should explore the scalability of these methods to larger areas and coarser resolutions, while investigating advanced approaches for integrating spatial and spectral data. Additionally, the influence of specific spectral bands and their interactions, along with agronomic and environmental factors, should be further examined. Incorporating spatial features such as soil characteristics and developing localized search algorithms for optimal neighborhood size could further refine predictive models and improve their robustness across diverse agricultural settings.

Author Contributions: Conceptualization, J.M.O.L. and E.N.-Y.; methodology, J.M.O.L. and E.N.-Y.; software, E.N.-Y.; validation, E.N.-Y.; formal analysis, J.M.O.L. and E.N.-Y.; investigation, E.N.-Y.; resources, J.M.O.L.; field work and data collection, K.A. and C.B.H.; writing—original draft preparation, E.N.-Y.; writing—review and editing, E.N.-Y., J.M.O.L., C.B.H., K.A. and D.R.S.; supervision, J.M.O.L. All authors have read and agreed to the published version of the manuscript.

Funding: This project was supported through the U.S. Department of Agriculture's Conservation Effects Assessment Project (CEAP), a multi-agency effort led by the Natural Resources Conservation Service (NRCS) to quantify the effects of voluntary conservation and strengthen data-driven management decisions across the nation's private lands, under Texas A&M AgriLife Cooperative agreement number NR213A750023C012.

Institutional Review Board Statement: Not applicable.

Informed Consent Statement: Not applicable.

Data Availability Statement: The data and code used in this study are available upon request from the corresponding author. Given that the study involves large volumes of aerial images and derived rasters, we are currently seeking a repository that can accommodate high-volume datasets.

Acknowledgments: The authors acknowledge the help of technical and non-technical persons involved in establishment and data collection of this study. This project was supported through the U.S. Department of Agriculture's Conservation Effects Assessment Project (CEAP), a multi-agency effort led by the Natural Resources Conservation Service (NRCS) to quantify the effects of voluntary conservation and strengthen data-driven management decisions across the nation's private lands, under Texas A&M AgriLife Cooperative agreement number NR213A750023C012. We would also like to express our sincere gratitude to ARS Grassland Soil and Water Research Laboratory for their support in conducting the fieldwork and maintaining the crop throughout the study, ensuring the quality and accuracy of the data collected. The authors acknowledge that the geospatial corrections for images using post-processing kinematics (PPK) and the creation of an orthomosaic of multispectral rasters were processed by Sayantan Sarkar at Texas A&M AgriLife Blackland Research and Extension Center, Temple, TX.

Conflicts of Interest: The authors declare no conflicts of interest.

References

1. Filippi, P.; Jones, E.J.; Wimalathunge, N.S.; Somarathna, P.D.S.N.; Pozza, L.E.; Ugbaje, S.U.; Jephcott, T.G.; Paterson, S.E.; Whelan, B.M.; Bishop, T.F.A. An approach to forecast grain crop yield using multi-layered, multi-farm data sets and machine learning. *Precis. Agric.* **2019**, *20*, 1015–1029. [CrossRef]
2. Lobell, D.B.; Roberts, M.J.; Schlenker, W.; Braun, N.; Little, B.B.; Rejesus, R.M.; Hammer, G.L. Greater Sensitivity to Drought Accompanies Maize Yield Increase in the U.S. Midwest. *Science* **2014**, *344*, 516–519. [CrossRef] [PubMed]

3. Hatfield, J.L.; Prueger, J.H. Temperature extremes: Effect on plant growth and development. *Weather. Clim. Extremes* **2015**, *10*, 4–10. [CrossRef]
4. Zhou, H.; Yang, J.; Lou, W.; Sheng, L.; Li, D.; Hu, H. Improving grain yield prediction through fusion of multi-temporal spectral features and agronomic trait parameters derived from UAV imagery. *Front. Plant Sci.* **2023**, *14*, 1217448. [CrossRef]
5. Noa-Yarasca, E.; Leyton, J.M.O.; Angerer, J. Biomass Time Series Forecasting Using Deep Learning Techniques. Is the Sophisticated Model Superior? In *Biometry and Statistical Computing*; ASA, CSSA, SSSA International Annual Meeting: St. Louis, MO, USA; Available online: https://scisoc.confex.com/scisoc/2023am/meetingapp.cgi/Paper/151648 (accessed on 10 April 2024).
6. Hunt, E.R., Jr.; Hively, W.D.; Fujikawa, S.J.; Linden, D.S.; Daughtry, C.S.T.; McCarty, G.W. Acquisition of NIR-Green-Blue Digital Photographs from Unmanned Aircraft for Crop Monitoring. *Remote Sens.* **2010**, *2*, 290–305. [CrossRef]
7. Cicek, H.; Sunohara, M.; Wilkes, G.; McNairn, H.; Pick, F.; Topp, E.; Lapen, D. Using vegetation indices from satellite remote sensing to assess corn and soybean response to controlled tile drainage. *Agric. Water Manag.* **2010**, *98*, 261–270. [CrossRef]
8. Killeen, P.; Kiringa, I.; Yeap, T.; Branco, P. Corn Grain Yield Prediction Using UAV-Based High Spatiotemporal Resolution Imagery, Machine Learning, and Spatial Cross-Validation. *Remote Sens.* **2024**, *16*, 683. [CrossRef]
9. Moeckel, T.; Dayananda, S.; Nidamanuri, R.R.; Nautiyal, S.; Hanumaiah, N.; Buerkert, A.; Wachendorf, M. Estimation of Vegetable Crop Parameter by Multi-temporal UAV-Borne Images. *Remote Sens.* **2018**, *10*, 805. [CrossRef]
10. Panda, S.S.; Ames, D.P.; Panigrahi, S. Application of Vegetation Indices for Agricultural Crop Yield Prediction Using Neural Network Techniques. *Remote Sens.* **2010**, *2*, 673–696. [CrossRef]
11. Lesage, J.; Pace, R.K. *Introduction to Spatial Econometrics*, 1st ed.; CRC Press: Boca Raton, FL, USA, 2023.
12. Lou, M.; Zhang, H.; Lei, X.; Li, C.; Zang, H. Spatial Autoregressive Models for Stand Top and Stand Mean Height Relationship in Mixed *Quercus mongolica* Broadleaved Natural Stands of Northeast China. *Forests* **2016**, *7*, 43. [CrossRef]
13. Fang, C.; Liu, H.; Li, G.; Sun, D.; Miao, Z. Estimating the Impact of Urbanization on Air Quality in China Using Spatial Regression Models. *Sustainability* **2015**, *7*, 15570–15592. [CrossRef]
14. Ponciano, P.F.; Scalon, J.D. Análise espacial da produção leiteira usando um modelo autoregressivo condicional. *Semin. Cienc. Agrar.* **2010**, *31*, 487–496. [CrossRef]
15. Ahn, K.-H.; Palmer, R. Regional flood frequency analysis using spatial proximity and basin characteristics: Quantile regression vs. parameter regression technique. *J. Hydrol.* **2016**, *540*, 515–526. [CrossRef]
16. Yoo, J.; Ready, R. The impact of agricultural conservation easement on nearby house prices: Incorporating spatial autocorrelation and spatial heterogeneity. *J. For. Econ.* **2016**, *25*, 78–93. [CrossRef]
17. Guo, L.; Zhao, C.; Zhang, H.; Chen, Y.; Linderman, M.; Zhang, Q.; Liu, Y. Comparisons of spatial and non-spatial models for predicting soil carbon content based on visible and near-infrared spectral technology. *Geoderma* **2017**, *285*, 280–292. [CrossRef]
18. Auffhammer, M.; Hsiang, S.; Schlenker, W.; Sobel, A. Using Weather Data and Climate Model Output in Economic Analyses of Climate Change. *Rev. Environ. Econ. Policy* **2013**, *7*, 181–198. [CrossRef]
19. Dell, M.; Jones, B.F.; Olken, B.A. What Do We Learn from the Weather? The New Climate-Economy Literature. *J. Econ. Lit.* **2014**, *52*, 740–798. [CrossRef]
20. Schlenker, W.; Roberts, M.J. Nonlinear temperature effects indicate severe damages to U.S. crop yields under climate change. *Proc. Natl. Acad. Sci. USA* **2009**, *106*, 15594–15598. [CrossRef]
21. Hawinkel, S.; De Meyer, S.; Maere, S. Spatial Regression Models for Field Trials: A Comparative Study and New Ideas. *Front. Plant Sci.* **2022**, *13*, 858711. [CrossRef]
22. Fischer, R.J.; Rekabdarkolaee, H.M.; Joshi, D.R.; Clay, D.E.; Clay, S.A. Soybean prediction using computationally efficient Bayesian spatial regression models and satellite imagery. *Agron. J.* **2024**, *116*, 2841–2849. [CrossRef]
23. Ward, M.; Gleditsch, K. *Spatial Regression Models*; SAGE Publications: Thousand Oaks, CA, USA, 2008. [CrossRef]
24. Rüttenauer, T. Spatial Regression Models: A Systematic Comparison of Different Model Specifications Using Monte Carlo Experiments. *Sociol. Methods Res.* **2022**, *51*, 728–759. [CrossRef]
25. Rehman, T.H.; Lundy, M.E.; Linquist, B.A. Comparative Sensitivity of Vegetation Indices Measured via Proximal and Aerial Sensors for Assessing N Status and Predicting Grain Yield in Rice Cropping Systems. *Remote Sens.* **2022**, *14*, 2770. [CrossRef]
26. Sarkar, S.; Leyton, J.M.O.; Noa-Yarasca, E.; Adhikari, K.; Hajda, C.B.; Smith, D.R. Integrating Remote Sensing and Soil Features for Enhanced Machine Learning-Based Corn Yield Prediction in the Southern US. *Sensors* **2025**, *25*, 543. [CrossRef] [PubMed]
27. Effrosynidis, D.; Sylaios, G.; Arampatzis, A. The Effect of Training Data Size on Disaster Classification from Twitter. *Information* **2024**, *15*, 393. [CrossRef]
28. Awan, F.M.; Saleem, Y.; Minerva, R.; Crespi, N. A Comparative Analysis of Machine/Deep Learning Models for Parking Space Availability Prediction. *Sensors* **2020**, *20*, 322. [CrossRef]
29. Noa-Yarasca, E.; Leyton, J.M.O.; Angerer, J.P. Deep Learning Model Effectiveness in Forecasting Limited-Size Aboveground Vegetation Biomass Time Series: Kenyan Grasslands Case Study. *Agronomy* **2024**, *14*, 349. [CrossRef]
30. Soil Survey Staff. Keys to Soil Taxonomy 11th Edition. Washington, DC, USA. Available online: https://www.nrcs.usda.gov/sites/default/files/2022-09/Keys-to-Soil-Taxonomy.pdf (accessed on 10 December 2024).

31. Adhikari, K.; Smith, D.R.; Hajda, C.; Kharel, T.P. Within-field yield stability and gross margin variations across corn fields and implications for precision conservation. *Precis. Agric.* **2023**, *24*, 1401–1416. [CrossRef]
32. FAO. Faostat: Crops and Livestock Products. Food and Agriculture Organization of the United Nations. Available online: https://www.fao.org/faostat/en/#data/QV (accessed on 19 July 2024).
33. USDA-NASS. Quick Stats. United States Department of Agriculture, National Agricultural Statistics Service. Available online: https://quickstats.nass.usda.gov/ (accessed on 19 July 2024).
34. Fu, W.J.; Jiang, P.K.; Zhou, G.M.; Zhao, K.L. Using Moran's I and GIS to study the spatial pattern of forest litter carbon density in a subtropical region of southeastern China. *Biogeosciences* **2014**, *11*, 2401–2409. [CrossRef]
35. O'sullivan, D.; Unwin, D. *Geographic Information Analysis*, 2nd ed.; John Wiley & Sons, Inc.: Hoboken, NJ, USA, 2010.
36. Noa-Yarasca, E. A Machine Learning Model of Riparian Vegetation Attenuated Stream Temperatures. Oregon State University, Corvallis, OR, USA. Available online: https://ir.library.oregonstate.edu/downloads/0r967b65c#page=137 (accessed on 10 April 2024).
37. Gitelson, A.A.; Viña, A.; Arkebauer, T.J.; Rundquist, D.C.; Keydan, G.; Leavitt, B. Remote estimation of leaf area index and green leaf biomass in maize canopies. *Geophys. Res. Lett.* **2003**, *30*, 1248. [CrossRef]
38. Peñuelas, J.; Gamon, J.; Fredeen, A.; Merino, J.; Field, C. Reflectance indices associated with physiological changes in nitrogen- and water-limited sunflower leaves. *Remote Sens. Environ.* **1994**, *48*, 135–146. [CrossRef]
39. Gitelson, A.A.; Merzlyak, M.N.; Zur, Y.; Stark, R.; Gritz, U. Non-Destructive and Remote Sensing Techniques for Estimation of Vegetation Status. Papers in Natural Resources, No. 273. 2001. Available online: https://digitalcommons.unl.edu/natrespapers/273/ (accessed on 18 November 2024).
40. Barnes, E.M.; Clarke, T.R.; Richards, S.E.; Colaizzi, P.D.; Haberland, J.; Kostrzewski, M.; Moran, M.S. Coincident Detection of Crop Water Stress, Nitrogen Status and Canopy Density Using Ground-Based Multi-spectral Data. In Proceedings of the 5th International Conference on Precision Agriculture, Bloomington, MN, USA, 16–19 July 2000; Robert, P.C., Rust, R.H., Larson, W.E., Eds.
41. Huete, A.; Didan, K.; Miura, T.; Rodriguez, E.P.; Gao, X.; Ferreira, L.G. Overview of the radiometric and biophysical performance of the MODIS vegetation indices. *Remote Sens. Environ.* **2002**, *83*, 195–213. [CrossRef]
42. Daughtry, C.S.T.; Walthall, C.L.; Kim, M.S.; De Colstoun, E.B.; McMurtrey, J.E., III. Estimating Corn Leaf Chlorophyll Concentration from Leaf and Canopy Reflectance. *Remote Sens. Environ.* **2000**, *74*, 229–239. [CrossRef]
43. Marcial-Pablo, M.d.J.; Gonzalez-Sanchez, A.; Jimenez-Jimenez, S.I.; Ontiveros-Capurata, R.E.; Ojeda-Bustamante, W. Estimation of vegetation fraction using RGB and multispectral images from UAV. *Int. J. Remote Sens.* **2019**, *40*, 420–438. [CrossRef]
44. Zarco-Tejada, P.J.; Berjón, A.; López-Lozano, R.; Miller, J.R.; Martín, P.; Cachorro, V.; González, M.R.; De Frutos, A. Assessing vineyard condition with hyperspectral indices: Leaf and canopy reflectance simulation in a row-structured discontinuous canopy. *Remote Sens. Environ.* **2005**, *99*, 271–287. [CrossRef]
45. Tucker, C.J. Red and photographic infrared linear combinations for monitoring vegetation. *Remote Sens. Environ.* **1979**, *8*, 127–150. [CrossRef]
46. Metternicht, G. Vegetation indices derived from high-resolution airborne videography for precision crop management. *Int. J. Remote Sens.* **2003**, *24*, 2855–2877. [CrossRef]
47. Merzlyak, M.N.; Gitelson, A.A.; Chivkunova, O.B.; Rakitin, V.Y. Non-destructive optical detection of pigment changes during leaf senescence and fruit ripening. *Physiol. Plant.* **1999**, *106*, 135–141. [CrossRef]
48. Broge, N.H.; Leblanc, E. Comparing prediction power and stability of broadband and hyperspectral vegetation indices for estimation of green leaf area index and canopy chlorophyll density. *Remote Sens. Environ.* **2001**, *76*, 156–172. [CrossRef]
49. Gitelson, A.A.; Kaufman, Y.J.; Merzlyak, M.N. Use of a green channel in remote sensing of global vegetation from EOS-MODIS. *Remote Sens. Environ.* **1996**, *58*, 289–298. [CrossRef]
50. Haboudane, D.; Tremblay, N.; Miller, J.R.; Vigneault, P. Remote Estimation of Crop Chlorophyll Content Using Spectral Indices Derived from Hyperspectral Data. *IEEE Trans. Geosci. Remote Sens.* **2008**, *46*, 423–437. [CrossRef]
51. Rouse, J.W., Jr.; Haas, R.H.; Schell, J.A.; Deering, D.W. Monitoring Vegetation Systems in the Great Plains with Erts. 1974. Washington, DC, USA. Available online: https://ui.adsabs.harvard.edu/abs/1974NASSP.351.309R/abstract (accessed on 18 November 2024).
52. Yang, C.; Everitt, J.H.; Bradford, J.M.; Murden, D. Airborne Hyperspectral Imagery and Yield Monitor Data for Mapping Cotton Yield Variability. *Precis. Agric.* **2004**, *5*, 445–461. [CrossRef]
53. Qi, J.; Chehbouni, A.; Huete, A.R.; Kerr, Y.H.; Sorooshian, S. A modified soil adjusted vegetation index. *Remote Sens. Environ.* **1994**, *48*, 119–126. [CrossRef]
54. Zhang, L.; Zhang, H.; Niu, Y.; Han, W. Mapping Maize Water Stress Based on UAV Multispectral Remote Sensing. *Remote Sens.* **2019**, *11*, 605. [CrossRef]

55. Baret, F.; Guyot, G.; Major, D. TSAVI: A Vegetation Index Which Minimizes Soil Brightness Effects On LAI And APAR Estimation. In Proceedings of the 12th Canadian Symposium on Remote Sensing Geoscience and Remote Sensing Symposium, Vancouver, BC, Canada, 10–14 July 1989; Institute of Electrical and Electronics Engineers (IEEE): Vancouver, BC, Canada, 1989; pp. 1355–1358. [CrossRef]
56. Breiman, L. Random forests. *Mach. Learn.* **2001**, *45*, 5–32. [CrossRef]
57. Chen, T.; Guestrin, C. "XGBoost". In Proceedings of the 22nd ACM SIGKDD International Conference on Knowledge Discovery and Data Mining, San Francisco, CA, USA, 13–17 August 2016; ACM: New York, NY, USA, 2016; pp. 785–794. [CrossRef]
58. Geurts, P.; Ernst, D.; Wehenkel, L. Extremely randomized trees. *Mach. Learn.* **2006**, *63*, 3–42. [CrossRef]
59. Natekin, A.; Knoll, A. Gradient boosting machines, a tutorial. *Front. Neurorobotics* **2013**, *7*, 21. [CrossRef]
60. Friedman, J.H. Greedy function approximation: A gradient boosting machine. *Ann. Stat.* **2001**, *29*, 1189–1232. [CrossRef]
61. Zhang, Z.; Zhao, Y.; Canes, A.; Steinberg, D.; Lyashevska, O. Predictive analytics with gradient boosting in clinical medicine. *Ann. Transl. Med.* **2019**, *7*, 152. [CrossRef]
62. Harris, C.R.; Millman, K.J.; van der Walt, S.J.; Gommers, R.; Virtanen, P.; Cournapeau, D.; Wieser, E.; Taylor, J.; Berg, S.; Smith, N.J.; et al. Array Programming with NumPy. *Nature* **2020**, *585*, 357–362. [CrossRef]
63. McKinney, W. Data Structures for Statistical Computing in Python. In Proceedings of the 9th Python in Science Conference, Austin, TX, USA, 28 June–3 July 2010; pp. 51–56.
64. Pedregosa, F.; Varoquaux, G.; Gramfort, A.; Michel, V.; Thirion, B.; Grisel, O.; Blondel, M.; Prettenhofer, P.; Weiss, R.; Dubourg, V.; et al. Scikit-Learn: Machine Learning in Python. *J. Mach. Learn. Res.* **2011**, *12*, 2825–2830.
65. Van Rossum, G.; Drake, F.L. *Python 3 Reference Manual*; CreateSpace: Scotts Valley, CA, USA, 2009.
66. Grossfeld, B. Geography and law. *Mich. Law Rev.* **1984**, *82*, 1510–1519. Available online: https://repository.law.umich.edu/mlr/vol82/iss5/25 (accessed on 22 November 2024). [CrossRef]
67. Bai, D.; Ye, L.; Yang, Z.; Wang, G. Impact of climate change on agricultural productivity: A combination of spatial Durbin model and entropy approaches. *Int. J. Clim. Chang. Strat. Manag.* **2024**, *16*, 26–48. [CrossRef]
68. Lichstein, J.W.; Simons, T.R.; Shriner, S.A.; Franzreb, K.E. Spatial Autocorrelation and Autoregressive Models In Ecology. *Ecol. Monogr.* **2003**, *72*, 445–463. [CrossRef]
69. Naher, A.; Almas, L.K.; Guerrero, B.; Shaheen, S. Spatiotemporal Economic Analysis of Corn and Wheat Production in the Texas High Plains. *Water* **2023**, *15*, 3553. [CrossRef]
70. Ayouba, K. Spatial dependence in production frontier models. *J. Prod. Anal.* **2023**, *60*, 21–36. [CrossRef]
71. Huo, X.-N.; Li, H.; Sun, D.-F.; Zhou, L.-D.; Li, B.-G. Combining Geostatistics with Moran's I Analysis for Mapping Soil Heavy Metals in Beijing, China. *Int. J. Environ. Res. Public Health* **2012**, *9*, 995–1017. [CrossRef]
72. Wu, G.; Fan, Y.; Riaz, N. Spatial Analysis of Agriculture Ecological Efficiency and Its Influence on Fiscal Expenditures. *Sustainability* **2022**, *14*, 9994. [CrossRef]
73. Sangoi, L. Understanding plant density effects on maize growth and development: An important issue to maximize grain yield. *Cienc. Rural.* **2001**, *31*, 159–168. [CrossRef]
74. Postma, J.A.; Hecht, V.L.; Hikosaka, K.; Nord, E.A.; Pons, T.L.; Poorter, H. Dividing the pie: A quantitative review on plant density responses. *Plant Cell Environ.* **2020**, *44*, 1072–1094. [CrossRef]
75. Trevisan, R.G.; Bullock, D.S.; Martin, N.F. Spatial variability of crop responses to agronomic inputs in on-farm precision experimentation. *Precis. Agric.* **2021**, *22*, 342–363. [CrossRef]
76. Noa-Yarasca, E.; Babbar-Sebens, M.; Jordan, C.E. Machine Learning Models for Prediction of Shade-Affected Stream Temperatures. *J. Hydrol. Eng.* **2025**, *30*, 04024058. [CrossRef]
77. Shrestha, A.; Bheemanahalli, R.; Adeli, A.; Samiappan, S.; Czarnecki, J.M.P.; McCraine, C.D.; Reddy, K.R.; Moorhead, R. Phenological stage and vegetation index for predicting corn yield under rainfed environments. *Front. Plant Sci.* **2023**, *14*, 1168732. [CrossRef]
78. Pinto, A.A.; Zerbato, C.; Rolim, G.d.S.; Júnior, M.R.B.; da Silva, L.F.V.; de Oliveira, R.P. Corn grain yield forecasting by satellite remote sensing and machine-learning models. *Agron. J.* **2022**, *114*, 2956–2968. [CrossRef]
79. Verma, B.; Prasad, R.; Srivastava, P.K.; Yadav, S.A.; Singh, P.; Singh, R. Investigation of optimal vegetation indices for retrieval of leaf chlorophyll and leaf area index using enhanced learning algorithms. *Comput. Electron. Agric.* **2022**, *192*, 106581. [CrossRef]
80. Radočaj, D.; Šiljeg, A.; Marinović, R.; Jurišić, M. State of Major Vegetation Indices in Precision Agriculture Studies Indexed in Web of Science: A Review. *Agriculture* **2023**, *13*, 707. [CrossRef]
81. Lawrence, R.L.; Ripple, W.J. Comparisons among Vegetation Indices and Bandwise Regression in a Highly Disturbed, Heterogeneous Landscape: Mount St. Helens, Washington. *Remote Sens. Environ.* **1998**, *64*, 91–102. [CrossRef]

AI

MDPI

Review

A Systematic Literature Review on Parameters Optimization for Smart Hydroponic Systems

Umar Shareef, Ateeq Ur Rehman and Rafiq Ahmad *

Aquaponics 4.0 Learning Factory (All Factory), Department of Mechanical Engineering, University of Alberta, 9211 116 St., Edmonton, AB T6G 2G8, Canada; ushareef@ualberta.ca (U.S.); ateequr@ualberta.ca (A.U.R.)
* Correspondence: rafiq.ahmad@ualberta.ca

Abstract: Hydroponics is a soilless farming technique that has emerged as a sustainable alternative. However, new technologies such as Industry 4.0, the internet of things (IoT), and artificial intelligence are needed to keep up with issues related to economics, automation, and social challenges in hydroponics farming. One significant issue is optimizing growth parameters to identify the best conditions for growing fruits and vegetables. These parameters include pH, total dissolved solids (TDS), electrical conductivity (EC), light intensity, daily light integral (DLI), and nutrient solution/ambient temperature and humidity. To address these challenges, a systematic literature review was conducted aiming to answer research questions regarding the optimal growth parameters for leafy green vegetables and herbs and spices grown in hydroponic systems. The review selected a total of 131 papers related to indoor farming, hydroponics, and aquaponics. The review selected a total of 123 papers related to indoor farming, hydroponics, and aquaponics. The majority of the articles focused on technology description (38.5%), artificial illumination (26.2%), and nutrient solution composition/parameters (13.8%). Additionally, remaining 10.7% articles focused on the application of sensors, slope, environment and economy. This comprehensive review provides valuable information on optimized growth parameters for smart hydroponic systems and explores future prospects and the application of digital technologies in this field.

Keywords: hydroponic; IoT; AI; growth parameters; optimal growth

1. Introduction

World Population and Food Insecurity

The world population is growing at an alarming rate, exerting tremendous pressure on the development of sophisticated and sustainable agricultural and food production systems and necessitating increased food production to adequately feed everyone. Unsustainable growth of the world population is causing resource and food scarcity globally [1]. Today, there are more than three times as many people in the world as there were in the mid-1900s. By mid-November 2022, the global human population had reached 8.0 billion, a significant increase from approximately 2.5 billion in 1950. This number is projected to rise to 8.5 billion by 2030 and 9.7 billion by 2050. Such sustained population growth presents substantial challenges to sustainable development, making accurate population data crucial for effective development planning [2].

As of 2024, the world's population is increasing by approximately 0.91% annually. This rate is slightly higher than the 0.88% rate in 2023, but lower than the 0.98% in 2020 and 1.06% in 2019, with around 73 million people being added each year [3].

The scarcity of land and water resources is pushing traditional agriculture to the limits [4]. To meet the growing need for food, global production will need to increase by approximately 70% from 2007 levels [5]. As more people move to cities, it is estimated that 75% of the world's population will live in urban areas. Currently, cities occupy just 3% of the Earth's land, but they consume 60–80% of the energy, produce 75% of the carbon emissions, and are home to 56% of the population [5]. In 2015, the United Nations General Assembly adopted the 2030 Agenda for Sustainable Development, with the second goal being zero hunger [6]. This goal focuses on establishing and developing climate-resilient, sustainable agricultural production systems to enhance productivity and efficiency while protecting ecosystems and combating climate change [7].

This paper presents a comprehensive review on the optimization of growth parameters for leafy greens, vegetables, and herbs cultivated in hydroponic systems. The focus is on identifying the key growth parameters, understanding their relative importance, and examining their effects on plant growth. Optimizing these parameters in smart hydroponic systems is crucial for achieving optimal plant growth, maximizing yields, and ensuring efficient use of resources.

2. Sustainable and Climate-Resilient Smart Agriculture and Food Production Systems

For an agricultural production system to be sustainable, it should be resource efficient, socially acceptable, and environment friendly [8]. Traditional soil-based agriculture is increasingly under threat due to unpredictable weather patterns driven by rising temperatures, reduced soil productivity, degradation, soil-borne diseases, and poor water management practices, which contribute to significant water wastage [9]. As the urban population continues to expand, meeting food needs with minimal environmental and energy impact can be achieved through sustainable food production systems [7]. In regions scarce in water and land, there has been a shift towards high-yield agricultural production technologies like hydroponics [10] to meet the growing demands for nutritious, locally grown food. Compared to traditional farming techniques, hydroponics produces higher yields through effective space management and efficient use of essential nutrients [5]. Moreover, hydroponics enables year-round harvesting of multiple crops and employs various technologies to regulate essential plant conditions. Hydroponics maximizes water and chemical usage efficiency, reducing the risk of hazardous waste and residues [5].

Since 1965, the global per-person availability of arable land has decreased from 0.39 hectares to 0.20 hectares [11] and rising urban populations are contributing to an increase in food miles. Landlocked or highly populated areas are particularly vulnerable. Countries like Singapore, Hong Kong, the UAE, Egypt, and Norway have less than 5% arable land. Vertical farming systems can substantially enhance yield per unit area, even with spatial variations in environmental growth conditions [12]. These urban farms could be located nearer to the point of consumption and can integrate seamlessly into the city's infrastructure rather than requiring dedicated structures [12]. For instance, a recent study demonstrated that a hydroponic rooftop greenhouse in Lisbon could reduce energy consumption by half compared to the current supply chain for tomatoes. Carbon dioxide enrichment in greenhouses has increased yields by 30% with CO_2 levels maintained at 700–1000 ppm [13].

2.1. *Smart Hydroponic Systems: An Efficient and Sustainable Alternative*

The initial approach to meet the food requirements of the 2050 population is to prioritize sustainability in key human activities, especially agriculture [7]. Feeding the human population by 2050 is not our only concern today; present-day global warming and widespread pollution are pressing urgent environmental and socioeconomic issues. Hydroponics could offer partial solutions to the drawbacks of traditional agriculture, including its significant contributions to CO_2 emissions and the loss of cultivable land due to outdated and unsustainable practices. By eliminating soil, hydroponics addresses issues like soil degradation, enables more efficient nutrient and pesticide management, and helps prevent soil-borne diseases [5,9]. Recirculation and the avoidance of evapotranspiration in certain agricultural practices can lead to a 95% reduction in water usage and an almost 100% reduction in the need for pesticides. A study in Japan suggested that utilizing waste heat from coal-fired power plants in nearby plant factories could result in a reduction of 1024 tons of CO_2 per year [14].

Hydroponics, derived from the Greek words 'hydro' (water) and 'ponos' (labor), is a form of horticulture that employs mineral nutrient solutions to grow plants instead of traditional soil cultivation methods [15]. This soilless growing technique provides plants with a nutrient-rich water solution for essential nutrients and minerals uptake, supported by a substrate (growing media) throughout their growth life [15].

Four thousand years ago, the Egyptians grew plants in big pots, as depicted in the wall painting of the Deir el Bahari temple. The earliest documented experiments on hydroponics started in 1699, and after 1929, commercial aspects of hydroponically grown crops started developing [15,16]. Advantages of hydroponics include efficient space utilization, climate-resilient sustainable agricultural and food production, quality assurance, and easy-to-manage technologies [15,17].

However, there are also disadvantages, such as the high initial capital investment that is required, the need for skilled labor to operate the system, and the potential environmental hazards posed by nutrient solution dumping if not handled properly [5,18,19]. Various types of hydroponic systems exist based on crop application, design, and water circulation systems; most of these systems consist of a nutrient solution reservoir and an aerator. These system types include:

1. Nutrient film technique (NFT) [5,19,20]
2. Aeroponics [5,19,21]
3. Deep water culture (DWC) [5,18,19,22]
4. Drip system [5,22]
5. Ebb and flow [5]
6. Aquaponics [5,7,23]

In the nutrient film technique (NFT), the roots of the plants are submerged in a flowing liquid nutrient film stream (Figure 1a). In aeroponics, roots are suspended in the air and receive nutrients through a periodic mist from a system of sprinklers (Figure 1b). In deep water culture (DWC), the plant's roots are submerged in the nutrient solution, with the rest of the plant supported by growing media or substrate above the water level (Figure 1c). The drip system directly pumps the nutrient solution to the roots of the plants with a regulated flow (Figure 1d). The ebb and flow system involves placing plants in a tray that is periodically flooded and drained with nutrient-rich water pumped from a reservoir below (Figure 1e).

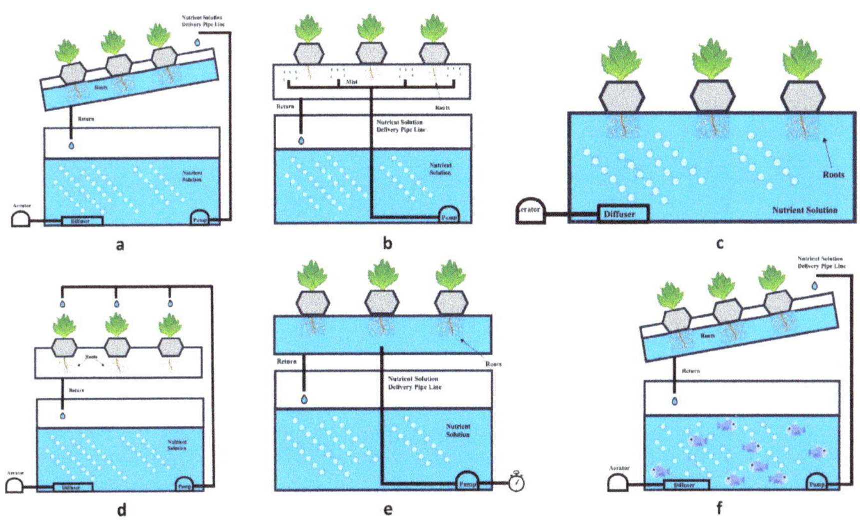

Figure 1. Different types of hydroponic systems. (**a**) NFT, (**b**) aeroponics, (**c**) deep water culture, (**d**) drip system, (**e**) ebb and flow, (**f**) aquaponics.

Aquaponics combines aquaculture (fish farming) and hydroponics. This technique utilizes the symbiotic relationship between plants and fish, where plants derive their nutritional requirements from fish waste. The plants absorb nutrients, and the microbial processes of nitrification and denitrification enable water recycling from the fish tank, establishing a balanced micro-ecosystem (Figure 1f).

Hydroponic systems are often referred to as plant factories because plants are cultivated under controlled conditions, such as pH, EC, TDS, etc. This approach not only improves productivity and efficiency but also enhances the nutritional value of the crops [24]. In a hydroponic system, plants receive necessary nutrients from mineral-rich solutions tailored to the specific needs of each plant species. Compared to traditional soil-based methods, this efficient nutrient delivery reduces losses from leaching, volatilization, and erosion. Industry 4.0 and the application of technologies like IoT, cloud computing, AI, and smart decision-making can make hydroponics even more efficient [22].

2.2. Challenges in Implementing Large-Scale Hydroponic Systems

When implementing hydroponic systems on a large scale, it is crucial to balance various key growth parameters due to the interdependence among different components of the systems [25]. Essential growth parameters, such as pH, EC, TDS, humidity, temperature (both nutrient solution and ambient), dissolved oxygen levels, and light intensity, must be closely monitored and managed [26]. When developed and adopted correctly, hydroponic systems can significantly enhance urban agriculture and contribute substantially to achieving Sustainable Development Goal 11 [12].

Hydroponic and fish production (aquaponics) are land-efficient food production systems that offer high yields and production benefits in urban areas with high population densities and limited, expensive farming land systems [7]. In these regions, where food demand is consistently high, aquaponics facilities can lower transportation costs and streamline supply chain management [27]. Adopting a decentralized production approach with small or medium-scale hydroponic systems in densely populated urban areas can minimize the ecological impact of agriculture, encourage self-employment and profitable business ventures, and foster community cooperation [5].

Recent studies in the hydroponic domain have focused on various aspects of the systems [23,24,28,29], management practices [30–33], types/species [22,34], energy use [7,35,36], sustainability, economy, and environment [37,38]. A complex system like hydroponics requires a deep understanding of multidisciplinary knowledge, encompassing agriculture, horticulture, mechanics, plant sciences, microbiology, chemistry, and more. Conducting a comprehensive review of hydroponics is a challenging task. Although the literature provides scattered data on important growth parameters for leafy greens and herbs, it lacks a single consolidated document consistent with all the available information [38–41].

2.3. Research Motivation and Paper Organization

Indoor farming requires a complete set of data for different species to be grown; this review, focusing on this aspect, provides a comprehensive data sheet to provide all the necessary data for important growth parameters for various leafy greens, herbs, and spices grown in a controlled environment.

This paper begins with an introduction and a comprehensive literature review. Section 3 describes the methodology and search criteria used to select papers for review. Section 4 discusses the key parameters that need to be monitored and controlled for optimal growth of leafy green vegetables and herbs in a hydroponic system. Section 5 outlines the application of AI and ML to smart hydroponic systems and future perspectives on existing research gaps and opportunities in the field and identifies key research areas for future work.

3. Research Methodology

The systematic literature review methodology can be divided into two main steps: (1) establishing the review protocol and (2) conducting the evaluation process.

3.1. Review Protocol

A review protocol outlines a clear strategy for conducting a literature review. This paper adheres to a systematic paper review process, which is a structured and methodical approach to analyzing and synthesizing existing literature on a particular topic. It involves defining specific criteria for selecting relevant papers, conducting a thorough search of academic databases, assessing the quality of the identified papers, and synthesizing the findings to draw conclusions or identify gaps in the existing research. This process aims to provide a comprehensive and unbiased overview of the current state of knowledge on the given subject.

The systematic review follows a well-defined protocol and aims to address a series of research questions. These questions cover various aspects and are designed to gain an understanding of the optimal growth parameters for different leafy greens, vegetables, and herbs in hydroponics. A preliminary search equation was formulated for the records search using Scopus, an online database for research articles. Based on this equation, a total of 1242 records were obtained. A year-wise comparison of the number of documents is presented in Figure 2, and the distribution of research article types is shown in Figure 3.

The majority of the records belonged to peer-reviewed journals and review papers. These records were further scrutinized based on the research questions and exclusion/inclusion criteria provided in Table 1.

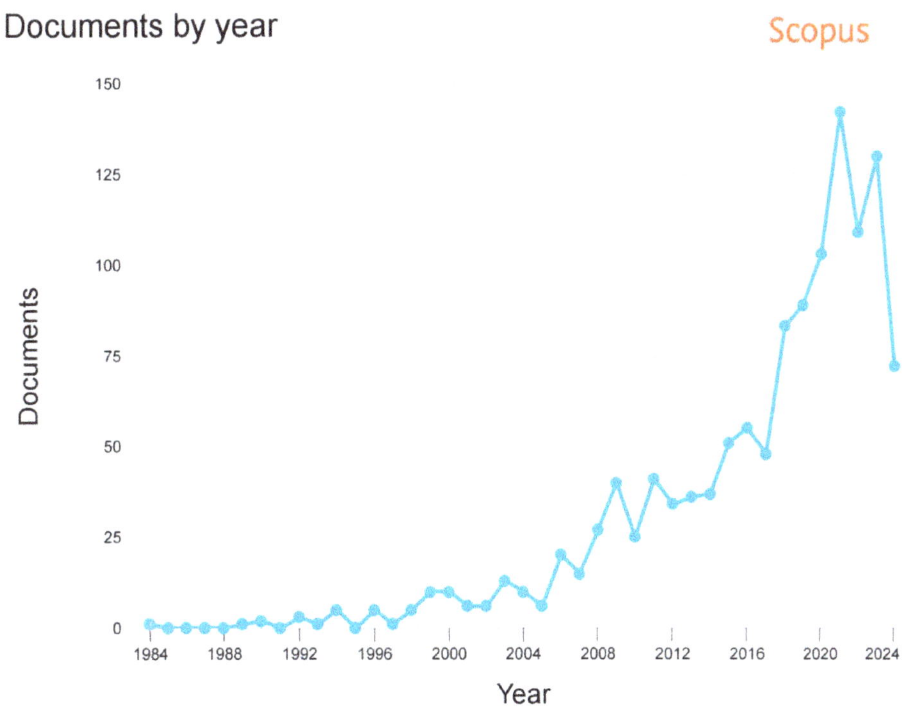

Figure 2. Year-wise data of records obtained from Scopus.

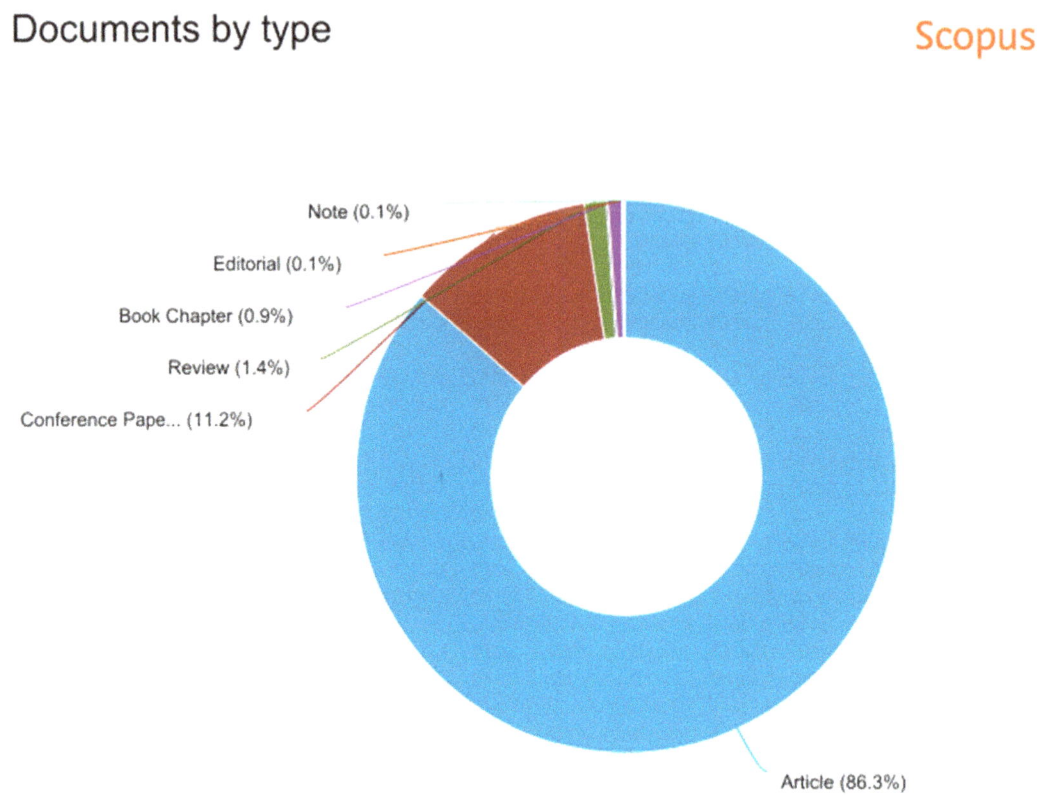

Figure 3. Research articles by their respective category.

Table 1. Review protocols, search questions, and inclusion/exclusion criteria.

Research Questions	• Classification of leafy green vegetables and herbs • What are growth parameters, and which growth parameters are most important (separately for leafy greens and herbs) for indoor/vertical/hydroponic farming? • What are the optimal ranges of the most important growth parameters for indoor farming, especially hydroponics (NFT and DWC in particular)? • How is growth affected by deviations from these optimal ranges? • How can IoT and automation be implemented to monitor, regulate, and correct these ranges? • Future directions considering sustainability and the Sustainable Development Goals (SDGs)
Search Equations	**Preliminary** • hydroponic AND growth AND parameters **Final Research Equation** • ALL (hydroponic, AND indoor AND farming, AND controlled AND environment, AND growth AND parameters, AND optimal AND growth) AND PUBYEAR > 2011 AND PUBYEAR < 2024 AND (LIMIT-TO (SUBJAREA, "AGRI") OR LIMIT-TO (SUBJAREA, "ENGI") OR LIMIT-TO (SUBJAREA, "ENVI") OR LIMIT-TO (SUBJAREA, "ENER")) AND (LIMIT-TO (DOCTYPE, "ar") OR LIMIT-TO (DOCTYPE, "re") OR LIMIT-TO (DOCTYPE, "bk") OR LIMIT-TO (DOCTYPE, "cp")) AND (LIMIT-TO (LANGUAGE, "English")).

Table 1. *Cont.*

Analysis Criteria	**Inclusion Criteria** • Conference papers and peer-reviewed journal articles • Studies published from 2011 to 2024 • Studies providing answers to research questions The article must include the title, abstract, source, DOI, and year of publication **Exclusion Criteria** • Summaries of editorials, seminars, events, and book reviews • Papers referring to optimal growth parameters or growth parameters but not applicable to aquaponics, hydroponic, or vertical farming • Papers published before 2011 • Non-English publications

Based on the above criteria and modified search equation to obtain more relevant and up-to-date literature, the final search equation yielded a total of 123 records, which were carried forward to the evaluation process.

3.2. Evaluation Method: PRISMA Approach

Once the review protocol is defined, the next step in the systematic analysis is to initiate the evaluation process. We utilized an evidence-based approach, PRISMA (Preferred Reporting Items for Systematic Reviews and Meta-Analysis) [42], which mapped out the number of records identified, screened, checked for eligibility, and eventually included in the systematic review. The modified search query yielded a total of 123 results. These were further evaluated by carefully examining the abstract of each paper. Papers that clearly addressed state-of-the-art hydroponic systems, optimization techniques, key parameters, optimal ranges, and the utilization of IoT and AI were retained for full-text eligibility checks, resulting in a total count of 65 (Figure 4).

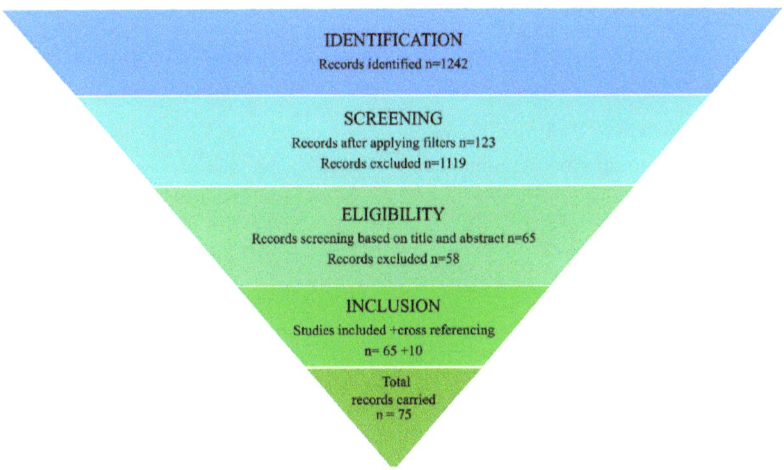

Figure 4. Evaluation of research articles based on PRISMA approach (adopted from [7]).

Subsequently, after a full-text review, 65 papers were selected for inclusion in this article. Among these, 25 papers primarily focused on growth parameters, optimization, and technology. The remaining papers discussed the fundamentals of hydroponic systems, key monitoring parameters, and optimal maintenance ranges. The systematic review yielded multidisciplinary articles. The majority of the articles were related to technology description (38.5%), artificial illumination (26.2%), and nutrient solution composition/parameters

(13.8%). Additionally, 10.8% of the articles pertained to the application of AI, digitization, and IoT, while remaining 10.7% articles focused on the application of sensors, slope, environment and economy (Figure 5).

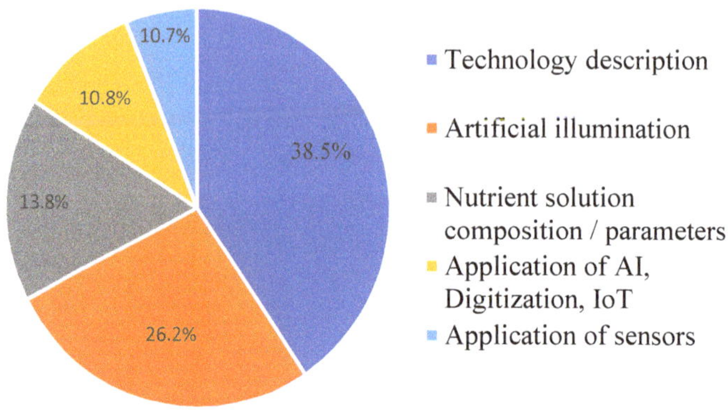

Figure 5. Percentage distribution of journal articles by field-related category.

This systematic review includes multidisciplinary articles from various subject areas, providing insights from different fields and contributing to a comprehensive summary of recent trends related to growth parameters in hydroponic systems. Major trends revealed through the systematic literature review included the effects of different growth parameters such as pH, EC, TDS, temperature (nutrient solution and ambient), substrate (growing/support media), dissolved oxygen, aeration, lighting conditions, and humidity on the growth of hydroponically grown plants. Other key areas included technology descriptions, growth parameters used for different types of hydroponic systems or indoor farming, and nutrient solution composition, delivery, and optimization. Additionally, artificial illumination and the application of emerging technologies in hydroponics were significant. These important parameters will be elaborated on in detail in the following sections.

4. Growth Parameters in Hydroponic Systems

The most important growth parameters to consider when monitoring and designing any controlled environment agriculture (CEA) system include water quality (nutrient solution), light intensity, nutrient availability, selected crop species, substrate/growing media, moisture availability, crop spacing, ambient temperature, air circulation and freshness, and relative humidity [43,44].

A systematic, careful, and balanced approach is necessary when dealing with the composition and management of the nutrient solution in any hydroponic system [45]. There are optimal ranges for these parameters for different plant species, which should be critically monitored and maintained to avoid resource wastage. The important growth parameters for any hydroponic system, and their ranges, measuring units, and relative importance and relevance, are presented in Table 2.

Table 2. Important growth parameters and ranges in hydroponic systems.

Sr No.	Parameter Description	Units	Linked to	Relative Importance/Significance	Optimal Range	References
1	pH	None	Nutrient solution	Highly significant	5–7.5	[4,5,15,21,22,34,44–84]
2	Electrical conductivity (EC)	Millisiemens per centimeter ($mScm^{-1}$)		Highly significant	0.8–3.5	[4,5,7,15,21–23,38,41,50,53,54,56,58,63–67,69–71,74–80,85]

Table 2. *Cont.*

Sr No.	Parameter Description	Units	Linked to	Relative Importance/Significance	Optimal Range	References
3	Temperature (nutrient solution)	Degrees centigrade °C		Highly significant	18–27	[21,23,26,29,37,41,43,54,57,58,60,62,72,78,85]
4	TDS	Parts per million (PPM)		Moderately significant	500–1800	[33,53,56]
5	Aeration/dissolved oxygen (DO)	Milligram per liter (mg L^{-1})		significant	3–10	[22,33,47,52,54,60,61,67,73,81,83,86]
6	Growing media/substrate		Plant growth/nutrient solution	Highly significant	Peat, mineral wool, rock wool, coconut coir etc.	[4,5,22,43,44,51–53,57,58,67,69–72,76,78–80,87–90]
7	Lighting/artificial illumination	PPFD (µmol $m^{-2}s^{-1}$)	Plant growth/overall system	Highly significant	80–600	[4,5,22,43,44,51–53,57,58,67,69–72,76,78–80,87–90]
8	Ambient temperature	Degrees centigrade °C	Overall system	Moderately significant	18–30	[13,21–23,33,38–41,43,47,51,54–57,60,61,63–65,67,71,75,78–80,82,83,91]
9	Relative humidity	None (percentage)	Overall system	Moderately significant	40–80	[4,13,22,33,38–41,43,47,51,54,55,57,60,63,65,67,70,71,75,78–80,85,91,92]
10	CO_2 dosing	Parts per million (PPM)	Plant growth	Optional supplementation	450–1200	[4,13,22,33,38–40,54,55,57,62–64,78,80,82,91,92]

4.1. Parameter Ranges in Hydroponic Systems

Crops can achieve higher quality through effective management of nutrient solutions [81]. Various physio-chemical phenomena can influence plant nutrient uptake [22,85]. The temperature of the nutrient solution can impact precipitation, co-precipitation, and complexation processes [85]. Maintaining an optimal temperature range of 18–27 °C [29,33,59,63,64,79] is crucial to prevent nutrient insolubility and ensure availability for plants.

The precipitation equilibrium is significantly influenced by pH variation. Continuous monitoring and control of pH are essential as it directly affects the nutrient uptake of the plants [4,23,33,74]. A pH above 7 can render zinc (Zn), iron (Fe), copper (Cu), manganese (Mg), and nickel (Ni) insoluble, while an acidic pH reduces the solubility of sulfur, phosphorus, and nitrogen. Even a minor change in pH has an apparent physiological response [81]. A pH range of 5–7.5 is generally recommended for most plant species in hydroponics [15,33,34,51,64].

Selective nutrient uptake in plant growth can deplete the solution of some of the nutrients, lowering the electrical conductivity (EC) of the nutrient solution. EC is an estimation of the ions present in the nutrient solution [35,85]; low EC indicates scarcity of nutrients in ionic form, and higher ranges indicate salt stress in plants. An EC range of 0.8–3.5 is suggested for smart hydroponic systems [15,33,76]. EC must be maintained at an optimal level for the healthy growth of the plants [4,5,22,76].

The choice of substrate or growing media depends on factors such as grower preference, nutrient solution pH, recyclability, and product shelf life [37,48]. An ideal substrate must have a balanced combination of these characteristics [76]: good aeration, durability, porosity, sterility, neutral pH, nutrient-holding capacity (cation exchange), lightweight, and ease of handling and reusability [5,16].

Light conditions significantly influence the nutritional and sensory qualities of plants [7,63,65]. For instance, lettuce shows optimal growth under artificial lighting with a photosynthetic photon flux density (PPFD) of 150–250 µmol$m^{-2}s^{-1}$ [7,62,64,93]. Hydroponics utilizes artificial illumination to supplement light requirements for the process of photosynthesis, a photochemical reaction that is crucial for converting CO_2 into carbohydrates [10,65]. Plant photoreceptors absorb this radiation and are responsible for the morphology and growth of the plant [7]. Light parameters, such as intensity, quality, wavelength, daily light integral (DLI), photosynthetically active radiation (PAR), and customized lighting recipes, play vital roles in plant chemical composition and sensory properties [86,88].

Maintaining an optimal humidity range is essential throughout the plant growth cycle, as humidity and temperature influence plant bio-metabolism [33,86,89]. A relative humidity range of 40–80% is considered to make a significant difference in plant growth in a hydroponic system [21,33,39,47,64]. Carbon dioxide (CO_2) concentration also affects plant growth significantly [23]. This is mainly because of related enhancements in photosynthesis and water use efficiency [13,90]. An ideal CO_2 concentration ranges from 450 to 1200 ppm, balancing higher yields without environmental harm [55,58,64,93]. Aerating the nutrient solution enhances plant growth, water uptake, and the leaf concentration of potassium, phosphorus, and magnesium [94]. There is a reasonable range within which plant growth is enhanced; higher aeration does not translate into a higher yield [95]. A range of 3–10 milligrams per liter (mg/L) is ideal for aeration to ensure the optimum growth of plants [34,49,61].

4.2. Plant Species-Based Growth Parameter Ranges for Hydroponic Systems

The previous section discussed the overall growth parameters for any hydroponic system, providing a wide range of parameters for optimal growth. This section focuses on the optimal growth parameters for specific plant species, as presented in Table 3.

Table 3. Plant specific (selected species) optimal growth parameters in hydroponic systems.

Plant Species	pH	EC (Milli Siemens cm^{-1})	Temperature (Nutrient Solution) (°C)	Growing Media/ Substrate	Lighting Intensity PPFD (μmol $m^{-2}s^{-1}$)	Photoperiod Hours (h) (Light/Dark)	Ambient Temperature (°C)	Relative Humidity (% Age)	References
Asparagus	6–6.8	1.4–1.8	20–28	Peat, mineral wool, perlite, rock wool, coconut coir etc.	150–200	8/16	18–30	45–80	[5,15,23,88]
Arugula	5.5–6.0	1.5–1.8	18–25		150–200	16/8	18–30	45–80	[1,7,23,67,91]
Basil	5.5–6.5	1.1–1.6	18–24		80–250	16/8	18–30	50–85	[1,4,5,7,15,21,50, 58,61,64,65,69,71, 89,96,97]
Celery	5.5–6.5	1.8–3	18–25		150	16/8	18–30	50–85	[5,15,23,90,93]
Kale	6–6.5	1.2–1.8	18–25		150–250	16/8	18–30	50–80	[7,60]
Leek	6.5–7	1.4–1.8	18–23		150–250	(12–14)/ (12–10)	18–30	60	[5,15,94,95]
Lettuce	5–7	1.5–2.5	18–25		150–250	14–17	18–27	45–80	[1,4,5,15,23,64,65, 70,80,91,93,96]
Parsley	6.0–6.5	1.8–2.2	18–25		150–200	16/8 or 14/8	18–30	45–80	[5,15,23]
Peppers	5.5–6	0.8–1.8	-		50–200	16/8	20–35	50–80	[1,5,15,23,65,77]
Strawberry	6.0	1.8–2.2	18–30		115–350	12–16	18–30	40–80	[1,5,7,15,23,64,84]
Spinach	6–6.5	1.8–2.2	20–30		100–150	13/11	20–35	20–30	[1,5,15,23,56,65, 72,96,98]
Tomato	5.5–6.5	2–4	-		50–200	15/9	-	-	[1,5,7,15,23,77,93]
Okra	5.5–6.5	2–2.4	-		-	-	20–35	50–80	[5,15,99]
Rhubarb	5.5–6.0	1.6–2.0	-		-	-	-	-	[5,15]
Rose	5.5–6.0	1.5–2.5	-		-	-	-	-	[5,15]
Sage	5.5–6.5	1.0–1.6	-		-	-	-	-	[5,15]

To achieve optimal growth in different plant species, different optimal levels of growth parameters are required, as presented in Table 3. Lettuce growth has been optimized [63] using 6.5, 2.3 mS/cm, 200 μmol $m^{-2}s^{-1}$, 16 h/day (light/dark), 24/21 °C day/night, 70/75% day/night, and 850 ppm ranges of pH, electrical conductivity, light intensity (PPFD), photoperiod, ambient temperature, relative humidity, and CO_2 level/supplementation, respectively. To obtain an optimal yield, strawberries and basil have been grown under pH 7, EC 2.3 mS cm^{-1}, ambient temperature 21 ± 2 °C, 55–70% relative humidity and 450 ppm CO_2, photoperiod 16/8 h/day (light/dark), and a light intensity of 200 μmol $m^{-2}s^{-1}$ [64]. A pH of 5.8, EC 1.85 mS/cm, ambient temperature 21 ± 2.5 °C, relative humidity 65 ± 5%, photoperiod hours/day (light/dark) and light intensity of 200 ± 5 μmol $m^{-2}s^{-1}$ have been used [65] to cultivate lettuce, spinach, kale, basil, and sweet Pepper in a controlled environment to check the growth effect of different light combinations on these species.

A recommended photoperiod of 16/8 h/day (light/dark) is generally used for most of the plant species grown hydroponically, and too much light is damaging to plants [96]. In one study, halophytes (salt-tolerant plants) were hydroponically grown using a photoperiod of 14/10 h/day (light/dark), PPFD 371.0 ±12.0 μmol $m^{-2}s^{-1}$, nutrient solution temperature of 22.9 ± 0.7 °C, pH of 7.8 ± 0.2, and a dissolved oxygen concentration of 6.7 ± 0.6 mg L^{-1} [60] to extract dissolved inorganic nutrients through an integrated multi-trophic aquaculture (IMTA). Additionally, Chinese kale was grown at five photo periods (12/12, 14/10, 16/8, 18/6, and 20/4 h/day (light/dark)) and five different light intensities (100, 130, 160, 190, and 220 μmol m^{-2} s^{-1}) [60] while other parameters were held at the same values (temperature, humidity, CO_2 concentration, photoperiod, light intensity, EC, and pH with values, 25 ± 3 °C, 65 ± 5%, 1000 ± 100 ppm, 16/8 h/day (light/dark), 150 μmol m^{-2} s^{-1}, 1.2 ± 0.1 (mS m^{-1}), and 6.5 ± 0.5, respectively). Better growth performance was observed under a light intensity of 160 and 190 μmol $m^{-2}s^{-1}$ with photoperiods of 16/8 and 18/6 h/day (light/dark), respectively [60].

In another study, arugula or rocket was grown at five different ECs (1.2, 1.5, 1.8, and 2.1 mS/cm) [67] while keeping other parameters like ambient temperature, humidity, light intensity, pH, and nutrient solution temperature at 18.5 °C, 50.85%, 200 μmol $m^{-2}s^{-1}$, 5.8, and 18.3 °C, respectively. An EC range of 15–1.8 mS/cm proved to be optimal for this experiment [67]. A group of researchers grew basil and lettuce under five different EC values, and it was concluded that 1.2 mS/cm is optimal for basil and 0.9 mS/cm is optimal for lettuce growth in an ebb and flow hydroponic system [4]. Three pH ranges (5.0–5.5, 5.5–6.0, 6.0–6.5) were shown to have significant effects on the physiological responses of lettuce in [80], with the lowest range causing a significant decrease in leaf area.

5. Discussion and Future Prospects

Further discussions and answers to the research questions regarding the application of smart technologies and tools, such as the Internet of Things (IoT), artificial intelligence (AI), machine learning (ML), and digital twinning in smart hydroponics, are detailed below.

5.1. Smart Hydroponic System

Smart agriculture relies on communication technologies and information for decision-making [75]. It utilizes smart monitoring devices and sensors to gather data from sources like IoT (Internet of Things) and ML (machine learning) to predict growth parameters for various plants, enabling precision farming [86,100]. In the smart agricultural domain, IoT devices collect data related to crop parameters and environmental conditions, such as pH, EC, humidity, temperature, leaf color, aeration, and lighting conditions. These data are then transmitted for further processing and action [20].

5.2. IoT-Assisted Parameters Control

In recent years, numerous efforts have been made to develop smart agricultural systems capable of monitoring and controlling various plant growth parameters such as light intensity, temperature, resource utilization, energy consumption, relative humidity, electrical conductivity, dissolved oxygen, and pH [7]. Utilizing IoT and automation techniques, a digital twin was developed for hydroponics that measures real-time parameters like pH, EC, humidity, and water temperature, and also predicts the growth and fresh weight of crops [23]. Furthermore, a model has been successfully developed to monitor growth parameters like pH, dissolved solids, fish feed rate, fish weight gain, nitrates, and plant growth. However, this model could not predict plant growth rate [101]. Through a manufacturing execution system (MES), Witzel et al. developed an aquaponics farm capable of capturing data from various sensors (relative humidity, light intensity, temperature, water level, energy consumption, and plant growth) to monitor and control real-time parameters of the system [102].

In another study, a digital twin based on a cyber-physical system (CPS) was developed, capable of gathering soil parameters data through probes and displaying the data on a dashboard, thus creating a fully functional digital twin [103]. Despite these advantages, IoT monitoring and parameter control face challenges such as a lack of standards, high initial costs, and security issues due to network encryption [7].

5.3. AI and ML for Parameters Optimization

Artificial intelligence (AI) and machine learning (ML) combined with big data and IoT are driving agricultural digitization. These technologies have the potential to improve real time monitoring, enhance production, and achieve greater control over the growth of crops [104]. ML algorithms and techniques are implemented in various aspects of agriculture, including crop management (yield prediction, species recognition, disease detection), livestock management, and water and soil management (type, pH, temperature, moisture content) [103].

One hydroponics system based on multi nodal data analysis demonstrated the ability to monitor and optimize six growth parameters (ambient temperature, air flow speed, relative humidity, CO_2 concentration, pH, EC, and nutrient solution temperature) related to plant growth in a hydroponic plant factory [79]. A group of researchers developed a control kit for application to any size hydroponic system in order to control some of the growth parameters (pH, EC, DO, temperature of nutrient solution) to an optimal level. This kit was able to increase lettuce yield by 21% as compared to uncontrolled growth [51]. Some novel KNN-algorithms have been shown to predict plant growth with 93% accuracy [52] in an NFT-hydroponic system by optimally controlling growth parameters. An Android-based app using cyber-physical system (CPS) was developed for real-time monitoring of vertical farms, with optimized control of parameters such as light, CO_2 volume, temperature, pH, and humidity, and live feedback through the app. This app also manages algae growth, which can pose a risk to plant growth [51]. A test bed was developed based on an open IoT platform for monitoring hydroponic systems. It optimizes and monitors parameters, sends alerts if parameters are not within optimal ranges, and estimates power consumption [49]. This test bed provides key insights into the optimization and correlation of different growth parameters.

5.4. Parameters Control in Hydroponic 4.0/5.0

Although the digitization of agriculture (Hydroponics 4.0) through the application of AI, ML, IoT, and similar technologies can help reduce food scarcity and increase production, several barriers exist, including technical, social, and economic challenges [24]. Agriculture 5.0 often lacks social fairness and a sustainable approach, focusing more on digitization and AI technologies for flexibility and enhanced productivity [32]. In 2021, the European Commission formally adopted Industry 5.0, which emphasizes sustainable, resilient, and human-centric production and technology systems [105,106]. Industry 5.0 aims to refine the interaction between humans and machines, creating robust systems and reducing environmental impacts through a circular economy. Key technologies of Industry 5.0 include collaborative robots (cobots), digital twins, smart materials, energy-efficient and secure data management, AI, and renewable energy resources [24,105].

Applying the principles of Industry 5.0 to hydroponics, termed Hydroponics 5.0, can significantly enhance agricultural production and quality. A cyber-physical cognitive system (CPCS) can be developed for hydroponic systems, capable of perceiving situations and taking remedial actions accordingly. This involves assigning repetitive tasks such as parameter monitoring, control, and optimization of subsequent actions to cobots, while humans handle critical thinking. Digital twins can help detect diseases and predict yield/growth, leading to better control over the growth period. This approach paves the way for sustainable, resilient, and climate-smart hydroponic systems [24,105,106].

6. Conclusions

Global food demand and food insecurity have paved the way for intensive agricultural production methods and techniques. Scientists and researchers are tirelessly working to integrate Industry 4.0 concepts into modern agriculture and farming to increase yields, reduce waste, minimize cost, and ensure sustainable production. Hydroponics offers a solution to food insecurity, enhances sustainability, and allows for the integration of modern technologies. However, it requires meticulous monitoring and control of various plant growth parameters to be economically and socially viable.

This paper provides insights into crucial growth parameters and their optimized ranges for hydroponic systems. Numerous research papers were analyzed to determine the optimal ranges for pH, electrical conductivity (EC), temperature (nutrient solution and ambient), aeration/dissolved oxygen (DO), growing media/substrate, lighting/artificial illumination, relative humidity, and CO_2 dosing, as well as plant-specific parameters. The concluded optimal ranges for these parameters are: pH, 5–7.5; EC, 0.8–3.5 mS/cm; temperature, 18–27 °C; aeration, 3–10 mg/L DO; growing media, peat, mineral wool, rock wool, or coconut coir; light intensity, 80–600 μmol $m^{-2}s^{-1}$; relative humidity, 40–80%; and CO_2 concentration, 450–1200 ppm. This paper also discusses the application and beneficial effects of smart monitoring, digital twins, machine learning (ML), artificial intelligence (AI), the Internet of Things (IoT), and Industry 4.0/5.0 in hydroponic systems. Additionally, potential roadblocks to adopting these technologies are highlighted.

Author Contributions: Conceptualization, U.S. and R.A.; methodology, R.A.; software, U.S. and A.U.R.; validation, A.U.R. and R.A.; formal analysis, R.A.; investigation, U.S. and A.U.R.; resources, U.S. and R.A.; data curation, R.A.; writing—original draft preparation, U.S.; writing—review and editing, A.U.R. visualization, R.A. and A.U.R.; supervision, R.A.; project administration, R.A.; funding acquisition, R.A. All authors have read and agreed to the published version of the manuscript.

Funding: The authors acknowledge the financial support for this work from the Natural Sciences and Engineering Research Council of Canada (NSERC) (Grant File No. ALLRP 545537-19 and RGPIN-2017-04516) and Higher Education Commission (HEC) of Pakistan.

Conflicts of Interest: The authors declare that they have no known competing financial interests or personal relationships that could have appeared to influence the work reported in this paper.

References

1. Nájera, C.; Gallegos-Cedillo, V.M.; Ros, M.; Pascual, J.A. Role of Spectrum-Light on Productivity, and Plant Quality over Vertical Farming Systems: Bibliometric Analysis. *Horticulturae* **2023**, *9*, 63. [CrossRef]
2. United Nations Department of Economic and Social Affairs, Population Division. *World Population Prospects 2022 Summary of Results*; UN DESA/POP/2022/TR/NO. 3; United Nations Department of Economic and Social Affairs, Population Division: New York, NY, USA, 2022.
3. World Population Clock: 8.1 Billion People (LIVE, 2024)—Worldometer. Available online: https://www.worldometers.info/world-population/ (accessed on 19 February 2024).
4. Hosseini, H.; Mozafari, V.; Roosta, H.R.; Shirani, H.; van de Vlasakker, P.C.H.; Farhangi, M. Nutrient Use in Vertical Farming: Optimal Electrical Conductivity of Nutrient Solution for Growth of Lettuce and Basil in Hydroponic Cultivation. *Horticulturae* **2021**, *7*, 283. [CrossRef]
5. Velazquez-Gonzalez, R.S.; Garcia-Garcia, A.L.; Ventura-Zapata, E.; Barceinas-Sanchez, J.D.O.; Sosa-Savedra, J.C. A Review on Hydroponics and the Technologies Associated for Medium- and Small-Scale Operations. *Agriculture* **2022**, *12*, 646. [CrossRef]

6. The Sustainable Development Agenda—United Nations Sustainable Development. Available online: https://www.un.org/sustainabledevelopment/development-agenda/ (accessed on 19 February 2024).
7. Gillani, S.A.; Abbasi, R.; Martinez, P.; Ahmad, R. Review on Energy Efficient Artificial Illumination in Aquaponics. *Clean. Circ. Bioeconomy* **2022**, *2*, 100015. [CrossRef]
8. Food and Agriculture Organization of the United Nations. *Sustainable Food Systems Concept and Framework: What Is a Sustainable Food System? Why Take a Food Systems Approach? Changing Food Systems*; FAO: Rome, Italy, 2018.
9. St. Clair, S.B.; Lynch, J.P. The Opening of Pandora's Box: Climate Change Impacts on Soil Fertility and Crop Nutrition in Developing Countries. *Plant Soil* **2010**, *335*, 101–115. [CrossRef]
10. Trivellini, A.; Toscano, S.; Romano, D.; Ferrante, A. The Role of Blue and Red Light in the Orchestration of Secondary Metabolites, Nutrient Transport and Plant Quality. *Plants* **2023**, *12*, 2026. [CrossRef] [PubMed]
11. World Bank Publications. *World Development Report 2016: Digital Dividends*; World Bank Publications: Chicago, IL, USA, 2016.
12. Buehler, D.; Junge, R. Global Trends and Current Status of Commercial Urban Rooftop Farming. *Sustainability* **2016**, *8*, 1108. [CrossRef]
13. Ward, R.; Jans-Singh, M.; Choudhary, R. Quantifying the Environmental and Energy Benefits of Food Growth in the Urban Environment. In *Smart Plant Factory: The Next Generation Indoor Vertical Farms*; Springer: Singapore, 2018; pp. 245–287; ISBN 9789811310652.
14. Togawa, T.; Fujita, T.; Dong, L.; Fujii, M.; Ooba, M. Feasibility Assessment of the Use of Power Plant-Sourced Waste Heat for Plant Factory Heating Considering Spatial Configuration. *J. Clean. Prod.* **2014**, *81*, 60–69. [CrossRef]
15. Electrical Conductivity and PH Guide for Hydroponics. Available online: https://www.researchgate.net/publication/309293679_Electrical_Conductivity_and_pH_Guide_for_Hydroponics?channel=doi&linkId=5808bddc08ae5ed04c012573&showFulltext=true (accessed on 21 August 2024). [CrossRef]
16. Torabi, M.; Mokhtarzadeh, A.; Mahlooji, M. The Role of Hydroponics Technique as a Standard Methodology in Various Aspects of Plant Biology Researches. In *Hydroponics—A Standard Methodology for Plant Biological Researches*; Asao, T., Ed.; InTech: Rijeka, Croatia, 2012.
17. Hydroponics | Oklahoma State University. Available online: https://extension.okstate.edu/fact-sheets/hydroponics.html (accessed on 21 August 2024).
18. Sharma, A.; Manpoong, C.; Devadas, V.S.; Kartha, B.D.; Pandey, H.; Wangsu, M. Crop Hydroponics, Phyto-Hydroponics, Crop Production, and Factors Affecting Soilless Culture. *ACS Agric. Sci. Technol.* **2022**, *2*, 1134–1150. [CrossRef]
19. Niu, G.; Masabni, J. Hydroponics. In *Plant Factory Basics, Applications and Advances*; Academic Press: Cambridge, MA, USA, 2022; pp. 153–166. [CrossRef]
20. Spehia, R.S.; Devi, M.; Singh, J.; Sharma, S.; Negi, A.; Singh, S.; Chauhan, N.; Sharma, D.; Sharma, J.C. Lettuce Growth and Yield in Hoagland Solution with an Organic Concoction. *Int. J. Veg. Sci.* **2018**, *24*, 557–566. [CrossRef]
21. Lakhiar, I.A.; Gao, J.; Syed, T.N.; Chandio, F.A.; Buttar, N.A. Modern Plant Cultivation Technologies in Agriculture under Controlled Environment: A Review on Aeroponics. *J. Plant Interact.* **2018**, *13*, 338–352. [CrossRef]
22. Ragaveena, S.; Shirly Edward, A.; Surendran, U. Smart Controlled Environment Agriculture Methods: A Holistic Review. *Rev. Environ. Sci. Biotechnol.* **2021**, *20*, 887–913. [CrossRef]
23. Reyes Yanes, A.; Abbasi, R.; Martinez, P.; Ahmad, R. Digital Twinning of Hydroponic Grow Beds in Intelligent Aquaponic Systems. *Sensors* **2022**, *22*, 7393. [CrossRef]
24. Abbasi, R.; Martinez, P.; Ahmad, R. The Digitization of Agricultural Industry—A Systematic Literature Review on Agriculture 4.0. *Smart Agric. Technol.* **2022**, *2*, 100042. [CrossRef]
25. Ahmed, N.; Turchini, G.M. Recirculating Aquaculture Systems (RAS): Environmental Solution and Climate Change Adaptation. *J. Clean. Prod.* **2021**, *297*, 126604. [CrossRef]
26. Reyes-Yanes, A.; Gelio, S.; Martinez, P.; Ahmad, R. Wireless Sensing Module for IoT Aquaponics Database Construction. *Int. J. Electron. Electr. Eng.* **2021**, *9*, 43–47. [CrossRef]
27. Oliver, L.P.; Coyle, S.D.; Bright, L.A.; Shultz, R.C.; Hager, J.V.; Tidwell, J.H. Comparison of Four Artificial Light Technologies for Indoor Aquaponic Production of Swiss Chard, Beta Vulgaris, and Kale, Brassica Oleracea. *J. World Aquac. Soc.* **2018**, *49*, 837–844. [CrossRef]
28. Abbasi, R.; Yanes, A.R.; Martinez Villanuera, E.; Ahmad, R. Real-Time Implementation of Digital Twin for Robot Based Production Line. In Proceedings of the Conference on Learning Factories (CLF) 2021, Edmonton, AB, Canada, 4 June 2021. [CrossRef]
29. Yang, T.; Kim, H.J. Comparisons of Nitrogen and Phosphorus Mass Balance for Tomato-, Basil-, and Lettuce-Based Aquaponic and Hydroponic Systems. *J. Clean. Prod.* **2020**, *274*, 122619. [CrossRef]
30. Pringle, A.M.; Handler, R.M.; Pearce, J.M. Aquavoltaics: Synergies for Dual Use of Water Area for Solar Photovoltaic Electricity Generation and Aquaculture. *Renew. Sustain. Energy Rev.* **2017**, *80*, 572–584. [CrossRef]
31. Ghamkhar, R.; Hartleb, C.; Wu, F.; Hicks, A. Life Cycle Assessment of a Cold Weather Aquaponic Food Production System. *J. Clean. Prod.* **2020**, *244*, 118767. [CrossRef]
32. Saiz-Rubio, V.; Rovira-Más, F. From Smart Farming towards Agriculture 5.0: A Review on Crop Data Management. *Agronomy* **2020**, *10*, 207. [CrossRef]
33. Yanes, A.R.; Martinez, P.; Ahmad, R. Towards Automated Aquaponics: A Review on Monitoring, IoT, and Smart Systems. *J. Clean. Prod.* **2020**, *263*, 121571. [CrossRef]

34. Maucieri, C.; Nicoletto, C.; Junge, R.; Schmautz, Z.; Sambo, P.; Borin, M. Hydroponic Systems and Water Management in Aquaponics: A Review. *Ital. J. Agron.* **2018**, *13*, 1–11. [CrossRef]
35. Silva, L.; Valdés-Lozano, D.; Escalante, E.; Gasca-Leyva, E. Dynamic Root Floating Technique: An Option to Reduce Electric Power Consumption in Aquaponic Systems. *J. Clean. Prod.* **2018**, *183*, 132–142. [CrossRef]
36. Kikuchi, Y.; Kanematsu, Y.; Yoshikawa, N.; Okubo, T.; Takagaki, M. Environmental and Resource Use Analysis of Plant Factories with Energy Technology Options: A Case Study in Japan. *J. Clean. Prod.* **2018**, *186*, 703–717. [CrossRef]
37. Arakkal Thaiparambil, N.; Radhakrishnan, V. Challenges in Achieving an Economically Sustainable Aquaponic System: A Review. *Aquac. Int.* **2022**, *30*, 3035–3066. [CrossRef]
38. Kelly, N.; Choe, D.; Meng, Q.; Runkle, E.S. Promotion of Lettuce Growth under an Increasing Daily Light Integral Depends on the Combination of the Photosynthetic Photon Flux Density and Photoperiod. *Sci. Hortic.* **2020**, *272*, 109565. [CrossRef]
39. Sapkota, S.; Sapkota, S.; Liu, Z. Effects of Nutrient Composition and Lettuce Cultivar on Crop Production in Hydroponic Culture. *Horticulturae* **2019**, *5*, 72. [CrossRef]
40. Johkan, M.; Shoji, K.; Goto, F.; Hahida, S.; Yoshihara, T. Effect of Green Light Wavelength and Intensity on Photomorphogenesis and Photosynthesis in Lactuca Sativa. *Environ. Exp. Bot.* **2012**, *75*, 128–133. [CrossRef]
41. Thakulla, D.; Dunn, B.; Hu, B.; Goad, C.; Maness, N. Nutrient Solution Temperature Affects Growth And∘ Brix Parameters of Seventeen Lettuce Cultivars Grown in an NFT Hydroponic System. *Horticulturae* **2021**, *7*, 321. [CrossRef]
42. Moher, D.; Liberati, A.; Tetzlaff, J.; Altman, D.G. Guidelines and Guidance Preferred Reporting Items for Systematic Reviews and Meta-Analyses: The PRISMA Statement. *Ann. Intern. Med.* **2009**, *151*, 264–269. [CrossRef] [PubMed]
43. Sarkar, A.; Majumder, M. Fuzzy Logic Approach in Prioritization of Crop Growing Parameters in Protected Farms: A Case in North East India. *CIGR J.* **2017**, *19*, 211–217.
44. Abbasi, R.; Martinez, P.; Ahmad, R. Data Acquisition and Monitoring Dashboard for IoT Enabled Aquaponics Facility. In Proceedings of the 2022 10th International Conference on Control, Mechatronics and Automation (ICCMA), Belval, Luxembourg, 9–12 November 2022; pp. 168–172. [CrossRef]
45. Sambo, P.; Nicoletto, C.; Giro, A.; Pii, Y.; Valentinuzzi, F.; Mimmo, T.; Lugli, P.; Orzes, G.; Mazzetto, F.; Astolfi, S.; et al. Hydroponic Solutions for Soilless Production Systems: Issues and Opportunities in a Smart Agriculture Perspective. *Front. Plant Sci.* **2019**, *10*, 923. [CrossRef] [PubMed]
46. Trejo-Téllez, L.I.; Gómez-Merino, F.C. Nutrient Solutions for Hydroponic Systems. In *Hydroponics—A Standard Methodology for Plant Biological Researches*; Asao, T., Ed.; InTech: Rijeka, Croatia, 2012.
47. Nguyen, T.P.D.; Jang, D.C.; Tran, T.T.H.; Nguyen, Q.T.; Kim, I.S.; Hoang, T.L.H.; Vu, N.T. Influence of Green Light Added with Red and Blue Leds on the Growth, Leaf Microstructure and Quality of Spinach (*Spinacia oleracea* L.). *Agronomy* **2021**, *11*, 1724. [CrossRef]
48. Baiyin, B.; Tagawa, K.; Yamada, M.; Wang, X.; Yamada, S.; Yamamoto, S.; Ibaraki, Y. Study on Plant Growth and Nutrient Uptake under Different Aeration Intensity in Hydroponics with the Application of Particle Image Velocimetry. *Agriculture* **2021**, *11*, 1140. [CrossRef]
49. Chowdhury, M.E.H.; Khandakar, A.; Ahmed, S.; Al-Khuzaei, F.; Hamdalla, J.; Haque, F.; Reaz, M.B.I.; Al Shafei, A.; Al-Emadi, N. Design, Construction and Testing of Iot Based Automated Indoor Vertical Hydroponics Farming Test-Bed in Qatar. *Sensors* **2020**, *20*, 5637. [CrossRef]
50. Yegül, U. Development of an Embedded Software and Control Kit to Be Used in Soilless Agriculture Production Systems. *Sensors* **2023**, *23*, 3706. [CrossRef]
51. Chuah, Y.D.; Lee, J.V.; Tan, S.S.; Ng, C.K. Implementation of Smart Monitoring System in Vertical Farming. *IOP Conf. Ser. Earth Environ. Sci.* **2019**, *268*, 012083. [CrossRef]
52. Mamatha, V.; Kavitha, J.C. Machine Learning Based Crop Growth Management in Greenhouse Environment Using Hydroponics Farming Techniques. *Meas. Sens.* **2023**, *25*, 100665. [CrossRef]
53. Fimbres-Acedo, Y.E.; Servín-Villegas, R.; Garza-Torres, R.; Endo, M.; Fitzsimmons, K.M.; Emerenciano, M.G.C.; Magallón-Servín, P.; López-Vela, M.; Magallón-Barajas, F.J. Hydroponic Horticulture Using Residual Waters from Oreochromis Niloticus Aquaculture with Biofloc Technology in Photoautotrophic Conditions with Chlorella Microalgae. *Aquac. Res.* **2020**, *51*, 4340–4360. [CrossRef]
54. Ibrahim, L.A.; Shaghaleh, H.; El-Kassar, G.M.; Abu-Hashim, M.; Elsadek, E.A.; Alhaj Hamoud, Y. Aquaponics: A Sustainable Path to Food Sovereignty and Enhanced Water Use Efficiency. *Water* **2023**, *15*, 4310. [CrossRef]
55. Noh, K.; Jeong, B.R. Increased Carbon Dioxide by Occupants Promotes Growth of Leafy Vegetables Grown in Indoor Cultivation System. *Sustainability* **2021**, *13*, 13288. [CrossRef]
56. Yessenamanova, M.; Yessenamanova, Z.; Tlepbergenova, A.; Moldir, M.; Nurgul, B. Optimization of Acidity and Electrical Conductivity in Hydroponic Vegetable Production: A STEM Educational Perspective. *Int. J. Des. Nat. Ecodynamics* **2023**, *18*, 1517–1524. [CrossRef]
57. Wang, L.; Ning, S.; Zheng, W.; Guo, J.; Li, Y.; Li, Y.; Chen, X.; Ben-Gal, A.; Wei, X. Performance Analysis of Two Typical Greenhouse Lettuce Production Systems: Commercial Hydroponic Production and Traditional Soil Cultivation. *Front. Plant Sci.* **2023**, 14. [CrossRef]
58. Priya, G.L.; Baskar, C.; Deshmane, S.S.; Adithya, C.; Das, S. Revolutionizing Holy-Basil Cultivation with AI-Enabled Hydroponics System. *IEEE Access* **2023**, *11*, 82624–82639. [CrossRef]

59. Atmaja, P.; Surantha, N. Smart Hydroponic Based on Nutrient Film Technique and Multistep Fuzzy Logic. *Int. J. Electr. Comput. Eng.* **2022**, *12*, 3146–3157. [CrossRef]
60. Custódio, M.; Villasante, S.; Calado, R.; Lillebø, A.I. Testing the Hydroponic Performance of the Edible Halophyte Halimione Portulacoides, a Potential Extractive Species for Coastal Integrated Multi-Trophic Aquaculture. *Sci. Total Environ.* **2021**, *766*, 144378. [CrossRef] [PubMed]
61. Sutulienė, R.; Laužikė, K.; Pukas, T.; Samuolienė, G. Effect of Light Intensity on the Growth and Antioxidant Activity of Sweet Basil and Lettuce. *Plants* **2022**, *11*, 1709. [CrossRef]
62. Boros, I.F.; Székely, G.; Balázs, L.; Csambalik, L.; Sipos, L. Effects of LED Lighting Environments on Lettuce (*Lactuca sativa* L.) in PFAL Systems—A Review. *Sci. Hortic.* **2023**, *321*, 112351. [CrossRef]
63. Carotti, L.; Pistillo, A.; Zauli, I.; Meneghello, D.; Martin, M.; Pennisi, G.; Gianquinto, G.; Orsini, F. Improving Water Use Efficiency in Vertical Farming: Effects of Growing Systems, Far-Red Radiation and Planting Density on Lettuce Cultivation. *Agric. Water Manag.* **2023**, *285*, 108365. [CrossRef]
64. Piovene, C.; Orsini, F.; Bosi, S.; Sanoubar, R.; Bregola, V.; Dinelli, G.; Gianquinto, G. Optimal Red:Blue Ratio in Led Lighting for Nutraceutical Indoor Horticulture. *Sci. Hortic.* **2015**, *193*, 202–208. [CrossRef]
65. Naznin, M.T.; Lefsrud, M.; Gravel, V.; Azad, M.O.K. Blue Light Added with Red LEDs Enhance Growth Characteristics, Pigments Content, and Antioxidant Capacity in Lettuce, Spinach, Kale, Basil, and Sweet Pepper in a Controlled Environment. *Plants* **2019**, *8*, 93. [CrossRef] [PubMed]
66. Custódio, M.; Cartaxana, P.; Villasante, S.; Calado, R.; Lillebø, A.I. Led Lighting and High-Density Planting Enhance the Cost-Efficiency of Halimione Portulacoides Extraction Units for Integrated Aquaculture. *Appl. Sci.* **2021**, *11*, 4995. [CrossRef]
67. Yang, T.; Samarakoon, U.; Altland, J.; Ling, P. Photosynthesis, Biomass Production, Nutritional Quality, and Flavor-Related Phytochemical Properties of Hydroponic-Grown Arugula (*Eruca sativa* Mill.) 'Standard' under Different Electrical Conductivities of Nutrient Solution. *Agronomy* **2021**, *11*, 1340. [CrossRef]
68. Carotti, L.; Potente, G.; Pennisi, G.; Ruiz, K.B.; Biondi, S.; Crepaldi, A.; Orsini, F.; Gianquinto, G.; Antognoni, F. Pulsed Led Light: Exploring the Balance between Energy Use and Nutraceutical Properties in Indoor-Grown Lettuce. *Agronomy* **2021**, *11*, 1106. [CrossRef]
69. Dou, H.; Niu, G.; Gu, M.; Masabni, J.G. Responses of Sweet Basil to Different Daily Light Integrals in Photosynthesis, Morphology, Yield, and Nutritional Quality. *HortScience* **2018**, *53*, 496–503. [CrossRef]
70. Kong, Y.; Nemali, A.; Mitchell, C.; Nemali, K. Spectral Quality of Light Can Affect Energy Consumption and Energy-Use Efficiency of Electrical Lighting in Indoor Lettuce Farming. *HortScience* **2019**, *54*, 865–872. [CrossRef]
71. Ercan Karaayak, P.; Inanoglu, S.; Karwe, M.V. Impact of Cold Plasma Treatment of Sweet Basil Seeds on the Growth and Quality of Basil Plants in a Lab-Scale Hydroponic System. *ACS Agric. Sci. Technol.* **2023**, *3*, 675–682. [CrossRef]
72. Farooq, A.; Verma, A.K.; Hittinahalli, C.M.; Varghese, T.; Pathak, M.S. Iron Supplementation in Aquaculture Wastewater and Its Impact on Osmoregulatory, Haematological, Blood Biochemical, and Stress Responses of Pangasius with Spinach in Nutrient Film Technique Based Aquaponics. *Aquaculture* **2023**, *567*, 739250. [CrossRef]
73. Nursyahid, A.; Agung Setyawan, T.; Sriyanto Nugroho, A.; Novitasari, M.; Pramono, S.; Dwi Wardihani, E. Performance of Acidity Nutrient Automation Using MQTT Protocol in Dutch Bucket System Hydroponic. *J. Eng. Sci. Technol.* **2023**, *18*, 214–228.
74. Langenfeld, N.J.; Pinto, D.F.; Faust, J.E.; Heins, R.; Bugbee, B. Principles of Nutrient and Water Management for Indoor Agriculture. *Sustainability* **2022**, *14*, 10204. [CrossRef]
75. Abd Ghani, R.; Omar, S.; Jolánkai, M.; Tarnawa, Á.; Khalid, N.; Kassai, M.K.; Kende, Z. Response of Shoot and Root Growth, Yield, and Chemical Composition to Nutrient Concentrations in Soybean Varieties Grown under Soilless and Controlled Environment Conditions. *Agriculture* **2023**, *13*, 1925. [CrossRef]
76. Stathopoulou, P.; Tsoumalakou, E.; Levizou, E.; Vanikiotis, T.; Zaoutsos, S.; Berillis, P. Iron and Potassium Fertilization Improve Rocket Growth without Affecting Tilapia Growth and Histomorphology Characteristics in Aquaponics. *Appl. Sci.* **2021**, *11*, 5681. [CrossRef]
77. Wei, Y.; Li, W.; An, D.; Li, D.; Jiao, Y.; Wei, Q. Equipment and Intelligent Control System in Aquaponics: A Review. *IEEE Access* **2019**, *7*, 169306–169326. [CrossRef]
78. Kim, H.; Oh, D.; Jang, H.; Koo, C.; Hong, T.; Kim, J. Development of a Multi-Node Monitoring System for Analyzing Plant Growth and Indoor Environment Interactions: An Empirical Study on a Plant Factory. *Comput. Electron. Agric.* **2023**, *214*, 108311. [CrossRef]
79. Lam, V.P.; Kim, S.J.; Park, J.S. Optimizing the Electrical Conductivity of a Nutrient Solution for Plant Growth and Bioactive Compounds of Agastache Rugosa in a Plant Factory. *Agronomy* **2020**, *10*, 76. [CrossRef]
80. Kudirka, G.; Viršilė, A.; Sutulienė, R.; Laužikė, K.; Samuolienė, G. Precise Management of Hydroponic Nutrient Solution PH: The Effects of Minor PH Changes and MES Buffer Molarity on Lettuce Physiological Properties. *Horticulturae* **2023**, *9*, 837. [CrossRef]
81. Fussy, A.; Papenbrock, J. An Overview of Soil and Soilless Cultivation Techniques—Chances, Challenges and the Neglected Question of Sustainability. *Plants* **2022**, *11*, 1153. [CrossRef]
82. Vatistas, C.; Avgoustaki, D.D.; Bartzanas, T. A Systematic Literature Review on Controlled-Environment Agriculture: How Vertical Farms and Greenhouses Can Influence the Sustainability and Footprint of Urban Microclimate with Local Food Production. *Atmosphere* **2022**, *13*, 1258. [CrossRef]

83. Sipos, L.; Balázs, L.; Székely, G.; Jung, A.; Sárosi, S.; Radácsi, P.; Csambalik, L. Optimization of Basil (*Ocimum basilicum* L.) Production in LED Light Environments—A Review. *Sci. Hortic.* **2021**, *289*, 110486. [CrossRef]
84. Warner, R.; Wu, B.S.; MacPherson, S.; Lefsrud, M. How the Distribution of Photon Delivery Impacts Crops in Indoor Plant Environments: A Review. *Sustainability* **2023**, *15*, 4645. [CrossRef]
85. Hendrickson, T.; Dunn, B.L.; Goad, C.; Hu, B.; Singh, H. Effects of Elevated Water Temperature on Growth of Basil Using Nutrient Film Technique. *HortScience* **2022**, *57*, 925–932. [CrossRef]
86. Naim Khalid, S.M. *FAO Crop Production Manual a Guide to Fruit and Vegetable Production in the Federated States of Micronesia*; FAO: Rome, Italy, 2020.
87. Lefsrud, M.G.; Kopsell, D.A.; Sams, C.E. Irradiance from Distinct Wavelength Light-Emitting Diodes Affect Secondary Metabolites in Kale. *HortScience* **2008**, *43*, 2243–2244. [CrossRef]
88. How to Grow Hydroponic Asparagus—Hydrobuilder Learning Center. Available online: https://hydrobuilder.com/learn/hydroponic-asparagus/ (accessed on 18 March 2024).
89. Pennisi, G.; Orsini, F.; Landolfo, M.; Pistillo, A.; Crepaldi, A.; Nicola, S.; Fernández, J.A.; Marcelis, L.F.M.; Gianquinto, G. Optimal Photoperiod for Indoor Cultivation of Leafy Vegetables and Herbs. *Eur. J. Hortic. Sci.* **2020**, *85*, 329–338. [CrossRef]
90. Yan, D.; Ding, K.; He, Y.; Fan, L.; Wang, S.; Jiang, X. Strontium Uptake and Effects in Hydroponically Cultivated Celery. *Water Air Soil Pollut.* **2020**, *231*, 279. [CrossRef]
91. Pennisi, G.; Pistillo, A.; Orsini, F.; Cellini, A.; Spinelli, F.; Nicola, S.; Fernandez, J.A.; Crepaldi, A.; Gianquinto, G.; Marcelis, L.F.M. Optimal Light Intensity for Sustainable Water and Energy Use in Indoor Cultivation of Lettuce and Basil under Red and Blue LEDs. *Sci. Hortic.* **2020**, *272*, 109508. [CrossRef]
92. Karimi, M.; Ahmadi, N.; Ebrahimi, M. Red LED Light Promotes Biomass, Flowering and Secondary Metabolites Accumulation in Hydroponically Grown *Hypericum perforatum* L. (Cv. Topas). *Ind. Crops Prod.* **2022**, *175*, 114239. [CrossRef]
93. Rahman, A.; Wahjuni, S.; Priandana, K. The Development of Hydroponic Nutrient Solutions Control Using Fuzzy and BPNN for Celery Plant. *Int. J. Adv. Sci. Eng. Inf. Technol.* **2022**, *12*, 431–436. [CrossRef]
94. Hydroponic Gardening—Leeks | Avocados Love Hydroponics. Available online: https://avocado99.wordpress.com/2008/07/02/hydroponic-gardening-leeks/ (accessed on 18 March 2024).
95. Growing Leeks Hydroponically—Nutrients, PH, and EC | Agri Farming. Available online: https://www.agrifarming.in/growing-leeks-hydroponically-nutrients-ph-and-ec (accessed on 18 March 2024).
96. Stamford, J.D.; Stevens, J.; Mullineaux, P.M.; Lawson, T. LED Lighting: A Grower's Guide to Light Spectra. *HortScience* **2023**, *58*, 180–196. [CrossRef]
97. IEEE Xplore Full-Text PDF. Available online: https://ieeexplore.ieee.org/stamp/stamp.jsp?tp=&arnumber=10198435 (accessed on 14 July 2024).
98. Rahman, M.M.; Field, D.L.; Ahmed, S.M.; Hasan, M.T.; Basher, M.K.; Alameh, K. LED Illumination for High-Quality High-Yield Crop Growth in Protected Cropping Environments. *Plants* **2021**, *10*, 2470. [CrossRef]
99. Rodrigues da Silva, A.A.; Francisco do Nascimento Sousa, P.; Soares de Lima, G.; Soares, L.A.d.A.; Gheyi, H.R.; de Andrade Silva, L.; Lopes Araújo Rocha, J.; Alberto Vieira de Azevedo, C. Salicylic Acid Alleviates the Effects of Salt Stress on the Physiology, Growth, and Production of Hydroponic Okra. *Arid. Land Res. Manag.* **2023**, *37*, 602–618. [CrossRef]
100. Mamledesai, H.; Soriano, M.A.; Ahmad, R. A Qualitative Tool Condition Monitoring Framework Using Convolution Neural Network and Transfer Learning. *Appl. Sci.* **2020**, *10*, 7298. [CrossRef]
101. Ahmed, A.; Zulfiqar, S.; Ghandar, A.; Chen, Y.; Hanai, M.; Theodoropoulos, G. Digital Twin Technology for Aquaponics: Towards Optimizing Food Production with Dynamic Data Driven Application Systems. *Commun. Comput. Inf. Sci.* **2019**, *1094*, 3–14. [CrossRef]
102. Witzel, O.; Wilm, S.; Karimanzira, D.; Baganz, D. Controlling and Regulation of Integrated Aquaponic Production Systems—An Approach for a Management Execution System (MES). *Inf. Process. Agric.* **2019**, *6*, 326–334. [CrossRef]
103. Liakos, K.G.; Busato, P.; Moshou, D.; Pearson, S.; Bochtis, D. Machine Learning in Agriculture: A Review. *Sensors* **2018**, *18*, 2674. [CrossRef]
104. Talaviya, T.; Shah, D.; Patel, N.; Yagnik, H.; Shah, M. Implementation of Artificial Intelligence in Agriculture for Optimisation of Irrigation and Application of Pesticides and Herbicides. *Artif. Intell. Agric.* **2020**, *4*, 58–73. [CrossRef]
105. Xu, X.; Lu, Y.; Vogel-Heuser, B.; Wang, L. Industry 4.0 and Industry 5.0—Inception, Conception and Perception. *J. Manuf. Syst.* **2021**, *61*, 530–535. [CrossRef]
106. Maddikunta, P.K.R.; Pham, Q.V.; Prabadevib, B.; Deepa, N.; Dev, K.; Gadekallu, T.R.; Ruby, R.; Liyanage, M. Industry 5.0: A Survey on Enabling Technologies and Potential Applications. *J. Ind. Inf. Integr.* **2022**, *26*, 100257. [CrossRef]

AI

Article

Application of Machine Learning for Insect Monitoring in Grain Facilities

Querriel Arvy Mendoza [1], Lester Pordesimo [2,*], Mitchell Neilsen [1], Paul Armstrong [2], James Campbell [2] and Princess Tiffany Mendoza [3]

1 Department of Computer Science, Kansas State University, Manhattan, KS 66506, USA; qasmendoza@ksu.edu (Q.A.M.); neilsen@ksu.edu (M.N.)
2 USDA-ARS Center for Grain and Animal Health Research, Stored Product Insect and Engineering Research Unit (SPIERU), Manhattan, KS 66502, USA; paul.armstrong@usda.gov (P.A.); james.campbell@usda.gov (J.C.)
3 Department of Grain Science and Industry, Kansas State University, Manhattan, KS 66506, USA; ptiffany@ksu.edu
* Correspondence: lester.pordesimo@usda.gov; Tel.: +1-785-776-2727

Abstract: In this study, a basic insect detection system consisting of a manual-focus camera, a Jetson Nano—a low-cost, low-power single-board computer, and a trained deep learning model was developed. The model was validated through a live visual feed. Detecting, classifying, and monitoring insect pests in a grain storage or food facility in real time is vital to making insect control decisions. The camera captures the image of the insect and passes it to a Jetson Nano for processing. The Jetson Nano runs a trained deep-learning model to detect the presence and species of insects. With three different lighting situations: white LED light, yellow LED light, and no lighting condition, the detection results are displayed on a monitor. Validating using F1 scores and comparing the accuracy based on light sources, the system was tested with a variety of stored grain insect pests and was able to detect and classify adult cigarette beetles and warehouse beetles with acceptable accuracy. The results demonstrate that the system is an effective and affordable automated solution to insect detection. Such an automated insect detection system can help reduce pest control costs and save producers time and energy while safeguarding the quality of stored products.

Keywords: deep learning; convolution neural network; insects; stored grain pests; insect detection; insect identification; insect monitoring; insect classification; machine learning

1. Introduction

While breeders and farmers strive for high-yielding grain varieties and increased crop production to meet the demand of the growing world population, up to 80% of total production is lost during harvesting, storage, and processing [1]. During storage, grain losses are estimated at 10–40% due to insect damage [2]. Insect infestation causes physical degradation (weight loss), grain quality and safety issues, and physical damage to storage structures and facilities, which all lead to health risks and economic loss [1–3].

Integrated Pest Management (IPM) has become the basis for insect management and control decisions in facilities where food and grain are processed or stored. IPM is based on insect ecology, and its operational plan has at least two key elements: monitoring-based decision-making and multiple pest control tactic applications. Given the challenges in finding and sampling insect populations directly in the temporal and spatial landscapes found in and around food facilities, monitoring relies on visual observation, or capture in traps, of dispersing adults. For stored insect pest prevention and control, various techniques and technologies have been explored and applied to identify, detect, and monitor insects in storage facilities. Commonly used are manual sampling traps and probes, manual inspection and sieving, the floatation method, kernel staining, uric acid measurement, the

Berlese funnel method, acoustic techniques, X-ray imaging, and the solid-phase micro-extraction method, among others [4]. Most of these methods are destructive, subjective, slow, labor-intensive, and time-consuming; usually have low accuracy and efficiency; could be limited to a few insect pests or specific insect growth stages; and may not be applicable to real-time applications [4]. Insect pests are small and, superficially, look very similar, so it takes highly trained people to accurately identify the species and age captured in traps and determine their risk level. Finally, most traps are selective in their captures since different pheromones and kairomones attract species, sexes, and developmental stages, making it difficult to measure populations accurately.

There is continued interest in the development of automated insect monitoring systems for both food production and the storage of raw and processed foods, but currently available technologies have limited ability to identify critical pests or require a person to do the actual insect identification. The use of artificial intelligence (AI) for the identification and counting of insects is a promising method to address the need for speed, timeliness, and reliability of insect information required for improving IPM methods. Cameras and other non-invasive sensors can continuously perform observations, and advances in computer vision technology, coupled with deep learning algorithms, provide methods to make AI insect identification possible. Images used to train deep-learning models could provide estimates of insect abundance, biomass, and diversity. Among the recent techniques for non-destructive, real-time, and rapid insect detection, near-infrared (NIR) spectroscopy and machine vision are promising. NIR spectroscopy is reliable and accurate but cannot detect low levels of insect infestation and differentiate between live and dead insects [3]. The machine vision system has been used for insect detection and monitoring, providing more benefits than the other methods. Aside from stored insect pest recognition and detection, it has also been applied in plant disease detection and crop quality grading [5]. Images and other outputs from these sensors used to train deep learning models could provide estimates of insect abundance, biomass, and diversity.

Previous work has focused on orchards and outdoor environmental pests, mostly flying, with some work conducted on crawling insects found in warehouses. Shuman et al. [6], Flinn et al. [7], and Banga et al. [4] looked at stored grain insects, while Potamitis et al. [8], Wen et al. [9], and Zhong et al. [10] focused on outdoor flying insects, and Eliopoulos et al. [11] examined crawling insects. Behavioral studies of insects could have benefited from automated insect enumeration and identification [12–14]. Shen et al. [15] recommended that mathematical modeling of insect movement continues to be a critical area of entomology research in which automated monitoring could play a significant role.

Machine vision has constantly been evolving as a technique for insect detection in stored grains. One of the first studies published was on detecting insects and body parts in bulk wheat samples, which used pattern recognition and multivariate analysis and obtained more than 90% accuracy [16]. Different algorithms and improved systems were developed to achieve high throughput [17], reduce instrument and computational costs, allow easy data access and real-time monitoring, and cover different insect pest species [18]. Some benefits of utilizing machine vision include: (a) the ability to analyze images and videos much faster and with higher accuracy than humans, which makes it suitable for high-speed quality control, inspection, and sorting tasks; (b) the capability to perform tasks consistently without errors or fatigue, ensuring consistent results; (c) the ability to analyze images and videos without physical contact, making it suitable for inspecting hazardous or fragile materials; (d) the flexibility of programming machine vision systems to analyze a wide variety of images and videos, enabling them to be used for various applications; and (e) the potential to reduce labor costs and enhance productivity, making it a cost-effective option for many applications.

However, the use of machine vision also has some drawbacks, which consist of: (a) the complexity of designing and implementing the systems, which require expertise in both hardware and software; (b) the cost of developing and implementing the systems, especially for specialized applications; (c) the systems' sensitivity to lighting conditions, which can affect the accuracy and consistency of their results; (d) the limited understanding of context and inability to make judgments based on experience or intuition of systems using machine vision; and (e) the limited applicability of the systems, especially for applications that require complex decision-making skills beyond image analysis.

More recently, machine vision was combined with machine learning for enhanced stored insect grain detection and classification accuracy [5]. One of the most widely used machine learning techniques is the convolutional neural network (CNN), which consists of a complex network structure and has the ability to execute convolution operations [19]. Shen et al. [15] applied a deep neural network to detect and identify six species of stored grain insects that were blended with grain and dockage materials. Although they achieved a high mean average precision (mAP) of 88%, they used different models for different insects and encountered issues such as one species being classified as two. Another insect detection system built on deep CNN focused on different insect sizes and obtained mAP up to 95%, but still, they only detected one insect species at a time [20].

Being able to detect, classify, and monitor insect pests present in a grain storage or food facility in real-time will be vital to decision-making and insect control. In our study, we aimed to develop a camera-based monitoring system to detect and identify crawling stored grain insect pests using CNN and validate the model using a live feed.

2. Materials and Methods

The list of hardware components for this project and the software solution to detect and identify insects in real time are described in the following sections, along with the testing. This basic system was tested on adult warehouse beetles (*Trogoderma variabile*) and cigarette beetles (*Lasioderma serricorne* (F.)) from a culture maintained at the USDA-ARS Center for Grain and Animal Health Research in Manhattan, KS, USA. These insect species were selected as the test cases because they are major insect pests in food processing facilities (equipment and structures) and food product warehouses, and they are morphologically different [21]. The adult cigarette beetle is 2 to 4 mm long, 1.25 to 1.5 mm wide, and light to dark brown. Their antennae are saw-like and have the same thickness from the base to the tip. The head is withdrawn under the insect when it is at rest or dead, giving the insect a characteristic humped appearance. Adult warehouse beetles are smaller than cigarette beetles, at about 2.7 to 3.5 mm in length, oval, and predominantly black, with dense, dark-brown scales or setae covering the hardened forewings or elytra.

2.1. Hardware Components

Raspberry Pi (RPi) version 4 (the mention of trade names or commercial products in this publication is solely for the purpose of providing specific information and does not imply recommendation or endorsement by the U.S. Department of Agriculture. USDA is an equal opportunity provider and employer) (Raspberry Pi Trading Ltd., Cambridge, UK) was used in the original design of this study. Following extensive testing, modifications, and the lack of computing resources, it was replaced with an NVIDIA Jetson Nano 4 GB developer kit (NVIDIA, Santa Clara, CA, USA) [22], which served as the primary Central Processing Unit (CPU). The Jetson Nano has a built-in, dedicated Graphical Processing Unit (GPU), and an external camera was mounted on the Camera Serial Interface (CSI) port. The Arducam IMX477 B0250 model (ArduCAM, Nanjing, China) [23] was used for capturing images that also served as a medium for detection and identification. To reduce the chance of overheating while the GPU and CPU were simultaneously processing, a constant, on-demand cooling system was mounted on the Jetson Nano. A Universal Serial Bus (USB) ring light was also used for illumination, and black, non-reflective paper was used to cover the part where the insects were placed. As for the base, a microscope stand was used to adjust the height settings to calibrate the focus. Figure 1 shows how the hardware products were connected.

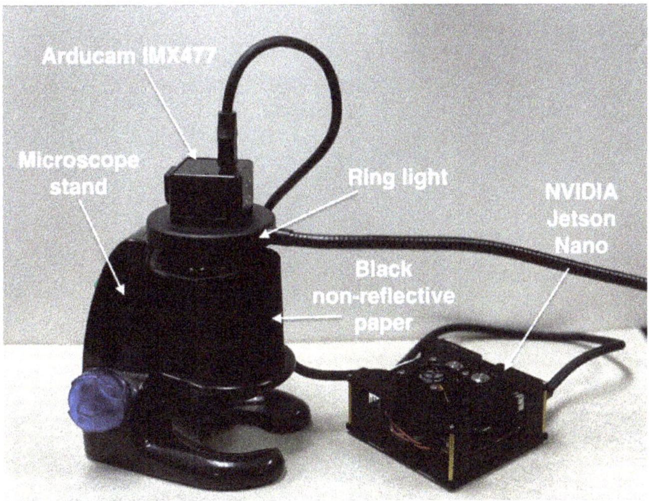

Figure 1. Connected hardware set-up.

A regular keyboard, mouse, and external monitor were also connected to the Jetson Nano for initial configuration.

2.2. Software Solution

As part of the software package for the Jetson Nano, the NVIDIA JetPack (SD card image) 4.2 was used, and the procedures listed on the website of NVIDIA were followed. After the first boot for the initialization phase, software drivers for the Arducam were downloaded from their website for primary configuration. The compatibility of the CSI camera with the Jetson Nano must require no connectivity problems since it serves as our eyes in the field.

2.3. Preparing the Model

The gathering of data was the most tedious part of this project. The summary of the steps is presented in Figure 2. First, a label file was created, which was a text file containing the list of names of our insects (cigarette and warehouse beetles).

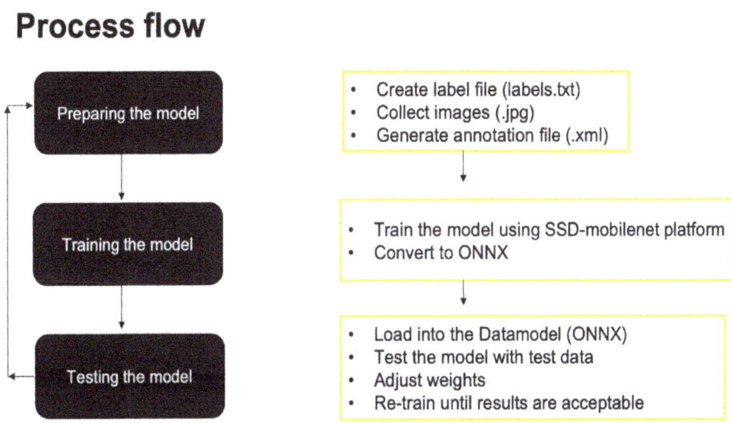

Figure 2. Model development process.

The next step was to capture images in JPG format. Two processes were used: first, images were downloaded from the internet (Figure 3) using basic data search, and second, using the CSI camera that was mounted on the Jetson Nano (Figure 4). Data augmentation was applied to each image. Data augmentation is a method of artificially expanding a dataset by applying various alterations to the existing samples. Data augmentation helps reduce overfitting in machine learning models and improves their ability to generalize to new data. In computer vision, popular data augmentation techniques include random cropping, flipping, rotations, translations, scaling, and color jittering. Training a model on a larger, augmented dataset can help it learn more robust features and generalize better to unseen data. It is frequently utilized when a model is trained to recognize objects or image patterns. The model may be taught to identify objects in various orientations, locations, and scales by applying random changes to existing photos. As a result, the model may be trained on a wider and more varied set of instances, which is particularly crucial for situations where there is little annotated data available.

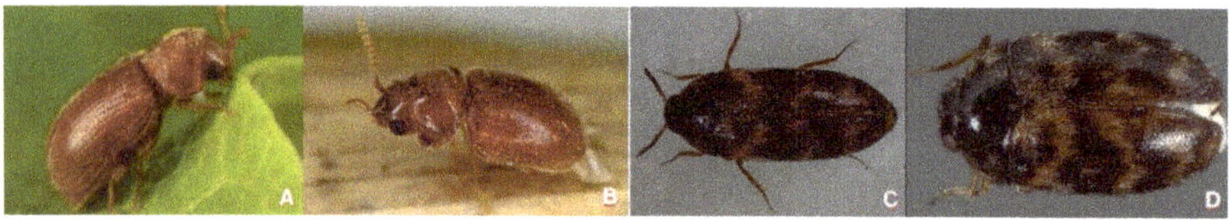

Figure 3. Downloaded sample images of insects used in the study. Cigarette beetle (**A**,**B**) and warehouse beetle (**C**,**D**).

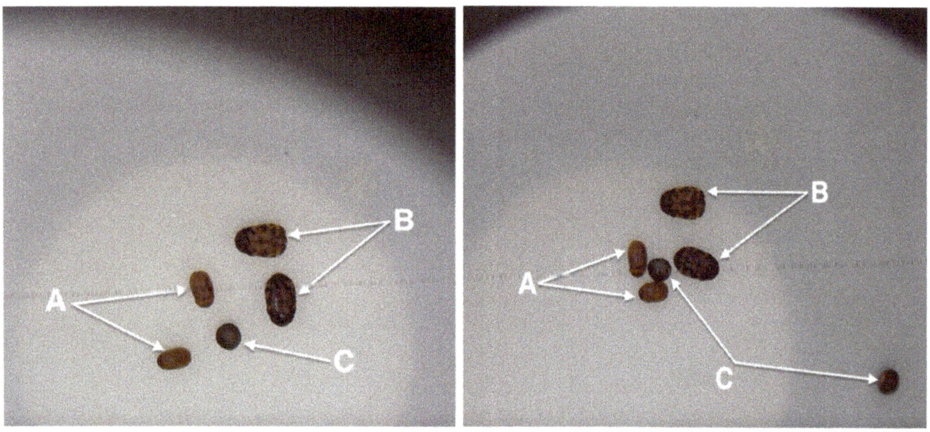

Figure 4. Two examples of actual images of the cigarette beetle (A), warehouse beetle (B), and black canola seeds (C) used in training the model.

More training data were produced by this process, which resulted in 1000 images (500 images for each insect classification), which were then segregated into 80–20 partitions: 80% for training data and 20% for test data (Table 1). However, it is important to apply augmentation techniques relevant to the problem domain and to use them judiciously to avoid over-augmentation because this can lead to a loss of information and decrease model performance.

Generating an annotation file was the next step in preparing the model. It is important in machine learning because it provides labeled data to train the model. The model uses this data to learn the relationships between inputs and outputs and to make predictions. The annotations provide information about features and classes of the data, which are critical for creating accurate and effective models. Without annotated data, the model would have no way of learning the desired relationships and predictions and would likely perform poorly. In this part, the LabelImg (https://github.com/tzutalin/labelImg (accessed on 10 August 2022) application was used for annotation. Steps on how to generate annotation files using LabelImg are listed on the GitHub page mentioned. By doing this, a new dataset was created that would provide the location of our subjects on the images captured by drawing a bounding box (with coordinates X-min, X-max, Y-min, and Y-max) (Figure 5). This annotation process was vital to data preparation since it served as the baseline of our training and detection phases; 800 annotated images were created based on this procedure.

Table 1. Partition of images per insect.

Insect Name	Scientific Name	Training Images	Test Images	Total
Cigarette beetle	*Lasioderma serricorne*	400	100	500
Warehouse beetle	*Trogoderma variabile*	400	100	500

```
<annotation>
  <folder>cigarette_beetle</folder>
  <filename>cigarette_beetle10.jpg</filename>
  <path>C:\Users\...\cigarette_beetle10.jpg</path>
  <source>
    <database>Unknown</database>
  </source>
  <size>
    <width>276</width>
    <height>183</height>
    <depth>3</depth>
  </size>
  <segmented>0</segmented>
  <object>
    <name>cigarette_beetle</name>
    <pose>Unspecified</pose>
    <truncated>0</truncated>
    <difficult>0</difficult>
    <bndbox>
      <xmin>157</xmin>
      <ymin>89</ymin>
      <xmax>241</xmax>
      <ymax>133</ymax>
    </bndbox>
  </object>
</annotation>
```


Figure 5. Annotation of the captured image.

2.4. Training the Model

The backbone used for this study was a single-shot multi-box detector (SSD), an algorithm that uses a multi-box that takes only one shot to detect multiple objects in an image. It uses a single deep neural network to accomplish this. It can also detect objects of different sizes in an image. SSD is independent of its base network, but some practical limitations will occur while deploying and running a complex neural network in real-time applications since it will require high computational power. For this reason, SSD was combined with MobileNet, which is one of the architecture models of the Convolution Neural Network (CNN) that mainly focuses on image classification for mobile applications. It uses depth-wise separable convolution layers instead of the standard convolution networks. One of the reasons why MobileNet was chosen is because of its architectural structure, which lessens the computational cost, and because it has very low computational power needed to run or apply transfer learning using a small computer such as the Jetson Nano.

When MobileNet V1 was used along with SSD (Figure 6), the computational process was reduced by omitting the last few layers, such as the FC, Maxpool, and Softmax. Therefore, the result from the last convolution layer in the MobileNet was used. Convoluting a stack of feature maps a few more times is needed to obtain a stack of feature maps. The result would then be used as input for its detection heads.

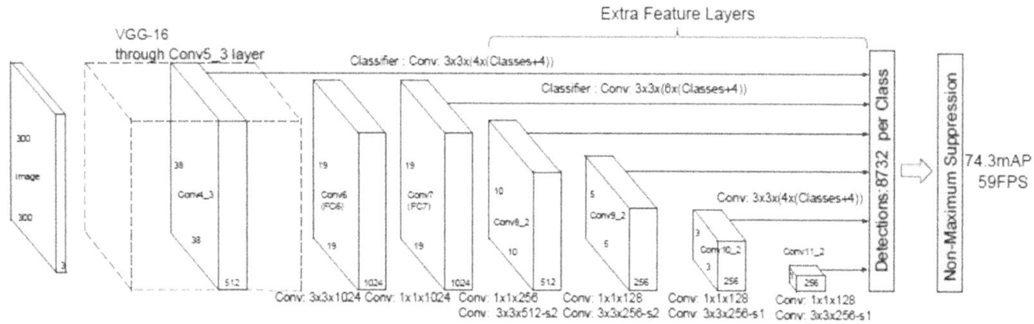

Figure 6. MobileNet-SSD architecture [24].

A resolution of 608 $\times$ 608 $\times$ 3 pixels was used to train these images, and they have been converted again to 300 $\times$ 300 $\times$ 3 pixels for validation since the latter was the default setting of SSD-MobileNet V1. This method produced better results than training them with 300 $\times$ 300 $\times$ 3 pixel images. The training images' greater quality improved accuracy for small items.

PyTorch is used to train the model. Pytorch is an open-source machine learning library for Python, primarily used for natural language processing and computer vision. It is based on the Torch library and provides a comfortable and flexible interface for building and training deep learning models. PyTorch also offers strong support for GPU acceleration, making it well-suited for training large-scale neural networks. The conversion of data from PyTorch format to Open Neural Network Exchange (ONNX) format was implemented. ONNX is an open standard for demonstrating deep learning models that enables interoperability between different AI frameworks, such as PyTorch, TensorFlow, Caffe2, and others. ONNX provides a common format for exchanging models between these frameworks, allowing researchers and developers to move models more easily between frameworks for training, inference, and experimentation. It also supports hardware acceleration on various platforms, including GPUs and custom hardware accelerators. By importing the ONNX data file to be used in TensorRT [25]. TensorRT is a Software Development Kit (SDK) that includes a deep learning inference optimizer and runtime library developed by NVIDIA. It is designed to optimize and accelerate the deployment of deep learning models on GPUs for real-time, high-performance inference, which the Jetson Nano uses.

3. Results and Discussion

Mean Average Precision (mAP) is a commonly used evaluation metric for information retrieval and computer vision tasks such as object detection and image classification [26]. It is the average precision value obtained at the first-ranked positive instances. It summarizes a ranked list of results by computing the average precision value for each item, where precision is the number of true positive instances divided by the number of positive instances predicted. The main use of mAP is to compare the performance of different algorithms and object detection models such as MobileNet-SSD, You Only Look Once (YOLO), and Faster R-CNN for a given task to assess the models in their pipelines.

Additionally, mAP is used in several benchmark challenges, such as the Common Objects in Context dataset (COCO) and Pascal Visual Object Classes. When computing for mAP, identifying a confusion matrix is needed (Table 2) along with Intersection over Union (IoU), Recall, and Precision. Generating a confusion matrix requires four attributes: True

Positive (TP)—the model properly predicted a label and matched the ground truth; True Negative (TN)—the model neither predicts the label nor comprises the ground truth; False Positive (FP)—a label was predicted by the model, although it is not part of the ground truth, and False Negative (FN)—the model did not predict a label, but it is part of the ground truth [26] as shown in Table 3.

Table 2. Confusion matrix with attributes.

Actual Result	Predicted No	Predicted Yes
Actual No	True Negative	False Positive
Actual Yes	False Negative	True Positive

Table 3. Confusion matrix with values from test images.

Insect	Predicted No	Predicted Yes
Cigarette Beetle	TN = 78	FP = 22
Warehouse Beetle	FN = 17	TP = 83

From here, substitute all our values from the test experiment to calculate these attributes. Since we utilized 100 test images for each insect, a total of 200 were used.

$$\text{Accuracy} = (\text{TP} + \text{TN})/\text{Total} = (83 + 78)/200 = 0.805$$

$$\text{Error Rate} = 1 - \text{Accuracy} = 1 - 0.805 = 0.195$$

$$\text{Precision} = \text{TP}/(\text{TP} + \text{FP}) = 83/(83 + 22) = 0.79$$

$$\text{Recall} = \text{TP}/(\text{TP} + \text{FN}) = 83/(83 + 17) = 0.83$$

Accuracy shows how often the classifier is correct overall, while Misclassification, also known as Error Rate, tells us how often the classifier is wrong. That is why calculating the error rate is one minus (−) the accuracy, because when you add them, the overall probability should be 1 or 100%. Knowing Precision and Recall is important since it will determine if the model is correctly implemented and to what extent. Precision is a useful indicator to determine when the costs of FP are high or simply answers the question: when it predicts "yes", how often is it correct? While Recall determines the number of actual Positives that our model successfully captures by classifying them as Positive (True Positive) or simply answering the question, when it is actually "yes", how often does it predict yes?

A balance between Precision and Recall requires an F1 score. The F1 score is a metric for machine learning that can be used in classification models. It is a proposed improvement to two simpler performance metrics. What is the difference between the F1 score and accuracy? The accuracy is the ratio of correctly predicted observations to the total number of observations. It is a simple metric that shows how well a model performs on a given task.

The F1 score is a harmonic mean of Precision and Recall, providing a single metric that summarizes the performance of the model. The F1 score is best used when you have an imbalanced class distribution, whereas accuracy works well if the class distribution is balanced. So, the F1 score considers the balance between Precision and Recall, while accuracy only considers the number of correct predictions. In other words, accuracy is sensitive to class imbalance, while the F1 score is not. The F1 score may be a better metric to use to find a balance between Precision and Recall, and there is an uneven class distribution.

$$\text{F1 score} = 2\,[(\text{Precision} \times \text{Recall})/(\text{Precision} + \text{Recall})] = 2\,[(0.79 \times 0.83)/(0.79 + 0.83)] = 0.81$$

When validating from our table below with 200 epochs on a 20-epoch increment, our best F1 score is 81% using white ring light within 120 epochs.

As shown in Table 4 and Figure 7, the validation loss decreases as the number of epochs increases. However, on a certain level, after the 120th epoch, it is slowly increasing again. This is called overfitting. Overfitting is a typical problem in machine learning, where a model is trained too successfully on the training data, leading to poor performance on new, unforeseen data. Overfitting occurs when a model learns the noise or random fluctuations in the training data rather than the underlying patterns. This leads to a model that fits the training data too closely but cannot generalize to new data. As a result, an overfit model will perform well on the training data but poorly on the test data or real-world data because it has not learned the true relationship between the input features and the output. To avoid this, an effective machine learning model should balance the ability to generalize to new data with the ability to fit the training data well to prevent overfitting. This can be accomplished by applying regularization techniques, simplifying the model, or providing more training data. Additionally, using a dedicated ring light greatly affects our accuracy. When comparing the white and yellow lights, there was a slight difference in detection, and most of the time, the white light gave the highest accuracy percentage. Then, epochs were increased from the usual 100 to 200. This did not improve the overall performance.

Table 4. Mean average precision table with values using different light sources.

Epoch	Validation Loss	No Ring Light	Yellow Ring Light	White Ring Light
20	2.9333	16%	23%	22%
40	2.2033	57%	68%	68%
60	1.9533	65%	68%	71%
80	1.7967	66%	70%	73%
100	1.5300	73%	78%	79%
120	1.5900	72%	76%	81%
140	1.6400	74%	77%	77%
160	1.6933	73%	76%	73%
180	1.7867	69%	73%	75%
200	1.7700	71%	73%	73%

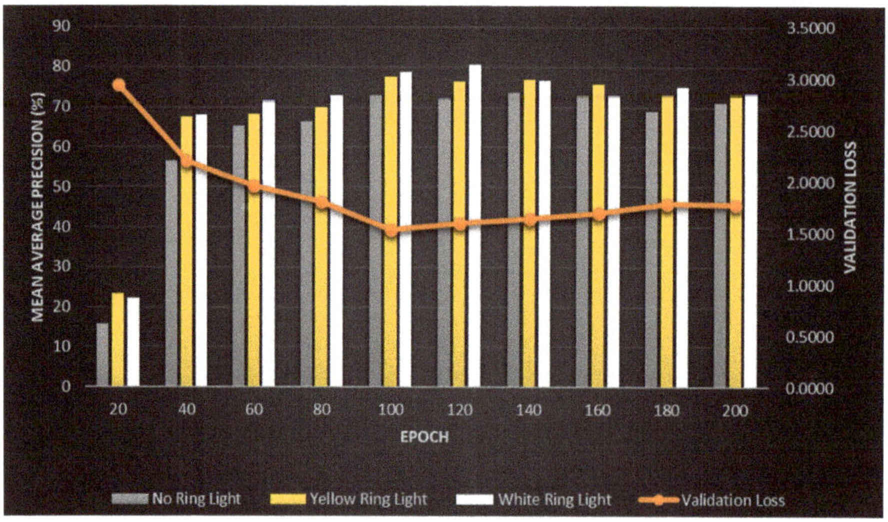

Figure 7. Bar chart comparison of mAP, including validation loss.

However, when the number of epochs was increased to around 120 epochs, the accuracy increased, and the validation loss decreased. After 125 epochs, the performance did not improve. From this result, the code was revised to take the lowest validation loss on a given epoch stream. Below is an example with 200 epochs, but the system uses the lowest validation loss at the 125th epoch (Figure 8).

```
2022-09-01 16:20:58 - Epoch: 199, Step: 140/170, Avg Loss: 2.4830, Avg Regression Loss: 0.8149, Avg Classification Loss: 1.6681
2022-09-01 16:21:16 - Epoch: 199, Step: 150/170, Avg Loss: 2.3959, Avg Regression Loss: 0.7244, Avg Classification Loss: 1.6751
2022-09-01 16:21:35 - Epoch: 199, Step: 160/170, Avg Loss: 2.1305, Avg Regression Loss: 0.7425, Avg Classification Loss: 1.3880
2022-09-01 16:22:51 - Epoch: 199, Validation Loss: 1.8761, Validation Regression Loss: 0.4292, Validation Classification Loss: 1.4469
2022-09-01 16:22:51 - Saved model models/insects/mb1-ssd-Epoch-199-Loss-1.8761031282018628.pth
2022-09-01 16:22:51 - Task done, exiting program.
root@desktop:/jetson-inference/python/training.../ssd# python onnx export.py --model-dir=models/insects
Namespace(batch_size=1, height=300, input=' ', labels='labels.txt', model_dir='models/insects', net='ssd-mobilenet', output=' ', width=300)
running on device cuda:0
Found best checkpoint with loss 1.533995 (models/insects/mb1-ssd-Epoch-124-Loss-1.5339945586475394.pth)
creating network: ssd-mobilenet
num classes: 3
loading checkpoint: models/insects/mb1-ssd-Epoch-124-Loss-1.5339945586475394.pth
exporting model to ONNX...
```

Figure 8. Completed 200 epochs with a validation loss of 1.876103 and still using the 125th epoch with a smaller validation loss.

Most published papers regarding insect detection and classification only use one insect per classification, but in this research paper, side-by-side testing was used with smaller-scaled insects. By doing this, a decrease in accuracy was noticed. When detecting and classifying one insect, accuracy is around 85–90% depending on the angle (our camera is in manual focus) and lighting condition. Then, doing it side-by-side, the accuracy is around 70–85% (Figure 9). One of the reasons is because the model is overfit. Increasing our training data set and re-training the model are needed to improve the results. Adjusting the annotation and providing additional spaces on the edges will also help.

Figure 9. The actual live camera shot detected and identified the beetle side-by-side.

4. Conclusions and Future Direction

This study successfully achieved its goal by developing a camera-based monitoring system to detect and identify small-scale insects, using a live feed video for the model to verify, and achieving a confidence level of at least 70% using a side-by-side comparison. There were some difficulties with the lighting and camera set-up since our camera is manual focus, and the total capacity of our designed model was limited. Adjusting the distance of the insects from the camera and manually adjusting the focal length of the camera to focus on the insects was a difficult and tedious task. This is not a reliable concept when applying it to a real-world scenario because, in the real world, insects are crawling and flying. This means losing valuable data while adjusting the settings on the camera since it is out of focus. Another challenge is to address the overfitting by providing additional training data.

Despite these challenges, the results were promising, and we believe the model has the potential for further improvement with the right modifications and technological advancements. To overcome the limitations of manual focus, we plan to integrate an auto-focus camera in the future. This will eliminate the need to manually adjust the camera settings now and then. Additionally, we will gather more diverse and larger data sets to increase the accuracy and robustness of the model. With these improvements, our camera-based monitoring system has the potential to become an effective tool for monitoring and identifying small-scale insects in various industries, such as agriculture and forestry. In conclusion, the camera-based monitoring system showed favorable results, and we believe that with further development, it can become a valuable tool for monitoring and identifying small-scale insects in a range of industries.

The ability of the system to identify insects accurately could lead to more effective pest management practices, reducing the reliance on harmful chemicals and improving crop yields. The system could also be used in environmental monitoring, providing valuable data on insect populations and their potential impacts on ecosystems. The potential applications of this technology are vast and exciting, and we are eager to continue its development and improve its capabilities. The technology could also have applications in areas such as public health by assisting in the early detection and monitoring of disease-carrying insects. It could also provide valuable information for scientific research, aiding in the study of insect behavior and its role in the food chain. Furthermore, the accuracy could greatly benefit farmers and food producers, leading to more efficient and cost-effective crop pest control. This technology can revolutionize how we approach insect detection and management, and its full potential is yet to be explored. It is important to point out that the test cases in this study involved insects that had stark morphological differences. Many insects are almost identical in features, so the technology would need much more refinement, or it could just serve as an initial screening tool for general identification.

Future research on machine vision is expected to investigate other aspects, such as the implications of using auto-focus cameras on image analysis. Performance comparisons of different algorithms are currently being carried out and will be written up in a new paper. Additionally, the impact of clear and blurred images on the analysis accuracy will be another area of focus. Although serving latency was deemed insignificant for two insects, further investigation will explore how it could become more relevant as the number of insects being analyzed rises. These additional factors will contribute to a more comprehensive understanding of the potential and constraints of machine vision in various applications.

For future work, the plan is to incorporate a cloud-based platform for storing and retrieving data for future analysis. Developing a stationary and autonomous mobile platform incorporating processing and communications capabilities to transfer information to a server for additional processing and user output is also on our list, as is implementation in a real-world scenario to help our farmers detect and identify these insects and provide the necessary actions to eliminate the level of infestation on their stored product facilities.

Author Contributions: Q.A.M.: Abstract, Introduction, Methodology, Investigation, Data Curation, Formal Analysis, Writing—Original draft preparation. L.P., P.A. and J.C.: Conceptualization, Funding Acquisition, Project Administration, Writing—Reviewing and Editing. L.P. and M.N.: Conceptualization, Research Supervision, Methodology, Resources, Writing—Reviewing and Editing. P.T.M.: Abstract, Introduction. All authors have read and agreed to the published version of the manuscript.

Funding: This research was supported by USDA-ARS (CRIS Project No. 3020-43440-008-00D). No other funding was used to support the research.

Institutional Review Board Statement: Not applicable.

Informed Consent Statement: Not applicable.

Data Availability Statement: The data collected in accomplishing the study is available upon request.

Acknowledgments: The research was supported by USDA (CRIS No. 3020-43440-008-00D) and by the Department of Computer Science of the College of Engineering, Kansas State University.

Conflicts of Interest: The authors declare no conflict of interest.

References

1. Kumar, D.; Kalita, P. Reducing Postharvest Losses during Storage of Grain Crops to Strengthen Food Security in Developing Countries. *Foods* **2017**, *6*, 8. [CrossRef] [PubMed]
2. Johnson, J.B. An overview of near-infrared spectroscopy (NIRS) for the detection of insect pests in stored grains. *J. Stored Prod. Res.* **2020**, *2020*, 101558. [CrossRef]
3. Srivastava, S.; Mishra, H.N. Detection of insect damaged rice grains using visible and near infrared hyperspectral imaging technique. *Chemom. Intell. Lab. Syst.* **2021**, *221*, 104489. [CrossRef]
4. Banga, K.S.; Kotwaliwale, N.; Mohapatra, D.; Giri, S.K. Techniques for insect detection in stored food grains: An overview. *Food Control* **2018**, *94*, 167–176. [CrossRef]
5. Kasinathan, T.; Singaraju, D.; Uyyala, S.R. Insect classification and detection in field crops using modern machine learning techniques. *Inf. Process. Agric.* **2021**, *8*, 446–457. [CrossRef]
6. Shuman, D.; Weaver, D.K.; Larson, R.G. Performance of an analytical, dual infrared-beam, stored-product insect monitoring system. *J. Econ. Entomol.* **2005**, *98*, 1723–1732. [CrossRef] [PubMed]
7. Flinn, P.W.; Opit, G.P.; Throne, J.E. Predicting Stored Grain Insect Population Densities Using an Electronic Probe Trap. *J. Econ. Entomol.* **2009**, *102*, 1696–1704. [CrossRef] [PubMed]
8. Potamitis, I.; Eliopoulos, P.; Rigakis, I. Automated remote insect surveillance at a global scale and the internet of things. *Robotics* **2017**, *6*, 19. [CrossRef]
9. Wen, C.; Guyer, D.E.; Lia, W. Local feature-based identification and classification for orchard insects. *Biosyst. Eng.* **2009**, *104*, 299–307. [CrossRef]
10. Zhong, Y.; Gao, J.; Lei, Q.; Zhou, Y. A Vision-Based Counting and Recognition System for Flying Insects in Intelligent Agriculture. *Sensors* **2018**, *18*, 1489. [CrossRef] [PubMed]
11. Eliopoulos, P.; Tatlas, N.-A.; Rigakis, I.; Potamitis, I. A "Smart" Trap Device for Detection of Crawling Insects and Other Arthropods in Urban Environments. *Electronics* **2018**, *7*, 161. [CrossRef]
12. Campbell, J.F.; Ching'oma, G.P.; Toews, M.D.; Ramaswamy, S.B. Spatial distribution and movement patterns of stored-product insects. In Proceedings of the 9th International Working Conference on Stored Product Protection, Campinas, Syo Paulo, Brazil, 15–18 October 2006; Lorini, I., Bacaltchuk, B., Beckel, H., Deckers, D., Sundfeld, E., dos Santos, J.P., Biagi, J.D., Celaro, J.C., FaroniL, R.D.A., de Bortolini, L.O.F., et al., Eds.; Brazilian Post-Harvest Association—ABRAPOS: Sao Paulo, Brazil, 2006; pp. 361–370.
13. Dowdy, A.K.; McGaughey, W.H. Seasonal activity of stored-product insects in and around farm-stored wheat. *J. Econ. Entomol.* **1994**, *93*, 1842–1847. [CrossRef]
14. Athanassiou, C.G.; Buchelos, C. Grain properties and insect distribution trends in silos of wheat. *J. Stored Prod. Res.* **2020**, *88*, 101632. [CrossRef]
15. Shen, Y.; Zhou, H.; Li, J.; Jian, F.; Javas, D.S. Detection of stored-grain insects using deep learning. *Comput. Electron. Agric.* **2018**, *145*, 319–325. [CrossRef]
16. Zayas, I.Y.; Flinn, P.W. Detection of insects in bulk wheat samples with machine vision. *Trans. ASAE* **1998**, *41*, 883–888. [CrossRef]
17. Ridgway, C.; Davies, E.R.; Chambers, J.; Mason, D.R.; Bateman, M. AE—Automation and Emerging Technologies: Rapid Machine VisionMethod for the Detection of Insects and other Particulate Bio-contaminants of Bulk Grain in Transit. *Biosyst. Eng.* **2002**, *83*, 21–30. [CrossRef]
18. Lima, M.C.F.; de Almeida Leandro, M.E.D.; Valero, C.; Coronel, L.C.P.; Bazzo, C.O.G. Automatic Detection and Monitoring of Insect Pests—A Review. *Agriculture* **2020**, *10*, 161. [CrossRef]
19. Liu, J.; Wang, X. Plant diseases and pests detection based on deep learning: A review. *Plant Methods* **2021**, *17*, 22. [CrossRef] [PubMed]
20. Li, J.; Zhou, H.; Wang, Z.; Jia, Q. Multi-scale detection of stored-grain insects for intelligent monitoring. *Comput. Electron. Agric.* **2020**, *168*, 105114. [CrossRef]
21. Edde, P.A.; Eaton, M.; Kells, S.A.; Philips, T.W. Biology, behavior, and ecology of pests in other durable commodities. In *Stored Product Protection*; Hagstrum, D.W., Phillips, T.W., Cuperus, G., Eds.; Kansas State University Agricultural Experiment Station and Cooperative Extension Service: Manhattan, KS, USA, 2012; pp. 45–62.

22. NVIDIA. NVIDIA® Jetson Nano™ 4GB Developer Kit. 2023. Available online: https://developer.nvidia.com/embedded/jetson-nano-developer-kit (accessed on 15 July 2022).
23. Arducam. Arducam®. 2020. Available online: https://www.arducam.com/product/arducam-complete-high-quality-camera-bundle-for-jetson-nano-xavier-nx/ (accessed on 1 August 2022).
24. Liu, W.; Anguelov, D.; Erhan, D.; Szegedy, C.; Reed, S.; Fu, C.; Berg, A.C. SSD: Single Shot MultiBox Detector. In *Computer Vision–ECCV 2016: 14th European Conference, Amsterdam, The Netherlands, 11–14 October 2016, Proceedings, Part I 14*; Lecture Notes in Computer Science; Leibe, B., Matas, J., Sebe, N., Welling, M., Eds.; Springer: Cham, Switzerland, 2016; Volume 9905. [CrossRef]
25. Bareeva, J. How to Convert a Model from PyTorch to TensorRT and Speed up Inference. 2020. Available online: https://learnopencv.com/how-to-convert-a-model-from-pytorch-to-tensorrt-and-speed-up-inference/ (accessed on 21 November 2022).
26. Shah, D. Mean Average Precision (mAP) Explained: Everything You Need to Know. 19 January 2023. Available online: https://www.v7labs.com/blog/mean-average-precision (accessed on 18 February 2023).

AI

MDPI

Article

FedBirdAg: A Low-Energy Federated Learning Platform for Bird Detection with Wireless Smart Cameras in Agriculture 4.0

Samy Benhoussa *, Gil De Sousa and Jean-Pierre Chanet

TSCF Research Unit, INRAE, Clermont Auvergne University, 63172 Aubière, France;
gil.de-sousa@inrae.fr (G.D.S.); jean-pierre.chanet@inrae.fr (J.-P.C.)
* Correspondence: samy.benhoussa@inrae.fr

Abstract: Birds can cause substantial damage to crops, directly affecting farmers' productivity and profitability. As a result, detecting bird presence in crop fields is crucial for effective crop management. Traditional agricultural practices have used various tools and techniques to deter pest birds, while digital agriculture has advanced these efforts through Internet of Things (IoT) and artificial intelligence (AI) technologies. With recent advancements in hardware and processing chips, connected devices can now utilize deep convolutional neural networks (CNNs) for on-field image classification. However, training these models can be energy-intensive, especially when large amounts of data, such as images, need to be transmitted for centralized model training. Federated learning (FL) offers a solution by enabling local training on edge devices, reducing data transmission costs and energy demands while also preserving data privacy and achieving shared model knowledge across connected devices. This paper proposes a low-energy federated learning framework for a compact smart camera network designed to perform simple image classification for bird detection in crop fields. The results demonstrate that this decentralized approach achieves performance comparable to a centrally trained model while consuming at least 8 times less energy. Further efficiency improvements, with a minimal tradeoff in performance reduction, are explored through early stopping.

Keywords: artificial intelligence; agriculture; autonomous things; energy-efficiency; federated learning; early stopping; bird detection; image classification

1. Introduction

Digital agriculture, also known as Agriculture 4.0, represents the latest evolution in agricultural practices, integrating advanced technologies and tools to develop sustainable agricultural systems able to address food security challenges posed by a rapidly growing global population [1]. This new era of agriculture leverages the latest digital technologies, particularly artificial intelligence (AI) [2,3], robotics [2,4], and the Internet of Things (IoT) [5], among others. These technologies rely on advanced, complexity-optimized software and lightweight, cutting-edge hardware to meet the performance demands of agricultural applications. Since digital agriculture devices must operate outdoors for extended periods, energy efficiency is essential for maintaining reliable and effective systems.

Numerous studies have focused on enhancing energy efficiency by reducing power consumption in energy-constrained connected devices such as wireless sensor networks (WSNs) when utilized in agricultural applications [6], with a focus on wireless transmission, which represents the most significant component of energy costs in WSNs [7]. These studies provided energy-aware networking techniques and protocols to reduce this cost [6,8,9].

As edge devices become increasingly capable of handling complex computations, autonomous things (AuTs), and largely the Internet of Autonomous Things (IoAT), are gaining traction in many fields [10], enabling enhanced decision-making by providing "in-situ" localized knowledge. These innovative systems deploy AI models directly in the field, enabling real-time decision-making and bringing the concept of Edge AI to life [11]. However, the deployment of autonomous things tends to increase power consumption due to the computational intensity of these models. Moreover, building effective knowledge models requires a training phase that can be energy-intensive, as large datasets often involve substantial computations and classically need to be transmitted to a remote server for AI model development before the model weights are sent back to edge devices for on-site inference.

Edge training enables the development of localized knowledge without requiring large data transfers to a remote server, thereby conserving energy by reducing transmission costs. However, training on edge devices poses challenges, as it demands substantial computational power and memory [12], which are often difficult to provide on these devices due to their limited processing and storage capacities. Also, training a network of connected things can cause communication traffic congestion due to increased interactions between the nodes [13]. From an energy perspective, training an AI model directly on the edge for subsequent on-device inference offers practical advantages, particularly in reducing energy costs associated with data transmission. However, this approach demands greater computational power in edge devices during the training phase. To optimize energy efficiency, a balance must be achieved between the computational costs of on-edge training and the transmission savings it provides.

Introduced by Google in 2016 [14], federated learning (FL) is a distributed learning paradigm for edge training, where a model is deployed across multiple devices to train locally on their data and share their knowledge with each other [15]. After being initialized in every node, local model weights are iteratively trained on local data, aggregated into a global model, and re-deployed for further local, on-device training until convergence is achieved (Figure 1). As the Figure depicts, Step 0 is performed only once, before the federated learning loop starts, while steps 1–3 are repeated in each federated round, which creates the federated learning loop. Federated rounds continue until the model converges, meaning local training no longer improves the global model. Images of a pigeon and a crow illustrating steps 2 and 3 exemplify non-independently and identically distributed (non-IID) data, as a camera may capture data with only specific features (e.g., only pigeons or only crows) during model training yet encounter previously unseen features during post-deployment inference (e.g., both pigeons and crows).

Training a wireless smart camera network (WSCN) using an FL training scheme enhances decision-making accuracy uniformly across the network's devices by enabling knowledge sharing among them. In cases where data is heterogeneous (IID) or non-IID, the local models address this issue through the global model aggregation process in FL. In other words, each camera learns from the collective data insights of other cameras without directly sharing their data. However, non-IID data can impede the performance of federated learning, making convergence more challenging [16]. Studies have been conducted to address these challenges through data-driven techniques [17] or optimizations of the federated aggregation (also referred to as averaging) scheme [18].

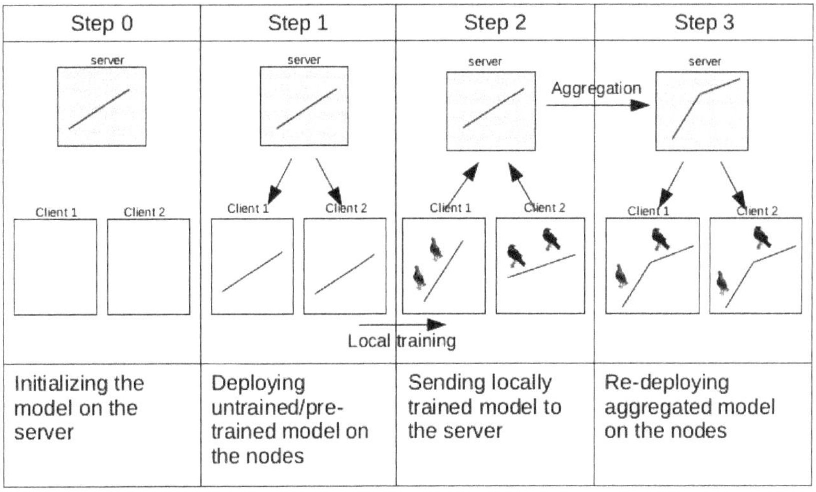

Figure 1. Illustration of FL training algorithm using our use case as an example.

In this paper, we propose an energy-efficient platform powered by a RaspBerry PI WSCN to detect bird presence in crop fields. The proposed system was designed for on-field training, guided by user supervision prior to deployment, and enabling adaptable and autonomous decision-making for optimizing agricultural profitability as an innovative technology in digital agriculture. Our work sets a primary objective of introducing an initial proof of concept of a low-energy platform for a lightweight smart camera network for bird detection in agricultural settings. The focus is to enhance the trade-off between energy efficiency of the post-deployment training phase and detection accuracy using FL through an on-field training simulation with a training dataset composed of real-world agricultural samples from a crop field vulnerable to pest bird attacks that was monitored from April 2021 to June 2021. We propose an FL framework for the on-site training of smart cameras, focusing on minimizing transmission energy costs while achieving equivalent model performance in inference compared to traditional remote training. Our approach strategically balances the computational energy required for edge-based training with the reduced energy demands for data transmission, creating a more efficient and adaptable system for field deployments. Our main contributions are as follows:

1. We introduce an energy-aware metric to assess the operational efficiency of AuTs in agricultural applications.
2. We develop a FL-based framework for the on-site training of a WSCN platform tailored for energy-efficient learning in the field. This framework is evaluated, using the introduced metric, against traditional centralized learning approaches in a bird detection use case for crop protection.
3. By examining trade-offs between computational energy and transmission energy in complex scenarios with highly non-IID data, we aim to provide insights into optimizing energy use for autonomous agricultural devices.

The paper is organized as follows. In Section 2, we review related work on AI and IoT applications in agriculture, focusing on training efficiency and FL applications with their advantages. We establish the novelty of our research by implementing FL specifically for the bird detection use case, with an emphasis on edge training efficiency. Section 3 outlines the methodology for implementing and analyzing our FL framework for on-site training by proposing a custom metric to assess the framework's efficiency. In Section 4, we implement this framework, assessing its efficiency through the previously introduced metric. Section 5 presents the results and compares our framework's efficiency to that of traditional centralized training methods. Section 6 provides conclusions on our findings, with insights into potential future work to enhance energy efficiency in edge AI, with a focus on optimizing inference-phase efficiency.

2. Related Work

Among the technologies driving digital agriculture, AI and IoT stand out as powerful tools for optimizing decision-making processes in farming. In training intelligent devices, FL, as previously introduced, offers several advantages, including enhanced privacy, energy efficiency, and knowledge sharing between trained devices. In this section, we review studies that integrate AI and IoT (or AIoT) applications in agriculture, emphasizing the applications and benefits of FL in the training phase of AIoT devices.

AI has found numerous applications in agriculture, whether integrated with computer vision for tasks like classification [19] or used for time-series forecasting [20]. Several platforms have developed AI-based automated decision-making systems to enhance agricultural efficiency and insight [21]. Additionally, the IoT has enabled the widespread use of connected devices to collect and process data directly at its source, allowing for more efficient and powerful agricultural systems [22]. Many agricultural use cases have harnessed AI and IoT technologies under the AIoT paradigm to boost performance and efficiency [23], utilizing FL, among other approaches, to train AuTs [24].

The application of AI and IoT in agriculture has become widespread and transformative, driving significant gains in efficiency and productivity across a range of agricultural use cases:

- **Crop monitoring**: IoT devices gather extensive data directly from the field, such as temperature, humidity, and soil moisture levels. AI models process this time-series data to predict crop growth and yield, enabling more informed decision-making for farmers [25–28].
- **Disease detection**: IoT-enabled cameras capture images of crops, which are analyzed via AI models using image recognition/classification techniques to identify symptoms of diseases early on. This allows for timely interventions, reducing the spread of diseases and improving crop health [29–33].
- **Soil conservation**: IoT sensors monitor key soil attributes like pH, nutrient levels, and moisture. AI algorithms analyze these data to provide recommendations for soil amendments and practices that enhance soil health and fertility [34,35].
- **Water conservation**: By tracking soil moisture and weather conditions, IoT devices feed data into AI systems that optimize irrigation schedules. This minimizes water usage while maintaining adequate hydration for crops, leading to better water conservation and reduced waste [36].
- **Smart harvesting**: IoT sensors continuously track environmental factors, providing real-time data that, when combined with AI models trained on historical datasets, can accurately identify the optimal time for harvesting. This approach maximizes yield quality while minimizing the risk of premature or delayed harvesting [37–39].
- **Animal monitoring**: IoT sensors, such as wearable devices, can monitor animal behavior, collecting data that track changes in key parameters indicative of health issues or other relevant conditions in livestock. These data, often represented as time series, can be used to train AI models capable of predicting diseases or classifying specific behaviors, enabling early intervention and improved herd management [40,41].
- **Supply chain management**: IoT devices monitor agricultural products throughout the supply chain, collecting data on conditions during transport and storage. AI ana-

lyzes this information to optimize logistics, manage inventory effectively, and reduce spoilage, ensuring that high-quality produce reaches consumers [42,43].

Overall, the integration of AI and IoT into agriculture is enabling smarter, data-driven practices, leading to improved sustainability, resource efficiency, and increased productivity.

In our study, we addressed the use case of detecting pest birds in crop fields using AI and IoT technologies. From an agricultural perspective, AIoT-enabled automated bird detection system can significantly enhance crop protection. Pest birds can cause extensive damage by eating seeds, fruits, or grains and spreading diseases. Detecting bird presence in real time helps prevent crop losses. Technically, detecting birds in a crop field is analogous to pest detection within the broader category of "disease detection" outlined above, as it also relies on image processing and classification techniques from camera-sensed data. The smart cameras capture field images, and AI models analyze these to identify distinct visual features indicative of birds, enabling timely interventions to protect crops.

Previous studies have focused on bird counting [44,45], species identification [46], or general bird recognition [47–50]. However, aside from [48], none of these works have specifically addressed the unique challenges of agriculture, such as detecting pest birds that pose a threat to crops. Moreover, they have not been trained on agricultural field images containing the specific pest bird species responsible for crop damage. Additionally, they do not consider the energy efficiency of the training process, a critical factor when deploying models on resource-constrained edge devices in agricultural environments, which is also lacking in the work of [48].

Unlike the works cited in this section on agricultural AI applications and bird detection, our goal is not to propose the most effective model in detecting pest birds accurately. Instead, we address the challenge of bird detection in agricultural crop fields through FL, with a strong emphasis on energy efficiency. Our approach prioritizes minimizing energy consumption during training, as our platform is designed to train directly in the field and quickly become functional for inference, reducing the energy demands of the training process. This work presents a simulation-based proof of concept, serving as a foundation for future on-field experiments with a platform of smart cameras capable of training efficiently in the field while maintaining reasonable inference performance. The study explores potential trade-offs between inference accuracy and the energy cost of training, which we aim to minimize using FL for initial convergence, followed by our proposed early stopping mechanism, LEFL.

Since its emergence, FL has been applied across various domains to train edge devices within the AIoT paradigm. Smart healthcare [51], smart industries [52], smart cities [53], and smart agriculture [54] are a few examples of domains that have already implemented FL frameworks to train their devices. The advantages of FL in these areas vary and include scalability, lower latency, data privacy, and energy efficiency, using fog aggregation to reduce frequent communication with a central server [55]. Also, FL offers significant advantages in environments with rapidly changing data, as training occurs directly at the edge, close to the data source [56]. This proximity allows the model to quickly adapt to new patterns and variations in real time, enhancing responsiveness and reducing latency compared to centralized training methods.

In our work, we explore the energy-efficiency advantage of FL in training agricultural connected devices, implementing a low-energy FL framework to train a MobileNet, a light convolutional neural network (CNN), in a WSCN for bird detection. Similar studies addressed the use of FL to train constrained devices in digital agriculture [54]. Recent studies highlighted several key advantages of applying FL in agriculture, such as data privacy and sovereignty [26,30–32,41,43,57,58], fault tolerance [32], reducing communication overhead [31], and improving model performance [28,33,59]. However, none of the surveyed works addressed the bird detection use case. Moreover, they did not place a primary emphasis on the energy efficiency of training agricultural AIoT devices.

Table 1 provides an overview of recent research papers on FL applied to agricultural AIoT systems, specifying the use case, the used IoT device(s), the AI model employed, and the key benefit gained from implementing FL.

Table 1. Previous works on FL applications in agricultural AIoT systems.

Paper	Use Case	IoT Device(s)/ Dataset(s)	AI Model(s)	FL Improvement(s)
[31]	Disease detection	UAVs	EfficientNet-B3	Communication overhead, data privacy
[32]	Weed detection	Hyperspectral Camera	Custom CNN	Fault tolerance, data sovereignty
[43]	Supply chain data management	Remote sensors, weather and soil data	Custom CNN, LSTM RNN	Data privacy
[26]	Crop yield prediction	Sensors, Cameras	ResNet-16, ResNet-28	Data privacy, data sovereignty
[30]	Maize leaf disease prediction	Cameras	AlexNet, SqueezeNet, ResNet-18, VGG-11, ShuffleNet	Data privacy
[57]	Intrusion detection	ToN-IoT dataset	GRU RNN	Data privacy
[41]	Automated animal activity recognition	Wearable sensors	CMI-Net	Data privacy
[58]	Securing agricultural IoT infrastructures	CSE-CIC-IDS2018, MQTTset, and InSDN datasets	Custom DNN, CNN, and RNN	Data privacy
[59]	Improving agricultural production	Smart sensors	Custom ML algorithm	Sensor control and adaptability
[28]	Crop classification	Climatic features	Gaussian Naive Bayes	Model accuracy
[33]	Disease and pest detection	Apple orchard images	ResNet-101	Model training speed
Our work	Bird detection	Smart cameras	MobileNetV2	Energy efficiency, knowledge sharing

3. Methodology

We present LEFL, a novel low-energy FL framework for the efficient training of a WSCN to detect bird presence in crop fields. The efficiency of a connected device can be measured as its performance relative to energy consumption. While device performance can vary, depending on the application, we focus on the quality of information (QoI) provided. By contrast to QoI used for basic IoT devices that cannot learn or adapt, such as wireless sensor networks (WSNs) [60], we define quality of knowledge (QoK) as the enhanced QoI for more advanced IoAT devices, such as WSCNs, capable of learning through local model training and developing knowledge. We use model accuracy—either of the locally trained models or the global shared model—as the primary metric to evaluate the QoK of the

WSCN. Achieving higher efficiency with LEFL involves achieving high QoK, reflected in accuracy enhancement by training the model while reducing the energy cost required for training through FL.

3.1. QoSAuT: Quality of Service of an Autonomous Thing

We introduce QoSAuT (quality of service of an autonomous thing) as an efficiency metric tailored for our use case. It is defined as the ratio of the shared model's accuracy to the energy expenditure incurred during the training process. This metric captures the balance between model performance in terms of QoK and the energy costs required to achieve that level of QoK, providing a holistic measure of the device's training efficiency. The metric can be expressed simply as follows:

$$QoSAuT(k, n_k, s_k) = \frac{\alpha.\Delta a_k}{n_k.(ct_k + \mu_k.s_k.x_k) + ci_k} \tag{1}$$

- k = AuT index.
- $QoSAuT_k$ = QoSAuT of the trained AuT k.
- Δa_k = accuracy gained via the model in k from the training, evaluated on unseen test data.
- n_k = number of training rounds of k.
- x_k = number of data frames sent via k in one training round.
- ct_k = computational cost * of one training round in k.
- ci_k = computational cost * of inference by the trained model in k.
- μ_k = transmission cost * of 1 byte by k.
- s_k = size (in bytes) of one data frame sent and/or received via k.
- α = the maximum portion of the battery capacity we are willing to allocate ** for training.

* The energy costs should be represented as a percentage of the total battery capacity of the smart camera during its operation, providing a clear metric for the device's energy consumption relative to its available power and making the QoSAuT a metric dependent on the capacity of the power storage used.

** The maximum percentage of the battery capacity allocated to the training phase is determined by the requirements of the subsequent inference phase—the more energy we allocate to training, the less we allocate to inference—which varies, depending on the specific application and its inference phase's requirements in terms of energy. Exceeding this limit during training can significantly impact QoSAuT, which is why this factor is incorporated into the formula.

It is important to note that the use of accuracy in evaluating QoSAuT is primarily intended to assess the overall quality of the training process, providing insight into the decision-making quality of the AuT independent of its specific use case. For a more adapted evaluation of decision-making quality emphasizing the user perspective, particularly in the context of bird detection, recall would be more appropriate, as false negatives are more detrimental than false positives (false negatives lead to crop damage, while false positives result in unnecessary alerts). However, in our work, accuracy serves as a general metric for assessing the training quality and energy-cost tradeoff of the trained AuT, irrespective of the deployment task.

Based on the definition of the QoSAuT metric above, we can interpret the quality of the autonomous thing's service by analyzing the quantitative metric's value. These interpretations are summarized in Table 2.

Table 2. Interpretation of the QoSAuT value.

QoSAuT	Quality of Service	Interpretation
$QoSAuT > 1$	Good	The AuT learned at a relatively low energy cost.
$QoSAuT = 1$	Average	The AuT learned near the limit of energy optimization.
$0 < QoSAuT < 1$	Bad	The AuT learned but at a relatively high energy cost.
$QoSAuT = 0$	Very bad	The AuT did not learn.
$QoSAuT < 0$	Disastrous	Theoretically impossible as the training must improve the model's accuracy.

3.2. LEFL: A Low-Energy Federated Learning Framework

We leverage FL to aggregate locally trained models into a global model without exchanging the vast amounts of raw data typically required during the training phase. This approach is particularly beneficial for smart cameras, for which sharing image data would be bandwidth-intensive and impractical. Instead, FL transmits model weight updates, which are significantly smaller in size, reducing the communication load and enabling more energy-efficient, scalable training across the network's nodes. While using FL for training helps alleviate network bottlenecks, allowing more data to be utilized effectively and improving the QoSAuT, our primary focus in this work is on the energy efficiency benefits, rather than the broader network-level advantages of FL. Federated learning (FL) can require numerous rounds for model convergence [14], and each round involves local computations on participating nodes. A high number of rounds can lead to increased energy consumption, presenting a challenge. Thus, a trade-off must be established between achieving model accuracy and minimizing energy costs. We propose LEFL, a low-energy federated learning framework that aims to address this by identifying the optimal point that balances consistent model accuracy with reduced energy expenditure. In our work, we leverage LEFL to improve the global QoSAuT of a WSCN.

3.3. Bird Detection Scenario

To implement LEFL the framework, we consider a bird detection use case for crop field protection using a WSCN. The user deploys the cameras across separate fields and instructs them to capture an image whenever they see a bird. After gathering a sufficient number of images, the user initiates the learning process. Once the training is complete, the camera network begins autonomous detection. The user can periodically check the system's performance and retrain the models if needed to improve autonomous decision-making.

Figure 2 presents the sequence diagram of the agricultural system to simulate. As shown in the Figure, the process begins with the farmer initiating a pre-training phase by instructing cameras to capture images of pest birds they detect in each camera's related field. Once sufficient images are collected, the training phase commences. After training, a post-training control phase tests the model. If the farmer approves the results, inference mode begins; otherwise, the pre-training phase is restarted for further improvement. While in inference mode, the farmer can still intervene at any time by restarting the process if deemed necessary. It is important to note that, at no point during the system's operation—whether before, during, or after training—does it transmit raw data to the user, only instructions/decisions that are lightweight messages. Keeping data entirely on the edge ensures enhanced privacy and minimizes transmission costs.

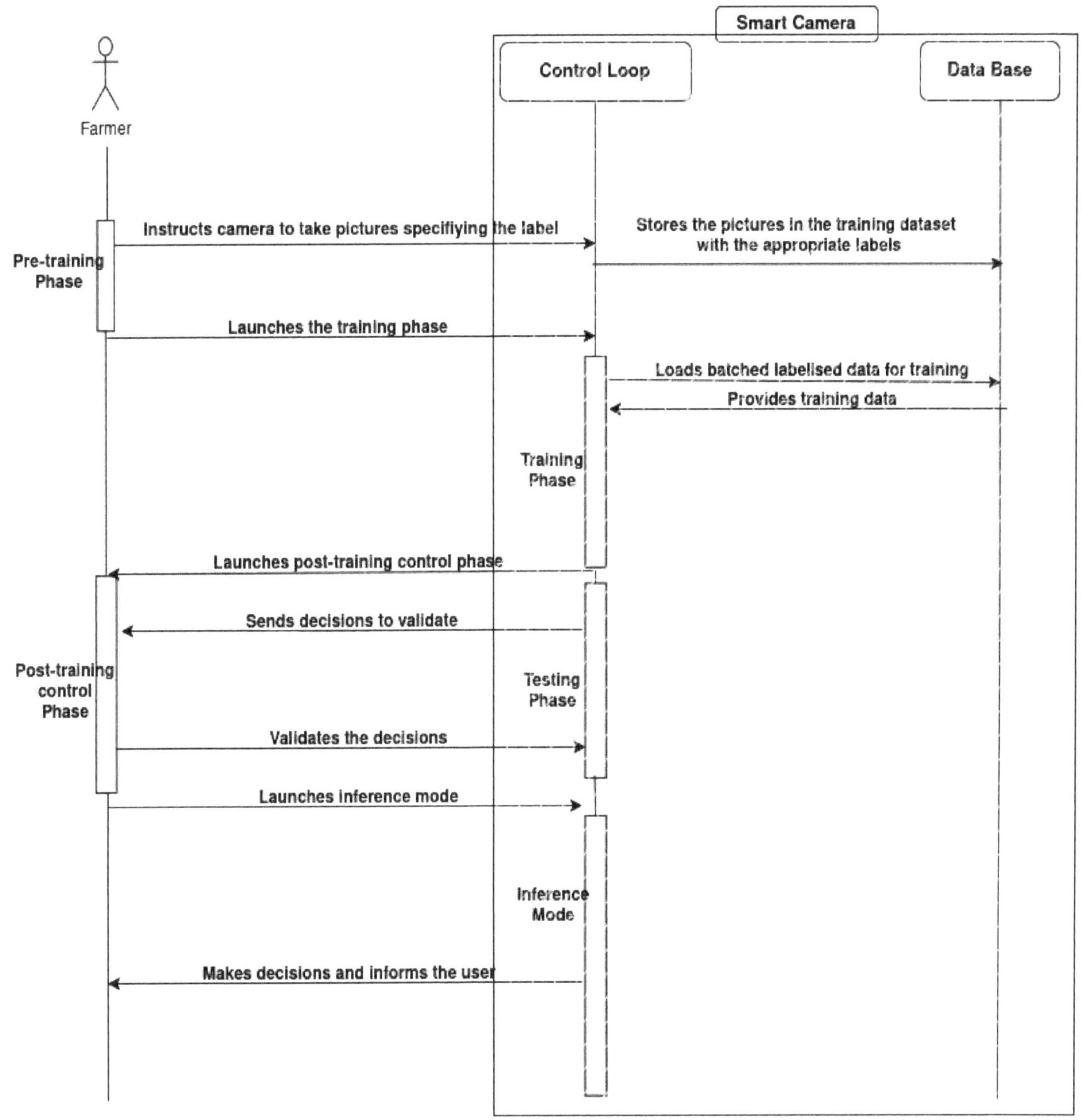

Figure 2. Sequence diagram illustrating the simulated agricultural system workflow.

4. Implementation

Our work presents a straightforward simulation of the previously described agricultural system. The implementation models a simple scenario where a farmer manages two similar crop fields requiring protection from pest birds. The simulation considers two pest bird species—pigeons and crows—highlighting the system's ability to detect different bird species to effectively safeguard the crops. To implement this simulation, we selected hardware likely to be optimal for powering a WSCN, prioritizing high computational capacity to handle the demanding processing tasks required for image classification (Section 4.1). Additionally, the training software and frameworks were carefully selected to optimize local training tasks, ensuring efficient and effective learning processes on edge devices

(Section 4.2). Finally, we gathered a dataset closely resembling real-world conditions to train the system effectively, ensuring its applicability and reliability in practical scenarios (Section 4.3).

4.1. Hardware

We opted for RaspBerry Pi 4 (Sony UK Technology Centre, Pencoed, UK) as a hardware implementation that delivers high-performance inference, supports efficient network communication, and maintains cost-effectiveness.

Table 3 offers a comparative overview of Raspberry Pi 4 against other market competitors designed for local image classification and training tasks. It highlights key specifications, including computational performance (focused on MobileNetV2), average training energy cost, price, and wireless communication capabilities with a focus on Wi-Fi compatibility.

Table 3. Comparative table of Raspberry Pi 4 and market competitors for local image classification and training.

Device	CPU	GPU/NPU	Memory	MobileNetV2 Performance (FPS)	Wireless Communication	Training Energy Cost (Power)	Price (USD)
Raspberry Pi 4	Quad-core Cortex-A72 @ 1.5 GHz	VideoCore VI GPU (32 GFLOPS)	1–8 GB LPDDR4	3–4	Wi-Fi 5 (802.11ac), Bluetooth 5.0	3.4 W	50
Jetson Nano	Quad-core Cortex-A57 @ 1.43 GHz	128 CUDA cores (1.8 TOPS)	4 GB LPDDR4	10–12	External USB Wi-Fi adapter required	5–10 W	99
Google Coral Dev Board	ARM Cortex-A53	Edge TPU (4 TOPS)	1 GB LPDDR4	50–60	Wi-Fi 5 (802.11ac), Bluetooth 4.1	2 W	129
Radxa Zero	Quad-core Cortex-A55 @ 1.4 GHz	Mali-G52 (0.6 TOPS NPU)	1–8 GB LPDDR4	2–3	Wi-Fi 5 (802.11ac) (only in advanced model), No wireless in basic model	3 W	40
BeagleBone AI	Dual-core Cortex-A15 @ 1.5 GHz	PowerVR SGX544 + C66x DSP	1 GB DDR3L	5–6	External USB Wi-Fi adapter required	7 W	120

Raspberry Pi 4 and the Radxa Zero stand out as the most accessible hardware options in terms of both affordability and energy efficiency, aligning well with the requirements of our system. Note that the Radxa Zero offers slightly lower performance compared to the Raspberry Pi 4 but demonstrates better energy efficiency. However, we excluded the Radxa Zero from consideration because our implementation requires hardware suitable for wireless communication, an important requirement for building a WSCN. Radxa Zero's basic model lacks integrated Wi-Fi capabilities. Considering these factors, we selected Raspberry Pi 4 as the most suitable choice for our implementation.

We focused on Wi-Fi, as it is the simplest and most practical communication technology for implementing our WSCN; Wi-Fi networks are already established in many agricultural settings, eliminating the need for additional communication infrastructure.

The computing unit selected for the system is Raspberry Pi 4, while the data acquisition component is the Raspberry Pi Camera Module. The simulated smart camera system (RaspBerry PI 4 + camera) is shown in Figure 3.

Figure 3. The designed system simulating a wireless smart camera for our platform implementation.

4.2. Software

As mentioned earlier, we opted for MobileNetV2 as the base model to equip our wireless smart cameras with AI-based classification capabilities. It is a lightweight, efficient CNN chosen for its suitability in performing image classification [61], which is our primary task.

Table 4 provides a comparative overview of our model alongside the most commonly used lightweight models for local edge image classification tasks. MobileNetV2 offers a strong balance between accuracy and computational efficiency, making it ideal for resource-constrained devices performing image classification.

Table 4. Comparison of lightweight image classification models for edge devices.

Model	Top-1 Accuracy	Model Size	Latency (CPU)	Energy Consumption	Notable Strengths
MobileNetV2	71.8%	14 MB	20 ms	Moderate	Lightweight, efficient for edge devices.
EfficientNet-B0	77.1%	20 MB	24 ms	Moderate-High	Higher accuracy; more resource-demanding.
SqueezeNet	58.1%	4.8 MB	18 ms	Low	Extremely lightweight, lower accuracy.
ShuffleNet V2	69.4%	8 MB	22 ms	Low	Faster on mobile CPUs, good trade-off.

To train our model, transfer learning was an effective technique, particularly in resource-constrained environments. This approach involves leveraging pre-trained models, which have already learned useful features on large datasets, such as ImageNet [62]. The model can be used in two ways: either without fine-tuning, with the pre-trained layers serving as feature extractors and leaving only the classification layers for adaptation to the new task, or with fine-tuning, with additional layers, beyond the classification ones, being updated to refine the model's performance for the specific task at hand.

Our model was designed specifically for mobile and embedded vision applications, and it utilizes depthwise separable convolutions, reducing the model size and computational complexity without compromising accuracy [61]. Transfer learning further enhances this by reusing its robust feature extraction layers while training only the final classification layers. This modular adaptability aligns with the requirements of applications like bird detection, for which the aim is to efficiently classify images into categories on constrained hardware performing in a changing environment. Figure 4 shows the plot of the architecture flow of our MobileNetV2-based model. As shown in the figure, our model contains a Tensorflow Keras MobileNetV2 as a base model, plus a fully connected GlobalAveragePooling2D layer as a classification head. The classification head converts the MobileNetV2's output features to a single 1280-element vector per image for further simplification to a scalar output (birds/no-birds) to be performed via the final dense layer. Note that, if fine-tuning is not used, only the weights of the classification head are updated during training. Otherwise, more layers of the "Functional" block are unfrozen for further training.

We trained our model using the TensorFlow framework. Developed by Google [63], TensorFlow is a popular framework for deep learning, known for its versatility and wide adoption in various edge devices, including Raspberry Pi. On its Lite version, it supports model training and inference on resource-constrained devices, making it suitable for edge AI applications.

Table 5 presents a comparison between TensorFlow 2.16 and PyTorch 2.4—another popular framework for deep learning (DL)—with a focus on their suitability for edge training and inference. It also highlights their resource efficiency, particularly in the context of power-constrained devices.

Table 5. Comparative table of TensorFlow and PyTorch for edge training and deployment.

Feature	TensorFlow	PyTorch
Ease of Use	High-level API with extensive documentation	Dynamic computational graph, more flexible
Performance on edge devices	Optimized for edge deployment with TensorFlow Lite	Requires optimizations for edge deployment
Model deployment	Easy integration with TensorFlow Lite for mobile/edge	Model conversion needed for deployment
Resource efficiency	Highly optimized for low-resource devices	May require more resources for similar tasks

We deploy our WSCN devices with the model illustrated in Figure 4, ensuring a unified approach to local training. The base model (MobileNetV2) is pre-trained on ImageNet dataset. After the classification head is added, the base model is frozen and the head is trained on each local device's data to develop a classification knowledge. To achieve global classification performance across all devices, we employed FL to facilitate global model convergence, ensuring that each device attains the same classification knowledge while preserving the uniqueness of local datasets.

Global model convergence in FL often requires numerous iterations [14], which can significantly impact energy consumption. Therefore, our focus is on optimizing the tradeoff between achieving a sufficient convergence rate and minimizing the number of iterations to conserve energy. This approach defines our low-energy FL framework, which we name LEFL, designed explicitly for energy-efficient training in resource-constrained environments like a WSCN.

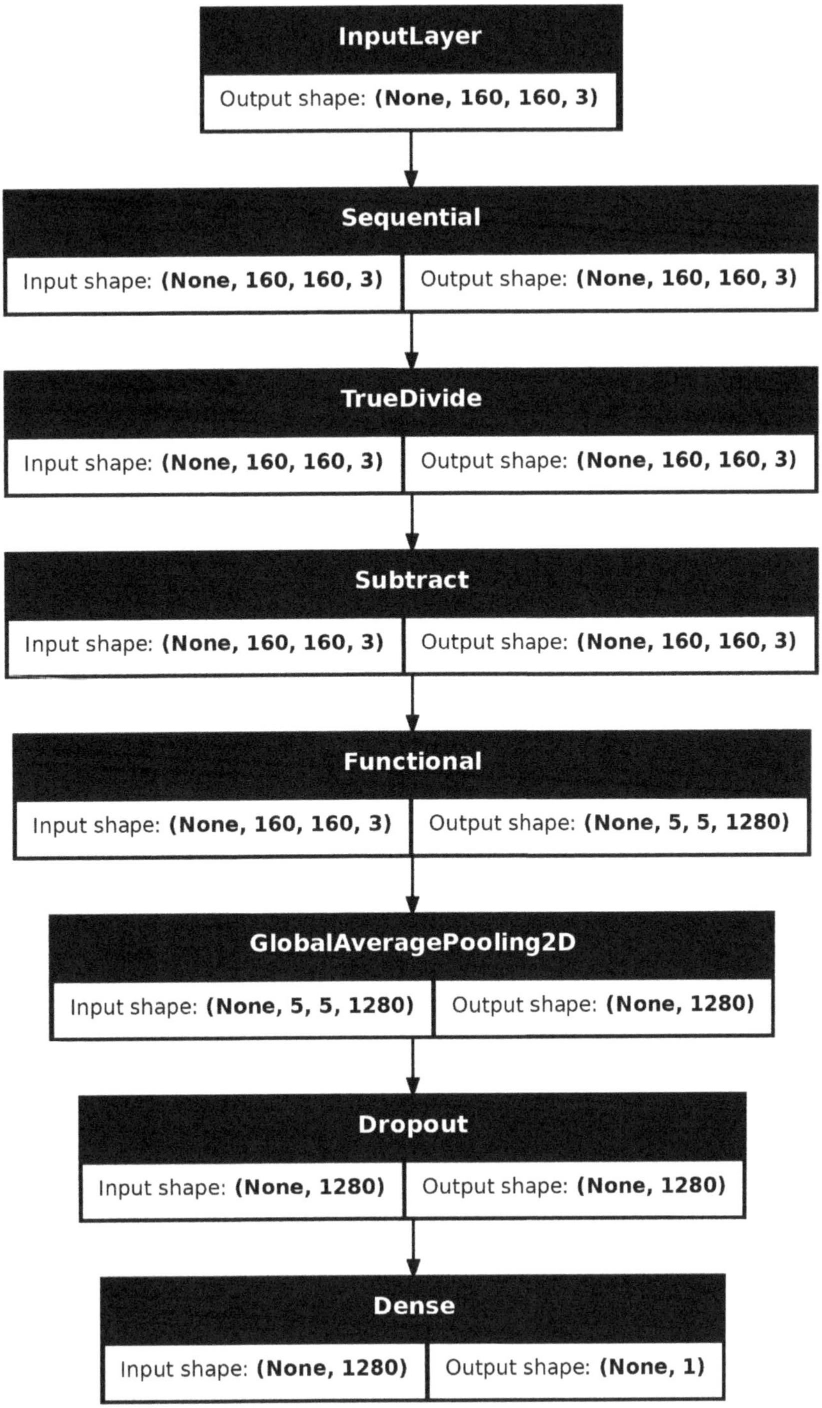

Figure 4. The architecture flow of the used model.

To implement our FL framework, we leveraged ready-to-use functionalities provided via FL libraries. Among the available options, we selected Flower [64], a robust and versatile framework tailored to establishing a client–server architecture and managing FL strategies. Flower simplifies the development and deployment of FL systems with its lightweight and adaptable design. Table 6 provides a comparative overview of various FL libraries [65], highlighting Flower's suitability for our specific requirements.

Table 6. Comparative table of Flower and other FL frameworks for edge training efficiency.

Framework	Ease of Use	Communication Efficiency	Energy Efficiency	Edge Adaptability	Observations
Flower	High	Optimized	High	Excellent	Lightweight, flexible, and edge-focused, DL framework-agnostic
TensorFLow FL (TFF)	Moderate	Moderate	Moderate	Good	Focused on TensorFlow-based implementations
PySyft	Moderate	Good	Moderate	Moderate	Emphasizes security but less edge-specific
FedML	Moderate	Optimized	High	Good	Versatile but more complex for lightweight systems
LEAF	Low	Basic	Low	Limited	Designed primarily for academic purposes where performance is not the primary focus

From the comparative table, it is evident that, while TFF aligns well with our choice of TensorFlow as a DL framework, its communication infrastructure does not prioritize wireless efficiency, which is crucial for our application. Flower, on the other hand, is designed with wireless communication efficiency in mind, making it more suitable for our system, which relies on wireless communication for federated training. While FedML offers optimized communication between devices during federated training, its adaptability to edge devices remains less refined compared to Flower. By leveraging Flower, we ensure an edge-optimized, scalable, communication-optimized, and energy-efficient implementation of our LEFL framework, which is essential for the resource-constrained devices in our WSCN.

The LEFL framework used in this work is a custom solution combining Flower for federated training and communication between devices and TensorFlow for efficient local model training on edge devices. This integration enables the collaborative training of a global model while maintaining energy efficiency.

The key contribution of our framework is an algorithm designed to determine the optimal stopping point for FL, an approach commonly referred to in the literature as an early stopping strategy. Resource-efficient early stopping strategies, like FLrce [66], ensure energy conservation without compromising model performance. In our work, unlike [66], where the stopping criterion was the number of conflicting clients, the stopping criterion of our strategy was guided by the $QoSAuT$ metric, defined in Section 3.1, which quantifies the trade-off between the trained model's acquired performance (accuracy) during training and its associated energy cost. We defined the partial $QoSAuT$, $QoSAuT_i$, as the trade-off between the model's acquired performance in one federated round, i, and its associated energy cost. Using $QoSAuT_i$, the algorithm dynamically evaluates the learning progress, halting the federated process when additional learning iterations yield diminishing returns in performance relative to energy expenditure. This adaptive strategy allows the LEFL framework to achieve both effective learning and operational sustainability through a novel energy-efficient early stopping strategy.

$QoSAuT_i$ is computed locally for each client and for each round. For instance, for the smart camera, k, in each round, i, we compute $Cr_{i,k}$, the stopping criteria of k at the round i with the following formula:

$$Cr_{i,k} = QoSAuT_i(k,1,8) = \frac{\alpha \Delta_i a_k}{ct_{i,k} + 8\mu_k x_{i,k} + ci_k} \quad (2)$$

We use, as fixed parameters, 1 as one training round (the federated round) and $8 = 4 \times 2$, as the size of a float is 4 bytes (transmitted model weights being floats), and 2 is the number of times a weight is sent across the network (sent from the client to the server and then from the server to the client after being updated).

And as variable parameters, we used the following:

- α = factor defined in Formula (1).
- $\Delta_i a_k$ = gained accuracy in the round, i, via the trained model in k.
- $x_{i,k}$ = number of weights sent in the round, i, via k.
- $ct_{i,k}$ = computational cost of the round, i, in k.
- ci_k = inference cost of the model in k.
- μ_k = transmission cost of 1 byte via k.

Since the number of weights transmitted in a federated round and the cost of sending 1 byte via the device are invariant data during the training process, and the inference cost of the model, ci_k, remains constant, regardless of the training process, the variant parameters of our early stopping strategy are as follows: $ct_{i,k}$ and $\Delta_i a_k$.

In each training round, we compute $ct_{i,k}$ directly, but computing $\Delta_i a_k$ requires the accuracy from the previous round. To address this, we maintain a global variable, $A[]$, which stores the achieved accuracy at the end of each round.

In FL, clients generally experience increasing accuracy during training. However, it may occasionally dip temporarily in a given round before resuming its upward trend. To address this, the early stopping mechanism is triggered if the $QoSAuT_i$ remains poor for two consecutive rounds, indicating a confirmed trend of declining training efficiency and justifying the need for early termination.

We implement the early stopping (ES) strategy of our LEFL framework based on the $QoSAuT_i$ values from the last two consecutive federated rounds. These values are calculated following Algorithm 1.

Algorithm 1 Computing $QoSAuT_i$.

Require: round i, accuracies table $A[]$, factor α, model's number of weights x, the cost of transmitting 1 byte μ, the cost of inference ci, the computational cost of the training round i $ct(i)$, as a percentage of the total available battery power.

```
if i = 0 then
    return False
end if
Δa ← A[i] − A[i − 1]
Q ← α.Δa / (ct(i)+8.μ.x+ci)
return Q
```

Our LEFL framework was designed to terminate federated learning rounds once the improvement in accuracy no longer justifies the corresponding energy expenditure, ensuring an optimal balance between performance—the accuracy of the model—gains and energy efficiency.

We conceptualized a function, **FL**(i), that performs the training for a federated round, i, updating the table with the accuracy achieved during the round. Specifically, **FL**(1) updates $A[0]$

with the initial accuracy before training and $A[1]$ with the model's accuracy upon the first round's completion. Additionally, we defined **CT**(i) as a function that computes the computational cost associated with the i-th federated round and **CI**() as a function to compute the computational cost of one inference via the model. Also, we assumed that the trained model had an attribute **trainable_weights** that returns the number of its trainable weights.

Using Algorithm 1 to compute $QoSAuT_i$, which we call "Q", and the computational cost functions CT and CI, we implemented the LEFL ES as presented in Algorithm 2.

Algorithm 2 LEFL ES Algorithm.

Require: factor α, total available battery power W_{max}, the cost of transmitting 1 byte to the server M

$\mu \leftarrow \frac{M}{W_{max}}$
$x \leftarrow model.trainable\_weights$
$ci \leftarrow \frac{CI()}{W_{max}}$
$A[] \leftarrow []$
$i \leftarrow 0$
$Cr \leftarrow$ **False**
while $Cr ==$ **False do**
  $i \leftarrow i+1$
  if $FL(i, A[])$ **then**
    $ct \leftarrow \frac{CT(i)}{W_{max}}$
    $Cr \leftarrow Q(i, A[], \alpha, x, \mu, ci, ct) < 1$ && $Q(i-1, A[], \alpha, x, \mu, ci, ct) < 1$
  else
    break
  end if
end while

Note that the LEFL's Early Stop algorithm was implemented on the client side (i.e., executed on each device k). If the stopping criterion is not met, the while loop continues until the server terminates the FL process upon achieving global convergence, and then every device k stops training.

4.3. Data

4.3.1. Dataset Overview

We trained our MobileNetV2 model, distributed across two smart cameras, as designed in Section 4.1, using a dataset of images taken directly from a field of germinating crops or newly sown seeds, which are highly attractive to pest birds. This dataset was provided by **Corentin Barbu** from the **Agronomie** Unit of **INRAE** and **Christophe Sausse** from **Terres Inovia** [67].

We annotated the data with two labels: *birds* (a pest bird species is present in the crop field) and *no_birds* (no pest bird species is present in the crop field). Two pest bird species are photographed on the monitored field: pigeons and crows. These species are particularly drawn to germinating crops and newly sown seeds, making them frequent targets for pest management strategies in agricultural environments.

The images were collected from a single plantation, with the field divided into two sections, each monitored via a separate camera, during a defined growth period (April to June) when grains attract pest birds. Due to seasonal constraints, the dataset does not include rainy or dusty conditions, which aligns with our objective of simulating on-field training during a consistent plantation growth period—where weather conditions typically remain stable throughout the training phase. The dataset includes bird samples captured at varying distances from the camera and at different times of the day (dawn, midday, and dusk), ensuring a balanced representation of lighting conditions. Additionally, it captures diverse bird sizes and behaviors, including birds flying close to the ground, resting on the soil, and appearing in various positions.

Figure 5 illustrates two samples from our dataset labeled as *birds*, showcasing the two pest bird species of our study: pigeons and crows. Typically, a sample from our dataset will feature both species, but one species can be more prominently represented in the crop field than the other.

(**a**) (**b**)

Figure 5. Two samples from our dataset depicting pest birds' presence in a crop field. (**a**) The pest birds are predominantly pigeons. (**b**) The pest birds are predominantly crows.

4.3.2. Data Distribution for Training

After duplicates, outliers, and unusable images were removed during the data-cleaning process, we obtained a curated dataset ready for training containing 115 images (57 *birds* + 58 *no_birds*).

Since the dataset is balanced, containing an equal number of "*birds*" and "*no_birds*" samples, accuracy is an appropriate metric for evaluating both training quality and overall model performance, eliminating the need for the F1 score.

To perform our training strategy, we split the dataset into two subsets: one for training (45 *birds* + 47 *no_birds*) and one for testing (12 *birds* + 11 *no_birds*). The training subset was then subdivided as follows: 75% for training (34 *birds* + 35 *no_birds*) and 25% for validation (11 *birds* + 12 *no_birds*). Table 7 provides an overview of the data distribution used in our training strategy.

Additionally, as a pre-processing step, we rescaled the images to 160 $\times$ 160 to match the input layer of our model to be trained (Figure 4). In real-world field deployment, the smart cameras capture images at an optimal resolution, providing 160 $\times$ 160-pixel images directly to the model for training. This approach significantly enhances energy efficiency by reducing both acquisition and computational costs, as it eliminates the need for additional preprocessing or resizing.

Table 7. Data distribution of our training strategy.

Subset	Birds	No_Birds	All
Training (60%)	34	35	69
Validation (20%)	11	12	23
Testing (20%)	12	11	23
Total (100%)	57	58	115

Given the small dataset size (115 samples), allocating 60% for training ensured that the model had enough examples to learn meaningful patterns while avoiding excessive overfitting. The 20% validation set is essential for tuning hyperparameters and preventing overfitting by evaluating model performance on unseen data during training. A validation set that is too small may lead to unstable results, while a larger one would reduce the already limited training set. With only 115 samples, more aggressive splits (e.g., 70/15/15 or 80/10/10) would either reduce validation robustness or limit the training set. The 60/20/20 split maintains a balance between learning, validation, and unbiased testing, which is particularly crucial in low-data scenarios.

As highlighted in Section 2, our addressed use case aligns closely with state-of-the-art pest and disease detection research. To contextualize the limitation of our dataset size, Table 8 presents a comparison of the sizes of the datasets employed in training classification models in similar studies, emphasizing the relative scale of data available for our training process.

Table 8. Dataset sizes of similar works (plant disease detection).

Paper	Dataset Size	Data Type
[29]	2029	Images
[30]	3852	Images
[31]	5400	Images
[32]	104,544	Hypervoxels
Our work	115	Images

However, in [31], 5400 samples were obtained from only 200 using data augmentation techniques, which included applying four different rotations and one mirroring transformation. In our work, we avoided using these techniques, as these augmentations could generate unrealistic or irrelevant data samples. Such transformations are not representative of real-world scenarios of a bird detection use case, given that our monitoring cameras remain fixed, capturing images consistently from the same angle of view. This fixed perspective makes it crucial to retain the natural orientation and positioning of elements within the captured scenes.

We mitigate our small dataset's limitation by employing two key techniques: First, transfer learning (Section 4.2), as a pre-trained model on a broader dataset, enables the system to perform effectively even with a small training dataset by leveraging pre-existing knowledge. Second, the operational design of our system ensures adaptability to new data. Users can intervene to halt inference mode if significant changes in the data are detected and initiate additional training phases (Section 3.3), allowing the model to continuously adapt to real-world variations. Finally, our application use case differs from plant disease detection in that our images involve simpler scenarios, such as birds present in a field, whereas identifying diseases often involves subtle and complex visual patterns. This makes the need for a larger dataset less critical in our case, as the task is less intricate.

To visualize our training data (Training + Validation subsets) distribution in the feature space and the separability of the two classes (*birds* and *no_birds*), we plot its feature space visualization (FSV). After reducing the data dimensionality, with principal component analysis (PCA), to two dimensions (two principal components), we obtained the FSV of our training data, as presented in Figure 6. Note that the FSV is represented in two dimensions, using only the first two principal components (PCs). This dimensionality reduction leads to overlapping representations, which might obscure potential separability observable in the original high-dimensional space. The overlap can also be attributed to the input image size (160×160), where images labeled as *birds* with relatively distant birds appear visually similar to images labeled as *no_birds*, as the low resolution fails to capture sufficient features of the far-spotted birds. Furthermore, images labeled as *no_birds* tend to cluster more tightly, reflecting the uniformity of crop field images without intruders, whereas *birds*-labeled images are more sparsely distributed. This sparsity arises from the inclusion of two pest bird species, pigeons and crows, each introducing distinct feature sets.

4.3.3. Data Distribution over Clients

To highlight the performance of our LEFL-based platform, we evaluated its effectiveness in addressing the bird detection task using two wireless smart cameras under two distinct scenarios:

1. During the training phase, both crop fields monitored via the cameras were equally likely to attract either bird species (homogeneous distribution). To replicate this scenario in our study, we split the training data between the two clients (cameras) in a balanced manner, ensuring an even representation of both bird species across the datasets. This data distribution over clients is called identically independent distribution (IID), meaning the data samples across all clients are independently drawn from the same probability distribution. (S_1).
2. During the training phase, one crop field is more likely to attract pigeons, while the other tends to attract crows. To simulate this scenario, we distribute the training data unequally between the two clients, assigning images showcasing more pigeons to the first client and images showcasing more crows to the second. This ensures a heterogeneous data distribution across the clients. Such a distribution is termed non-identically independent distribution (non-IID), indicating that data samples on each client are drawn from distinct probability distributions (S_2).

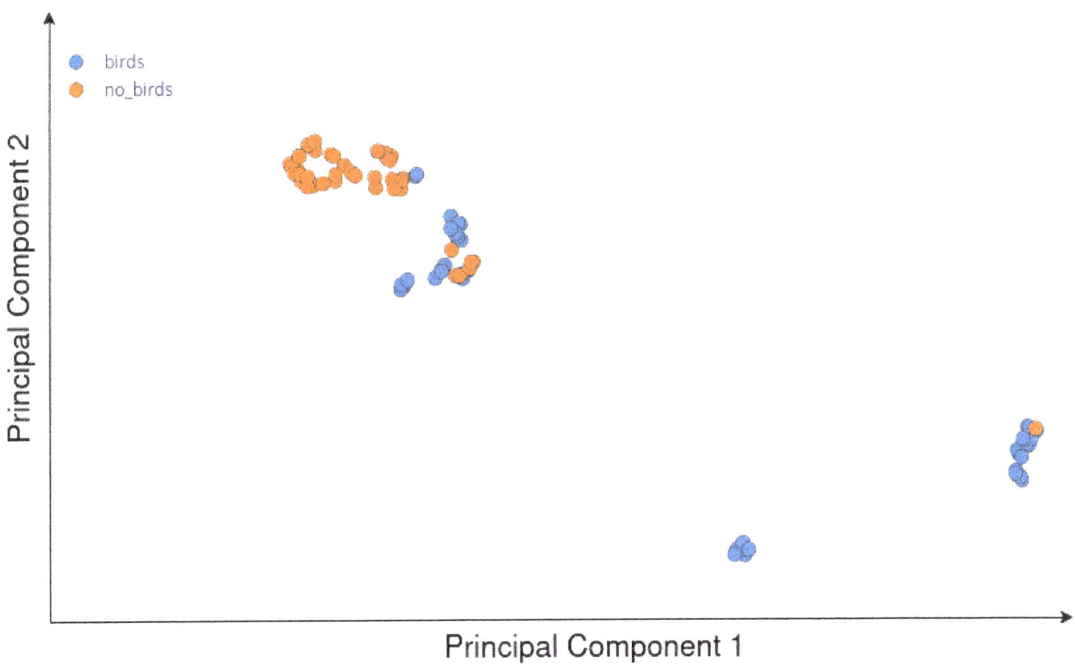

Figure 6. The feature space visualization of our training dataset (92 samples).

Our training dataset consists of 92 images, out of which 45 are labeled as *birds*. Among these, 26 images predominantly feature pigeons, while 19 images primarily contain crows.

In both scenarios, we split the training set (comprising the training and validation subsets) across the clients while keeping the test subset identical for both by duplicating it. This approach highlights the heterogeneity that may arise during training, as the data distribution varies between clients. However, during the testing phase, we evaluate performance on the same data to emphasize the model's ability to generalize and adapt to a more balanced data distribution, which is representative of real-world scenarios during inference. In terms of data size, both clients in both scenarios have an equal share of the training dataset, with each receiving 46 samples, which constitutes half of the total training dataset (92 samples).

In Scenario S_1, the 45 images labeled as *birds* are distributed evenly between the two clients. Client 1 receives 13 images predominantly featuring pigeons and 10 images predominantly featuring crows, while Client 2 receives 13 images with more pigeons and 9 images with more crows. The *no_bird*-labeled images are allocated randomly across both clients to simulate a balanced and homogeneous distribution for this category. Table 9 provides an overview of the training data distribution across the clients in this scenario.

Table 9. Training data distribution over the clients in scenario S_1 (IID).

Client	Birds (Mostly Pigeons)	Birds (Mostly Crows)	No_Birds	Total
Client 1	13	10	23	46
Client 2	13	9	24	46

In Scenario S_2, the 45 images labeled as *birds* are distributed non-identically between the two clients. Client 1 receives the 26 images predominantly featuring pigeons, while Client 2 is assigned the 19 images predominantly featuring crows. The *no_birds*-labeled images are allocated randomly but unequally across both clients, compensating for the smaller number of *birds*-labeled images in Client 2. This setup reflects the possibility that a client may have fewer *birds*-labeled samples during the training phase—i.e., fewer birds intruded on the monitored crop field during the training phase—simulating real-world situations where more birds can intrude a crop field than the other one during the training phase. Table 10 provides an overview of the training data distribution across the clients in this scenario.

Table 10. Training data distribution over the clients in scenario S_2 (non-IID).

Client	Birds (Mostly Pigeons)	Birds (Mostly Crows)	No_Birds	Total
Client 1	26	0	20	46
Client 2	0	19	27	46

5. Results

We implemented a standard centralized training strategy for MobileNetV2 using transfer learning, which serves as our benchmark for comparison (Section 5.1).

To evaluate our LEFL framework's efficiency, we carefully distinguished between the two previously discussed scenarios, treating them separately to distinctly assess our framework's efficiency in both the IID (Section 5.3.1) and non-IID (Section 5.3.2) training data cases.

The Shift Project—Lean ICT [68] demonstrated that transmitting 1 byte of data over a Wi-Fi network consumes energy in the range of 10^{-10} kWh, which is equivalent to a range of 0.3–1 µJ. For the sake of simplifying calculations and obtaining a general estimate of the energy range, we set $M = 1$ µJ.

In our simulation, we set the maximum power capacity to $W_{max} = 5 \times 10^7 \mu J$, which is equivalent to 14 Wh. This power capacity can be provided via a **RaspBerry PI Power Pack** with a capacity of 3800 mAh, assuming a nominal voltage of 3.7 V, typical for lithium-ion batteries.

We estimated the energy consumption of a single training round by summing the energy cost of training the classification head (the trainable layers without fine-tuning) for one epoch and the energy cost of feature extraction performed using the frozen MobileNetV2. The first component was derived from the number of arithmetic operations executed during one epoch [69], as the second was deduced from the MobileNetV2's inference cost [70].

The classification head comprised 42,599,680 connections (Figure 4). Using this method, we determined that each training round required approximately 1.2×10^{10} FLOPs (floating point operations) to train it.

Raspberry Pi 4 features a quad-core ARM Cortex-A72 CPU, with each core running at up to 1.5 GHz. Assuming each core executes two double-precision FLOPs per clock cycle, the theoretical peak performance for all four cores is calculated as follows: $2 \times 1.5 \times 4 = 12$ GFLOPS (FLOPs per second).

Training consumes approximately 3.4 W (Table 3). This translates into an energy cost as follows: $3.4/12 = 0.283$ W/GFLOP, or, with W converted to J/s, approximately 283 pJ/FLOP (or 2.8×10^{-10} J/FLOP). By multiplying this value by the number of FLOPs required for training the classification head for one epoch (calculated earlier), the estimated computational energy cost is as follows: 2.8×10^{-10} J/FLOP $\times$ 1.2×10^{10} FLOPs = 3.4 J.

Performing feature extraction with the frozen MobileNetV2 on one input image of 160×160 requires 2.5×10^8 FLOPs [70]. Consequently, the energy consumed via the frozen MobileNetV2 for feature extraction in one epoch (42 images) is as follows: $2.5 \times 10^8 \times 42 \times 2.8 \times 10^{-10} = 2.9$ J.

From the previous calculations, the computational energy cost for a single training epoch is as follows: $2.9 + 3.4 = 6.3$ J. Since the same layers are trained throughout the entire FL process with identical parameters, we set the computational cost for a single federated round i as follows: $ct_i = 6.3 \times 10^3$ µJ.

One inference via our model involves feature extraction using MobileNetV2, followed by classification through a classification head with 1280 parameters (Figure 4). The global average pooling layer processes the 5×5 feature maps by performing 25 additions and 1 division per feature map across 1280 feature maps. This results in $(25 + 1) \times 1280 = 33{,}280$ FLOPs. The dense layer processes the 1280-dimensional input by performing two operations (multiplication and addition) per weight, resulting in $2 \times 1280 = 2560$ FLOPs. The total inference cost is, therefore, 33,280 + 2560 = 35,840. At an energy cost of 283 pJ/FLOP, this translates to $283 \times 35{,}840$ = 10.15 µJ. Note that the Dropout layer is only active during training and incurs no computational overhead during inference.

Based on the previous calculations, the computational energy cost for an inference is as follows: $2.9.10^3 + 10.15 = 2.9.10^3$ µJ. We set $ci = 2.9.10^3$ µJ.

Given our use case, which relies on small training datasets but requires a new training phase after every significant change in the data, it is essential to minimize the energy cost of this phase. In doing so, the training process can be repeated as often as needed while conserving battery power for the inference mode. We primarily set $\alpha = 1\%$, as training had to incur the lowest possible energy cost compared to inference, enabling additional training sessions, as demonstrated in Section 3.3, without depleting the battery required for inference.

5.1. Benchmark Preparation

To obtain comparable results in terms of training efficiency, we set the same optimal hyperparameters for the benchmark, as well as for both the IID and non-IID FL scenarios.

This approach allows us to directly compare the performance of the different training strategies under uniform conditions, ensuring a fair evaluation across all training setups.

We imported the pre-trained MobileNetV2 as a base model from **Keras** using **keras.applications.MobileNetV2** and specifying the parameter *weights* to *'imagenet'*. We added the classification head as illustrated in Figure 4, and we froze the base model layers to use them as a feature extractor while training the head.

For our training, we chose the following hyperparameters:

- **Batch size:** For our simulation, we adopted a classic batch learning scheme, also referred to as offline learning. In this approach, the training dataset is divided into smaller batches. The model processes one batch at a time, updating its weights only after completing the forward and backward passes for each batch. A large batch size helps the model learn faster but may require more memory and computational resources. Given the constraints of working on resource-limited devices and a small training dataset of 92 samples, we selected a small batch size of 16 samples.
- **Base learning rate:** The learning rate controls how much to change the model in response to the error each time the model weights are updated. A low learning rate allows the model to learn more fine-grained patterns in the data, but it requires more iterations to converge, leading to higher computational costs. Conversely, a high learning rate can accelerate convergence, but it may risk overshooting optimal values and potentially leading to poor model performance. Given the small size of our dataset, which is more prone to quick convergence, we used a low base learning rate of 0.001 in conjunction with the Adaptive Moment Estimation (Adam) Optimizer.
- **Optimizer:** We employed the Adam optimizer, which dynamically adjusts the learning rate for each parameter during training. The optimizer reduces the risk of overfitting on our small dataset while maintaining good generalization performance. Small updates to the weights ensure that the model does not memorize training examples too quickly.
- **Number of epochs:** In one epoch, the model trains on the entire dataset. Too few epochs may lead to underfitting, while too many can result in overfitting. Given our small dataset, which has limited variability and is not very sparse in the feature space (Figure 6), a small number of epochs is sufficient to achieve model convergence. As shown in Figure 7, the model begins to overfit after the 11th epoch, with the validation accuracy plateauing while the training accuracy continues to rise. Based on this observation, we set the number of epochs to 11 to balance sufficient training while preventing overfitting. The model begins to overfit after a relatively small number of epochs, likely due to the limited size of the training dataset and the low diversity of samples, particularly those labeled as *no_birds*.

To establish our benchmark performance, we first trained the model without fine-tuning and then fine-tuned the last 100 layers of MobileNetV2 for four more epochs.

Figure 8 presents the model confusion matrices in the test data set after training without fine-tuning and with fine-tuning, respectively. Fine-tuning offers limited advantages due to the simplicity of the classification task and the nature of the dataset used for training and testing. Utilizing the pre-trained MobileNetV2 for feature extraction is sufficient, as it effectively captures the necessary features. Training focuses solely on classifying the extracted features, enabling rapid learning. Fine-tuning additional layers in the base model does not result in significant improvements.

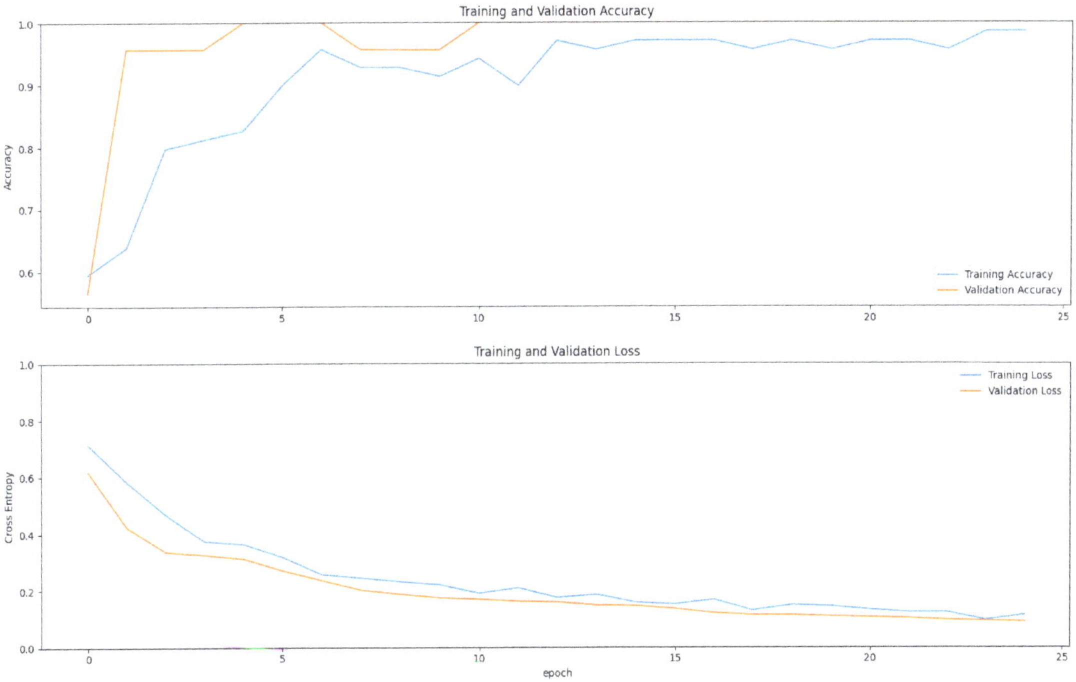

Figure 7. Training and validation loss and accuracy curves for the model trained with the previously set hyperparameters for 25 epochs.

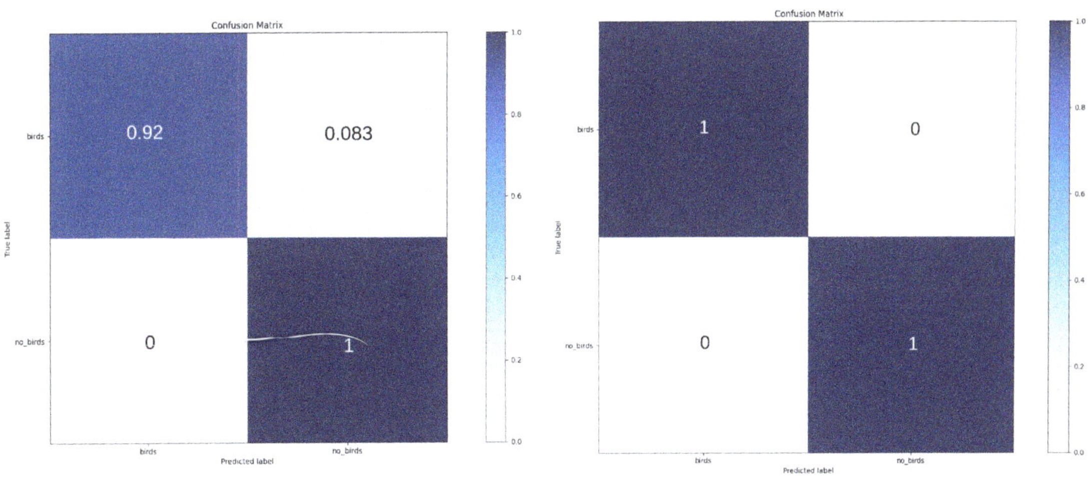

(a) (b)

Figure 8. Normalized confusion matrices of the model after the benchmark training with and without fine-tuning. The vertical axis represents the true labels, ordered as "birds" and "no_birds", while the horizontal axis represents the predicted labels, also ordered as "birds" and "no_birds", respectively. (**a**) Normalized confusion matrix of the model trained without fine-tuning. (**b**) Normalized confusion matrix of the model trained with fine-tuning.

5.2. FL Preparation

To test our LEFL framework, we implemented an FL setup using hyperparameters derived from the benchmark configuration (Section 5.1) but adapted to fit the specific requirements of FL. These adaptations ensure compatibility with the federated environment while maintaining consistency with the benchmarking conditions.

- **Batch size per client:** In our FL scenario, each client is assigned 42 samples for training, which represents half of the base training dataset, as there are two clients. To accommodate this, we adjust the batch size by halving it. This results in a batch size of 8 samples per client.
- **Base learning rate:** Since we halved the batch size, we also reduced the learning rate by half to maintain learning behavior aligning with our benchmark. We set the base learning rate to 0.0005 for each client.
- **Optimizer:** We kept the same optimizer as the benchmark (Adam) for both clients.
- **Number of epochs per federated round:** To maintain global control over the learning convergence, we ensured that each client performed one training pass per federated round. Therefore, we set the number of epochs to 1 for each client.
- **Aggregation algorithm:** Since we simulated a use case where both cameras monitor crop fields with equal importance and the training data are distributed equally across the clients, ensuring that no client has more weight than the other, we used the classic FedAvg algorithm with equal weights for all clients. The formula below illustrates the algorithm, with w^{t+1} being the global weight at round $t+1$ and w_1^t and w_2^t being the local weights after round t of clients 1 and 2, respectively.

$$w^{t+1} = \frac{1}{2}(w_1^t + w_2^t) \tag{3}$$

FedAvg stands out as the most suited algorithm for our study. FedSGD (federated stochastic gradient descent), for instance, transmits model gradients at every iteration, ensuring a more immediate synchronization of updates. However, this approach results in significant communication overhead, which is impractical in our low-energy, on-field training setup. FedProx, an extension of FedAvg that introduces a proximal term to stabilize updates in heterogeneous environments, could be beneficial in scenarios with highly diverse datasets, but given that our two cameras operate in similar agricultural conditions, such complexity is unnecessary. Scaffold, another alternative, attempts to correct model drift in non-IID settings by maintaining control variates but at the cost of additional computation and storage requirements, making it less suitable for our lightweight edge devices. FedOpt and other adaptive optimization-based methods, while promising for large-scale networks with diverse devices, introduce hyperparameter tuning challenges and increased computational costs that do not align with our goal of minimal energy expenditure. In contrast, FedAvg strikes the optimal balance for our scenario—it efficiently aggregates local model updates by averaging them at periodic intervals, significantly reducing communication overhead while maintaining convergence stability. Given that both cameras operate under similar conditions, the risk of model divergence is minimal, allowing FedAvg to perform effectively without additional complexity.

5.3. LEFL's Performance

5.3.1. LEFL's Performance on IID Data

To demonstrate the performance of our LEFL approach on an IID training dataset, we first trained our platform to achieve FL convergence, followed by training with LEFL.

Convergence is typically indicated when accuracies of local models converge, showing minimal improvement over subsequent training rounds. Figure 9 illustrates the training and validation accuracy for both clients 1 and 2 during 25 federated rounds within the IID data scenario. In this case, convergence was achieved in the 12th round, as shown in the figure. This result aligns with the benchmark's convergence point (11th epoch) because the data are IID, ensuring that both clients (Client 1 and Client 2) share identical distributions and replicating the benchmark data distribution.

Figure 10 plots the $QoSAuT_i(k)$ (as defined in Section 3.1) in each round, i, for both clients ($k = \{1,2\}$) in the FL-to-convergence scenario, highlighting the specific round where the LEFL ES mechanism terminates the training. Each client exhibits a unique progression in its $QoSAuT_i$, driven by the inter-client diversity of the data. This diversity leads to distinct learning patterns as the model adapts locally within each client. The variation in learning patterns explains why the LEFL ES mechanism is triggered at different times for each client. It is important to note that the stopping criterion used in our LEFL ES mechanism, relying here on two consecutive $QoSAuT$ values (<1), can be adapted to the complexity of the considered models. This criterion should theoretically indicate that further training is no longer energy-efficient. In other use cases involving significantly larger datasets (e.g., 1000+ samples), our approach may prove suboptimal. In such cases, convergence may occur much later in higher rounds, and the $QoSAuT$ metric may experience temporary declines for several rounds ($\geq$2) before subsequently improving and compensating for the earlier poor values. This is not the case in our use case, as convergence is reached quickly, causing the QoSAuT to decrease despite experiencing the fluctuations shortly after early stopping.

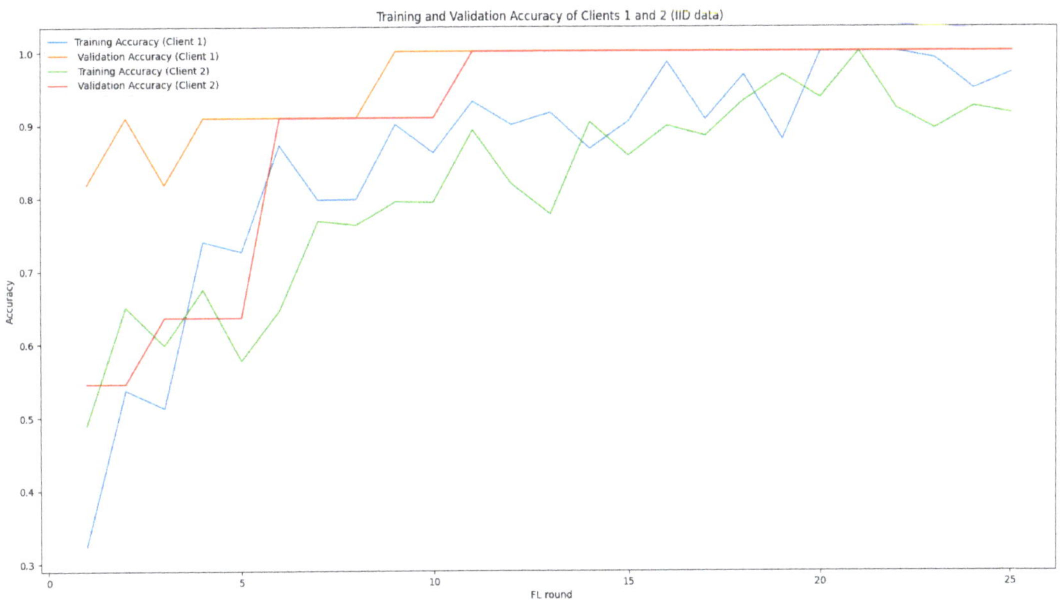

Figure 9. Training and validation loss and accuracy curves for both clients 1 and 2 with previously set parameters, trained for 25 FL rounds on IID data.

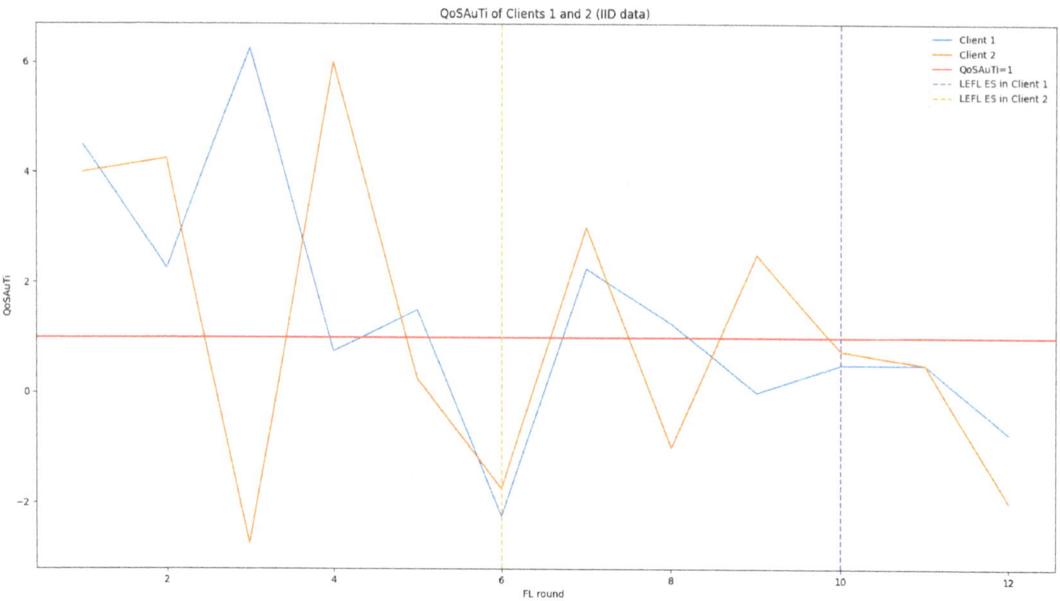

Figure 10. $QoSAuT_i$ for both clients 1 and 2 trained to convergence, with the activation of LEFL early stopping (ES) in IID data scenario.

Figure 11 plots the confusion matrices of the global final model trained with LEFL and FL-to-convergence, respectively. In FL-to-convergence, we achieved results comparable to those of benchmark centralized learning without fine-tuning. However, in LEFL, accuracy is slightly reduced in favor of energy efficiency. This highlights the tradeoff enabled by LEFL between inference performance and energy-conscious training. It is worth noting that the accuracy of the model regarding the *no_birds* label did not decrease with LEFL. This is because partial convergence for this label was achieved earlier, given the limited diversity of its samples.

Table 11 summarizes the $QoSAuT$ (as defined in Section 3.1) for each device across three scenarios: LEFL, FL-to-convergence, and a classic remote benchmark training scheme. In the remote benchmark scheme, devices transmit their data to a remote server for centralized training under benchmark conditions (Section 5.1), with the resulting weights then deployed back to the devices for inference. All scenarios were evaluated under an IID data distribution.

Table 11. Test accuracy and $QoSAuT$ of each device trained on IID data in three scenarios: LEFL, FL-to-convergence, and remote benchmark.

Scenario	LEFL	FL-to-Convergence	Remote Benchmark
Test Accuracy	0.91	0.96	0.96
$QoSAuT(1)$	1.55	1.40	0.09
$QoSAuT(2)$	2.56	1.40	0.09

FL improved the $QoSAuT$ of both clients by at least a factor of 15, primarily due to the significant reduction in the amount of data transmitted to the central server for training. Instead of transmitting large datasets as in the benchmark, FL keeps the data on edge devices, minimizing communication overhead. However, this advantage in reduced data transmission can be offset by increased computational costs if FL takes longer to converge, as is often the case in more complex scenarios. In our bird detection use case, FL appears particularly advantageous compared to traditional centralized training leaving limited room for further improvement via LEFL. As a result, LEFL slightly enhanced the $QoSAuT$. This enhancement varies between clients because the stopping mechanism was triggered at different rounds for each client.

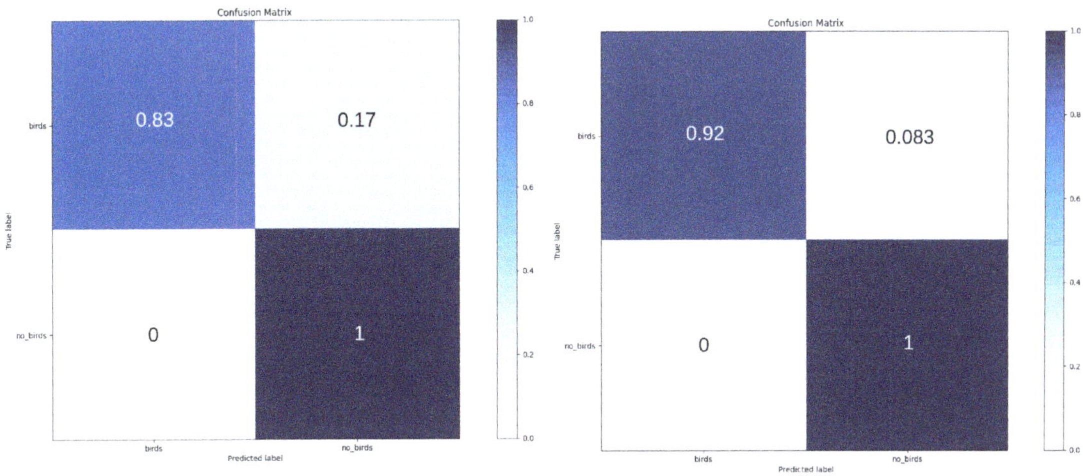

(**a**) (**b**)

Figure 11. Normalized confusion matrices of the model within the IID data scenario trained using LEFL and FL-to-convergence schemes. The vertical axis represents the true labels, ordered as "birds" and "no_birds," while the horizontal axis represents the predicted labels, also ordered as "birds" and "no_birds," respectively. (**a**) Normalized confusion matrix of the model trained with LEFL in the IID data scenario. (**b**) Normalized confusion matrix of the model trained with FL-to-convergence in the IID data scenario.

It is important to note that the $QoSAuT$ of client 2 is influenced by client 1. After client 2 stopped training, it continued receiving updated weights from client 1, which was still learning, while maintaining its own training energy. This observation leads us to conclude that, in the context of FL, the $QoSAuT$ should be considered a global metric for the entire network, representing the average $QoSAuT$ of all clients. This approach accounts for the fact that each client's performance is influenced by the participation of others in the training process, impacting their respective $QoSAuT$ values.

5.3.2. LEFL's Performance on Non-IID Data

To demonstrate the performance of our LEFL approach on a non-IID training dataset, we first trained our platform to achieve FL convergence, followed by training with LEFL.

Figure 12 illustrates the training and validation accuracy for both clients 1 and 2 during 25 federated rounds within the non-IID data scenario. In this case, convergence occurred later compared to the IID data scenario, at the 22th round. This delayed convergence is attributed to the heterogeneous data distribution, which causes the client models to require more rounds to align and achieve consensus.

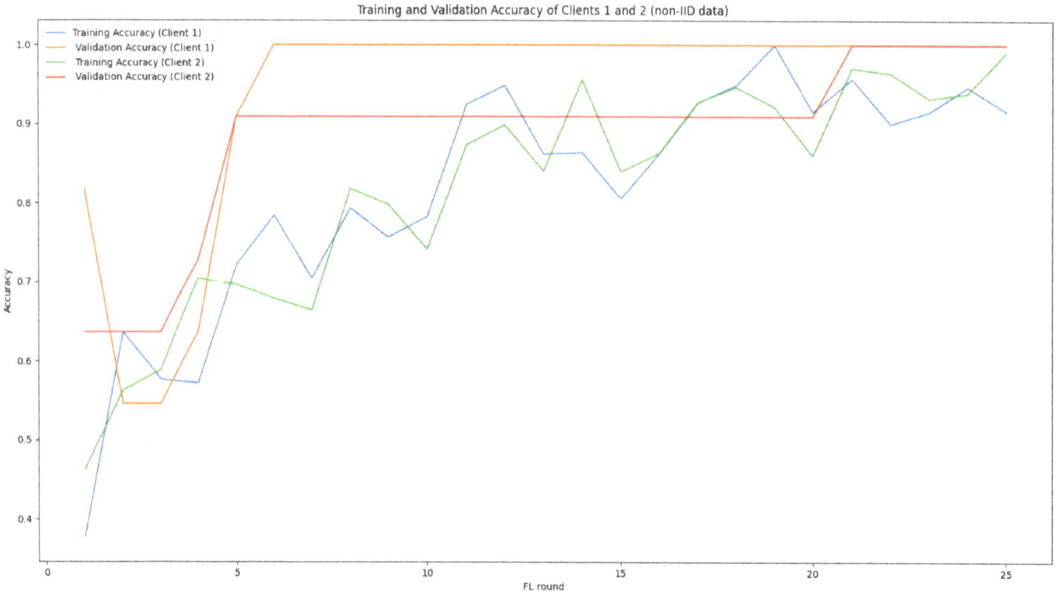

Figure 12. Training and validation loss and accuracy curves for both clients 1 and 2 with previously set parameters, trained for 25 FL rounds on non-IID data.

Figure 13 plots the $QoSAuT_i(k)$ in each round i for both clients ($k = \{1,2\}$) in the FL-to-convergence scenario, highlighting the specific round where LEFL ES mechanism terminates the training. $QoSAuT_i$ exhibits similar patterns to the IID data scenario, though it converges a few rounds later. However, the LEFL ES mechanism triggers as early as in the IID scenario, causing the model to learn less compared to the IID case.

Figure 14 plots the confusion matrices of the global final model trained with LEFL and FL-to-convergence, respectively. In the FL-to-convergence scenario, we achieved results comparable to those of FL-to-convergence and benchmark centralized learning with IID data. However, LEFL achieved lower accuracy compared to its performance in the IID data scenario due to the ES mechanism being triggered several rounds earlier than convergence, as convergence occurred later in the non-IID data scenario. It is important to note that the LEFL performance drop observed in the non-IID scenario compared to the IID scenario does not affect the *no_bird* label. This is because the samples for this label remain IID, even in the non-IID scenario, due to their lack of diversity in the dataset.

Table 12 summarizes the $QoSAuT$ for each device across 3 scenarios: LEFL, FL-to-convergence, and a classic remote benchmark training scheme. All scenarios are evaluated under a non-IID data distribution. However, it is important to highlight that, in the remote benchmark scenario, the data are inherently IID since they are centralized and aggregated on a server for training.

Table 12. Test accuracy and $QoSAuT$ of each device trained on non-IID data in three scenarios: LEFL, FL-to-convergence, and remote benchmark.

Scenario	LEFL	FL-to-Convergence	Remote Benchmark
Test Accuracy	0.87	0.96	0.96
$QoSAuT(1)$	1.44	0.77	0.09
$QoSAuT(2)$	1.80	0.77	0.09

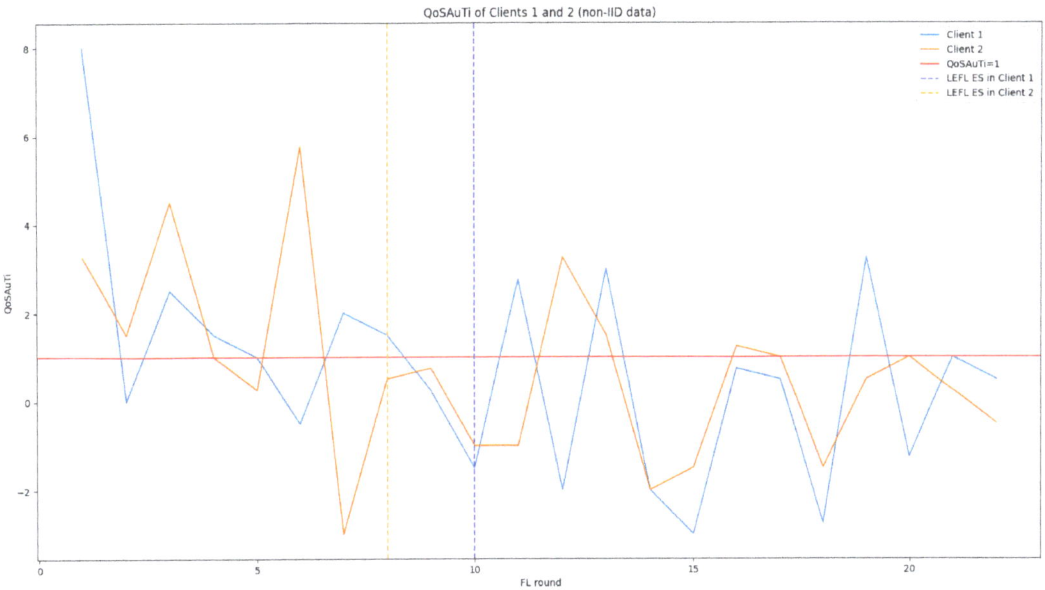

Figure 13. $QoSAuT_i$ for both clients 1 and 2 trained to convergence, with the activation of LEFL early stopping (ES), in the non-IID data scenario.

FL improved the $QoSAuT$ of both clients by at least a factor of 8. In the FL-to-convergence scenario, the improvement was smaller compared to the IID data scenario. This is because convergence occurred later with non-IID data, resulting in the $QoSAuT$ being reduced to half of that in the IID scenario to achieve the same model accuracy level. With LEFL, the model's accuracy was lower due to the previously mentioned reasons (convergence occurred later in non-IID data). However, while the $QoSAuT$ was slightly lower than in the IID scenario, it remained at a comparable level (lower only by a factor of 1.1 to 1.4). Within the same non-IID data scenario, it is important to note, the $QoSAuT$ doubles in LEFL compared to FL-to-convergence. Comparing these two scenarios reveals that the model achieves only 9% less accuracy but doubles the $QoSAuT$. This highlights the effectiveness of LEFL in optimizing the trade-off between accuracy and energy consumption. From a user's perspective, agricultural devices that operate longer with slightly less accurate decisions can be more beneficial than devices that deliver higher accuracy but have a lifespan that is half as long.

The significant $QoSAuT$ improvements observed with FL-to-convergence compared to the benchmark—far exceeding the gains achieved using LEFL relative to FL-to-convergence—can be attributed to the simplicity of the classification task, the limited dataset diversity, and the small number of clients. These factors facilitate quicker convergence, even in non-IID data scenarios, compared to more complex tasks where data are more diverse and require more FL rounds to reach convergence. In more diverse non-IID data scenarios, LEFL is likely to exhibit even greater advantages in terms of $QoSAuT$ over the traditional FL-to-convergence training scheme.

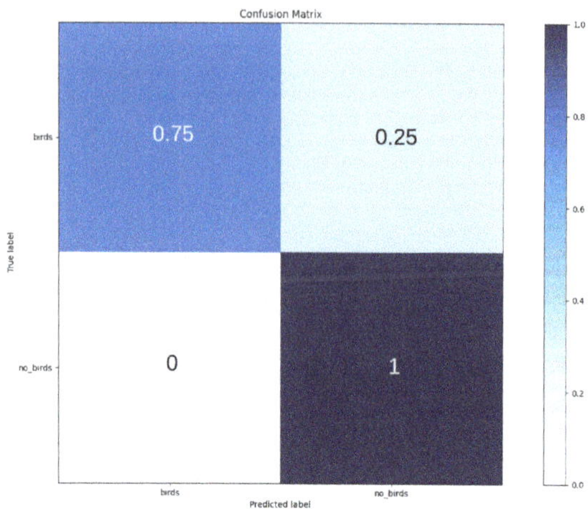

(**a**)

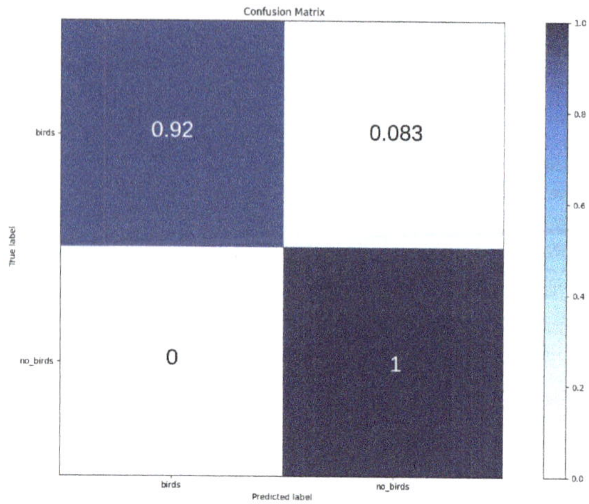

(**b**)

Figure 14. Normalized confusion matrices of the model within the non-IID data scenario trained with LEFL and FL-to-convergence schemes. The vertical axis represents the true labels, ordered as "birds" and "no_birds," while the horizontal axis represents the predicted labels, also ordered as "birds" and "no_birds," respectively. (**a**) Normalized confusion matrix of the model trained with LEFL in non-IID data scenario. (**b**) Normalized confusion matrix of the model trained with FL-to-convergence in non-IID data scenario.

6. Conclusions and Future Work

In this paper, we have proposed an Artificial Intelligence of Things (AIoT) system for bird detection in agricultural applications, with a strong focus on energy efficiency, particularly in terms of training costs. By leveraging federated learning (FL), we enhanced the energy efficiency of the system, reducing its energy consumption while maintaining the same decision-making quality. Furthermore, we also proposed a low-energy FL system (LEFL) in order to achieve a balance between decision quality and energy sobriety by stopping the FL training process earlier. To quantify decision-making quality, we used the model's accuracy, and to evaluate energy efficiency, we introduced a novel metric, $QoSAuT$, which measures the quality of the service provided via our system in terms of decision-making. We tested our approach on a lightweight but realistic dataset obtained from real-world applications. This dataset consisted of around one hundred samples categorized into *birds* and *no_birds* tailored for a simple classification task, which we distributed across our system in both IID and non-IID configurations to simulate potential scenarios that could occur during the training phase.

Training our designed system on the previously described dataset demonstrated a 15-fold improvement in energy efficiency using FL in the IID configuration and an eight-fold improvement in the non-IID configuration. With LEFL, the performance also improved slightly in IID scenarios with a low reduction in accuracy. However, a more significant improvement was achieved in non-IID scenarios with an also limited reduction in accuracy.

Our work presents a comprehensive approach to designing an energy-efficient, AIoT-based system for bird detection in agriculture 4.0. In future work, its robustness and scalability performances could be evaluated in more complex scenarios by diversifying the dataset, incorporating additional bird species, and broadening the feature diversity, such as including elements in the *no_bird* category like human presence or the presence of other animals and objects. However, composing the WSCN with two smart cameras is sufficient as a proof of concept, effectively demonstrating the LEFL advantages while remaining generalizable to networks with more nodes.

Additionally, the early stopping criteria used in our LEFL approach can be adapted to more complex tasks where FL convergence occurs significantly later than in our case. Future studies could explore alternative early stopping mechanisms, including methods for function approximation predicting the evolution of $QoSAuT$ to identify the point at which further rounds no longer contribute to enhancing the overall efficiency of the training process. Such forecasting could provide a more precise and effective stopping criterion.

Another strategy for improving the energy efficiency of our AIoT system is to optimize the inference phase by employing technologies with lower computational demands. This entails making thoughtful choices regarding both the hardware and the AI model in order to minimize resource consumption during inference. One possible path for hardware improvement is using neuromorphic vision sensors, also known as event-based or dynamic vision sensors (DVSs) [71]. These sensors capture only changes in the visual scene, enabling significant energy savings compared to the continuous frame-by-frame image capture of traditional cameras used in our system. From a model improvement perspective, spiking neural networks (SNNs) [72] are inherently event-driven, optimizing energy efficiency in local computations. This makes them particularly well-suited for processing event-based data from DVSs, enabling energy-efficient decision-making at the edge.

To further advance the bird detection use case and align it more closely with real-world agricultural/environmental requirements, semantic segmentation can be incorporated to not only improve the precision of bird detection but also quantify birds' presence by accurately counting them in crop fields, thus estimating the scale of potential losses caused. An energy-efficient approach to addressing this involves both optimized hardware and an adapted model, as discussed earlier, by training a wireless DVS network powered via an SNN model for semantic segmentation. Additionally, integrating LEFL could further improve training efficiency, making the system energy-efficient in both the training and inference phases.

EvSegSNN [73] successfully combines SNNs with DVSs to perform semantic segmentation, demonstrating promising results from an energy-efficient standpoint. Implementing this approach in the bird detection scenario, while incorporating LEFL for energy-efficient training, could be an exciting research direction.

Author Contributions: Conceptualization, S.B.; methodology, S.B.; software, S.B.; validation, S.B., G.D.S. and J.-P.C.; formal analysis, S.B., G.D.S. and J.-P.C.; investigation, S.B.; resources, S.B., G.D.S. and J.-P.C.; data curation, S.B.; writing—original draft preparation, S.B.; writing—review and editing, S.B., G.D.S. and J.-P.C.; supervision, G.D.S. and J.-P.C.; project administration, G.D.S. and J.-P.C.; funding acquisition, G.D.S. and J.-P.C. All authors have read and agreed to the published version of the manuscript.

Funding: This research was funded by the Agence Nationale de la Recherche (ANR) of the French Government through the program "Investissements d'Avenir" (16-IDEX-0001 CAP 20-25), Clermont Auvergne Metropole, and the INRAE MathNum Scientific Department.

Institutional Review Board Statement: Not applicable.

Informed Consent Statement: Not applicable.

Data Availability Statement: The data presented in this study are available upon request from Corentin Barbu (INRAE Agronomie) and Christophe Sausse (Terres Inovia).

Acknowledgments: We sincerely thank the funding organizations—The Agence Nationale de la Recherche (ANR) of the French Government, Clermont Auvergne Métropole, and the INRAE MathNum Department—for their support, which made it possible to explore innovative AIoT-based solutions for bird detection in agriculture 4.0. We also express our gratitude to Corentin Barbu and Christophe Sausse for providing access to the dataset. This invaluable resource, derived from real-world data, was instrumental in conducting our experiments and validating our approach.

Conflicts of Interest: The authors declare no conflicts of interest.

References

1. Basso, B.; Antle, J. Digital agriculture to design sustainable agricultural systems. *Nat. Sustain.* **2020**, *3*, 254–256. [CrossRef]
2. Wakchaure, M.; Patle, B.; Mahindrakar, A. Application of AI techniques and robotics in agriculture: A review. *Artif. Intell. Life Sci.* **2023**, *3*, 100057. [CrossRef]
3. Eli-Chukwu, N.C. Applications of Artificial Intelligence in Agriculture: A Review. *Eng. Technol. Appl. Sci. Res.* **2019**, *9*, 4377–4383. [CrossRef]
4. Lytridis, C.; Kaburlasos, V.G.; Pachidis, T.; Manios, M.; Vrochidou, E.; Kalampokas, T.; Chatzistamatis, S. An Overview of Cooperative Robotics in Agriculture. *Agronomy* **2021**, *11*, 1818. [CrossRef]
5. Farooq, M.S.; Riaz, S.; Abid, A.; Umer, T.; Zikria, Y.B. Role of IoT Technology in Agriculture: A Systematic Literature Review. *Electronics* **2020**, *9*, 319. [CrossRef]
6. Banđur, D.; Jakšić, B.; Banđur, M.; Jović, S. An analysis of energy efficiency in Wireless Sensor Networks (WSNs) applied in smart agriculture. *Comput. Electron. Agric.* **2019**, *156*, 500–507. [CrossRef]
7. Raghunathan, V.; Schurgers, C.; Park, S.; Srivastava, M. Energy-aware wireless microsensor networks. *IEEE Signal Process. Mag.* **2002**, *19*, 40–50. [CrossRef]
8. Jawad, H.; Nordin, R.; Gharghan, S.; Jawad, A.; Ismail, M. Energy-Efficient Wireless Sensor Networks for Precision Agriculture: A Review. *Sensors* **2017**, *17*, 1781. [CrossRef]
9. Sahota, H.; Kumar, R.; Kamal, A.; Huang, J. An energy-efficient wireless sensor network for precision agriculture. In *Proceedings of the IEEE Symposium on Computers and Communications, Riccione, Italy, 22–25 June 2010*; IEEE: Piscataway, NJ, USA, 2010; pp. 347–350. [CrossRef]
10. Hemmati, A.; Rahmani, A.M. The Internet of Autonomous Things applications: A taxonomy, technologies, and future directions. *Internet Things* **2022**, *20*, 100635. [CrossRef]
11. Singh, R.; Gill, S.S. Edge AI: A survey. *Internet Things Cyber-Phys. Syst.* **2023**, *3*, 71–92. [CrossRef]
12. Kukreja, N.; Shilova, A.; Beaumont, O.; Huckelheim, J.; Ferrier, N.; Hovland, P.; Gorman, G. Training on the Edge: The why and the how. *arXiv* **2019**, arXiv.1903.03051.
13. Shi, Y.; Yang, K.; Jiang, T.; Zhang, J.; Letaief, K.B. Communication-Efficient Edge AI: Algorithms and Systems. *arXiv* **2020**, arXiv.2002.09668.
14. McMahan, H.B.; Moore, E.; Ramage, D.; Hampson, S.; Arcas, B.A.y. Communication-Efficient Learning of Deep Networks from Decentralized Data. *arXiv* **2016**, arXiv.1602.05629.
15. Zhang, C.; Xie, Y.; Bai, H.; Yu, B.; Li, W.; Gao, Y. A survey on federated learning. *Knowl.-Based Syst.* **2021**, *216*, 106775. [CrossRef]
16. Zhu, H.; Xu, J.; Liu, S.; Jin, Y. Federated Learning on Non-IID Data: A Survey. *arXiv* **2021**, arXiv:cs.LG/2106.06843.
17. Zhao, Y.; Li, M.; Lai, L.; Suda, N.; Civin, D.; Chandra, V. Federated Learning with Non-IID Data. *arXiv* **2018**, arXiv.1806.00582.
18. Zhao, Z.; Feng, C.; Hong, W.; Jiang, J.; Jia, C.; Quek, T.Q.S.; Peng, M. Federated Learning with Non-IID Data in Wireless Networks. *IEEE Trans. Wirel. Commun.* **2022**, *21*, 1927–1942. [CrossRef]
19. Restrepo-Arias, J.F.; Branch-Bedoya, J.W.; Awad, G. Image classification on smart agriculture platforms: Systematic literature review. *Artif. Intell. Agric.* **2024**, *13*, 1–17. [CrossRef]
20. Liu, G.; Zhong, K.; Li, H.; Chen, T.; Wang, Y. A state of art review on time series forecasting with machine learning for environmental parameters in agricultural greenhouses. *Inf. Process. Agric.* **2024**, *11*, 143–162. [CrossRef]
21. Jha, K.; Doshi, A.; Patel, P.; Shah, M. A comprehensive review on automation in agriculture using artificial intelligence. *Artif. Intell. Agric.* **2019**, *2*, 1–12. [CrossRef]
22. Xu, J.; Gu, B.; Tian, G. Review of agricultural IoT technology. *Artif. Intell. Agric.* **2022**, *6*, 10–22. [CrossRef]
23. Qazi, S.; Khawaja, B.A.; Farooq, Q.U. IoT-Equipped and AI-Enabled Next Generation Smart Agriculture: A Critical Review, Current Challenges and Future Trends. *IEEE Access* **2022**, *10*, 21219–21235. [CrossRef]
24. Muhammed, D.; Ahvar, E.; Ahvar, S.; Trocan, M.; Montpetit, M.J.; Ehsani, R. Artificial Intelligence of Things (AIoT) for smart agriculture: A review of architectures, technologies and solutions. *J. Netw. Comput. Appl.* **2024**, *228*, 103905. [CrossRef]
25. Kumar, S.; Chowdhary, G.; Udutalapally, V.; Das, D.; Mohanty, S.P. gCrop: Internet-of-Leaf-Things (IoLT) for Monitoring of the Growth of Crops in Smart Agriculture. In Proceedings of the 2019 IEEE International Symposium on Smart Electronic Systems (iSES) (Formerly iNiS), Rourkela, India, 16–18 December 2019; pp. 53–56. [CrossRef]

26. T, M.; Makkithaya, K.; G, N.V. A Federated Learning-Based Crop Yield Prediction for Agricultural Production Risk Management. In *Proceedings of the 2022 IEEE Delhi Section Conference (DELCON), New Delhi, India, 11–13 February 2022;* IEEE: Piscataway, NJ, USA, 2022. [CrossRef]
27. Yu, C.; Shen, S.; Zhang, K.; Zhao, H.; Shi, Y. Energy-Aware Device Scheduling for Joint Federated Learning in Edge-assisted Internet of Agriculture Things. In *Proceedings of the 2022 IEEE Wireless Communications and Networking Conference (WCNC), Austin, TX, USA, 10–13 April 2022;* IEEE: Piscataway, NJ, USA, 2022. [CrossRef]
28. Idoje, G.; Dagiuklas, T.; Iqbal, M. Federated Learning: Crop classification in a smart farm decentralised network. *Smart Agric. Technol.* **2023**, *5*, 100277. [CrossRef]
29. Jiang, P.; Chen, Y.; Liu, B.; He, D.; Liang, C. Real-Time Detection of Apple Leaf Diseases Using Deep Learning Approach Based on Improved Convolutional Neural Networks. *IEEE Access* **2019**, *7*, 59069–59080. [CrossRef]
30. Antico, T.M.; Moreira, L.F.R.; Moreira, R. Evaluating the Potential of Federated Learning for Maize Leaf Disease Prediction. In Proceedings of the Anais do XIX Encontro Nacional de Inteligência Artificial e Computacional (ENIAC 2022), Sociedade Brasileira de Computaçăo—SBC, 2022, ENIAC 2022, Campinas, Brazil, 28 November–1 December 2022. [CrossRef]
31. Khan, F.S.; Khan, S.; Mohd, M.N.H.; Waseem, A.; Khan, M.N.A.; Ali, S.; Ahmed, R. Federated learning-based UAVs for the diagnosis of Plant Diseases. In *Proceedings of the 2022 International Conference on Engineering and Emerging Technologies (ICEET), Kuala Lumpur, Malaysia, 27–28 October 2022;* IEEE: Piscataway, NJ, USA, 2022. [CrossRef]
32. Patros, P.; Ooi, M.; Huang, V.; Mayo, M.; Anderson, C.; Burroughs, S.; Baughman, M.; Almurshed, O.; Rana, O.; Chard, R.; et al. Rural AI: Serverless-Powered Federated Learning for Remote Applications. *IEEE Internet Comput.* **2023**, *27*, 28–34. [CrossRef]
33. Deng, F.; Mao, W.; Zeng, Z.; Zeng, H.; Wei, B. Multiple Diseases and Pests Detection Based on Federated Learning and Improved Faster R-CNN. *IEEE Trans. Instrum. Meas.* **2022**, *71*, 3523811. [CrossRef]
34. Vincent, D.R.; Deepa, N.; Elavarasan, D.; Srinivasan, K.; Chauhdary, S.H.; Iwendi, C. Sensors Driven AI-Based Agriculture Recommendation Model for Assessing Land Suitability. *Sensors* **2019**, *19*, 3667. [CrossRef]
35. Murugamani, C.; Shitharth, S.; Hemalatha, S.; Kshirsagar, P.R.; Riyazuddin, K.; Naveed, Q.N.; Islam, S.; Mazher Ali, S.P.; Batu, A. Machine Learning Technique for Precision Agriculture Applications in 5G-Based Internet of Things. *Wirel. Commun. Mob. Comput.* **2022**, *2022*, 6534238. [CrossRef]
36. Dahane, A.; Benameur, R.; Kechar, B.; Benyamina, A. An IoT Based Smart Farming System Using Machine Learning. In Proceedings of the 2020 International Symposium on Networks, Computers and Communications (ISNCC), Montreal, QC, Canada, 20–22 October 2020; pp. 1–6. [CrossRef]
37. Cheng, Z.; Zhang, F. Flower End-to-End Detection Based on YOLOv4 Using a Mobile Device. *Wirel. Commun. Mob. Comput.* **2020**, *2020*, 1–9. [CrossRef]
38. Hsu, C.W.; Huang, Y.H.; Huang, N.F. Real-time Dragonfruit's Ripeness Classification System with Edge Computing Based on Convolution Neural Network. In *Proceedings of the 2022 International Conference on Information Networking (ICOIN), Jeju-si, Republic of Korea, 12–15 January 2022;* IEEE: Piscataway, NJ, USA, 2022; pp. 177–182. [CrossRef]
39. Paul, P.B.; Biswas, S.; Bairagi, A.K.; Masud, M. Data-Driven Decision Making for Smart Cultivation. In Proceedings of the 2021 IEEE International Symposium on Smart Electronic Systems (iSES), Jaipur, India, 18–22 December 2021; IEEE: Piscataway, NJ, USA, 2021; pp. 249–254. [CrossRef]
40. Arablouei, R.; Wang, L.; Currie, L.; Yates, J.; Alvarenga, F.A.; Bishop-Hurley, G.J. Animal behavior classification via deep learning on embedded systems. *Comput. Electron. Agric.* **2023**, *207*, 107707. [CrossRef]
41. Mao, A.; Huang, E.; Gan, H.; Liu, K. FedAAR: A Novel Federated Learning Framework for Animal Activity Recognition with Wearable Sensors. *Animals* **2022**, *12*, 2142. [CrossRef] [PubMed]
42. Aliahmadi, A.; Nozari, H.; Ghahremani-Nahr, J. AIoT-based Sustainable Smart Supply Chain Framework. *Int. J. Innov. Manag. Econ. Soc. Sci.* **2022**, *2*, 28–38. [CrossRef]
43. Durrant, A.; Markovic, M.; Matthews, D.; May, D.; Enright, J.; Leontidis, G. The role of cross-silo federated learning in facilitating data sharing in the agri-food sector. *Comput. Electron. Agric.* **2022**, *193*, 106648. [CrossRef]
44. Chabot, D.; Francis, C.M. Computer-automated bird detection and counts in high-resolution aerial images: A review. *J. Field Ornithol.* **2016**, *87*, 343–359. [CrossRef]
45. Akçay, H.G.; Kabasakal, B.; Aksu, D.; Demir, N.; Öz, M.; Erdoğan, A. Automated Bird Counting with Deep Learning for Regional Bird Distribution Mapping. *Animals* **2020**, *10*, 1207. [CrossRef]
46. Wäldchen, J.; Mäder, P. Machine learning for image based species identification. *Methods Ecol. Evol.* **2018**, *9*, 2216–2225. [CrossRef]
47. Hong, S.J.; Han, Y.; Kim, S.Y.; Lee, A.Y.; Kim, G. Application of Deep-Learning Methods to Bird Detection Using Unmanned Aerial Vehicle Imagery. *Sensors* **2019**, *19*, 1651. [CrossRef]
48. Lee, S.; Lee, M.; Jeon, H.; Smith, A. Bird Detection in Agriculture Environment using Image Processing and Neural Network. In Proceedings of the 2019 6th International Conference on Control, Decision and Information Technologies (CoDIT), Paris, France, 23–26 April 2019; pp. 1658–1663. [CrossRef]

49. Li, C.; Zhang, B.; Hu, H.; Dai, J. Enhanced Bird Detection from Low-Resolution Aerial Image Using Deep Neural Networks. *Neural Process. Lett.* **2018**, *49*, 1021–1039. [CrossRef]
50. Mashuk, F.; Sattar, A.; Sultana, N. Machine Learning Approach for Bird Detection. In *Proceedings of the 2021 Third International Conference on Intelligent Communication Technologies and Virtual Mobile Networks (ICICV), Tirunelveli, India, 4–6 February 2021;* IEEE: Piscataway, NJ, USA, 2021; pp. 818–822. [CrossRef]
51. Nguyen, D.C.; Pham, Q.V.; Pathirana, P.N.; Ding, M.; Seneviratne, A.; Lin, Z.; Dobre, O.; Hwang, W.J. Federated Learning for Smart Healthcare: A Survey. *ACM Comput. Surv.* **2022**, *55*, 1–37. [CrossRef]
52. Zhou, J.; Zhang, S.; Lu, Q.; Dai, W.; Chen, M.; Liu, X.; Pirttikangas, S.; Shi, Y.; Zhang, W.; Herrera-Viedma, E. A Survey on Federated Learning and its Applications for Accelerating Industrial Internet of Things. *arXiv* **2021**, arXiv.2104.10501.
53. Pandya, S.; Srivastava, G.; Jhaveri, R.; Babu, M.R.; Bhattacharya, S.; Maddikunta, P.K.R.; Mastorakis, S.; Piran, M.J.; Gadekallu, T.R. Federated learning for smart cities: A comprehensive survey. *Sustain. Energy Technol. Assess.* **2023**, *55*, 102987. [CrossRef]
54. Žalik, K.R.; Žalik, M. A Review of Federated Learning in Agriculture. *Sensors* **2023**, *23*, 9566. [CrossRef] [PubMed]
55. Saha, R.; Misra, S.; Deb, P.K. FogFL: Fog-Assisted Federated Learning for Resource-Constrained IoT Devices. *IEEE Internet Things J.* **2021**, *8*, 8456–8463. [CrossRef]
56. Kumar, A.; Srirama, S.N., Fog Enabled Distributed Training Architecture for Federated Learning. In *Big Data Analytics;* Springer International Publishing: Berlin/Heidelberg, Germany, 2021; pp. 78–92. [CrossRef]
57. Kumar, P.; Gupta, G.P.; Tripathi, R. PEFL: Deep Privacy-Encoding-Based Federated Learning Framework for Smart Agriculture. *IEEE Micro* **2022**, *42*, 33–40. [CrossRef]
58. Friha, O.; Ferrag, M.A.; Shu, L.; Maglaras, L.; Choo, K.K.R.; Nafaa, M. FELIDS: Federated learning-based intrusion detection system for agricultural Internet of Things. *J. Parallel Distrib. Comput.* **2022**, *165*, 17–31. [CrossRef]
59. Abu-Khadrah, A.; Ali, A.M.; Jarrah, M. An Amendable Multi-Function Control Method using Federated Learning for Smart Sensors in Agricultural Production Improvements. *ACM Trans. Sens. Netw.* **2023**. . [CrossRef]
60. Sachidananda, V.; Khelil, A.; Suri, N. Quality of information in wireless sensor networks: A survey. *Citeseer* **2010**.
61. Dong, K.; Zhou, C.; Ruan, Y.; Li, Y. MobileNetV2 Model for Image Classification. In *Proceedings of the 2020 2nd International Conference on Information Technology and Computer Application (ITCA), Guangzhou, China, 18–20 December 2020;* IEEE: Piscataway, NJ, USA, 2020; pp. 476–480. [CrossRef]
62. Gupta, J.; Pathak, S.; Kumar, G. Deep Learning (CNN) and Transfer Learning: A Review. *J. Phys. Conf. Ser.* **2022**, *2273*, 012029. [CrossRef]
63. Abadi, M.; Barham, P.; Chen, J.; Chen, Z.; Davis, A.; Dean, J.; Devin, M.; Ghemawat, S.; Irving, G.; Isard, M.; et al. TensorFlow: A system for large-scale machine learning. *arXiv* **2016**, arxiv.1605.08695.
64. Beutel, D.J.; Topal, T.; Mathur, A.; Qiu, X.; Fernandez-Marques, J.; Gao, Y.; Sani, L.; Li, K.H.; Parcollet, T.; de Gusmão, P.P.B.; et al. Flower: A Friendly Federated Learning Research Framework. *arXiv* **2020**, arXiv.2007.14390.
65. Stuecke, J. Top 7 Open-Source Frameworks for Federated Learning. 2024. Available online: https://www.apheris.com/resources/blog/top-7-open-source-frameworks-for-federated-learning/ (accessed on 13 December 2024).
66. Niu, Z.; Dong, H.; Qin, A.K.; Gu, T. FLrce: Resource-Efficient Federated Learning with Early-Stopping Strategy. *arXiv* **2023**, arXiv.2310.09789.
67. Sausse, C.; Barbu, C.; Bertrand, M.; Thibord, J.B. Dégâts d'oiseaux à la levée: Vers un changement de méthode? *Phytoma* **2022**, 35–38.
68. Ferreboeuf, H. The Shift Project, Lean ICT Report. 2018. Available online: https://theshiftproject.org/en/article/lean-ict-our-new-report/ (accessed on 12 December 2024).
69. Sevilla, J.; Heim, L.; Hobbhahn, M.; Besiroglu, T.; Ho, A.; Villalobos, P. Estimating Training Compute of Deep Learning Models. 2022. Available online: https://epoch.ai/blog/estimating-training-compute (accessed on 18 December 2024).
70. Sandler, M.; Howard, A.; Zhu, M.; Zhmoginov, A.; Chen, L.C. MobileNetV2: Inverted Residuals and Linear Bottlenecks. *arXiv* **2018**, arXiv.1801.04381.
71. Liao, F.; Zhou, F.; Chai, Y. Neuromorphic vision sensors: Principle, progress and perspectives. *J. Semicond.* **2021**, *42*, 013105. [CrossRef]
72. Ghosh-Dastidar, S.; Adeli, H. Spiking neural networks. *Int. J. Neural Syst.* **2009**, *19*, 295–308. [CrossRef]
73. Hareb, D.; Martinet, J. EvSegSNN: Neuromorphic Semantic Segmentation for Event Data. *arXiv* **2024**, arXiv.2406.14178.

AI

MDPI

Article

Adapting a Large-Scale Transformer Model to Decode Chicken Vocalizations: A Non-Invasive AI Approach to Poultry Welfare

Suresh Neethirajan [1,2]

[1] Faculty of Agriculture, Dalhousie University, 6050 University Ave, Halifax, NS B3H 1W5, Canada; sneethir@gmail.com
[2] Faculty of Computer Science, Dalhousie University, 6050 University Ave, Halifax, NS B3H 1W5, Canada

Abstract: Natural Language Processing (NLP) and advanced acoustic analysis have opened new avenues in animal welfare research by decoding the vocal signals of farm animals. This study explored the feasibility of adapting a large-scale Transformer-based model, OpenAI's Whisper, originally developed for human speech recognition, to decode chicken vocalizations. Our primary objective was to determine whether Whisper could effectively identify acoustic patterns associated with emotional and physiological states in poultry, thereby enabling real-time, non-invasive welfare assessments. To achieve this, chicken vocal data were recorded under diverse experimental conditions, including healthy versus unhealthy birds, pre-stress versus post-stress scenarios, and quiet versus noisy environments. The audio recordings were processed through Whisper, producing text-like outputs. Although these outputs did not represent literal translations of chicken vocalizations into human language, they exhibited consistent patterns in token sequences and sentiment indicators strongly correlated with recognized poultry stressors and welfare conditions. Sentiment analysis using standard NLP tools (e.g., polarity scoring) identified notable shifts in "negative" and "positive" scores that corresponded closely with documented changes in vocal intensity associated with stress events and altered physiological states. Despite the inherent domain mismatch—given Whisper's original training on human speech—the findings clearly demonstrate the model's capability to reliably capture acoustic features significant to poultry welfare. Recognizing the limitations associated with applying English-oriented sentiment tools, this study proposes future multimodal validation frameworks incorporating physiological sensors and behavioral observations to further strengthen biological interpretability. To our knowledge, this work provides the first demonstration that Transformer-based architectures, even without species-specific fine-tuning, can effectively encode meaningful acoustic patterns from animal vocalizations, highlighting their transformative potential for advancing productivity, sustainability, and welfare practices in precision poultry farming.

Keywords: bioacoustics; poultry welfare; chicken vocalizations; natural language processing; animal welfare; acoustic analysis; precision livestock farming; animal vocal communication; sound classification; acoustic feature encoding

1. Introduction

The application of Artificial Intelligence (AI) to the study of animal vocalizations has transformed how researchers and farmers assess the health and welfare of livestock. In particular, acoustic analysis combined with advanced machine learning (ML) techniques offers a promising non-invasive strategy for real-time monitoring of poultry welfare [1,2].

This approach addresses several limitations of traditional observation-based methods, which are often subjective, time-consuming, and potentially stressful for the animals [3,4]. By leveraging AI-driven acoustic analysis within precision livestock farming practices, researchers have the opportunity to enhance productivity while simultaneously improving ethical standards in poultry farming [5,6].

Traditional welfare assessments in poultry rely heavily on physical and behavioral observations, such as feather condition, posture, vocal frequency, or changes in feeding activity. While these qualitative assessments are widely used, they are prone to variability among observers and lack the continuous monitoring necessary for timely interventions [7]. In contrast, AI models designed for vocal signal detection provide automated surveillance capable of identifying subtle changes in vocal patterns associated with stress, fear, disease, or distress. These systems reduce human–animal interactions during monitoring, thereby minimizing stress responses triggered by observers and preserving natural bird behavior [8,9].

1.1. AI for Acoustic Analysis in Poultry Welfare

Recent advances in machine learning—particularly deep learning—have enabled significant breakthroughs in detecting and classifying animal vocalizations [10]. Historically, much of this work has relied on specialized Convolutional Neural Networks (CNNs) or Recurrent Neural Networks (RNNs), which operate on time–frequency representations (spectrograms) of recorded calls [11,12]. For example:

(a) Successful identification of bird species from field recordings [13,14].
(b) Real-time monitoring of farm animal distress signals [15,16].
(c) Early detection of poultry diseases through cough-like vocal patterns [17,18].

Despite their effectiveness in classification tasks such as identifying distress calls or categorizing alarm calls [19,20], these models often require carefully curated datasets labeled by domain experts. The reliance on large labeled datasets remains a bottleneck for broader adoption. Transfer learning techniques have emerged as a solution to this challenge by repurposing pretrained models originally developed for human speech recognition to detect specific avian vocal patterns [21,22]. For instance, CNNs trained on thousands of hours of human speech have been successfully adapted to identify avian vocalizations with reduced domain-specific training requirements.

While these methods are powerful for classification tasks like detecting distress calls or categorizing alarm signals, they fail to capture the semantic complexity inherent in animal communication [23]. Chickens produce a variety of vocalizations—such as alarm calls, food calls, comfort calls, and distress calls—that reflect not only immediate threats or resource availability but also underlying emotional or physiological states [24,25]. Deciphering these nuanced signals is critical for enabling farmers to respond appropriately and improve both animal welfare and farm efficiency [26].

1.2. State of the Art in Bioacoustic Methods

The field of bioacoustics encompasses research not only on poultry but also on wildlife and other livestock species. CNNs remain central to many state-of-the-art approaches, particularly for species identification in acoustic ecology [27,28]. These methods frequently achieve performance metrics exceeding 90% accuracy for species detection tasks and demonstrate robustness against environmental noise when using advanced architectures like hybrid CNN-RNN models or Attention-based Inception networks [29].

More recently, researchers have explored Transformer architectures capable of processing sequences of acoustic frames over extended time windows—a feature that is particularly relevant for analyzing social and emotional contexts in vocalizations [30,31]. However, applying these methods to poultry welfare remains limited. Current classifiers primarily distinguish between "distress" and "non-distress" calls or "healthy" versus "unhealthy" sounds without providing deeper semantic interpretations or emotional insights into the vocalizations [32].

Large-scale Automatic Speech Recognition (ASR) models like DeepSpeech, wav2vec, and Whisper have revolutionized human speech recognition by leveraging massive amounts of textual and audio data. Whisper stands out for its ability to handle multilingual inputs and noisy environments with high accuracy [33,34]. Although Whisper was designed specifically for human language processing, its Transformer-based architecture may be capable of detecting structural similarities in non-human vocalizations—such as pitch changes or rhythmic patterns—if suitably fine-tuned or guided [35]. This raises an intriguing question: can a model trained purely on human language be adapted to extract meaningful text-like representations from chicken vocalizations?

1.3. The Potential of Large-Scale ASR Models (Whisper) for Poultry

Whisper's Transformer architecture excels at capturing long-range dependencies within sequential data through its self-attention mechanism. While its front-end acoustic processing is optimized for human phonetics and lexicons, there is potential for its high-level representations to encode non-linguistic information such as pitch transitions or amplitude envelopes. If these patterns emerge consistently as pseudo-linguistic tokens during transcription processes—despite being nonsensical—they could still provide meaningful insights into stress or contentment levels in chickens when analyzed using sentiment tools originally designed for textual data [36,37].

Early proofs-of-concept have demonstrated partial success in applying large-scale Transformers to other animal communication systems such as whale calls [38,39]. While direct translation is not feasible due to the absence of linguistic meaning in animal sounds, hidden layers within these models appear capable of clustering similar call types based on consistent acoustic features. This phenomenon highlights an important insight: advanced ASR systems can act as generalized audio encoders rather than strictly speech decoders [40].

1.4. Gaps, Challenges, and Ethical Considerations

Despite promising results from early studies applying AI to animal vocalizations, several significant challenges remain:

(a) Data Requirements: Training large-scale Transformers like Whisper requires extensive high-quality labeled recordings annotated by poultry experts—a time-intensive process that limits scalability [41,42].

(b) Background Noise: Commercial farms are inherently noisy environments with overlapping bird calls and mechanical sounds such as ventilation fans. While Whisper is robust against noise in human speech contexts, farm-specific noise poses unique challenges that complicate analysis [43].

(c) Ethical Constraints: Inducing stress or disease states in flocks solely for data collection purposes raises ethical concerns. Additionally, variations in farm layouts and climates reduce the feasibility of standardized sampling protocols across different facilities [44].

(d) Interpretation vs. Translation: Even if Whisper generates consistent transcriptions like "nihai nihai", these strings lack semantic meaning in human language. Extracting actionable insights requires careful correlation with validated welfare metrics such as physiological indicators or behavioral observations [45].

Successfully addressing these challenges would advance animal welfare science significantly by creating pipelines for continuous real-time monitoring that reduce labor demands while improving early detection capabilities and overall welfare standards.

1.5. Objectives and Contributions

Building on the above context, the present study aims to bridge state-of-the-art AI approaches with practical poultry welfare needs. Specifically, we investigate whether OpenAI's Whisper—a large-scale model trained on massive human speech datasets—can be harnessed or adapted to interpret chicken vocalizations. The overarching hypotheses are that certain repetitive or distinctive token sequences in Whisper's output will align with known behavioral states (e.g., pre- vs. post-stress, healthy vs. unhealthy, quiet vs. noisy environments). By applying standard sentiment analysis methods to these automatically generated texts, we explore the potential to gauge changes in emotional valence or stress. Four key objectives guide this work:

(1) Adapt Whisper for Non-Human Audio—Assess if Whisper can produce stable "transcriptions" of chicken vocalizations. Even if the "words" do not match any natural language, their consistency across multiple audio samples could encode stress or welfare signals.
(2) Sentiment and Emotional Analysis—Apply standard NLP sentiment tools—originally built for English text—to the token sequences, checking for systematic shifts that map onto conditions like disease, fear, or contentment.
(3) Compare Across Conditions—Measure how the distribution of these token-based "sentiments" diverges between healthy vs. unhealthy birds, or pre- vs. post-stress. If the approach is sensitive, it may capture early welfare changes more effectively than visual inspection.
(4) Generate Insights for Future Fine-Tuning—Evaluate how domain mismatch impacts performance and highlight strategies—such as domain-specific lexicon augmentation or partial re-training—to refine such models for poultry bioacoustics.

In the broader context of AI and agriculture, this research offers conceptual and practical benefits. Conceptually, it expands the potential of large-scale Transformer-based ASR models beyond human speech, suggesting a path to "universal" acoustic recognition systems that can decode communication in numerous species [46]). Practically, real-time welfare monitoring reduces reliance on invasive tests, ensures earlier interventions for disease or stress, and fosters a more humane environment in large-scale poultry operations [47,48].

Overall, the intersection of machine listening, bioacoustics, and poultry welfare stands at a pivotal juncture. By harnessing the latest AI techniques—both specialized CNN-based classifiers and emerging Transformer-based ASR—we can move closer to a future where farm management decisions are informed by continuous, data-driven, and empathetic insights into the lives of the animals under our care.

2. Materials and Methods

2.1. Datasets

Two open-access datasets were used to evaluate the feasibility of applying Whisper to poultry vocalizations. The first dataset (Dataset 1) was collected at the CARUS animal facility at Wageningen University and comprises audio recordings from fifty-two Super Nick chickens housed in three separate cages. Mild stressors, such as an umbrella opening or pre-recorded dog barking, were introduced at controlled intervals, enabling the collection of both baseline and post-stress vocalizations. These recordings, detailed in [49], are publicly available through Zenodo. The dataset captures a range of stress responses, offering both controlled and ecologically relevant vocal samples for analysis.

The second dataset (Dataset 2), described in [50], contains 346 audio recordings obtained at the Bowen University poultry research farm. Each file is categorized as healthy, noisy, or unhealthy, based on observed behaviors and acoustic characteristics. Healthy recordings consist of typical chicken vocalizations without audible distress, while noisy recordings include background disruptions such as vehicular sounds or human chatter. In contrast, unhealthy recordings feature vocal indicators of potential health issues, including coughing and snoring. The recordings, which range in duration from 5 to 60 s, represent a diverse set of environmental and health-related conditions. This dataset is freely accessible on Mendeley Data. Details on the characteristics, experimental conditions, and composition of both datasets used in this study are summarized in Table 1.

Table 1. Overview and characteristics of chicken vocalization datasets used in the study.

Dataset	Source	Number of Birds	Experimental Conditions	Number of Recordings	Duration of Recordings	Recording Characteristics
Dataset 1	CARUS facility, Wageningen University [49]	52	Pre-stress and post-stress groups (acute stress induced via umbrella opening), and a non-stressed control group.	>900 audio files (available via Zenodo)	1 h recordings per session, segmented into shorter clips (1–5 s)	Controlled experimental setting; mild acute stressors regularly introduced; minimal environmental noise.
Dataset 2	Bowen University poultry research farm [50]	100	Healthy (treated) and unhealthy (untreated) groups monitored over 65 days; unhealthy group developed respiratory illness naturally.	346 audio files categorized as - Healthy (139) - Unhealthy (121) - Noise (86) (available via Mendeley Data)	5–60 s per recording	Field environment; systematic recording (morning, afternoon, night); high-resolution audio (24-bit, 96 kHz); abnormal vocalizations captured (e.g., coughing, snoring); background noise recorded separately; spectral de-noising applied

The classification of stress-related vocalizations in this study was based on prior literature rather than direct physiological measurements. Existing research has established well-documented categorizations of poultry distress calls, linking distinct vocal patterns—such as increased frequency, amplitude, and duration—to stressors like predator threats, handling, and environmental disturbances [6,45]. Additionally, previous studies have identified characteristic alarm calls, discomfort vocalizations, and illness-related sounds in chickens, which have been manually annotated in earlier datasets [9,25]. Given these established associations, our labeling process adheres to previously validated classification frameworks rather than relying on new expert annotations or physiological markers.

However, we recognize that directly associating acoustic patterns with welfare states requires further validation through behavioral observations and physiological monitoring. Future work should incorporate expert-annotated datasets and sensor-based welfare indicators to enhance the accuracy and reliability of automated stress detection models. All experimental protocols related to these datasets were approved by the respective institutional review boards. In Dataset 1, stress induction was conducted under strict ethical guidelines, with measures such as controlled exposure to stimuli implemented to minimize discomfort to the chickens.

2.2. Analysis Pipeline

To unify both datasets under a single processing strategy, we designed a multi-step pipeline that addresses audio preparation, transcription via Whisper, and downstream text-based analyses.

2.2.1. Preprocessing

All audio files were converted to a standardized format using FFmpeg (16 kHz sampling rate, single-channel mono). Background noise in the audio files was meticulously filtered out using the Spectral De-noise module of the Izotope RX software (Sound on Sound, UK, Version 11.2.0), a critical step in preprocessing to ensure clarity in the vocalization data. Subsequently, to achieve consistent intensity across all recordings, the sounds were normalized using Sound Forge Pro 11 (Magix Software GmbH, Germany, version 17.0.3.177), further refining the quality of the audio data for analysis. Each file was then segmented into shorter clips of 1–5 s to facilitate computational efficiency. These segments were chosen because abrupt changes in vocalization patterns often occur within a few seconds, making shorter clips more amenable to detailed, per-clip analysis.

2.2.2. Transcription with Whisper

The preprocessed audio segments were fed into the base version of OpenAI's Whisper model, which generates text tokens traditionally intended to represent words in human speech. Although Whisper's training data derive solely from human languages, we hypothesized that stable patterns or repeated tokens might emerge in the output when presented with non-human sounds. In practice, Whisper often produced "nonsensical" strings—for instance, repeated fragments of text or unusual characters. However, the consistency of these fragments across segments with similar acoustic properties was of particular interest, as it could reflect underlying acoustic structures such as intensity, pitch, and harmonic content.

2.2.3. Sentiment and Textual Analysis

Following transcription, all text outputs were structured and analyzed in Python (Version 3.13.1). Using the SentimentIntensityAnalyzer from the NLTK library, each transcription was assigned a sentiment score (positive, negative, neutral). While these lexicons were originally developed for English text, we examined whether relative shifts in positivity or negativity correlated with changes in vocal intensity or distress levels.

The rationale for sentiment analysis stems from the hypothesis that Whisper's phonetic patterns and structured token sequences may encode latent acoustic features linked to poultry welfare. Although sentiment scores do not directly reflect emotional states in chickens, they serve as proxies for stress-induced vocal changes. Prior research has established a connection between increased vocalization intensity, higher pitch, and repetitive distress calls with heightened physiological arousal [50]. By analyzing sentiment shifts across experimental conditions, we aim to capture variations in acoustic expression that align with known behavioral indicators of welfare.

To further validate this approach, future iterations should integrate direct acoustic feature extraction methods—such as Mel-frequency cepstral coefficients (MFCCs) and spectral entropy measures—alongside Whisper's sentiment-based outputs. Additionally, correlating sentiment scores with external welfare metrics, such as heart rate, body temperature, or corticosterone levels, would provide stronger biological validation for AI-driven poultry welfare monitoring.

Beyond sentiment analysis, additional textual features—including n-gram frequencies (unigram, bigram, trigram) and part-of-speech tags—were extracted to explore potential syntactic or "vocabulary-like" structures in Whisper's outputs. These analyses were conducted using NumPy, Pandas, and Scikit-learn, while Matplotlib (version 3.9) and Seaborn (Version 0.9.1) were employed for visualization through distribution plots, boxplots, and heatmaps.

2.2.4. Statistical and Visualization Tools

To gauge the similarity of transcribed tokens across different experimental conditions (e.g., pre- vs. post-stress, healthy vs. unhealthy, or quiet vs. noisy), we computed cosine similarity scores between the text embeddings. This approach allowed us to determine how closely the "transcriptions" aligned or diverged within each category. Furthermore, we used Latent Dirichlet Allocation (LDA) topic modeling to discern whether any overarching "topics" or clusters of tokens emerged consistently in subsets of the data (e.g., all post-stress clips forming one distinct cluster). This higher-level structure helps identify patterns that sentiment scores alone may not capture.

2.3. *Comparisons with Prior Acoustic Analysis*

Whisper's outputs are not literal chicken "words", making standard speech recognition metrics such as word error rate (WER) inapplicable. Instead, we compared our approach to prior research that employed CNNs, Hidden Markov Models (HMMs), and random forests to detect chicken distress calls or diagnose avian diseases. These studies report classification accuracies ranging from 67% to over 90%, depending on task complexity [15,16]. However, our objective is not to surpass these models in word recognition but to assess whether Whisper-generated "transcriptions" exhibit sentiment and textual patterns that correlate with known welfare conditions (healthy vs. unhealthy, pre- vs. post-stress). Therefore, our performance evaluation prioritizes coherence and interpretability of sentiment shifts over literal transcription accuracy.

While Whisper shows promise in extracting meaningful patterns from non-human vocalizations, a direct benchmark against domain-specific avian models would enhance its comparative value. However, such a comparison presents inherent challenges. CNNs and RNN-based models for bioacoustics rely on spectrogram-based feature extraction and require large-scale labeled datasets for supervised learning. In contrast, Whisper operates as a sequence-to-sequence ASR model, producing text-like outputs that do not directly align with conventional classification metrics such as accuracy, precision, recall, or F1-score. Consequently, standard speech recognition performance evaluations do not apply to non-linguistic vocalizations.

Despite these differences, prior studies using CNNs and HMMs for poultry distress calls have reported 67–90% classification accuracy [15,16]. Given that Whisper has not been fine-tuned for avian data, direct comparison would require a domain adaptation step, which falls beyond the scope of this feasibility study. Instead, our analysis focuses on interpreting Whisper's token outputs and their correlation with welfare indicators. Future work should explore a hybrid approach, integrating CNN-based acoustic features with Whisper's text outputs to balance interpretability and classification accuracy in poultry vocalization analysis.

2.4. *Limitations and Potential Biases*

The domain mismatch between human speech and chicken vocalizations stands out as the most significant limitation. Whisper's underlying weights prioritize human phonetic and linguistic structures, thus producing text tokens (e.g., "nihai", "kichi") that have no direct meaning in poultry communication. The practical assumption here is that consistent token repetition or sentiment skew could correlate with acoustic features indicative of stress or well-being. Another limitation pertains to annotation. The datasets used in this study provide broad labels such as "healthy", "unhealthy", and "noisy", but do not include granular annotations that differentiate alarm calls, threat calls, or various social calls. Future work would benefit from more extensive labeling efforts to precisely link model output to each vocalization's ecological or emotional function.

Background noise remains a technical challenge in large-scale operations. Despite applying basic noise reduction, farm environments can contain overlapping calls and irregular mechanical sounds that potentially reduce the model's sensitivity to subtle changes in chicken vocalizations. Addressing this would require domain-adaptive noise filtering or targeted fine-tuning of Whisper's acoustic front-end. Finally, any sentiment analysis performed with an English-based lexicon inherently lacks domain-specific calibration for animal calls. Consequently, the positive or negative sentiment scores reported here should be interpreted with caution. They do not directly map onto the emotional states of the chickens but serve as proxies for changes in vocal intensity or patterning, potentially linked to welfare conditions. Despite these caveats, the results offer an intriguing proof-of-concept for deploying large-scale ASR models in non-human acoustic domains.

3. Results

The core aim of this research was to investigate whether OpenAI's Whisper model—originally trained on large-scale human speech corpora—could generate text outputs that meaningfully capture acoustic variations in chicken vocalizations. Two main datasets served as testbeds: one focusing on chickens before and after stress induction (Dataset 1), and the other comparing healthy, noisy, and unhealthy conditions (Dataset 2). Throughout the experiments, Whisper frequently produced "nonsensical" text tokens (e.g., "nihai", "kichi", "going") when transcribing chicken calls. Despite the apparent domain mismatch, careful analysis of token patterns, sentiment scores, linguistic metrics, and topic clusters revealed consistent correlations with known stressors or health statuses.

3.1. Whisper Output Characteristics and Consistency

Whisper's outputs for chicken vocalizations typically comprised repetitive tokens or fragments that bore no linguistic resemblance to human words. Nevertheless, these tokens emerged in a highly consistent manner across recordings with similar acoustic features. Figure 1 illustrates how loud or urgent calls—often occurring in post-stress or unhealthy contexts—regularly generated repeated sequences (for instance, "nihai nihai"), whereas calmer baseline calls produced a broader, less repetitive token distribution. Figure 1 illustrates how repeated tokens and character usage reflect distinct acoustic profiles across conditions. This consistency suggests that Whisper's latent feature extraction was sensitive to amplitude, pitch, or call duration patterns, even though it was never trained on avian data.

In Dataset 1, chickens recorded immediately after mild stressors (e.g., an opened umbrella) exhibited more frequent repeated tokens and stronger amplitude signals. Meanwhile, pre-stress recordings yielded token outputs that were more varied, suggesting lower overall vocal intensity. A similar trend emerged in Dataset 2: "unhealthy" chickens that coughed or exhibited rale-like sounds tended to produce repetitive token strings, whereas healthy calls were more diverse. Importantly, the presence of random punctuation and foreign characters rose significantly in "noisy" environments with overlapping background sounds. Although this phenomenon contributed additional "nonsense" to the transcripts, the patterns themselves remained stable within each noisy clip, implying that Whisper's decoder was reacting systematically to the acoustic confusion.

(A)

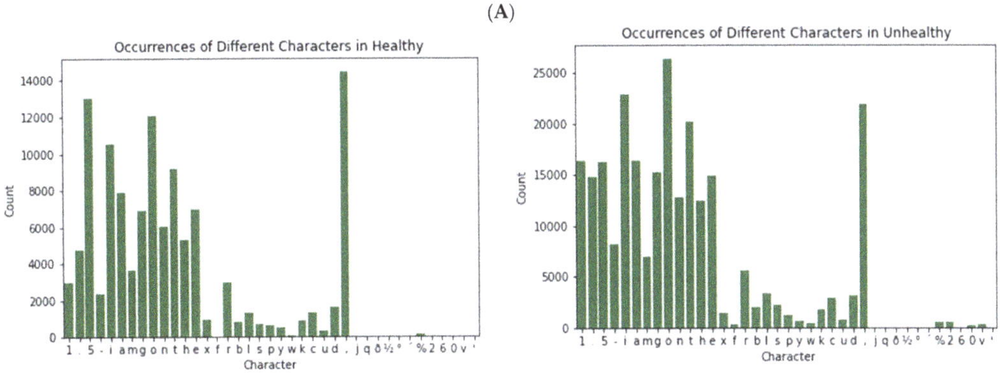

(B)

Figure 1. Unigram word cloud (**A**) and character frequency analysis (**B**) demonstrating consistent token patterns correlated with chicken welfare conditions.

3.2. Sentiment Analysis Across Experimental Conditions

To probe whether these domain-mismatched transcriptions might reflect actual differences in welfare states, we applied an English-oriented sentiment scoring tool (NLTK's SentimentIntensityAnalyzer) to each token sequence.

3.2.1. Pre- vs. Post-Stress (Dataset 1)

Segments recorded in the immediate aftermath of stress induction displayed a pronounced increase in "negative" sentiment scores, consistent with heightened vocal urgency or agitation. Meanwhile, targeted interventions, such as partial covers that calm the flock, occasionally led to lower negative sentiment, presumably reflecting more subdued call patterns. Although these "negative" labels do not literally map to avian distress, they highlight an alignment between token repetition and the sentiment analyzer's classification of urgency or intensity.

3.2.2. Healthy vs. Noisy vs. Unhealthy (Dataset 2)

Recordings taken in noisy environments exhibited higher negative polarity scores than healthy ones, suggesting that Whisper's repeated punctuation or foreign glyphs—prompted by background chaos—triggered the analyzer's negative weighting. In "unhealthy" segments, the sentiment distribution was more variable; some clips approached strongly negative scores, possibly from coughing or labored breathing, while others remained closer to neutral. This variation (Figure 2) underscores the complexity of linking raw token repetition to highly heterogeneous physiological conditions. Still, across both datasets, elevated "negative" scores loosely tracked the presence of disruptive noise, stress, or health problems. It should be noted that the absolute sentiment values may be misleading but relative differences are informative.

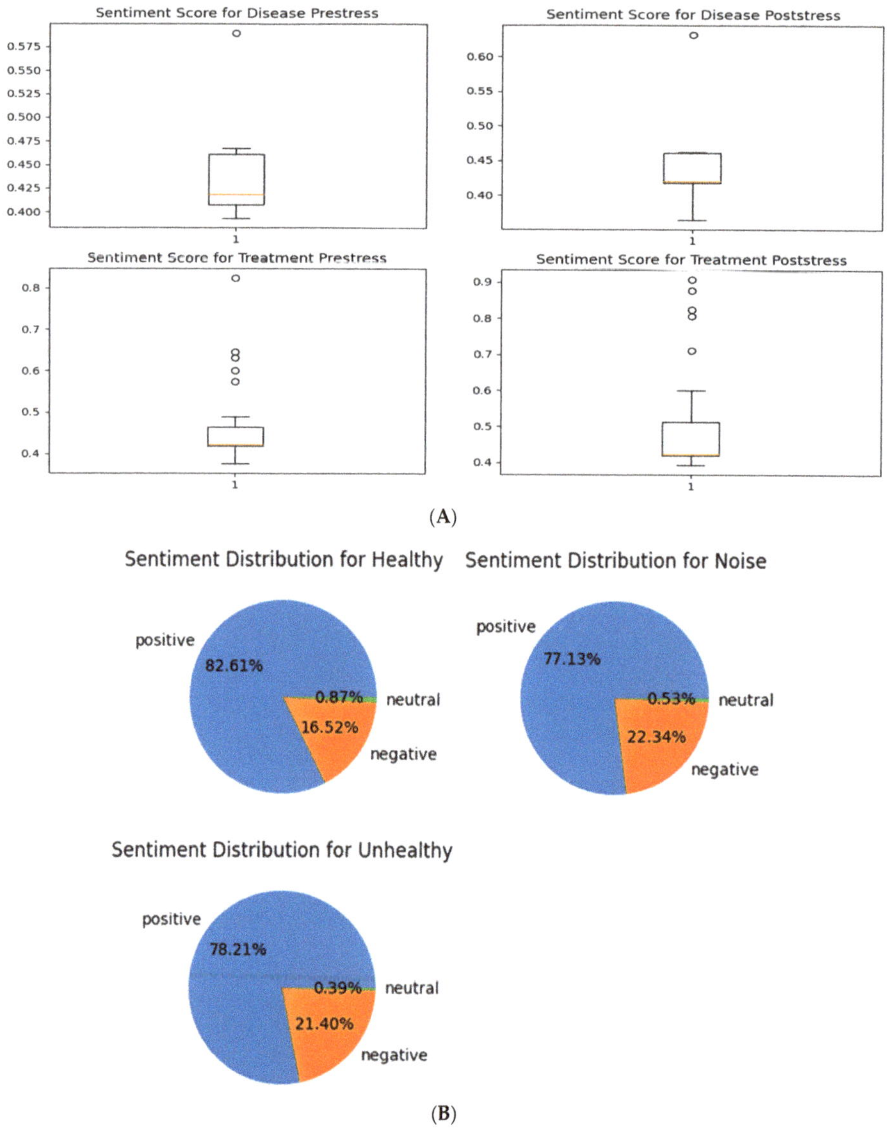

Figure 2. Comparative sentiment analysis showing differences between pre-stress and post-stress conditions (**A**) and among healthy, noisy, and unhealthy contexts (**B**), highlighting shifts in negativity corresponding to increased vocal intensity or distress.

3.3. Linguistic Metrics and Topic Modeling Insights

Despite the absence of genuine semantics, several conventional textual analyses uncovered systematic differences tied to each experimental condition. Character and word frequency counts, for example, highlighted how tokens like "nihai", "kichi", "going", or "room" disproportionately clustered in high-distress or high-volume recordings. Figures 3 and 4 explore these frequencies further, revealing that "nihai" often topped the token list in post-stress calls, while calmer sessions showed a more balanced vocabulary. When part-of-speech (POS) tags were assigned—albeit with an English-trained NLP tagger—post-stress transcripts contained higher counts of "adjectives" or "present participles", presumably because repeated tokens or morphological variants (e.g., "nihaiii", "goinggg") triggered these labels.

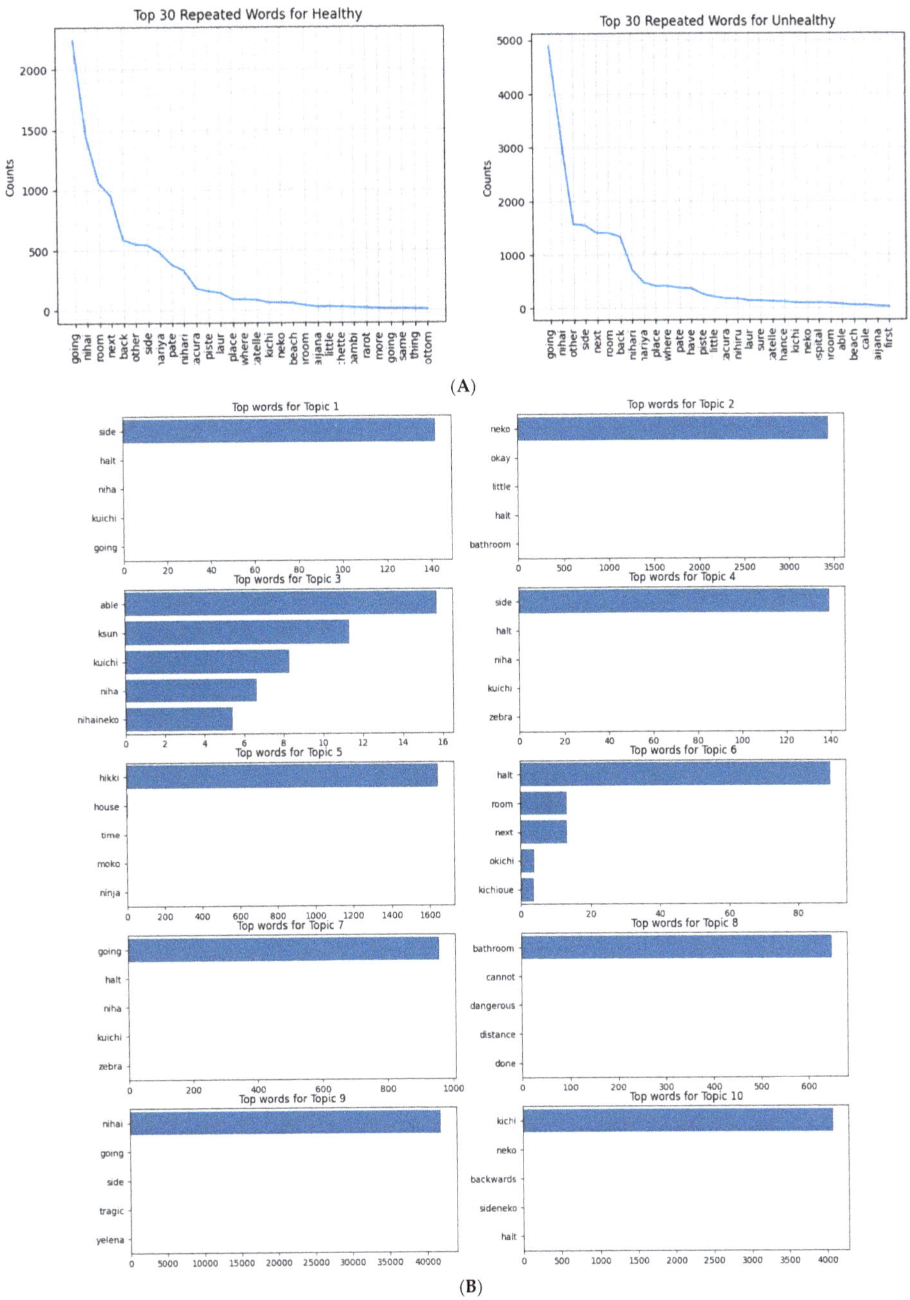

Figure 3. Token frequency rankings (**A**) and topic modeling (**B**) demonstrating recurrent Whisper-generated tokens (e.g., "nihai", "going") that distinguish between healthy and distressed poultry vocalizations.

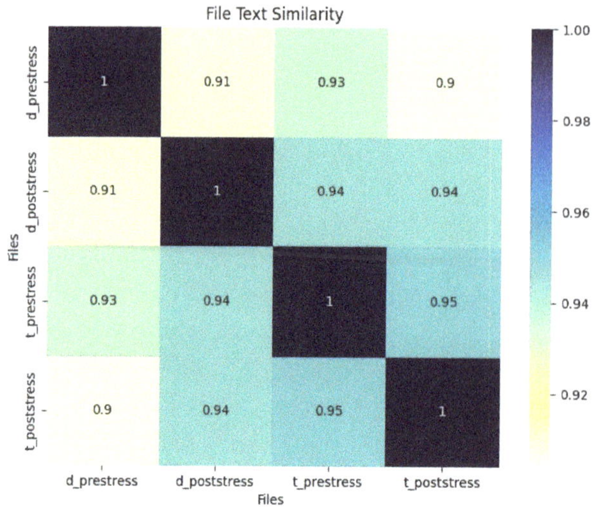

(**A**)

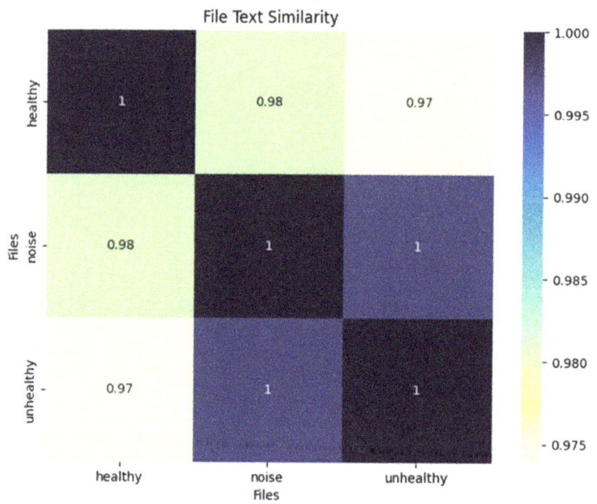

(**B**)

Figure 4. Cosine similarity heatmaps illustrating distinct clustering of Whisper-generated transcripts: Panel (**A**) differentiates post-stress or noisy vocalizations from baseline calls; Panel (**B**) demonstrates tight clustering among healthy recordings, confirming Whisper's sensitivity to condition-specific acoustic patterns.

To determine whether these token patterns might form distinct clusters, we employed Latent Dirichlet Allocation (LDA). Topic modeling identified recurring sets of tokens that separated pre-stress from post-stress segments and healthy from unhealthy vocalizations. These "topics" significantly overlapped with repeated token fragments often associated with negative sentiment. Such clustering supports the notion that the same acoustic indicators leading to repeated text outputs also cause the model to group these transcripts into distinct "themes". Even though the tokens themselves have no dictionary meaning, their distribution across topics reflects underlying acoustic conditions. Figure 3 (Panel B) shows an example of how repeated "nihai" sequences coalesce into a post-stress cluster, while more diverse tokens populate pre-stress or healthy topics.

3.4. Text Similarity and Condition Separation

To quantify the similarity of transcripts, we vectorized each token sequence (for instance, via TF-IDF weighting) and calculated cosine similarity. Figure 4 depicts the resulting heatmaps, in which segments from the same condition (e.g., all healthy recordings) consistently group together, while segments from opposing conditions (e.g., healthy vs. unhealthy) occupy distinct regions. This aligns with prior findings that repeated or intense calls yield characteristic textual proxies, whereas quieter, stable vocalizations map onto more varied or sparse outputs. Post-stress or unhealthy states thus exhibit lower textual similarity to calmer, baseline conditions.

These clustering results parallel the phenomenon observed in CNN-based spectrogram classification, where distress calls cluster together in high-dimensional feature space. Here, however, the separation arises from a Transformer-based ASR converting amplitude and pitch features into repeated tokens that NLP similarity metrics interpret as "closely related documents." This underscores a key point: large-scale speech models can function as generalized acoustic encoders, capturing enough detail to differentiate stress or illness conditions in poultry even without specialized training on avian calls.

Extended visualizations—covering bigram/trigram word clouds, correlation matrices, and more detailed LDA plots—appear in the Supplemental File (Figures S1–S9). Those additional figures illustrate how character-level distributions, parts-of-speech ratios, and multi-dimensional scaling corroborate the main findings presented here.

All extended analyses that support the quantitative and qualitative observations above are detailed in the Supplemental Material. For example, Figures S1 and S2 expand on the token distributions across different stress levels, while Figures S3–S9 illustrate additional topic modeling results, multimodal correlation plots, and character-level frequency charts. These supplemental items enrich the core findings, providing finer-grained evidence of how Whisper's transcriptions reflect acoustic changes in multiple poultry contexts.

4. Discussion

The human-oriented ASR model (Whisper) has the potential to generate text outputs consistently correlated with variations in chicken vocal patterns across stress, noise, and health conditions. Despite the clear mismatch in language domain, the model's acoustic front-end appears adept at encoding pitch, amplitude, and spectral cues into stable sequences of tokens. When these tokens are analyzed using standard NLP sentiment and clustering tools, they reveal differences that align with known events (e.g., stress induction, background chaos) or health issues (e.g., coughing, rales).

One of the most striking outcomes is that Whisper, trained solely on human speech, repeatedly produced token strings like "nihai nihai" whenever birds emitted louder or more rapid calls. Although these strings hold no genuine meaning in English, their repetitive nature effectively captures "signature" acoustic features typical of urgent or distressed vocalizations. Similar phenomena were noted in pilot studies applying speech encoders to whale calls or bat echolocation signals. The ability of large-scale ASR models to act as generic feature extractors challenges conventional assumptions that these models are confined to processing human phonemes and syntax. Instead, they appear to extract high-level temporal or frequency cues that can generalize to non-human vocal repertoires, at least for the purpose of classification or monitoring.

From a practical standpoint, such an approach offers a non-invasive, real-time monitoring strategy for poultry farms. Rather than relying on labor-intensive visual inspections or invasive physiological sampling, farmers could place microphones to record calls continuously and run Whisper locally. Sudden spikes in negative sentiment or repeated tokens might alert them to health problems, predatory threats, or environmental stressors (e.g., equipment failure). This could enable early intervention—potentially mitigating large-scale losses and reducing antibiotic use. Moreover, this method avoids the need for extensive custom labeling of chicken calls before initial deployment. The unscrupulous textual outputs become relevant only insofar as they are stable or repeated under certain conditions.

Despite these benefits, several constraints and areas for improvement arise. Domain mismatch remains the largest hurdle. Because Whisper is not trained on avian calls, it occasionally "hallucinates" random punctuation, foreign letters, or morphological expansions. In quiet or intermediate conditions, these artifacts can reduce the clarity of the textual signals. Additionally, the broad "negative", "positive", and "neutral" labels from standard English-based sentiment dictionaries do not necessarily map onto genuine emotional states in chickens. Such labels are better interpreted as proxies for amplitude or repetition in the calls. A domain-specific lexicon or direct correlation with validated ethograms would likely yield a more accurate measure of real distress or well-being.

Noise contamination in commercial barns also poses a challenge. While Whisper handles moderate noise well for human speech, the chaotic overlaps of machinery, ventilation, and hundreds of clucking birds may degrade recognition. Preliminary filtering or specialized microphone arrays might isolate individual calls more effectively, boosting interpretive accuracy. Additionally, multi-speaker diarization—commonly used in human teleconferencing—could help segregate overlapping calls from different birds, further refining the textual output.

Comparison with prior poultry bioacoustics research underscores the novelty of this approach. Traditional methods typically employ Convolutional Neural Networks (CNNs), Hidden Markov Models (HMMs), or random forests trained on spectrograms or carefully annotated data. These strategies can achieve high accuracy (often above 90%) for specific tasks (e.g., distress-call detection), but they often require substantial domain-specific labeling. By contrast, the Whisper-based pipeline leverages an enormous pretrained model "as-is", with minimal or no additional training. It does not compete directly with fine-grained classification accuracy, yet it provides a fast route to real-time detection of broad changes in vocalization patterns. Looking ahead, partial fine-tuning of Whisper's acoustic layers on a smaller, labeled set of avian calls could improve the clarity of the textual tokens, potentially bridging some of the gap to specialized models.

A final point of interest is the concept of a future "chicken translator". While genuine translation is unlikely—chickens do not produce language in the human sense—these results hint at advanced acoustic systems that might map repeated calls to well-defined categories such as alarm calls, comfort calls, or social chatter. Integrating other data streams (camera-based posture analysis, physiological sensors) could enrich the interpretability of repeated tokens and provide robust multimodal indicators of a flock's welfare. For instance, consistent spikes in "nihai nihai" tokens plus decreased feeding activity might indicate significant stress or illness, prompting a more targeted veterinary check.

Taken as a whole, this proof-of-concept demonstrates that a large-scale human ASR model can be repurposed—without specialized domain re-training—to capture meaningful patterns in chicken vocalizations. While certainly not a replacement for domain-tuned architecture, it opens new directions for rapid, non-invasive acoustic surveillance in precision livestock farming. The next phase of research might incorporate partial fine-tuning, advanced noise mitigation, and more precise linking of token repetition to specific behavioral or physiological states. In time, these innovations could substantially improve early-warning systems, reduce mortality through prompt intervention, and support a more humane and effective management of poultry welfare.

5. Conclusions

The findings demonstrate that OpenAI's Whisper, a Transformer-based model originally developed for human speech recognition, can effectively decode acoustic features from chicken vocalizations. Despite inherent domain differences, Whisper reliably extracted consistent patterns from poultry vocalizations associated with specific emotional and physiological states. Sentiment analysis further revealed systematic changes in acoustic signals corresponding to recognized stress and welfare conditions, effectively distinguishing periods of distress from states of calmness. The observed correlation between Whisper-generated sentiment shifts and known stress indicators indicates that Transformer-based architectures, even without species-specific fine-tuning, can serve as powerful generalized acoustic analyzers in animal welfare research.

Several promising areas emerge for future research to enhance model accuracy and robustness. Firstly, fine-tuning the Whisper model using annotated chicken vocalization datasets could significantly improve sensitivity to poultry-specific acoustic cues. Creating carefully annotated vocalization databases, including diverse environmental conditions, varied flock sizes, and different breeds—would further refine acoustic feature detection capabilities. Additionally, developing a multimodal validation framework that integrates physiological markers, such as heart rate variability, corticosterone levels, and thermographic imaging, along with behavioral observations through video analysis, could significantly improve the biological interpretability of acoustic findings. Another promising direction is the application of anomaly detection techniques, such as Isolation Forest or Dynamic Time Warping (DTW), to detect unusual vocal patterns that conventional sentiment-based methods might overlook. Moreover, conducting rigorous null-model analyses with synthetic or shuffled vocalizations would help establish clearer baseline comparisons.

Ultimately, combining Transformer-based acoustic analysis with these complementary approaches holds substantial potential to revolutionize precision poultry farming, enabling real-time monitoring systems capable of significantly enhancing welfare, operational efficiency, productivity, and overall sustainability.

Supplementary Materials: The following supporting information can be downloaded at: https://www.mdpi.com/article/10.3390/10.3390/ai6040065/s1, Figure S1: Comparative analysis of text metrics: text length, number of vowels, number of consonants, and number of stopwords across disease pre-stress, disease post-stress, treatment pre-stress, and treatment post-stress; Figure S2: Distribution of vowel frequencies across different stages of stress and treatment; Figure S3: Consonant frequency analysis in various phases of stress and treatment; Figure S4: Emotional landscape of laying hens—pie chart analysis across phases; Figure S5: Comparative distribution of parts of speech in chicken vocalizations across disease and treatment phases; Figure S6: Comparative analysis of text metrics—healthy vs. noise vs. unhealthy phases; Figure S7: Distribution of vowel frequencies across different stages of chicken vocalization—healthy, noise, and unhealthy dataset analysis; Figure S8: frequency analysis in various phases of chicken vocalizations—healthy, noise, and unhealthy dataset comparisons; Figure S9: Distribution of sentiment scores and emotional profiles in healthy vs. noise vs. unhealthy phases; Figure S10: Extended correlation and emotional state analyses across chicken health phases; Figure S11: Additional unigram (A), bigram (B), and trigram (C) word clouds for healthy, noise, and unhealthy phases; Figure S12: Comparative distribution of parts of speech in chicken vocalizations—healthy, noise, and unhealthy phases; Table S1: Comparative sentiment analysis of poultry vocalizations under varied stress and environmental conditions. Dataset can be downloaded from Zenodo [51]—Neethirajan, S. (2023). Vocalization Patterns in Laying Hens—An Analysis of Stress-Induced Audio Responses [Data set]. https://doi.org/10.5281/zenodo.10433023.

Funding: This research was funded by the Natural Sciences and Engineering Research Council of Canada (Project R37245).

Institutional Review Board Statement: The data used in this study were obtained from previously published work and all the animal experiments in the respective studies were approved by the ethics boards of the respective universities.

Data Availability Statement: The data will be made available upon reasonable request.

Conflicts of Interest: The author declares no conflicts of interest.

References

1. Neethirajan, S. Digital phenotyping: A game changer for the broiler industry. *Animals* **2023**, *13*, 2585. [CrossRef] [PubMed]
2. Ferreira, J.C.; Campos, A.T.; Ferraz, P.F.P.; Bahuti, M.; Junior, T.Y.; Silva, J.P.D.; Ferreira, S.C. Dynamics of the Thermal Environment in Climate-Controlled Poultry Houses for Broiler Chickens. *AgriEngineering* **2024**, *6*, 3891–3911. [CrossRef]
3. Marino, L. Thinking chickens: A review of cognition, emotion, and behavior in the domestic chicken. *Anim. Cogn.* **2017**, *20*, 127–147. [CrossRef]
4. Hernandez, E.; Llonch, P.; Turner, P.V. Applied animal ethics in industrial food animal production: Exploring the role of the veterinarian. *Animals* **2022**, *12*, 678. [CrossRef]
5. Rollin, B. Why is agricultural animal welfare important? The social and ethical context. In *Improving Animal Welfare: A Practical Approach*, 3rd ed.; CABI: Wallingford, UK, 2021; pp. 46–59.
6. Neethirajan, S. Automated tracking systems for the assessment of farmed poultry. *Animals* **2022**, *12*, 232. [CrossRef] [PubMed]
7. Neethirajan, S.; Kemp, B. Digital livestock farming. *Sens. Bio-Sens. Res.* **2021**, *32*, 100408. [CrossRef]
8. Neethirajan, S. Transforming the adaptation physiology of farm animals through sensors. *Animals* **2020**, *10*, 1512. [CrossRef]
9. Vasdal, G.; Muri, K.; Stubsjøen, S.M.; Moe, R.O.; Kittelsen, K. Qualitative behaviour assessment as part of a welfare assessment in flocks of laying hens. *Appl. Anim. Behav. Sci.* **2022**, *246*, 105535. [CrossRef]
10. Cai, J.; Yan, Y.; Cheok, A. Deciphering Avian Emotions: A Novel AI and Machine Learning Approach to Understanding Chicken Vocalizations. *Res. Sq.* **2023**. [CrossRef]
11. Ghani, B. Machine Learning-Based Analysis of Bird Vocalizations. Ph.D. Thesis, University of Göttingen, Göttingen, Germany, 2022. Available online: https://ediss.uni-goettingen.de/handle/11858/13959 (accessed on 15 September 2023).
12. Ranjard, L.; Ross, H.A. Unsupervised bird song syllable classification using evolving neural networks. *J. Acoust. Soc. Am.* **2008**, *123*, 4358–4368. [CrossRef]
13. Piczak, K.J. Recognizing bird species in audio recordings using deep convolutional neural networks. In *CLEF (Working Notes)*; Sun SITE Central Europe: London, UK, 2016; pp. 534–543. Available online: https://ceur-ws.org/Vol-1609/16090534.pdf (accessed on 16 September 2023).
14. Gupta, G.; Kshirsagar, M.; Zhong, M.; Gholami, S.; Ferres, J.L. Comparing recurrent convolutional neural networks for large scale bird species classification. *Sci. Rep.* **2021**, *11*, 17085. [CrossRef]
15. Mao, A.; Giraudet, C.S.; Liu, K.; De Almeida Nolasco, I.; Xie, Z.; Xie, Z.; Gao, Y.; Theobald, J.; Bhatta, D.; Stewart, R.; et al. Automated identification of chicken distress vocalizations using deep learning models. *J. R. Soc. Interface* **2022**, *19*, 20210921. [CrossRef] [PubMed]
16. Jung, D.H.; Kim, N.Y.; Moon, S.H.; Kim, H.S.; Lee, T.S.; Yang, J.S.; Lee, J.Y.; Han, X.; Park, S.H. Classification of vocalization recordings of laying hens and cattle using convolutional neural network models. *J. Biosyst. Eng.* **2021**, *46*, 217–224. [CrossRef]
17. Cheng, B.; Zhang, S. A novel chicken voice recognition method using the orthogonal matching pursuit algorithm. In Proceedings of the 2015 8th International Congress on Image and Signal Processing (CISP), Shenyang, China, 14–16 October 2015; IEEE: Piscataway, NJ, USA, 2016; pp. 1266–1271.
18. Sadeghi, M.; Banakar, A.; Khazaee, M.; Soleimani, M.R. An intelligent procedure for the detection and classification of chickens infected by clostridium perfringens based on their vocalization. *Braz. J. Poult. Sci.* **2015**, *17*, 537–544.
19. Lokhandwala, S.; Sinha, R.; Ganji, S.; Pailla, B. Decoding Asian elephant vocalisations: Unravelling call types, context-specific behaviors, and individual identities. In Proceedings of the International Conference on Speech and Computer, Hubli, India, 29 November–1 December 2023; Springer Nature: Cham, Switzerland, 2023; pp. 367–379.
20. Zhang, L.; Liu, J.; Zhang, Z.; Liang, W. Tit alarm calls trigger anti-predator behavior in free-range domestic chickens. *Appl. Anim. Behav. Sci.* **2023**, *265*, 106009.
21. Guo, X. UL-net: Fusion spatial and temporal features for bird voice detection. In Proceedings of the 2022 IEEE 2nd International Conference on Electronic Technology, Communication and Information (ICETCI), Changchun, China, 27–29 May 2022; IEEE: Piscataway, NJ, USA, 2022; pp. 1009–1013.
22. Rajan, R.; Noumida, A. Multi-label bird species classification using transfer learning. In Proceedings of the 2021 International Conference on Communication, Control and Information Sciences (ICCISc), Idukki, India, 16–18 June 2021; IEEE: Piscataway, NJ, USA, 2021; Volume 1, pp. 1–5.
23. Nicol, C.J. *The Behavioural Biology of Chickens*; CABI: Wallingford, UK, 2015.

24. Sibley, D.A. *What It's Like to Be a Bird: From Flying to Nesting, Eating to Singing—What Birds Are Doing, and Why*; Knopf: New York, NY, USA, 2020.
25. Herborn, K.A.; McElligott, A.G.; Mitchell, M.A.; Sandilands, V.; Bradshaw, B.; Asher, L. Spectral entropy of early-life distress calls as an iceberg indicator of chicken welfare. *J. R. Soc. Interface* **2020**, *17*, 20200086.
26. Burnham, R. Animal calling behaviours and what this can tell us about the effects of changing soundscapes. *Acoustics* **2023**, *5*, 631–652. [CrossRef]
27. Potamitis, I. Deep learning for detection of bird vocalisations. *arXiv* **2016**, arXiv:1609.08408. [CrossRef]
28. Wang, H.; Xu, Y.; Yu, Y.; Lin, Y.; Ran, J. An efficient model for a vast number of bird species identification based on acoustic features. *Animals* **2022**, *12*, 2434. [CrossRef]
29. Noumida, A.; Rajan, R. Multi-label bird species classification from audio recordings using attention framework. *Appl. Acoust.* **2022**, *197*, 108901.
30. Fundel, F.; Braun, D.A.; Gottwald, S. Automatic bat call classification using transformer networks. *Ecol. Inform.* **2023**, *78*, 102288. [CrossRef]
31. Yang, J.; Carstens, B.C.; Provost, K.L. Machine learning reveals relationships between song, climate, and migration in coastal *Zonotrichia leucophrys*. *bioRxiv* **2023**. [CrossRef]
32. de Carvalho, P.S.; Grzywalski, T.; Hou, Y.; Thomas, P.; Dedeurwaerder, A.; De Gussem, M.; Tuyttens, F.; Devos, P.; Botteldooren, D.; Antonissen, G. Automated detection of broiler vocalizations. A Machine Learning Approach for Broiler Chicken Vocalization Monitoring. *Poult. Sci.* **2025**, *104*, 104962.
33. Vachhani, B.; Singh, D.; Lawyer, R. Multi-resolution Approach to Identification of Spoken Languages and To Improve Overall Language Diarization System Using Whisper Model. In Proceedings of the Interspeech 2023, Dublin, Ireland, 20–24 August 2023; Volume 2023, pp. 1993–1997.
34. Yeo, J.H.; Kim, M.; Watanabe, S.; Ro, Y.M. Visual Speech Recognition for Low-resource Languages with Automatic Labels From Whisper Model. *arXiv* **2023**, arXiv:2309.08535.
35. Stowell, D.; Wood, M.D.; Pamuła, H.; Stylianou, Y.; Glotin, H. Automatic acoustic detection of birds through deep learning: The first bird audio detection challenge. *Methods Ecol. Evol.* **2019**, *10*, 368–380. [CrossRef]
36. Dongan, G.; Akbulut, F.P. Multi-modal fusion learning through biosignal, audio, and visual content for detection of mental stress. *Neural Comput. Appl.* **2023**, *35*, 24435–24454. [CrossRef]
37. Webster, J.; Margerison, J. (Eds.) *Management and Welfare of Farm Animals: The UFAW Farm Handbook*; John Wiley & Sons: Hoboken, NJ, USA, 2022.
38. Andreas, J.; Beguš, G.; Bronstein, M.M.; Diamant, R.; Delaney, D.; Gero, S.; Goldwasser, S.; Gruber, D.F.; de Haas, S.; Malkin, P.; et al. Cetacean translation initiative: A roadmap to deciphering the communication of sperm whales. *arXiv* **2021**, arXiv:2104.08614.
39. Kather, S. Analysis of Sperm Whale (*Physeter macrocephalus*) Vocalisations in the Azores: Coda Repertoires and Their Behavioural Context. Master's Thesis, The University of the Azores, Ponta Delgada, Portugal, 2023.
40. Malik, M.; Malik, M.K.; Mehmood, K.; Makhdoom, I. Automatic speech recognition: A survey. *Multimed. Tools Appl.* **2021**, *80*, 9411–9457. [CrossRef]
41. Neethirajan, S.; Reimert, I.; Kemp, B. Measuring farm animal emotions—Sensor-based approaches. *Sensors* **2021**, *21*, 553. [CrossRef]
42. Bessa Ferreira, V.H.; Dutour, M.; Oscarsson, R.; Gjøen, J.; Jensen, P. Effects of domestication on responses of chickens and red junglefowl to conspecific calls: A pilot study. *PLoS ONE* **2022**, *17*, e0279553. [CrossRef]
43. Pijpers, N.; van den Heuvel, H.; Duncan, I.H.; Yorzinski, J.; Neethirajan, S. Understanding chicks' emotions: Are eye blinks & facial temperatures reliable indicators? *bioRxiv* **2022**. [CrossRef]
44. Neethirajan, S. Artificial intelligence and sensor innovations: Enhancing livestock welfare with a human-centric approach. *Hum.-Centric Intell. Syst.* **2024**, *4*, 77–92.
45. Marler, P. Bird calls: Their potential for behavioral neurobiology. *Ann. N. Y. Acad. Sci.* **2004**, *1016*, 31–44. [PubMed]
46. Vernes, S.C.; Kriengwatana, B.P.; Beeck, V.C.; Fischer, J.; Tyack, P.L.; Ten Cate, C.; Janik, V.M. The multi-dimensional nature of vocal learning. *Philos. Trans. R. Soc. B* **2021**, *376*, 20200236.
47. Papageorgiou, M.; Goliomytis, M.; Tzamaloukas, O.; Miltiadou, D.; Simitzis, P. Positive Welfare Indicators and Their Association with Sustainable Management Systems in Poultry. *Sustainability* **2023**, *15*, 10890. [CrossRef]
48. van den Heuvel, H.; Youssef, A.; Grat, L.M.; Neethirajan, S. Quantifying the Effect of an Acute Stressor in Laying Hens using Thermographic Imaging and Vocalisations. *bioRxiv* **2022**. [CrossRef]
49. Adebayo, S.; Aworinde, H.O.; Akinwunmi, A.O.; Alabi, O.M.; Ayandiji, A.; Sakpere, A.B.; Adeyemo, A.; Oyebamiji, A.K.; Olaide, O.; Kizito, E. Enhancing poultry health management through machine learning-based analysis of vocalization signals dataset. *Data Brief* **2023**, *50*, 109528.

50. Griebel, U.; Oller, D.K. From emotional signals to symbols. *Front. Psychol.* **2024**, *15*, 1135288.
51. Neethirajan, S. Vocalization Patterns in Laying Hens—An Analysis of Stress-Induced Audio Responses [Data Set]. 2023. Available online: https://zenodo.org/records/10433023 (accessed on 15 September 2023).

MDPI AG
Grosspeteranlage 5
4052 Basel
Switzerland
Tel.: +41 61 683 77 34

AI Editorial Office
E-mail: ai@mdpi.com
www.mdpi.com/journal/ai

www.ingramcontent.com/pod-product-compliance
Lightning Source LLC
Chambersburg PA
CBHW061605220926
51850CB00005B/431
9783725871001